Student Solutions Manual
for
Tipler and Mosca's
Physics for Scientists and Engineers
Fifth Edition

DAVID MILLS
Professor Emeritus
College of the Redwoods

CHARLES L. ADLER
Saint Mary's College of Maryland

W. H. Freeman and Company
New York

Copyright © 2003 by W. H. Freeman and Company

All rights reserved.

Printed in the United States of America

ISBN: 0-7167-8333-9

First printing 2003

W. H. Freeman and Company
41 Madison Avenue
New York, NY 10010
Houndmills, Basingstoke RG21 6XS, England

CONTENTS

To the Student

This solution manual accompanies *Physics for Scientists and Engineers, 5e*, by Paul Tipler and Gene Mosca. Following the structure of the solutions to the Worked Examples in the text, we begin the solutions to the back-of-the-chapter numerical problems with a brief discussion of the physics of the problem, represent the problem pictorially whenever appropriate, express the physics of the solution in the form of a mathematical model, fill in any intermediate steps as needed, make the appropriate substitutions and algebraic simplifications, and complete the solution with the substitution of numerical values (including their units) and the evaluation of whatever physical quantity was called for in the problem. This is the problem-solving strategy used by experienced learners of physics and it is our hope that you will see the value in such an approach to problem solving and learn to use it consistently.

Believing that it will maximize your learning of physics, we encourage you to create your own solution before referring to the solutions in this manual. You may find that, by following this approach, you will find different, but equally valid, solutions to some of the problems. In any event, studying the solutions contained herein without having first attempted the problems will do little to help you learn physics.

You'll find that nearly all problems with numerical answers have their answers given to three significant figures. Most of the exceptions to this rule are in the solutions to the problems for Section 1-5 on Significant Figures and Order of Magnitude. When the nature of the problem makes it desirable to do so, we keep more than three significant figures in the answers to intermediate steps and then round to three significant figures for the final answer. Some of the Estimation and Approximation Problems have answers to fewer than three significant figures.

Physics for Scientists and Engineers, 5e includes numerous spreadsheet problems. Most of them call for the plotting of one or more graphs. The solutions to these problems were also generated using Microsoft Excel and its "paste special" feature, so that you can easily make changes to the graphical parts of the solutions.

Acknowledgments

Charles L. Adler (Saint Mary's College of Maryland) is the author of the new problems appearing in the Fifth Edition. Chuck saved me (dm) many hours of work by providing rough-draft solutions to these new problems, and I thank him for this help. Gene Mosca (United States Naval Academy and the co-author of the Fifth Edition) helped me tremendously by reviewing my early work in great detail, helping me clarify many of my solutions, and providing solutions when I was unsure how best to proceed. It was a pleasure to collaborate with both Chuck and Gene in the creation of this solutions manual. They share my hope that you will find these solutions useful in learning physics.

We want to thank Lay Nam Chang (Virginia Polytechnic Institute), Brent A. Corbin (UCLA), Alan Cresswell (Shippensburg University), Ricardo S. Decca (Indiana University–Purdue University), Michael Dubson (The University of Colorado at Boulder), David Faust (Mount Hood Community College), Philip Fraundorf (The University of Missouri–Saint Louis), Clint Harper (Moorpark College), Kristi R. G. Hendrickson (University of Puget Sound), Michael Hildreth (The University of Notre Dame), David Ingram (Ohio University), James J. Kolata (The University of Notre Dame), Eric Lane (The University of Tennessee–Chattanooga), Jerome Licini (Lehigh University), Laura McCullough (The University of Wisconsin–Stout), Carl Mungan (United States Naval Academy), Jeffrey S. Olafsen (University of Kansas), Robert Pompi (The State University of New York at Binghamton), R. J. Rollefson (Wesleyan University), Andrew Scherbakov (Georgia Institute of Technology), Bruce A. Schumm (University of Chicago), Dan Styer (Oberlin College), Daniel Marlow (Princeton University), Jeffrey Sundquist (Palm Beach Community College–South), Cyrus Taylor (Case Western Reserve University), and Fulin Zuo (University of Miami), for their reviews of the problems and their solutions.

Jerome Licini (Lehigh University), Michael Crivello (San Diego Mesa College), Paul Quinn (University of Kansas), and Daniel Lucas (University of Wisconsin–Madison) error checked the solutions. Without their thorough and critical work, many errors would have remained to be discovered by the users of this solutions manual. Their assistance is greatly appreciated. In spite of their best efforts, there may still be errors in some of the solutions, and for those I (dm) assume full responsibility. Should you find errors or think of alternative solutions that you would like to call to my attention, I would appreciate it if you would communicate them to me by sending them to asktipler@whfreeman.com.

It was a pleasure to work with Brian Donnellan, Media and Supplements Editor for Physics, who guided us through the creation of this solutions manual. Our thanks to Amanda McCorquodale and Eileen McGinnis for organizing the reviewing and error-checking process.

February 2003

David Mills
Professor Emeritus
College of the Redwoods

Charles L. Adler
Saint Mary's College of Maryland

Chapter 1
Systems of Measurement

Conceptual Problems

*1 • Which of the following is *not* one of the fundamental physical quantities in the SI system? (*a*) mass. (*b*) length. (*c*) force. (*d*) time. (*e*) All of the above are fundamental physical quantities.

Determine the Concept The fundamental physical quantities in the SI system include mass, length, and time. Force, being the product of mass and acceleration, is not a fundamental quantity. $\boxed{(c) \text{ is correct.}}$

*5 • The prefix pico means (*a*) 10^{-12}, (*b*) 10^{-6}, (*c*) 10^{-3}, (*d*) 10^{6}, (*e*) 10^{9}.

Determine the Concept Consulting Table 1-1 we note that the prefix pico means 10^{-12}. $\boxed{(a) \text{ is correct.}}$

Estimation and Approximation

*10 •• The angle subtended by the moon's diameter at a point on the earth is about 0.524°. Use this and the fact that the moon is about 384 Mm away to find the diameter of the moon. (The angle θ subtended by the moon is approximately D/r_m, where D is the diameter of the moon and r_m is the distance to the moon.)

Figure 1-2 Problem 10

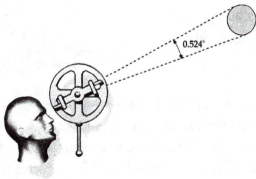

Picture the Problem Because θ is small, we can approximate it by $\theta \approx D/r_m$ provided that it is in radian measure. We can solve this relationship for the diameter of the moon.

Express the moon's diameter D $D = \theta r_m$
in terms of the angle it subtends

at the earth θ and the earth-moon
distance r_m:

Find θ in radians:

$$\theta = 0.524° \times \frac{2\pi \, \text{rad}}{360°} = 0.00915 \, \text{rad}$$

Substitute and evaluate D:

$$D = (0.00915 \, \text{rad})(384 \, \text{Mm})$$
$$= \boxed{3.51 \times 10^6 \, \text{m}}$$

*11 ••   The sun has a mass of 1.99×10^{30} kg and is composed mostly of
hydrogen, with only a small fraction being heavier elements. The hydrogen atom
has a mass of 1.67×10^{-27} kg. Estimate the number of hydrogen atoms in the sun.

Picture the Problem We'll assume that the sun is made up entirely of hydrogen.
Then we can relate the mass of the sun to the number of hydrogen atoms and the
mass of each.

Express the mass of the sun M_S
as the product of the number of
hydrogen atoms N_H and the mass
of each atom
M_H:

$$M_S = N_H M_H$$

Solve for N_H:

$$N_H = \frac{M_S}{M_H}$$

Substitute numerical values and
evaluate N_H:

$$N_H = \frac{1.99 \times 10^{30} \, \text{kg}}{1.67 \times 10^{-27} \, \text{kg}} = \boxed{1.19 \times 10^{57}}$$

*14 ••   (*a*) Estimate the number of gallons of gasoline used per day by
automobiles in the United States and the total amount of money spent on it. (*b*)
If 19.4 gallons of gasoline can be made from one barrel of crude oil, estimate the
total number of barrels of oil imported into the United States per year to make
gasoline. How many barrels per day is this?

Picture the Problem The population of the United States is roughly 3×10^8
people. Assuming that the average family has four people, with an average of two
cars per family, there are about 1.5×10^8 cars in the United States. If we double
that number to include trucks, cabs, etc., we have 3×10^8 vehicles. Let's assume
that each vehicle uses, on average, about 12 gallons of gasoline per week.

(a) Find the daily consumption of gasoline G:

$$G = (3 \times 10^8 \text{ vehicles})(2 \text{ gal/d})$$
$$= 6 \times 10^8 \text{ gal/d}$$

Assuming a price per gallon $P = \$1.50$, find the daily cost C of gasoline:

$$C = GP = (6 \times 10^8 \text{ gal/d})(\$1.50/\text{gal})$$
$$= \$9 \times 10^8 /\text{d} \approx \boxed{\$1 \text{ billion dollars/d}}$$

(b) Relate the number of barrels N of crude oil required annually to the yearly consumption of gasoline Y and the number of gallons of gasoline n that can be made from one barrel of crude oil:

$$N = \frac{Y}{n} = \frac{G\Delta t}{n}$$

Substitute numerical values and estimate N:

$$N = \frac{(6 \times 10^8 \text{ gal/d})(365.24 \text{ d/y})}{19.4 \text{ gal/barrel}}$$
$$\approx \boxed{10^{10} \text{ barrels/y}}$$

*17 •• Estimate the yearly toll revenue of the George Washington Bridge in New York. At last glance, the toll is \$6 to go into New York from New Jersey; going from New York into New Jersey is free. There are a total of 14 lanes.

Picture the Problem Assume that, on average, four cars go through each toll station per minute. Let R represent the yearly revenue from the tolls. We can estimate the yearly revenue from the number of lanes N, the number of cars per minute n, and the \$6 toll per car C.

$$R = NnC = 14 \text{ lanes} \times 4 \frac{\text{cars}}{\text{min}} \times 60 \frac{\text{min}}{\text{h}} \times 24 \frac{\text{h}}{\text{d}} \times 365.24 \frac{\text{d}}{\text{y}} \times \frac{\$6}{\text{car}} = \boxed{\$177\text{M}}$$

Units

*20 • Write out the following (which are not SI units) without using abbreviations. For example, 10^3 meters = 1 kilometer: (a) 10^{-12} boo, (b) 10^9 low, (c) 10^{-6} phone, (d) 10^{-18} boy, (e) 10^6 phone, (f) 10^{-9} goat, (g) 10^{12} bull.

Picture the Problem We can use the definitions of the metric prefixes listed in Table 1-1 to express each of these quantities without abbreviations.

(a) 10^{-12} boo = $\boxed{1 \text{ picoboo}}$ (e) 10^{6} phone = $\boxed{1 \text{ megaphone}}$

(b) 10^{9} low = $\boxed{1 \text{ gigalow}}$ (f) 10^{-9} goat = $\boxed{1 \text{ nanogoat}}$

(c) 10^{-6} phone = $\boxed{1 \text{ microphone}}$ (g) 10^{12} bull = $\boxed{1 \text{ terabull}}$

(d) 10^{-18} boy = $\boxed{1 \text{ attoboy}}$

Conversion of Units

*25 •   A basketball player is $6 \text{ ft } 10\frac{1}{2} \text{ in}$ tall. What is his height in centimeters?

Picture the Problem We'll first express his height in inches and then use the conversion factor 1 in = 2.54 cm.

Express the player's height in inches:

$$h = 6\,\text{ft} \times \frac{12\,\text{in}}{\text{ft}} + 10.5\,\text{in} = 82.5\,\text{in}$$

Convert h into cm:

$$h = 82.5\,\text{in} \times \frac{2.54\,\text{cm}}{\text{in}} = \boxed{210\,\text{cm}}$$

*28 •   Find the conversion factor to convert from miles per hour into kilometers per hour.

Picture the Problem Let v be the speed of an object in mi/h. We can use the conversion factor 1 mi = 1.61 km to convert this speed to km/h.

Multiply v mi/h by 1.61 km/mi to convert v into km/h:

$$v\,\frac{\text{mi}}{\text{h}} = v\,\frac{\text{mi}}{\text{h}} \times \frac{1.61\,\text{km}}{\text{mi}} = \boxed{1.61v\,\text{km/h}}$$

*33 ••   In the following, x is in meters, t is in seconds, v is in meters per second, and the acceleration a is in meters per second squared. Find the SI units of each combination: (a) v^2/x, (b) $\sqrt{x/a}$, (c) $\frac{1}{2}at^2$.

Picture the Problem We can treat the SI units as though they are algebraic quantities to simplify each of these combinations of physical quantities and constants.

(a) Express and simplify the units of v^2/x:

$$\frac{(m/s)^2}{m} = \frac{m^2}{m \cdot s^2} = \boxed{\frac{m}{s^2}}$$

(b) Express and simplify the units of $\sqrt{x/a}$:

$$\sqrt{\frac{m}{m/s^2}} = \sqrt{s^2} = \boxed{s}$$

(c) Noting that the constant factor $\frac{1}{2}$ has no units, express and simplify the units of $\frac{1}{2}at^2$:

$$\left(\frac{m}{s^2}\right)(s)^2 = \left(\frac{m}{s^2}\right)(s^2) = \boxed{m}$$

Dimensions of Physical Quantities

*36 •• The SI unit of force, the kilogram-meter per second squared (kg·m/s²) is called the newton (N). Find the dimensions and the SI units of the constant G in Newton's law of gravitation $F = Gm_1m_2/r^2$.

Picture the Problem We can solve Newton's law of gravitation for G and substitute the dimensions of the variables. Treating them as algebraic quantities will allow us to express the dimensions in their simplest form. Finally, we can substitute the SI units for the dimensions to find the units of G.

Solve Newton's law of gravitation for G to obtain:

$$G = \frac{Fr^2}{m_1m_2}$$

Substitute the dimensions of the variables:

$$G = \frac{\frac{ML}{T^2} \times L^2}{M^2} = \boxed{\frac{L^3}{MT^2}}$$

Use the SI units for L, M, and T:

Units of G are $\boxed{\dfrac{m^3}{kg \cdot s^2}}$

*41 •• When an object falls through air, there is a drag force that depends on the product of the surface area of the object and the square of its velocity, that is, $F_{air} = CAv^2$, where C is a constant. Determine the dimensions of C.

Picture the Problem We can find the dimensions of C by solving the drag force equation for C and substituting the dimensions of force, area, and velocity.

Solve the drag force equation for the constant C:

$$C = \frac{F_{air}}{Av^2}$$

Express this equation
dimensionally:

$$[C] = \frac{[F_{air}]}{[A][v]^2}$$

Substitute the dimensions of
force, area, and velocity and
simplify to obtain:

$$[C] = \frac{\dfrac{ML}{T^2}}{L^2\left(\dfrac{L}{T}\right)^2} = \boxed{\dfrac{M}{L^3}}$$

Scientific Notation and Significant Figures

*43 •   Express as a decimal number without using powers of 10 notation:
(a) 3×10^4, (b) 6.2×10^{-3}, (c) 4×10^{-6}, (d) 2.17×10^5.

Picture the Problem We can use the rules governing scientific notation to
express each of these numbers as a decimal number.

(a) $3 \times 10^4 = \boxed{30,000}$ (c) $4 \times 10^{-6} = \boxed{0.000004}$

(b) $6.2 \times 10^{-3} = \boxed{0.0062}$ (d) $2.17 \times 10^5 = \boxed{217,000}$

*47 •   A cell membrane has a thickness of about 7 nm. How many cell
membranes would it take to make a stack 1 in high?

Picture the Problem Let N represent the required number of membranes and
express N in terms of the thickness of each cell membrane.

Express N in terms of the
thickness of a single membrane:

$$N = \frac{1\,\text{in}}{7\,\text{nm}}$$

Convert the units into SI units
and simplify to obtain:

$$N = \frac{1\,\text{in}}{7\,\text{nm}} \times \frac{2.54\,\text{cm}}{\text{in}} \times \frac{1\,\text{m}}{100\,\text{cm}} \times \frac{1\,\text{nm}}{10^{-9}\,\text{m}}$$

$$= \boxed{4 \times 10^6}$$

*49 •   Perform the following calculations and round off the answers to the
correct number of significant figures: (a) $3.141592654 \times (23.2)^2$, (b) $2 \times 3.141592654 \times 0.76$, (c) $(4/3)\, \pi \times (1.1)^3$, (d) $(2.0)^5/3.141592654$.

Picture the Problem Apply the general rules concerning the multiplication,
division, addition, and subtraction of measurements to evaluate each of the

given expressions.

(*a*) The second factor and the result have three significant figures:

$$3.141592654 \times (23.2)^2 = \boxed{1.69 \times 10^3}$$

(*b*) We'll assume that 2 is exact. Therefore, the result will have two significant figures:

$$2 \times 3.141592654 \times 0.76 = \boxed{4.8}$$

(*c*) We'll assume that 4/3 is exact. Therefore the result will have two significant figures:

$$\frac{4}{3}\pi \times (1.1)^3 = \boxed{5.6}$$

(*d*) Because 2.0 has two significant figures, the result has two significant figures:

$$\frac{(2.0)^5}{3.141592654} = \boxed{10}$$

General Problems

*51 •   If you could count $1 per second, how many years would it take to count 1 billion dollars (1 billion = 10^9)?

Picture the Problem We can use a series of conversion factors to convert 1 billion seconds into years.

Multiply 1 billion seconds by the appropriate conversion factors to convert it into years:

$$10^9\,\text{s} = 10^9\,\text{s} \times \frac{1\text{h}}{3600\,\text{s}} \times \frac{1\text{d}}{24\,\text{h}} \times \frac{1\text{y}}{365.24\,\text{d}} = \boxed{31.7\,\text{y}}$$

*57 ••   The astronomical unit (AU) is defined in terms of the distance from the earth to the sun, namely 1.496×10^{11} m. The parsec is the radius of a circle for which a central angle of 1 s intercepts an arc of length 1 AU. The light-year is the distance that light travels in one year. (*a*) How many parsecs are there in one astronomical unit? (*b*) How many meters are in a parsec? (*c*) How many meters in a light-year? (*d*) How many astronomical units in a light-year? (*e*) How many light-years in a parsec?

Picture the Problem We can use the relationship between an angle *θ*, measured in radians, subtended at the center of a circle, the radius *R* of the circle, and the length L of the arc to answer these questions concerning the astronomical units of measure.

(*a*) Relate the angle θ subtended by an arc of length S to the distance R:

$$\theta = \frac{S}{R} \qquad (1)$$

Solve for and evaluate S:

$$S = R\theta$$

$$= (1\,\text{parsec})(1\,\text{s})\left(\frac{1\,\text{min}}{60\,\text{s}}\right)$$

$$\times \left(\frac{1°}{60\,\text{min}}\right)\left(\frac{2\pi\,\text{rad}}{360°}\right)$$

$$= \boxed{4.85 \times 10^{-6}\,\text{parsec}}$$

(*b*) Solve equation (1) for and evaluate R:

$$R = \frac{S}{\theta}$$

$$= \frac{1.496 \times 10^{11}\,\text{m}}{(1\,\text{s})\left(\dfrac{1\,\text{min}}{60\,\text{s}}\right)\left(\dfrac{1°}{60\,\text{min}}\right)\left(\dfrac{2\pi\,\text{rad}}{360°}\right)}$$

$$= \boxed{3.09 \times 10^{16}\,\text{m}}$$

(*c*) Relate the distance D light travels in a given interval of time Δt to its speed c and evaluate D for $\Delta t = 1$ y:

$$D = c\Delta t$$

$$= \left(3 \times 10^{8}\,\frac{\text{m}}{\text{s}}\right)(1\,\text{y})\left(3.156 \times 10^{7}\,\frac{\text{s}}{\text{y}}\right)$$

$$= \boxed{9.47 \times 10^{15}\,\text{m}}$$

(*d*) Use the definition of 1 AU and the result from part (*c*) to obtain:

$$1\,c \cdot \text{y} = \left(9.47 \times 10^{15}\,\text{m}\right)\left(\frac{1\,\text{AU}}{1.496 \times 10^{11}\,\text{m}}\right)$$

$$= \boxed{6.33 \times 10^{4}\,\text{AU}}$$

(*e*) Combine the results of parts (*b*) and (*c*) to obtain:

$$1\,\text{parsec} = \left(3.08 \times 10^{16}\,\text{m}\right)$$

$$\times \left(\frac{1\,c \cdot \text{y}}{9.47 \times 10^{15}\,\text{m}}\right)$$

$$= \boxed{3.25\,c \cdot \text{y}}$$

*59 ••   The Super-Kamiokande neutrino detector in Japan is a large transparent cylinder filled with ultra-pure water. The height of the cylinder is 41.4 m and the diameter is 39.3 m. Calculate the mass of the water in the cylinder. Does this match the claim posted on the official Super-K website that the detector uses 50,000 tons of water? The density of water is 1000 kg/m³.

Picture the Problem We can use the definition of density to relate the mass of the water in the cylinder to its volume and the formula for the volume of a cylinder to express the volume of water used in the detector's cylinder. To convert our answer in kg to lb, we can use the fact that 1 kg weighs about 2.205 lb.

Relate the mass of water contained in the cylinder to its density and volume:

$$m = \rho V$$

Express the volume of a cylinder in terms of its diameter d and height h:

$$V = A_{\text{base}} h = \frac{\pi}{4} d^2 h$$

Substitute to obtain:

$$m = \rho \frac{\pi}{4} d^2 h$$

Substitute numerical values and evaluate m:

$$m = \left(10^3 \text{ kg/m}^3\right)\left(\frac{\pi}{4}\right)(39.3\,\text{m})^2 (41.4\,\text{m})$$

$$= 5.02 \times 10^7 \text{ kg}$$

Convert 5.02×10^7 kg to tons:

$$m = 5.02 \times 10^7 \text{ kg} \times \frac{2.205\,\text{lb}}{\text{kg}} \times \frac{1\,\text{ton}}{2000\,\text{lb}}$$

$$= 55.4 \times 10^3 \text{ ton}$$

The 50,000 - ton claim is conservative. The actual weight is closer to 55,000 tons.

*62 •••   The period T of a simple pendulum depends on the length L of the pendulum and the acceleration of gravity g (dimensions L/T^2). (*a*) Find a simple combination of L and g that has the dimensions of time. (*b*) Check the dependence of the period T on the length L by measuring the period (time for a complete swing back and forth) of a pendulum for two different values of L. (*c*) The correct formula relating T to L and g involves a constant that is a multiple of π, and cannot be obtained by the dimensional analysis of part (*a*). It can be found by experiment as in Part (*b*) if g is known. Using the value $g = 9.81$ m/s² and your experimental results from Part (*b*), find the formula relating T to L and g.

Picture the Problem We can express the relationship between the period T of the pendulum, its length L, and the acceleration of gravity g as $T = CL^a g^b$ and perform dimensional analysis to find the values of a and b and, hence, the function relating these variables. Once we've performed the experiment called for in part (b), we can determine an experimental value for C.

(a) Express T as the product of L and g raised to powers a and b:

$$T = CL^a g^b \qquad (1)$$

where C is a dimensionless constant.

Write this equation in dimensional form:

$$[T] = [L]^a [g]^b$$

Noting that the symbols for the dimension of the period and length of the pendulum are the same as those representing the physical quantities, substitute the dimensions to obtain:

$$T = L^a \left(\frac{L}{T^2} \right)^b$$

Because L does not appear on the left-hand side of the equation, we can write this equation as:

$$L^0 T^1 = L^{a+b} T^{-2b}$$

Equate the exponents to obtain:

$$a + b = 0 \text{ and } -2b = 1$$

Solve these equations simultaneously to find a and b:

$$a = \tfrac{1}{2} \text{ and } b = -\tfrac{1}{2}$$

Substitute in equation (1) to obtain:

$$T = CL^{1/2} g^{-1/2} = \boxed{C\sqrt{\frac{L}{g}}} \qquad (2)$$

(b) If you use pendulums of lengths 1 m and 0.5 m; the periods should be about:

$$T(1\,\text{m}) = \boxed{2\,\text{s}}$$

and

$$T(0.5\,\text{m}) = \boxed{1.4\,\text{s}}$$

(c) Solve equation (2) for C:

$$C = T\sqrt{\frac{g}{L}}$$

Evaluate C with $L = 1$ m and $T = 2$ s:

$$C = (2\,\mathrm{s})\sqrt{\frac{9.81\,\mathrm{m/s}^2}{1\,\mathrm{m}}} = 6.26 \approx 2\pi$$

Substitute in equation (2) to obtain:

$$\boxed{T = 2\pi\sqrt{\frac{L}{g}}}$$

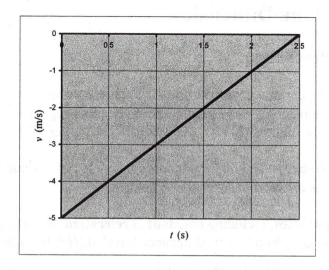

*8   ••   Draw careful graphs of the position and velocity and acceleration over the time period $0 \leq t \leq 25$ s for a cart that

(*a*) moves away from the origin at a slow and steady (constant) velocity for the first 5 s;
(*b*) moves away at a medium-fast, steady (constant) velocity for the next 5 s;
(*c*) stands still for the next 5 s;
(*d*) moves toward the origin at a slow and steady (constant) velocity for the next 5 s;
(*e*) stands still for the last 5 s.

Determine the Concept Velocity is the slope of the position versus time curve and acceleration is the slope of the velocity versus time curve. See the graphs below.

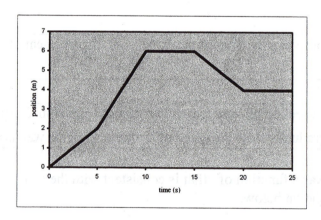

Evaluate C with $L = 1$ m and $T = 2$ s:

$$C = (2\,s)\sqrt{\frac{9.81\,m/s^2}{1\,m}} = 6.26 \approx 2\pi$$

Substitute in equation (2) to obtain:

$$T = \boxed{2\pi\sqrt{\frac{L}{g}}}$$

Chapter 2
Motion in One Dimension

Conceptual Problems

*2 • An object thrown straight up falls back to the ground. Its time of flight is T, its maximum height is H, and its height at release is negligible. Its average speed for the entire flight is (a) H/T, (b) 0, (c) $H/(2T)$, (d) $2H/T$.

Determine the Concept The important concept here is that "average speed" is being requested as opposed to "average velocity".

Under all circumstances, including **constant acceleration**, the definition of the average speed is the ratio of the total distance traveled $(H + H)$ to the total time elapsed, in this case $2H/T$. $\boxed{(d) \text{ is correct.}}$

Remarks: Because this motion involves a round trip, if the question asked for "average velocity," the answer would be zero.

*5 • Stand in the center of a large room. Call movement to your right "positive," and movement to your left "negative." Walk across the room along a straight line in such a way that, after getting started, your velocity is negative but your acceleration is positive. (a) Is your displacement initially positive or negative? Explain. (b) Describe how you vary your speed as you walk. (c) Sketch a graph of v versus t for your motion.

Determine the Concept The important concept is that when both the acceleration and the velocity are in the same direction, the speed increases. On the other hand, when the acceleration and the velocity are in opposite directions, the speed decreases.

(a) | Because your velocity remains negative, your displacement must be *negative*.

(b) | Define the direction of your trip as the negative direction. During the last five steps gradually slow the speed of walking, until the wall is reached.

(c) A graph of v as a function of t that is consistent with the conditions stated in the problem is shown below.

13

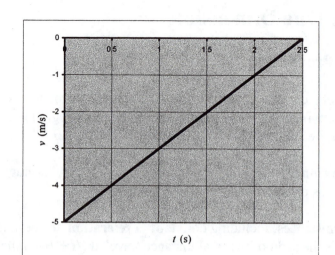

*8 ••   Draw careful graphs of the position and velocity and acceleration over the time period $0 \leq t \leq 25$ s for a cart that

(*a*) moves away from the origin at a slow and steady (constant) velocity for the first 5 s;
(*b*) moves away at a medium-fast, steady (constant) velocity for the next 5 s;
(*c*) stands still for the next 5 s;
(*d*) moves toward the origin at a slow and steady (constant) velocity for the next 5 s;
(*e*) stands still for the last 5 s.

Determine the Concept Velocity is the slope of the position versus time curve and acceleration is the slope of the velocity versus time curve. See the graphs below.

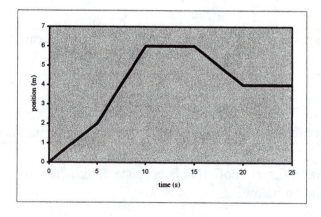

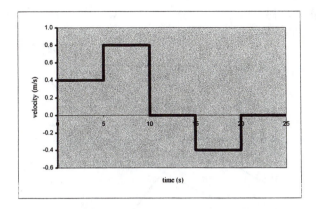

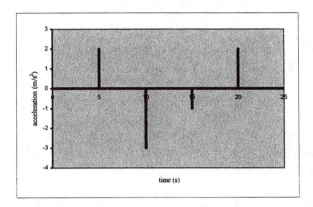

*11 •• Dr. Josiah S. Carberry stands at the top of the Sears tower in Chicago. Wanting to emulate Galileo, and ignoring the safety of the pedestrians below, he drops a bowling ball from the top of the tower. One second later, he drops a second bowling ball. While the balls are in the air, does their separation (*a*) increase over time, (*b*) decrease, or (*c*) stay the same? Ignore any effects that air resistance may have.

Determine the Concept Neglecting air resistance, the balls are in free fall, each with the same free-fall acceleration, which is a constant.

At the time the second ball is released, the first ball is already moving. Thus, during any time interval their velocities will increase by exactly the same amount. What can be said about the speeds of the two balls? *The first ball will always be moving faster than the second ball.*

This being the case, what happens to the separation of the two balls while they are both falling? *Their separation increases.* | (*a*) is correct. |

*13 • Which of the velocity-versus-time curves in Figure 2-24 best describes the motion of an object with constant positive acceleration?

Figure 2-24 Problem 13

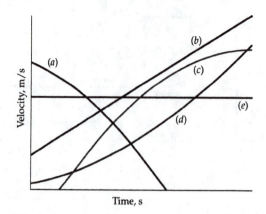

Determine the Concept The slope of a $v(t)$ curve at any point in time represents the acceleration at that instant. Only one curve has a constant and positive slope.

(b) is correct.

*15 • Is it possible for the average velocity of an object to be zero during some interval, even though its average velocity for the first half of the interval is not zero? Explain.

Determine the Concept Note that the "average velocity" is being requested as opposed to the "average speed."

Yes. In any roundtrip, A to B, and back to A, the average velocity is zero.

$$v_{av(A \to B \to A)} = \frac{\Delta x}{\Delta t} = \frac{\Delta x_{AB} + \Delta x_{BA}}{\Delta t}$$

$$= \frac{\Delta x_{AB} + (-\Delta x_{BA})}{\Delta t} = \frac{0}{\Delta t}$$

$$= \boxed{0}$$

On the other hand, the average velocity between A and B is not generally zero.

$$v_{av(A \to B)} = \frac{\Delta x_{AB}}{\Delta t} \neq \boxed{0}$$

Remarks: Consider an object launched up in the air. Its average velocity on the way up is NOT zero. Neither is it zero on the way down. However, over the round trip, it is zero.

*20 •• For each of the four graphs of x versus t in Figure 2-27, answer the following questions: (a) Is the velocity at time t_2 greater than, less than, or equal to the velocity at time t_1? (b) Is the speed at time t_2 greater than, less than, or equal to the speed at time t_1?

Figure 2-27 Problem 20

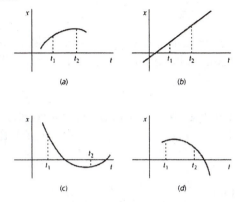

Determine the Concept In one-dimensional motion, the velocity is the slope of a position-versus-time plot and can be either positive or negative. On the other hand, the speed is the magnitude of the velocity and can only be positive. We'll use v to denote velocity and the word "speed" for how fast the object is moving.

(a)

curve *a*: $v(t_2) < v(t_1)$

curve *b*: $v(t_2) = v(t_1)$

curve *c*: $v(t_2) > v(t_1)$

curve *d*: $v(t_2) < v(t_1)$

(b)

curve *a*: $\text{speed}(t_2) < \text{speed}(t_1)$

curve *b*: $\text{speed}(t_2) = \text{speed}(t_1)$

curve *c*: $\text{speed}(t_2) < \text{speed}(t_1)$

curve *d*: $\text{speed}(t_2) > \text{speed}(t_1)$

*27 •• Assume that a Porsche accelerates uniformly from 80.5 km/h (50 mi/h) at $t = 0$ to 113 km/h (70 mi/h) at $t = 9$ s. Which graph in Figure 2-28 best describes the motion of the car?

Figure 2-28 Problem 27

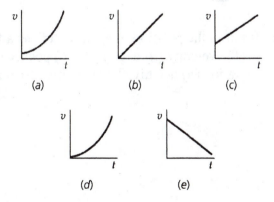

Determine the Concept Because the Porsche accelerates uniformly, we need to look for a graph that represents constant acceleration. We are told that the Porsche has a constant acceleration that is positive (the velocity is increasing); therefore we must look for a velocity-versus-time curve with a positive constant slope and a nonzero intercept. $\boxed{(c)\,\text{is correct.}}$

*28 ••   An object is dropped from rest and falls a distance D in a given time. If the time during which it falls is doubled, the distance it falls will be (a) 4D, (b) 2D, (c) D, (d) D/2, (e) D/4.

Determine the Concept In the absence of air resistance, the object experiences constant acceleration. Choose a coordinate system in which the downward direction is positive.

Express the distance D that an object, released from rest, falls in time t:

$$D = \tfrac{1}{2}gt^2$$

Because the distance fallen varies with the square of the time, during the first two seconds it falls four times the distance it falls during the first second.

$\boxed{(a)\,\text{is correct.}}$

*31 •   If an object is moving at constant acceleration in a straight line, its instantaneous velocity halfway through any time interval is (a) greater than its average velocity, (b) less than its average velocity, (c) equal to its average velocity, (d) half its average velocity, (e) twice its average velocity.

Determine the Concept Because the acceleration of the object is constant, the constant acceleration equations can be used to describe its motion. The special expression for average velocity for constant acceleration

is $v_{av} = \dfrac{v_i + v_f}{2}$. $\boxed{(c)\,\text{is correct.}}$

*39 ••   Figure 2-30 shows the position of a car plotted as a function of time. At which times t_0 to t_7 is the velocity (a) negative? (b) positive? (c) zero? At which times is the acceleration (a) negative? (b) positive? (c) zero?

Figure 2-30 Problem 39

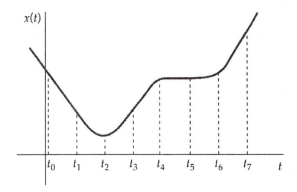

Determine the Concept The velocity is the slope of a position versus time curve and the acceleration is the rate at which the velocity, and thus the slope, changes.

Velocity	(a) Negative at t_0 and t_1.
	(b) Positive at t_3, t_4, t_6, and t_7.
	(c) Zero at t_2 and t_5.
Acceleration	(a) Negative at t_4.
	(b) Positive at t_2 and t_6.
	(c) Zero at t_0, t_1, t_3, t_5, and t_7.

Remarks: The acceleration is positive at points where the slope increases as you move toward the right.

Estimation and Approximation

*43 ••   Occasionally, people can survive after falling large distances if the surface they fall on is soft enough. During a traverse of the Eiger's infamous Nordvand, mountaineer Carlos Ragone's rock anchor pulled out and he plummeted 500 feet to land in snow. Amazingly, he suffered only a few bruises and a wrenched shoulder. (a) What final speed did he reach before impact? Ignore air resistance. (b) Assuming that his impact left a hole in the snow four feet deep, estimate his acceleration as he slowed to a stop. Assume that the acceleration was constant. Express this as a multiple of g (the magnitude of free-fall acceleration at the surface of the earth).

Picture the Problem In the absence of air resistance, Carlos' acceleration is constant. Because all the motion is downward, let's use a coordinate system in which downward is positive and the origin is at the point at which the fall began.

(a) Using a constant-acceleration equation, relate Carlos' final velocity to his initial velocity,

$$v^2 = v_0^2 + 2a\Delta y$$

and, because $v_0 = 0$ and $a = g$,

| Using the definition of average speed, express the travel time for the nerve impulse: | $\Delta t = \dfrac{\Delta x}{v_{av}}$ |

Substitute numerical values and evaluate Δt:

$$\Delta t = \frac{1.7\,\mathrm{m}}{120\,\mathrm{m/s}} = \boxed{14.2\,\mathrm{ms}}$$

Speed, Displacement, and Velocity

*49 •   A runner runs 2.5 km, in a straight line, in 9 min and then takes 30 min to walk back to the starting point. (*a*) What is the runner's average velocity for the first 9 min? (*b*) What is the average velocity for the time spent walking? (*c*) What is the average velocity for the whole trip? (*d*) What is the average speed for the whole trip?

Picture the Problem In this problem the runner is traveling in a straight line but not at constant speed - first she runs, then she walks. Let's choose a coordinate system in which her initial direction of motion is taken as the positive *x* direction.

(*a*) Using the definition of average velocity, calculate the average velocity for the first 9 min:

$$v_{av} = \frac{\Delta x}{\Delta t} = \frac{2.5\,\mathrm{km}}{9\,\mathrm{min}} = \boxed{0.278\,\mathrm{km/min}}$$

(*b*) Using the definition of average velocity, calculate her average speed for the 30 min spent walking:

$$v_{av} = \frac{\Delta x}{\Delta t} = \frac{-2.5\,\mathrm{km}}{30\,\mathrm{min}}$$
$$= \boxed{-0.0833\,\mathrm{km/min}}$$

(*c*) Express her average velocity for the whole trip:

$$v_{av} = \frac{\Delta x_{\text{round trip}}}{\Delta t} = \frac{0}{\Delta t} = \boxed{0}$$

(*d*) Finally, express her average speed for the whole trip:

$$\text{Average speed} = \frac{\text{distance traveled}}{\text{elapsed time}}$$
$$= \frac{2(2.5\,\mathrm{km})}{30\,\mathrm{min} + 9\,\mathrm{min}}$$
$$= \boxed{0.128\,\mathrm{km/min}}$$

*52 •   The speed of light, *c*, is 3×10^8 m/s. (*a*) How long does it take for light to travel from the sun to the earth, a distance of 1.5×10^{11} m? (*b*) How long does it take light to travel from the moon to the earth, a distance of 3.84×10^8 m? (*c*) A light-year is a unit of distance equal to that traveled by light in 1 year. Convert 1 light-year into kilometers and miles.

Figure 2-30 Problem 39

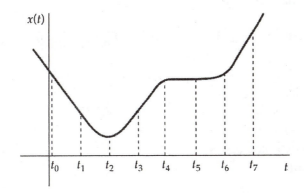

Determine the Concept The velocity is the slope of a position versus time curve and the acceleration is the rate at which the velocity, and thus the slope, changes.

Velocity

(a) Negative at t_0 and t_1.
(b) Positive at t_3, t_4, t_6, and t_7.
(c) Zero at t_2 and t_5.

Acceleration

(a) Negative at t_4.
(b) Positive at t_2 and t_6.
(c) Zero at t_0, t_1, t_3, t_5, and t_7.

Remarks: The acceleration is positive at points where the slope increases as you move toward the right.

Estimation and Approximation

*43 ••   Occasionally, people can survive after falling large distances if the surface they fall on is soft enough. During a traverse of the Eiger's infamous Nordvand, mountaineer Carlos Ragone's rock anchor pulled out and he plummeted 500 feet to land in snow. Amazingly, he suffered only a few bruises and a wrenched shoulder. (a) What final speed did he reach before impact? Ignore air resistance. (b) Assuming that his impact left a hole in the snow four feet deep, estimate his acceleration as he slowed to a stop. Assume that the acceleration was constant. Express this as a multiple of g (the magnitude of free-fall acceleration at the surface of the earth).

Picture the Problem In the absence of air resistance, Carlos' acceleration is constant. Because all the motion is downward, let's use a coordinate system in which downward is positive and the origin is at the point at which the fall began.

(a) Using a constant-acceleration equation, relate Carlos' final velocity to his initial velocity,

$v^2 = v_0^2 + 2a\Delta y$

and, because $v_0 = 0$ and $a = g$,

acceleration, and distance fallen and solve for his final velocity:

$$v = \sqrt{2g\Delta y}$$

Substitute numerical values and evaluate v:

$$v = \sqrt{2(9.81\,\text{m/s}^2)(150\,\text{m})} = \boxed{54.2\,\text{m/s}}$$

(b) While his acceleration by the snow is not constant, solve the same constant- acceleration equation to get an estimate of his average acceleration:

$$a = \frac{v^2 - v_0^2}{2\Delta y}$$

Substitute numerical values and evaluate a:

$$a = \frac{-(54\,\text{m/s}^2)^2}{2(1.22\text{m})} = -1.20 \times 10^3\,\text{m/s}^2$$

$$= \boxed{-123g}$$

Remarks: The final velocity we obtained in part (a), approximately 121 mph, is about the same as the terminal velocity for an "average" man. This solution is probably only good to about 20% accuracy.

*46 ••   The photograph in Figure 2-32 is a short-time exposure (1/30 s) of a juggler with two tennis balls in the air. The tennis ball near the top of its trajectory is less blurred than the lower one. Why is that? Can you estimate the speed of the lower ball from the picture?

Figure 2-32 Problem 46

Determine the Concept This is a constant-acceleration problem with $a = -g$ if we take upward to be the positive direction.

At the maximum height the ball will reach, its speed will be near zero and when the ball has just been tossed in the air its speed is

Because the ball is moving slowly its blur is relatively short (i.e., there is less blurring).

near its maximum value. What conclusion can you draw from the image of the ball near its maximum height?

To estimate the initial speed of the ball:

a) Estimate how far the ball being tossed moves in 1/30 s:

The ball moves about 3 ball diameters in 1/30 s.

b) Estimate the diameter of a tennis ball:

The diameter of a tennis ball is approximately 5 cm.

c) Now one can calculate the approximate distance the ball moved in 1/30 s:

$$\text{Distance traveled} = (3 \text{ diameters})$$
$$\times (5 \text{ cm/diameter})$$
$$= 15 \text{ cm}$$

d) Calculate the average speed of the tennis ball over this distance:

$$\text{Average speed} = \frac{15\,\text{cm}}{\frac{1}{30}\text{s}} = 450 \text{ cm/s}$$
$$= 4.50 \text{ m/s}$$

e) Because the time interval is very short, the average speed of the ball is a good approximation to its initial speed:

$$\therefore v_0 = 4.5 \text{ m/s}$$

f) Finally, use the constant acceleration equation $v^2 = v_0^2 + 2a\Delta y$ to solve for and evaluate Δy:

$$\Delta y = \frac{-v_0^2}{2a} = \frac{-(4.5\,\text{m/s})^2}{2(-9.81\,\text{m/s}^2)} = \boxed{1.03\,\text{m}}$$

Remarks: This maximum height is in good agreement with the height of the higher ball in the photograph.

*47 ••   Look up the speed at which a nerve impulse travels through the body. Estimate the time between stubbing your toe on a rock and feeling the pain due to this.

Picture the Problem The average speed of a nerve impulse is approximately 120 m/s. Assume an average height of 1.7 m and use the definition of average speed to estimate the travel time for the nerve impulse.

Using the definition of average speed, express the travel time for the nerve impulse:

$$\Delta t = \frac{\Delta x}{v_{av}}$$

Substitute numerical values and evaluate Δt:

$$\Delta t = \frac{1.7\,\text{m}}{120\,\text{m/s}} = \boxed{14.2\,\text{ms}}$$

Speed, Displacement, and Velocity

*49 •   A runner runs 2.5 km, in a straight line, in 9 min and then takes 30 min to walk back to the starting point. (*a*) What is the runner's average velocity for the first 9 min? (*b*) What is the average velocity for the time spent walking? (*c*) What is the average velocity for the whole trip? (*d*) What is the average speed for the whole trip?

Picture the Problem In this problem the runner is traveling in a straight line but not at constant speed - first she runs, then she walks. Let's choose a coordinate system in which her initial direction of motion is taken as the positive x direction.

(*a*) Using the definition of average velocity, calculate the average velocity for the first 9 min:

$$v_{av} = \frac{\Delta x}{\Delta t} = \frac{2.5\,\text{km}}{9\,\text{min}} = \boxed{0.278\,\text{km/min}}$$

(*b*) Using the definition of average velocity, calculate her average speed for the 30 min spent walking:

$$v_{av} = \frac{\Delta x}{\Delta t} = \frac{-2.5\,\text{km}}{30\,\text{min}}$$
$$= \boxed{-0.0833\,\text{km/min}}$$

(*c*) Express her average velocity for the whole trip:

$$v_{av} = \frac{\Delta x_{\text{round trip}}}{\Delta t} = \frac{0}{\Delta t} = \boxed{0}$$

(*d*) Finally, express her average speed for the whole trip:

$$\text{Average speed} = \frac{\text{distance traveled}}{\text{elapsed time}}$$
$$= \frac{2(2.5\,\text{km})}{30\,\text{min} + 9\,\text{min}}$$
$$= \boxed{0.128\,\text{km/min}}$$

*52 •   The speed of light, c, is 3×10^8 m/s. (*a*) How long does it take for light to travel from the sun to the earth, a distance of 1.5×10^{11} m? (*b*) How long does it take light to travel from the moon to the earth, a distance of 3.84×10^8 m? (*c*) A light-year is a unit of distance equal to that traveled by light in 1 year. Convert 1 light-year into kilometers and miles.

Picture the Problem In free space, light travels in a straight line at constant speed, c.

(a) Using the definition of average speed, solve for and evaluate the time required for light to travel from the sun to the earth:

$$\text{average speed} = \frac{s}{t}$$

and

$$t = \frac{s}{\text{average speed}} = \frac{1.5 \times 10^{11} \, \text{m}}{3 \times 10^8 \, \text{m/s}}$$

$$= 500 \, \text{s} = \boxed{8.33 \, \text{min}}$$

(b) Proceed as in (a) this time using the moon-earth distance:

$$t = \frac{3.84 \times 10^8 \, \text{m}}{3 \times 10^8 \, \text{m/s}} = \boxed{1.28 \, \text{s}}$$

(c) One light-year is the distance light travels in a vacuum in one year:

$$1 \, \text{light-year} = 9.48 \times 10^{15} \, \text{m} = \boxed{9.48 \times 10^{12} \, \text{km}}$$

$$= \left(9.48 \times 10^{12} \, \text{km}\right)\left(1 \, \text{mi}/1.61 \, \text{km}\right)$$

$$= \boxed{5.89 \times 10^{12} \, \text{mi}}$$

*55 •• An archer fires an arrow, which produces a muffled "thwok" as it hits a target. If the archer hears the "thwok" exactly 1 s after firing the arrow and the average speed of the arrow was 40 m/s, what was the distance separating the archer and the target? Use 340 m/s for the speed of sound.

Picture the Problem Note that both the arrow and the sound travel a distance d. We can use the relationship between distance traveled, the speed of sound, the speed of the arrow, and the elapsed time to find the distance separating the archer and the target.

Express the elapsed time between the archer firing the arrow and hearing it strike the target:

$$\Delta t = 1 \, \text{s} = \Delta t_{\text{arrow}} + \Delta t_{\text{sound}}$$

Express the transit times for the arrow and the sound in terms of the distance, d, and their speeds:

$$\Delta t_{\text{arrow}} = \frac{d}{|v_{\text{arrow}}|} = \frac{d}{40 \, \text{m/s}}$$

and

$$\Delta t_{\text{sound}} = \frac{d}{|v_{\text{sound}}|} = \frac{d}{340 \, \text{m/s}}$$

Substitute these two relationships in the expression obtained in step 1 and solve for d:

$$\frac{d}{40 \, \text{m/s}} + \frac{d}{340 \, \text{m/s}} = 1 \, \text{s}$$

and $d = \boxed{35.8 \, \text{m}}$

*59 ••   The cheetah can run as fast as $v_1 = 113$ km/h, the falcon can fly as fast as $v_2 = 161$ km/h, and the sailfish can swim as fast as $v_3 = 105$ km/h. The three of them run a relay with each covering a distance L at maximum speed. What is the average speed v of this relay team? Compare this with the average of the three speeds.

Picture the Problem Ignoring the time intervals during which members of this relay time get up to their running speeds, their accelerations are zero and their average speed can be found from its definition.

<table>
<tr><td>Using its definition, relate the average speed to the total distance traveled and the elapsed time:</td><td>$$|v_{av}| = \frac{\text{distance traveled}}{\text{elapsed time}}$$</td></tr>
<tr><td>Express the time required for each animal to travel a distance L:</td><td>$$t_{cheetah} = \frac{L}{v_{cheetah}},$$
$$t_{falcon} = \frac{L}{v_{falcon}},$$
and
$$t_{sailfish} = \frac{L}{v_{sailfish}}$$</td></tr>
<tr><td>Express the total time, Δt:</td><td>$$\Delta t = L\left(\frac{1}{v_{cheetah}} + \frac{1}{v_{falcon}} + \frac{1}{v_{sailfish}}\right)$$</td></tr>
</table>

Use the total distance traveled by the relay team and the elapsed time to calculate the average speed:

$$|v_{av}| = \frac{3L}{L\left(\dfrac{1}{113\,\text{km/h}} + \dfrac{1}{161\,\text{km/h}} + \dfrac{1}{105\,\text{km/h}}\right)} = \boxed{122\,\text{km/h}}$$

Calculate the average of the three speeds:

$$\text{Average}_{\text{three speeds}} = \frac{113\,\text{km/h} + 161\,\text{km/h} + 105\,\text{km/h}}{3} = \boxed{126\,\text{km/h} = 1.03 v_{av}}$$

*61 ••   A car traveling at a constant speed of 20 m/s passes an intersection at time $t = 0$, and 5 s later another car traveling at a constant speed of 30 m/s passes the same intersection in the same direction. (*a*) Sketch the position functions $x_1(t)$ and $x_2(t)$ for the two cars. (*b*) Determine when the second car will overtake the first. (*c*) How far from the intersection will the two cars be when they pull even? (*d*) Where is the first car when the second car passes the intersection?

Picture the Problem One way to solve this problem is by using a graphing calculator to plot the positions of each car as a function of time. Plotting these positions as functions of time allows us to visualize the motion of the two cars relative to the (fixed) ground. More importantly, it allows us to see the motion of the two cars relative to each other. We can, for example, tell how far apart the cars are at any given time by determining the length of a vertical line segment from one curve to the other.

(*a*) Letting the origin of our coordinate system be at the intersection, the position of the slower car, $x_1(t)$, is given by:

$x_1(t) = 20t$
where x_1 is in meters if t is in seconds.

Because the faster car is also moving at a constant speed, we know that the position of this car is given by a function of the form:

$x_2(t) = 30t + b$

We know that when $t = 5$ s, this second car is at the intersection (i.e., $x_2(5\text{ s}) = 0$). Using this information, you can convince yourself that:

$b = -150$ m

Thus, the position of the faster car is given by:

$x_2(t) = 30t - 150$

One can use a graphing calculator, graphing paper, or a spreadsheet to obtain the graphs of $x_1(t)$ (the solid line) and $x_2(t)$ (the dashed line) shown to the right:

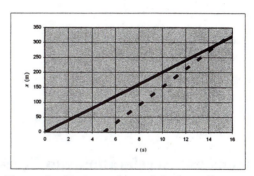

(*b*) Use the time coordinate of the intersection of the two lines to determine the time at which the second car overtakes the first:

From the intersection of the two lines, one can see that the second car will "overtake" (catch up to) the first car at $\boxed{t = 15\text{ s}}$.

(*c*) Use the position coordinate of the intersection of the two lines to determine the distance from the intersection at which the

From the intersection of the two lines, one can see that the distance from the intersection is $\boxed{300\text{ m}}$.

second car catches up to the first car:

(d) Draw a vertical line from $t = 5$ s to the red line and then read the position coordinate of the intersection of this line and the red line to determine the position of the first car when the second car went through the intersection:

From the graph, when the second car passes the intersection, the first car was $\boxed{100 \text{ m ahead}}$.

Acceleration

*68 •• The position of an object is related to time by $x = At^2 - Bt + C$, where $A = 8$ m/s², $B = 6$ m/s, and $C = 4$ m. Find the instantaneous velocity and acceleration as functions of time.

Picture the Problem The instantaneous velocity is dx/dt and the acceleration is dv/dt .

Using the definitions of instantaneous velocity and acceleration, determine v and a:

$$v = \frac{dx}{dt} = \frac{d}{dt}\left[At^2 - Bt + C\right] = 2At - B$$

and

$$a = \frac{dv}{dr} = \frac{d}{dt}\left[2At - B\right] = 2A$$

Substitute numerical values for A and B and evaluate v and a:

$$v = 2\left(8\text{m/s}^2\right)t - 6\text{ m/s}$$
$$= \boxed{\left(16 \text{ m/s}^2\right)t - 6\text{m/s}}$$

and

$$a = 2\left(8 \text{ m/s}^2\right) = \boxed{16.0 \text{ m/s}^2}$$

Constant Acceleration and Free-Fall

*70 • An object projected upward with initial velocity v_0 attains a height h. Another object projected up with initial velocity $2v_0$ will attain a height of (a) 4h, (b) 3h, (c) 2h, (d) h.

Picture the Problem Because the acceleration is constant ($-g$) we can use a constant-acceleration equation to find the height of the projectile.

Using a constant-acceleration equation, express the height of

$$v^2 = v_0^2 + 2a\Delta y$$

the object as a function of its initial velocity, the acceleration due to gravity, and its displacement:

Solve for $\Delta y_{max} = h$:

Because $v(h) = 0$,

$$h = \frac{-v_0^2}{2(-g)} = \frac{v_0^2}{2g}$$

From this expression for h we see that the maximum height attained is proportional to the square of the launch speed:

$$h \propto v_0^2$$

Therefore, doubling the initial speed gives four times the height:

$$h_{2v_0} = \frac{(2v_0)^2}{2g} = 4\left(\frac{v_0^2}{2g}\right) = 4h_{v_0}$$

and (a) is correct.

*73 • An object with constant acceleration has a velocity of 10 m/s when it is at $x = 6$ m and of 15 m/s when it is at $x = 10$ m. What is its acceleration?

Picture the Problem Because the acceleration of the object is constant we can use constant-acceleration equations to describe its motion.

Using a constant-acceleration equation, relate the velocity to the acceleration and the displacement:

$$v^2 = v_0^2 + 2a\,\Delta x$$

Solve for the acceleration:

$$a = \frac{v^2 - v_0^2}{2\,\Delta x}$$

Substitute numerical values and evaluate a:

$$a = \frac{(15^2 - 10^2)\,\text{m}^2/\text{s}^2}{2(4\,\text{m})} = \boxed{15.6\,\text{m/s}^2}$$

*77 •• A load of bricks is being lifted by a crane at a steady velocity of 5 m/s when one brick falls off 6 m above the ground. (a) Sketch $x(t)$ to show the motion of the free brick. (b) What is the greatest height the brick reaches above the ground? (c) How long does it take to reach the ground? (d) What is its speed just before it hits the ground?

Picture the Problem In the absence of air resistance, the brick experiences constant acceleration and we can use constant-acceleration equations to describe its motion. Constant acceleration implies a parabolic position-versus-time curve.

(*a*) Using a constant-acceleration equation, relate the position of the brick to its initial position, initial velocity, acceleration, and time into its fall:

The graph of
$$y = y_0 + v_0 t + \tfrac{1}{2}(-g)t^2$$
$$= 6\,\text{m} + (5\,\text{m/s})t - (4.91\,\text{m/s}^2)t^2$$
was plotted using a spreadsheet program.

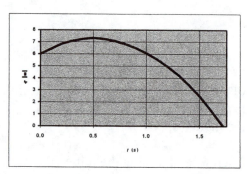

(*b*) Relate the greatest height reached by the brick to its height when it falls off the load and the additional height it rises Δy_{max}:

$$h = y_0 + \Delta y_{\text{max}}$$

Using a constant-acceleration equation, relate the height reached by the brick to its acceleration and initial velocity:

$$v_{\text{top}}^2 = v_0^2 + 2(-g)\Delta y_{\text{max}}$$
or, because $v_{\text{top}} = 0$,
$$0 = v_0^2 + 2(-g)\Delta y_{\text{max}}$$

Solve for Δy_{max}:

$$\Delta y_{\text{max}} = \frac{v_0^2}{2g}$$

Substitute numerical values and evaluate Δy_{max}:

$$\Delta y_{\text{max}} = \frac{(5\,\text{m/s})^2}{2(9.81\,\text{m/s}^2)} = 1.27\,\text{m}$$

Substitute to obtain:

$$h = y_0 + \Delta y_{\text{max}} = 6\,\text{m} + 1.27\,\text{m} = \boxed{7.27\,\text{m}}$$

Note: The graph shown above confirms this result.

(c) Using the quadratic formula, solve for t in the equation obtained in part (a):

$$t = \frac{-v_0 \pm \sqrt{v_0^2 - 4\left(\frac{-g}{2}\right)(-\Delta y)}}{2\left(\frac{-g}{2}\right)}$$

$$= \left(\frac{v_0}{g}\right)\left(1 \pm \sqrt{1 - \frac{2g(\Delta y)}{v_0^2}}\right)$$

With $y_{bottom} = 0$ and $y_0 = 6$ m or $\Delta y = -6$ m, we have $t = \boxed{1.73\,\text{s}}$ and $t = -0.708$ s. Note: The second solution is nonphysical.

(d) Using a constant-acceleration equation, relate the speed of the brick on impact to its acceleration and displacement, and solve for its speed:

$$v = \sqrt{2gh}$$

Substitute numerical values and evaluate v:

$$v = \sqrt{2(9.81\,\text{m/s}^2)(7.27\,\text{m})} = \boxed{11.9\,\text{m/s}}$$

*79 ••   An object is dropped from rest at a height of 120 m. Find the distance it falls during its final second in the air.

Picture the Problem In the absence of air resistance, the object's acceleration is constant. Choose a coordinate system in which downward is positive and the origin is at the point of release. In this coordinate system, $a = g$ and $y = 120$ m at the bottom of the fall.

Express the distance fallen in the last second in terms of the object's position at impact and its position 1 s before impact:

$$\Delta y_{last\,second} = 120\,\text{m} - y_{1\,s\,before\,impact} \quad (1)$$

Using a constant-acceleration equation, relate the object's position upon impact to its initial position, initial velocity, and fall time:

$$y = y_0 + v_0 t + \tfrac{1}{2}gt^2$$
or, because $y_0 = 0$ and $v_0 = 0$,
$$y = \tfrac{1}{2}gt_{fall}^2$$

Solve for the fall time:

$$t_{fall} = \sqrt{\frac{2y}{g}}$$

Substitute numerical values and evaluate t_{fall}:

$$t_{fall} = \sqrt{\frac{2(120\,m)}{9.81\,m/s^2}} = 4.95\,s$$

We know that, one second before impact, the object has fallen for 3.95 s. Using the same constant-acceleration equation, calculate the object's position 3.95 s into its fall:

$$y(3.95\,s) = \tfrac{1}{2}(9.81\ m/s^2)(3.95\,s)^2$$
$$= 76.4\,m$$

Substitute in equation (1) to obtain:

$$\Delta y_{last\ second} = 120\,m - 76.4\,m = \boxed{43.6\,m}$$

*81 •   A stone is thrown vertically from a 200-m tall cliff. During the last half second of its flight the stone travels a distance of 45 m. Find the initial speed of the stone.

Picture the Problem In the absence of air resistance, the acceleration of the stone is constant. Choose a coordinate system with the origin at the bottom of the trajectory and the upward direction positive. Let $v_{f\text{-}1/2}$ be the speed one-half second before impact and v_f the speed at impact.

Using a constant-acceleration equation, express the final speed of the stone in terms of its initial speed, acceleration, and displacement:

$$v_f^2 = v_0^2 + 2a\Delta y$$

Solve for the initial speed of the stone:

$$v_0 = \sqrt{v_f^2 + 2g\Delta y} \qquad (1)$$

Find the average speed in the last half second:

$$v_{av} = \frac{v_{f\text{-}1/2} + v_f}{2} = \frac{\Delta x_{last\ half\ second}}{\Delta t} = \frac{45\,m}{0.5\,s}$$
$$= 90\,m/s$$
and
$$v_{f\text{-}1/2} + v_f = 2(90\,m/s) = 180\,m/s$$

Using a constant-acceleration equation, express the change in speed of the stone in the last half second in terms of the acceleration and the elapsed time; solve for the change in its speed:

$$\Delta v = v_f - v_{f\text{-}1/2} = g\Delta t$$
$$= (9.81\,m/s^2)(0.5\,s)$$
$$= 4.91\,m/s$$

Add the equations that express the sum and difference of $v_{f-\frac{1}{2}}$ and v_f and solve for v_f:

$$v_f = \frac{180\,\text{m/s} + 4.91\,\text{m/s}}{2} = 92.5\,\text{m/s}$$

Substitute in equation (1) and evaluate v_0:

$$v_0 = \sqrt{(92.5\text{m/s})^2 + 2(9.81\text{m/s}^2)(-200\text{m})}$$

$$= \boxed{68.1\,\text{m/s}}$$

Remarks: The stone may be thrown either up or down from the cliff and the results after it passes the cliff on the way down are the same.

*84 •• It is relatively easy to use a spreadsheet program such as Microsoft Excel to solve certain types of physics problems. For example, you probably solved Problem 75 using algebra. Let's solve Problem 75 in a different way, this time using a spreadsheet program. While we can solve this problem using algebra, there are many places in physics where we can't get an alternative solution so easily, and have to rely on numerical methods like the one shown here. (*a*) Using Microsoft Excel or some other spreadsheet program, generate a graph of the height versus time for the ball in Problem 75 (thrown upwards with an initial velocity of 20 m/s). Determine the maximum height, the time it was in the air, and the time(s) when the ball is 15 m above the ground by inspection (i.e., look at the graph and find them.) (*b*) Now change the initial velocity to 10 m/s, and find the maximum height the ball reaches and the time the ball spends in the air.

Picture the Problem While we can solve this problem analytically, there are many physical situations in which it is not easy to do so and one has to rely on numerical methods; for example, see the spreadsheet solution shown below. Because we're neglecting the height of the release point, the position of the ball as a function of time is given by $y = v_0 t - \frac{1}{2}gt^2$. The formulas used to calculate the quantities in the columns are as follows:

Cell	Content/Formula	Algebraic Form
B1	20	v_0
B2	9.81	g
B5	0	t
B6	B5 + 0.1	$t + \Delta t$
C6	$B$1*B6 − 0.5*$B$2*B6^2	$v_0 t - \frac{1}{2}gt^2$

(*a*)

	A	B	C
1	v0 =	20	m/s
2	g =	9.81	m/s^2
3		t	height
4		(s)	(m)

5		0.0	0.00
6		0.1	1.95
7		0.2	3.80
8		0.3	5.56
9		0.4	7.22
10		0.5	8.77
40		3.5	9.91
41		3.6	8.43
42		3.7	6.85
43		3.8	5.17
44		3.9	3.39
45		4.0	1.52
46		4.1	−0.45

The graph shown below was generated from the data in the previous table. Note that the maximum height reached is a little more than 20 m and the time of flight is about 4 s.

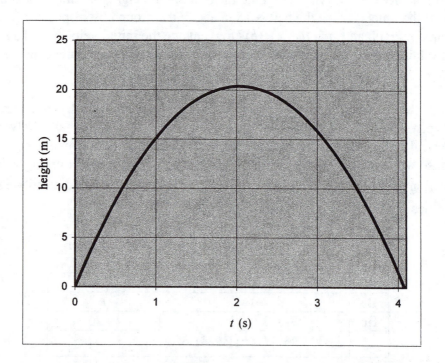

(b) In the spreadsheet, change the value in cell B1 from 20 to 10. The graph should automatically update. With an initial velocity of 10 m/s, the maximum height achieved is approximately 5 m and the time-of-flight is approximately 2 s.

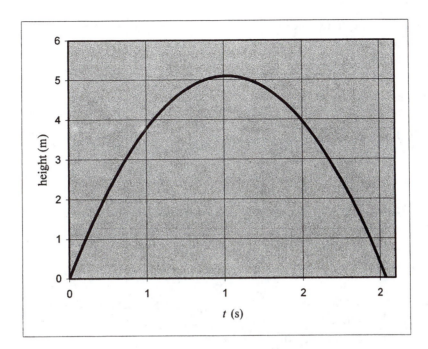

*85 •• Al and Bert are jogging side-by-side on a trail in the woods at a speed of 0.75 m/s. Suddenly Al sees the end of the trail 35 m ahead and decides to speed up to reach it. He accelerates at a constant rate of 0.5 m/s², leaving Bert behind, who continues on at a constant speed. (*a*) How long does it take Al to reach the end of the trail? (*b*) Once he reaches the end of the trail, he immediately turns around and heads back along the trail with a constant speed of 0.85 m/s. How long does it take him to reach Bert? (*c*) How far are they from the end of the trail when they meet?

Picture the Problem Because the accelerations of both Al and Bert are constant, constant-acceleration equations can be used to describe their motions. Choose the origin of the coordinate system to be where Al decides to begin his sprint.

(*a*) Using a constant-acceleration equation, relate Al's initial velocity, his acceleration, and the time to reach the end of the trail to his displacement in reaching the end of the trail:

$$\Delta x = v_0 t + \tfrac{1}{2}at^2$$

Substitute numerical values to obtain:

$$35\,\text{m} = (0.75\,\text{m/s})t + \tfrac{1}{2}(0.5\,\text{m/s}^2)t^2$$

Solve for the time required for Al to reach the end of the trail:

$$t = \boxed{10.4\,\text{s}}$$

(*b*) Using constant-acceleration equations, express the positions

$$x_{\text{Bert}} = x_{\text{Bert},0} + (0.75\,\text{m/s})t$$

and

of Bert and Al as functions of time. At the instant Al turns around at the end of the trail, $t = 0$. Also, $x = 0$ at a point 35 m from the end of the trail:

$$x_{Al} = x_{Al,0} - (0.85\,\text{m/s})t$$
$$= 35\,\text{m} - (0.85\,\text{m/s})t$$

Calculate Bert's position at $t = 0$. At that time he has been running for 10.4 s:

$$x_{Bert,0} = (0.75\,\text{m/s})(10.4\,\text{s}) = 7.80\,\text{m}$$

Because Bert and Al will be at the same location when they meet, equate their position functions and solve for t:

$$7.80\,\text{m} + (0.75\,\text{m/s})t = 35\,\text{m} - (0.85\,\text{m/s})t$$
and
$$t = 17.0\,\text{s}$$

To determine the elapsed time from when Al began his accelerated run, we need to add 10.4 s to this time:

$$t_{start} = 17.0\,\text{s} + 10.4\,\text{s} = \boxed{27.4\,\text{s}}$$

(c) Express Bert's distance from the end of the trail when he and Al meet:

$$d_{end\,of\,trail} = 35\,\text{m} - x_{Bert,0}$$
$$- d_{Bert\,runs\,until\,he\,meets\,Al}$$

Substitute numerical values and evaluate $d_{end\,of\,trail}$:

$$d_{end\,of\,trail} = 35\,\text{m} - 7.80\,\text{m}$$
$$- (17\,\text{s})(0.75\,\text{m/s}$$
$$= \boxed{14.5\,\text{m}}$$

*89 ••   In a classroom demonstration, a glider moves along an inclined air track with constant acceleration. It is projected from the start of the track with an initial velocity. After 8 s have elapsed, it is 100 cm from its starting point and is moving along the track at a velocity of –15 cm/s. Find the initial velocity and the acceleration.

Picture the Problem The acceleration of the glider on the air track is constant. Its average acceleration is equal to the instantaneous (constant) acceleration. Choose a coordinate system in which the initial direction of the glider's motion is the positive direction.

Using the definition of acceleration, express the average acceleration of the glider in terms of the glider's velocity change and the elapsed time:

$$a = a_{av} = \frac{\Delta v}{\Delta t}$$

Using a constant-acceleration equation, express the average velocity of the glider in terms of the displacement of the glider and the elapsed time:

$$v_{av} = \frac{\Delta x}{\Delta t} = \frac{v_0 + v}{2}$$

Solve for and evaluate the initial velocity:

$$v_0 = \frac{2\Delta x}{\Delta t} - v = \frac{2(100\,\text{cm})}{8\,\text{s}} - (-15\,\text{cm/s})$$

$$= \boxed{40.0\,\text{cm/s}}$$

Substitute this value of v_0 and evaluate the average acceleration of the glider:

$$a = \frac{-15\,\text{cm/s} - (40\,\text{cm/s})}{8\,\text{s}}$$

$$= \boxed{-6.88\,\text{cm/s}^2}$$

*94 ••   A motorcycle officer hidden at an intersection observes a car that ignores a stop sign, crosses the intersection, and continues on at constant speed. The police officer takes off in pursuit 2.0 s after the car has passed the stop sign, accelerates at 6.2 m/s² until her speed is 110 km/h, and then continues at this speed until she catches the car. At that instant, the car is 1.4 km from the intersection. How fast was the car traveling?

Picture the Problem The acceleration of the police officer's car is positive and constant and the acceleration of the speeder's car is zero. Choose a coordinate system such that the direction of motion of the two vehicles is the positive direction and the origin is at the stop sign.

Express the velocity of the car in terms of the distance it will travel until the police officer catches up to it and the time that will elapse during this chase:

$$v_{car} = \frac{d_{caught}}{t_{car}}$$

Letting t_1 be the time during which she accelerates and t_2 the time of travel at $v_1 = 110$ km/h, express the time of travel of the police officer:

$$t_{officer} = t_1 + t_2$$

Convert 110 km/h into m/s:

$$v_1 = (110\,\text{km/h})(10^3\,\text{m/km})(1\,\text{h}/3600\,\text{s})$$
$$= 30.6\,\text{m/s}$$

Express and evaluate t_1:

$$t_1 = \frac{v_1}{a_{motorcycle}} = \frac{30.6\,\text{m/s}}{6.2\,\text{m/s}^2} = 4.94\,\text{s}$$

Express and evaluate d_1:

$$d_1 = \tfrac{1}{2}v_1 t_1 = \tfrac{1}{2}(30.6\,\text{m/s})(4.94\,\text{s})$$
$$= 75.6\,\text{m}$$

Determine d_2:

$$d_2 = d_{\text{caught}} - d_1 = 1400\,\text{m} - 75.6\,\text{m}$$
$$= 1324.4\,\text{m}$$

Express and evaluate t_2:

$$t_2 = \frac{d_2}{v_1} = \frac{1324.4\,\text{m}}{30.6\,\text{m/s}} = 43.3\,\text{s}$$

Express the time of travel of the car:

$$t_{\text{car}} = 2.0\,\text{s} + 4.93\,\text{s} + 43.3\,\text{s} = 50.2\,\text{s}$$

Finally, find the speed of the car:

$$v_{\text{car}} = \frac{d_{\text{caught}}}{t_{\text{car}}} = \frac{1400\,\text{m}}{50.2\,\text{s}} = 27.9\,\text{m/s}$$

$$= (27.9\,\text{m/s})\left(\frac{1\,\text{mi/h}}{0.447\,\text{m/s}}\right)$$

$$= \boxed{62.4\,\text{mi/h}}$$

*98 •   At the end of *Charlie and the Chocolate Factory*, Willie Wonka presses a button that shoots the great glass elevator through the roof of his chocolate factory. (*a*) If the elevator reaches a maximum height of 10 km above the roof, what was its speed immediately after crashing through the roof? Ignore air resistance, even though in this case it makes little sense to ignore it. (*b*) Assume that the elevator's speed just after it crashes through the roof was half of what it was just before its impact with the roof. Assuming that it started from rest on the ground floor of the chocolate factory, and that the height of the roof is 150 m above the ground floor, what uniform acceleration is needed for it to reach this high speed?

Picture the Problem This is a composite of two constant accelerations with the acceleration equal to one constant prior to the elevator hitting the roof, and equal to a different constant after crashing through it. Choose a coordinate system in which the upward direction is positive and apply constant-acceleration equations.

(*a*) Using a constant-acceleration equation, relate the velocity to the acceleration and displacement:

$$v^2 = v_0^2 + 2a\,\Delta y$$
or, because $v = 0$ and $a = -g$,
$$0 = v_0^2 - 2g\,\Delta y$$

Solve for v_0:

$$v_0 = \sqrt{2g\Delta y}$$

Substitute numerical values and evaluate v_0:

$$v_0 = \sqrt{2(9.81\,\text{m/s}^2)(10^4\,\text{m})} = \boxed{443\,\text{m/s}}$$

(b) Find the velocity of the elevator just before it crashed through the roof:

$$v_f = 2 \times 443 \text{ m/s} = 886 \text{ m/s}$$

Using the same constant-acceleration equation, this time with $v_0 = 0$, solve for the acceleration:

$$v^2 = 2a\Delta y \Rightarrow a = \frac{v^2}{2\Delta y}$$

Substitute numerical values and evaluate a:

$$a = \frac{(886 \text{ m/s})^2}{2(150 \text{ m})} = 2.62 \times 10^3 \text{ m/s}^2$$

$$= \boxed{267g}$$

*101 •• Consider the motion of a particle that experiences free fall with a constant acceleration. Before the advent of computer-driven data-logging software, we used to do a free-fall experiment in which a coated tape was placed vertically next to the path of a dropped conducting puck. A high-voltage spark generator would cause an arc to jump between two vertical wires through the falling puck and through the tape, thereby marking the tape at fixed time intervals Δt. Show that the change in height in successive time intervals for an object falling from rest follows *Galileo's Rule of Odd Numbers*: $\Delta y_{21} = 3 \Delta y_{10}$, $\Delta y_{32} = 5 \Delta y_{10}$, . . ., where Δy_{10} is the change in y during the first interval of duration Δt, Δy_{21} is the change in y during the second interval of duration Δt, etc.

Picture the Problem In the absence of air resistance, the puck experiences constant acceleration and we can use constant-acceleration equations to describe its position as a function of time. Choose a coordinate system in which downward is positive, the particle starts from rest ($v_0 = 0$), and the starting height is zero ($y_0 = 0$).

Using a constant-acceleration equation, relate the position of the falling puck to the acceleration and the time. Evaluate the y-position at successive equal time intervals Δt, $2\Delta t$, $3\Delta t$, etc:

$$y_1 = \frac{-g}{2}(\Delta t)^2 = \frac{-g}{2}(\Delta t)^2$$

$$y_2 = \frac{-g}{2}(2\Delta t)^2 = \frac{-g}{2}(4)(\Delta t)^2$$

$$y_3 = \frac{-g}{2}(3\Delta t)^2 = \frac{-g}{2}(9)(\Delta t)^2$$

$$y_4 = \frac{-g}{2}(4\Delta t)^2 = \frac{-g}{2}(16)(\Delta t)^2$$

etc.

Evaluate the changes in those positions in each time interval:

$$\Delta y_{10} = y_1 - 0 = \left(\frac{-g}{2}\right)(\Delta t)^2$$

$$\Delta y_{21} = y_2 - y_1 = 3\left(\frac{-g}{2}\right)(\Delta t)^2 = 3\Delta y_{10}$$

$$\Delta y_{32} = y_3 - y_2 = 5\left(\frac{-g}{2}\right)(\Delta t)^2 = 5\Delta y_{10}$$

$$\Delta y_{43} = y_4 - y_3 = 7\left(\frac{-g}{2}\right)(\Delta t)^2 = 7\Delta y_{10}$$

etc.

*103 ••   A plane landing on a small tropical island has just 70 m of runway on which to stop. If its initial speed is 60 m/s, (a) what is the minimum acceleration of the plane during landing, assuming it to be constant? (b) How long does it take for the plane to stop with this acceleration?

Picture the Problem We can use constant-acceleration equations with the final velocity $v = 0$ to find the acceleration and stopping time of the plane.

(a) Using a constant-acceleration equation, relate the known velocities to the acceleration and displacement:

$$v^2 = v_0^2 + 2a\,\Delta x$$

Solve for a:

$$a = \frac{v^2 - v_0^2}{2\,\Delta x} = \frac{-v_0^2}{2\,\Delta x}$$

Substitute numerical values and evaluate a:

$$a = \frac{-(60\,\text{m/s})^2}{2(70\,\text{m})} = \boxed{-25.7\,\text{m/s}^2}$$

(b) Using a constant-acceleration equation, relate the final and initial speeds of the plane to its acceleration and stopping time:

$$v = v_0 + a\Delta t$$

Solve for and evaluate the stopping time:

$$\Delta t = \frac{v - v_0}{a} = \frac{0 - 60\,\text{m/s}}{-25.7\,\text{m/s}^2} = \boxed{2.33\,\text{s}}$$

*105 ••   If it were possible for a spacecraft to maintain a constant acceleration indefinitely, trips to the planets of the Solar System could be undertaken in days or weeks, while voyages to the nearer stars would only take a few years. (a) Show that g, the magnitude of free-fall acceleration on earth, is approximately 1 $c\cdot y/y^2$

(See problem 52 for the definition of a light-year.) (b) Using data from the tables at the back of the book, find the time it would take for a one-way trip from Earth to Mars (at Mars' closest approach to Earth). Assume that the spacecraft starts from rest, travels along a straight line, accelerates halfway at 1 g, and then flips around and decelerates at 1 g for the rest of the trip.

Picture the Problem Note: No material body can travel at speeds faster than light. When one is dealing with problems of this sort, the kinematic formulae for displacement, velocity and acceleration are no longer valid, and one must invoke the special theory of relativity to answer questions such as these. For now, ignore such subtleties. Although the formulas you are using (i.e., the constant-acceleration equations) are not quite correct, your answer to part (b) will be wrong by about 1%.

(a) This part of the problem is an exercise in the conversion of units. Make use of the fact that $1 \, c \cdot y = 9.47 \times 10^{15}$ m and $1 \, y = 3.16 \times 10^7$ s:

$$g = \left(9.81 \text{m/s}^2\right)\left(\frac{1 c \cdot y}{9.47 \times 10^{15} \text{m}}\right)\left(\frac{\left(3.16 \times 10^7 \text{s}\right)^2}{\left(1 \text{y}\right)^2}\right) = \boxed{1.03 c \cdot y / y^2}$$

(b) Let $t_{1/2}$ represent the time it takes to reach the halfway point. Then the total trip time is:

$$t = 2 \, t_{1/2} \qquad (1)$$

Use a constant- acceleration equation to relate the half-distance to Mars Δx to the initial speed, acceleration, and half-trip time $t_{1/2}$:

$$\Delta x = v_0 t + \tfrac{1}{2} a t_{1/2}^2$$

Since $v_0 = 0$ and $a = g$:

$$t_{1/2} = \sqrt{\frac{2\Delta x}{a}}$$

The distance from Earth to Mars at closest approach is 7.8×10^{10} m. Substitute numerical values and evaluate $t_{1/2}$:

$$t_{1/2} = \sqrt{\frac{2\left(3.9 \times 10^{10} \text{ m}\right)}{9.81 \text{m/s}^2}} = 8.92 \times 10^4 \text{ s}$$

Substitute for $t_{1/2}$ in equation (1) to obtain:

$$t = 2\left(8.92 \times 10^4 \text{ s}\right) = 1.78 \times 10^5 \text{ s} \approx \boxed{2 \text{d}}$$

Remarks: Our result in part (b) seems remarkably short, considering how far Mars is and how low the acceleration is.

*110 •• Starting at one station, a subway train accelerates from rest at a constant rate of 1.0 m/s² for half the distance to the next station, then slows down at the same rate for the second half of the journey. The total distance between stations is 900 m. (*a*) Sketch a graph of the velocity *v* as a function of time over the full journey. (*b*) Sketch a graph of the distance covered as a function of time over the full journey. Place appropriate numerical values on both axes.

Determine the Concept The problem describes two intervals of constant acceleration; one when the train's velocity is increasing, and a second when it is decreasing.

(*a*) Using a constant-acceleration equation, relate the half-distance Δx between stations to the initial speed v_0, the acceleration a of the train, and the time-to-midpoint Δt:

$$\Delta x = v_0 \Delta t + \tfrac{1}{2} a (\Delta t)^2$$
or, because $v_0 = 0$,
$$\Delta x = \tfrac{1}{2} a (\Delta t)^2$$

Solve for Δt:

$$\Delta t = \sqrt{\frac{2\Delta x}{a}}$$

Substitute numerical values and evaluate the time-to-midpoint Δt:

$$\Delta t = \sqrt{\frac{2(450\,\text{m})}{1\,\text{m/s}^2}} = 30.0\,\text{s}$$

Because the train accelerates uniformly and from rest, the first part of its velocity graph will be linear, pass through the origin, and last for 30 s. Because it slows down uniformly and at the same rate for the second half of its journey, this part of its graph will also be linear but with a negative slope. The graph of *v* as a function of *t* is shown to the right.

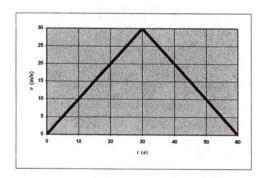

(*b*) The graph of *x* as a function of *t* is obtained from the graph of *v* as a function of *t* by finding the area under the velocity curve. Looking at the velocity graph, note that when the train has been in motion for 10 s, it will have traveled a distance of

$$\tfrac{1}{2}(10\,\text{s})(10\,\text{m/s}) = 50\,\text{m}$$

Selecting additional points from the velocity graph and calculating the areas under the curve will confirm the graph

and that this distance is plotted above 10 s on the graph to the right.

of x as a function of t that is shown.

Integration of the Equations of Motion

*115 •   The velocity of a particle is given by $v(t) = (6 \text{ m/s}^2)\, t + (3 \text{ m/s})$. (*a*) Sketch v versus t, and find the area under the curve for the interval $t = 0$ to $t = 5$ s. (*b*) Find the position function $x(t)$. Use it to calculate the displacement during the interval $t = 0$ to $t = 5$ s.

Picture the Problem The integral of a function is equal to the "area" between the curve for that function and the independent-variable axis.

(*a*) The graph is shown to the right. The distance is found by determining the area under the curve. You can accomplish this easily Because the shape of the area under the curve is a trapezoid.

$$A = \left(\frac{33\,\text{m/s} + 3\,\text{m/s}}{2}\right)(5\text{s} - 0\text{s}) = 90\,\text{m}$$

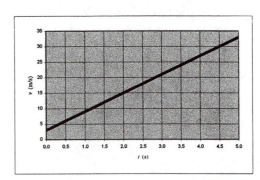

$A = (36 \text{ blocks})(2.5 \text{ m/block}) = \boxed{90 \text{ m}}$

Alternatively, we could just count the blocks and fractions thereof.

There are approximately 36 blocks each having an area of
$(5 \text{ m/s})(0.5 \text{ s}) = 2.5 \text{ m}$.

(*b*) To find the position function $x(t)$, we integrate the velocity function $v(t)$ over the time interval in question:

$$x(t) = \int_0^t v(t')dt'$$

$$= \int_0^t \left[(6\,\text{m/s}^2)t' + (3\,\text{m/s})\right]dt'$$

and

$$\boxed{x(t) = (3\,\text{m/s}^2)t^2 + (3\,\text{m/s})t}$$

Now evaluate $x(t)$ at 0 s and 5 s respectively and subtract to obtain Δx:

$\Delta x = x(5\text{s}) - x(0\text{s}) = 90\,\text{m} - 0\,\text{m}$

$= \boxed{90.0 \text{ m}}$

*117 ••   The velocity of a particle is given by $v = (7 \text{ m/s}^3)t^2 - 5 \text{ m/s}$, where t is in seconds and v is in meters per second. If the particle starts from the origin, $x_0 = 0$, at $t_0 = 0$, find the general position function $x(t)$.

Picture the Problem Because the velocity of the particle varies with the square of the time, the acceleration is not constant. The displacement of the particle is found by integration.

Express the velocity of a particle as the derivative of its position function:

$$v(t) \equiv \frac{dx(t)}{dt}$$

Separate the variables to obtain:

$$dx(t) = v(t)dt$$

Express the integral of x from $x_0 = 0$ to x and t from $t_0 = 0$ to t:

$$x(t) = \int_{t_0=0}^{x(t)} dx' = \int_{t_0=0}^{t} v(t')dt'$$

Substitute for $v(t')$ to obtain:

$$x(t) = \int_{t_0=0}^{t} \left[(7 \text{ m/s}^3)t'^2 - (5 \text{ m/s})\right] dt'$$

$$= \boxed{(\tfrac{7}{3} \text{ m/s}^3)t^3 - (5 \text{ m/s})t}$$

*121 ••   Figure 2-39 shows a plot of x versus t for a body moving along a straight line. Sketch rough graphs of v as a function of t and a as a function of t for this motion.

Figure 2-39 Problem 121

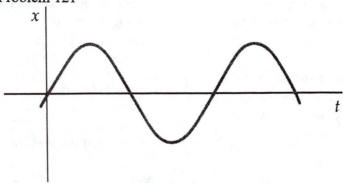

Picture the Problem Because the position of the body is not described by a parabolic function, the acceleration is not constant.

Select a series of points on the graph of $x(t)$ (e.g., at the extreme values and where the graph crosses the t axis), draw tangent lines at those points, and measure their slopes. In doing this, you are evaluating $v = dx/dt$ at these points. Plot these slopes above the times at which you measured the slopes. Your graph should closely resemble the one to the right.

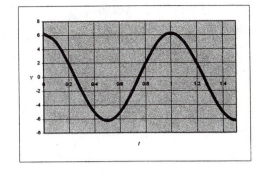

Select a series of points on the graph of $v(t)$ (e.g., at the extreme values and where the graph crosses the t axis), draw tangent lines at those points, and measure their slopes. In doing this, you are evaluating $a = dv/dt$ at these points. Plot these slopes above the times at which you measured the slopes. Your graph should closely resemble the one at the right.

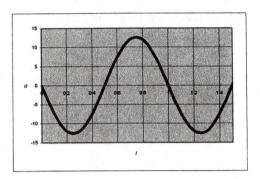

General Problems

*126 ••• The position of a body oscillating on a spring is given by $x = A \sin \omega t$, where A and ω are constants with values $A = 5$ cm and $\omega = 0.175 \text{ s}^{-1}$. (a) Plot x as a function of t for $0 \le t \le 36$ s. (b) Measure the slope of your graph at $t = 0$ to find the velocity at this time. (c) Calculate the average velocity for a series of intervals beginning at $t = 0$ and ending at $t = 6, 3, 2, 1, 0.5$, and 0.25 s. (d) Compute dx/dt and find the velocity at time $t = 0$. (e) Compare your results in parts (c) and (d).

Picture the Problem We can obtain an average velocity, $v_{av} = \Delta x/\Delta t$, over fixed time intervals. The instantaneous velocity, $v = dx/dt$ can only be obtained by differentiation.

(*a*) The graph of *x* versus *t* is shown to the right:

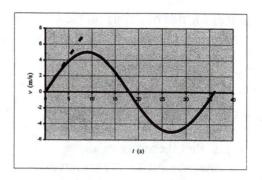

(*b*) Draw a tangent line at the origin and measure its rise and run. Use this ratio to obtain an approximate value for the slope at the origin:

The tangent line appears to, at least approximately, pass through the point (5, 4). Using the origin as the second point,

$$\Delta x = 4 \text{ cm} - 0 = 4 \text{ cm}$$

and

$$\Delta t = 5 \text{ s} - 0 = 5 \text{ s}$$

Therefore, the slope of the tangent line and the velocity of the body as it passes through the origin is approximately

$$v(0) = \frac{\text{rise}}{\text{run}} = \frac{\Delta x}{\Delta t} = \frac{4\,\text{cm}}{5\,\text{s}} = \boxed{0.8\,\text{cm/s}}$$

(*c*) Calculate the average velocity for the series of time intervals given by completing the table shown below:

t_0	t	Δt	x_0	x	Δx	$v_{av}=\Delta x/\Delta t$
(s)	(s)	(s)	(cm)	(cm)	(cm)	(m/s)
0	6	6	0	4.34	4.34	0.723
0	3	3	0	2.51	2.51	0.835
0	2	2	0	1.71	1.71	0.857
0	1	1	0	0.871	0.871	0.871
0	0.5	0.5	0	0.437	0.437	0.874

(*d*) Express the time derivative of the position:

$$\frac{dx}{dt} = A\omega\cos\omega t$$

Substitute numerical values and evaluate $\dfrac{dx}{dt}$ at $t = 0$:

$$\dfrac{dx}{dt} = A\omega\cos 0 = A\omega$$

$$= (0.05\,\mathrm{m})(0.175\,\mathrm{s}^{-1})$$

$$= \boxed{0.875\,\mathrm{cm/s}}$$

(e) Compare the average velocities from part (c) with the instantaneous velocity from part (d):

As Δt, and thus Δx, becomes small, the value for the average velocity approaches that for the instantaneous velocity obtained in part (d). For $\Delta t = 0.25$ s, they agree to three significant figures.

*131 ••• In Problem 130, a rock falls through water with a continuously decreasing acceleration of the form $a(t) = ge^{-bt}$, where b is a positive constant. In physics, we are not often given acceleration directly as a function of time, but usually either as a function of position or of velocity. Assume that the rock's acceleration as a function of *velocity* has the form $a = g - bv$ where g is the magnitude of free-fall acceleration and v is the rock's speed. Prove that if the rock has an initial velocity $v = 0$ at time $t = 0$, it will have the dependence on *time* given above.

Picture the Problem Because the acceleration of the rock is a function of its velocity, it is not constant. Choose a coordinate system in which downward is positive and the origin is at the point of release of the rock.

Rewrite $a = g - bv$ explicitly as a differential equation:

$$\dfrac{dv}{dt} = g - bv$$

Separate the variables, v on the left, t on the right:

$$\dfrac{dv}{g - bv} = dt$$

Integrate the left-hand side of this equation from 0 to v and the right-hand side from 0 to t:

$$\int_{0}^{v} \dfrac{dv'}{g - bv'} = \int_{0}^{t} dt'$$

and

$$-\dfrac{1}{b}\ln\!\left(\dfrac{g - bv}{g}\right) = t$$

Solve this expression for v.

$$v = \dfrac{g}{b}\left(1 - e^{-bt}\right)$$

Finally, differentiate this expression with respect to time to obtain an expression for the acceleration and to complete the proof.

$$a = \frac{dv}{dt} = \boxed{ge^{-bt}}$$

Chapter 3
Motion in Two and Three Dimensions

Conceptual Problems

*1 • Can the magnitude of the displacement of a particle be less than the distance traveled by the particle along its path? Can its magnitude be more than the distance traveled? Explain.

Determine the Concept The distance traveled along a path can be represented as a sequence of displacements.

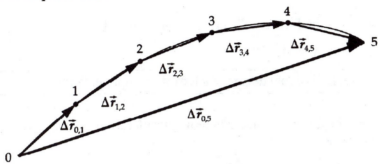

Suppose we take a trip along some path and consider the trip as a sequence of many very small displacements. The net displacement is the vector sum of the very small displacements, and the total distance traveled is the sum of the magnitudes of the very small displacements. That is,

$$total \ distance = \left| \Delta \vec{r}_{0,1} \right| + \left| \Delta \vec{r}_{1,2} \right| + \left| \Delta \vec{r}_{2,3} \right| + ... + \left| \Delta \vec{r}_{N-1,N} \right|$$

where N is the number of very small displacements. (For this to be exactly true we have to take the limit as N goes to infinity and each displacement magnitude goes to zero.) Now, using "the shortest distance between two points is a straight line," we have

$$\left| \Delta \vec{r}_{0,N} \right| \leq \left| \Delta \vec{r}_{0,1} \right| + \left| \Delta \vec{r}_{1,2} \right| + \left| \Delta \vec{r}_{2,3} \right| + ... + \left| \Delta \vec{r}_{N-1,N} \right|,$$

where $\left| \Delta \vec{r}_{0,N} \right|$ is the magnitude of the net displacement.

Hence, we have shown that the magnitude of the displacement of a particle is less than or equal to the distance it travels along its path.

*6 •   Can a vector be equal to zero and still have one or more components not equal to zero?

Determine the Concept The diagram shows a vector $\vec{A}$ and its components A_x and A_y. We can relate the magnitude of $\vec{A}$ is related to the lengths of its components through the Pythagorean theorem.

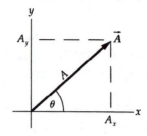

Suppose that $\vec{A}$ is equal to zero. Then $A^2 = A_x^2 + A_y^2 = 0$.
But $A_x^2 + A_y^2 = 0 \Rightarrow A_x = A_y = 0$.

> No. If a vector is equal to zero, each of its components must be zero too.

*8 •   True or false: The instantaneous-acceleration vector is *always* in the direction of motion.

Determine the Concept The *instantaneous acceleration* is the limiting value, as Δt approaches zero, of $\Delta \vec{v}/\Delta t$. Thus, the acceleration vector is in the same direction as $\Delta \vec{v}$.

False. Consider a ball that has been thrown upward near the surface of the earth and is slowing down. *The direction of its motion is upward.*

The diagram shows the ball's velocity vectors at two instants of time ... as well as the determination of $\Delta \vec{v}$. Note that because $\Delta \vec{v}$ is downward so is the *acceleration* of the ball.

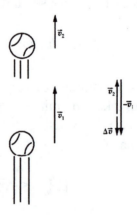

*12 •   Assuming constant acceleration, if you know the position vectors of a particle at two points on its path and also know the time it took to move from one point to the other, you can then compute (*a*) the particle's average velocity, (*b*) the particle's average acceleration, (*c*) the particle's instantaneous velocity, (*d*) the particle's instantaneous acceleration, (*e*) insufficient information is given to describe the particle's motion.

Determine the Concept The average velocity of a particle, $\vec{v}_{av}$, is the ratio of the particle's displacement to the time required for the displacement.

(*a*) We can calculate $\Delta \vec{r}$ from the given information and Δt is known.

$\boxed{(a)\text{ is correct.}}$

(*b*) We do not have enough information to calculate $\Delta \vec{v}$ and cannot compute the particle's average acceleration.

(*c*) We would need to know how the particle's velocity varies with time in order to compute its instantaneous velocity.

(*d*) We would need to know how the particle's velocity varies with time in order to compute its instantaneous acceleration.

*15 • Give examples of motion in which the directions of the velocity and acceleration vectors are (*a*) opposite, (*b*) the same, and (*c*) mutually perpendicular.

Determine the Concept The velocity vector is defined by $\vec{v} = d\vec{r}/dt$, while the acceleration vector is defined by $\vec{a} = d\vec{v}/dt$.

(*a*) A car moving along a straight road while braking.

(*b*) A car moving along a straight road while speeding up.

(*c*) A particle moving around a circular track at constant speed.

*18 •• As a bungee jumper approaches the lowest point in her descent, the rubber band holding her stretches and she loses speed as she continues to move downward. Assuming that she is dropping straight downward, make a motion diagram to find the direction of her acceleration vector as she slows down by drawing her velocity vectors at times t_1 and t_2, where $\Delta t = t_2 - t_1$ is small. From your drawing find the direction of the change in velocity $\Delta \vec{v} = \vec{v}_2 - \vec{v}_1$, and thus the direction of the acceleration vector.

Determine the Concept The acceleration vector is in the same direction as the *change in velocity vector*, $\Delta \vec{v}$.

The drawing is shown to the right.

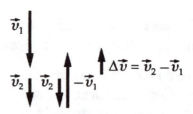

*21 •   True or false: When a projectile is fired horizontally, it takes the same amount of time to reach the ground as an identical projectile dropped from rest from the same height. Ignore the effects of air resistance.

Determine the Concept True. In the absence of air resistance, both projectiles experience the same downward acceleration. Because both projectiles have initial vertical velocities of zero, their vertical motions must be identical.

*28 ••   A vector $\vec{A}(t)$ has a constant magnitude but is changing direction. (*a*) Find $d\vec{A}/dt$ in the following manner: draw the vectors $\vec{A}(t+\Delta t)$ and $\vec{A}(t)$ for a small time interval Δt, and find the difference $\Delta\vec{A} = \vec{A}(t+\Delta t) - \vec{A}(t)$ graphically. How is the direction of $\Delta\vec{A}$ related to $\vec{A}$ for small time intervals? (*b*) Interpret this result for the special cases where $\vec{A}$ represents the position of a particle with respect to some coordinate system. (*c*) Could $\vec{A}$ represent a velocity vector? Explain.

(*a*) The vectors $\vec{A}(t)$ and $\vec{A}(t+\Delta t)$ are of equal length but point in slightly different directions. $\Delta\vec{A}$ is shown in the diagram below. Note that $\Delta\vec{A}$ is nearly perpendicular to $\vec{A}(t)$. For very small time intervals, $\Delta\vec{A}$ and $\vec{A}(t)$ are perpendicular to one another. Therefore, $d\vec{A}/dt$ is perpendicular to $\vec{A}$.

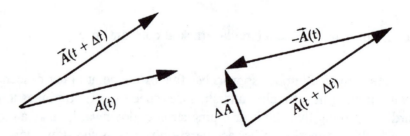

(*b*) If $\vec{A}$ represents the position of a particle, the particle must be undergoing circular motion (i.e., it is at a constant distance from some origin). The velocity vector is tangent to the particle's trajectory; in the case of a circle, it is perpendicular to the circle's radius.

(*c*) Yes, it could in the case of uniform circular motion. The speed of the particle is constant, but its heading is changing constantly. The acceleration vector in this case is always perpendicular to the velocity vector.

*30 •• Two cannons are pointed directly toward each other as shown in Figure 3-43. When fired, the cannonballs will follow the trajectories shown - P is the point where the trajectories cross each other. If we want the cannonballs to hit each other, should the gun crews fire cannon A first, cannon B first, or should they fire simultaneously? Ignore the effects of air resistance.

Figure 3-43 Problem 30

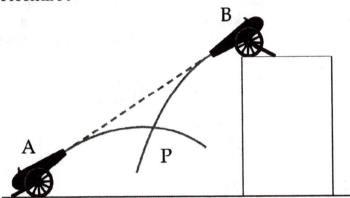

Determine the Concept We'll assume that the cannons are identical and use a constant acceleration equation to express the displacement of each cannonball as a function of time. Having done so, we can then establish the condition under which they will have the same vertical position at a given time and, hence, collide. The modified diagram shown below shows the displacements of both cannonballs.

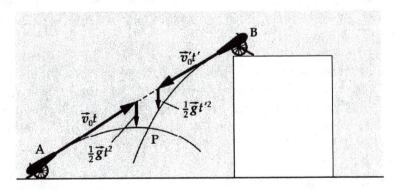

Express the displacement of the cannonball from cannon A at any time t after being fired and before any collision:

$$\Delta \vec{r} = \vec{v}_0 t + \tfrac{1}{2} \vec{g} t^2$$

Express the displacement of the cannonball from cannon A at any time t' after being fired and before any collision:

$$\Delta \vec{r}' = \vec{v}_0' t' + \tfrac{1}{2} \vec{g} t'^2$$

> If the guns are fired simultaneously, $t = t'$ and the balls are the same distance $\frac{1}{2}gt^2$ below the line of sight at all times. Therefore, they should fire the guns simultaneously.

Remarks: This is the "monkey and hunter" problem in disguise. If you imagine a monkey in the position shown below, and the two guns are fired simultaneously, and the monkey begins to fall when the guns are fired, then the monkey and the two cannonballs will all reach point P at the same time.

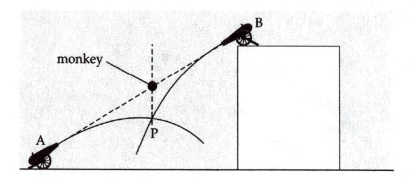

Estimation and Approximation

*36 •• Estimate how far you can throw a ball if you throw it (a) horizontally while standing on level ground, (b) at $\theta = 45°$ while standing on level ground, (c) horizontally from the top of a building 12 m high, (d) at $\theta = 45°$ from the top of a building 12 m high.

Picture the Problem During the flight of the ball the acceleration is constant and equal to 9.81 m/s² directed downward. We can find the flight time from the vertical part of the motion, and then use the horizontal part of the motion to find the horizontal distance. We'll assume that the release point of the ball is 2 m above your feet.

Make a sketch of the motion. Include coordinate axes, initial and final positions, and initial velocity components:

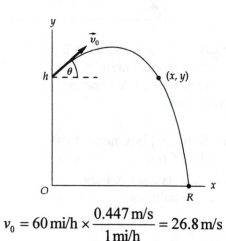

Obviously, how far you throw the ball will depend on how fast you can throw it. A major league

$$v_0 = 60\,\text{mi/h} \times \frac{0.447\,\text{m/s}}{1\,\text{mi/h}} = 26.8\,\text{m/s}$$

baseball pitcher can throw a fastball at 90 mi/h or so. Assume that you can throw a ball at two-thirds that speed to obtain:

There is no acceleration in the x direction, so the horizontal motion is one of constant velocity. Express the horizontal position of the ball as a function of time:

$$x = v_{0x}t \qquad (1)$$

Assuming that the release point of the ball is a distance h above the ground, express the vertical position of the ball as a function of time:

$$y = h + v_{0y}t + \tfrac{1}{2}a_y t^2 \qquad (2)$$

(a) For $\theta = 0$ we have:

$$v_{0x} = v_0 \cos\theta_0 = (26.8\,\text{m/s})\cos 0°$$
$$= 26.8\,\text{m/s}$$
and
$$v_{0y} = v_0 \sin\theta_0 = (26.8\,\text{m/s})\sin 0° = 0$$

Substitute in equations (1) and (2) to obtain:

$$x = (26.8\,\text{m/s})t$$
and
$$y = 2\,\text{m} + \tfrac{1}{2}(-9.81\,\text{m/s}^2)t^2$$

Eliminate t between these equations to obtain:

$$y = 2\,\text{m} - \frac{4.91\,\text{m/s}^2}{(26.8\,\text{m/s})^2}x^2$$

At impact, $y = 0$ and $x = R$:

$$0 = 2\,\text{m} - \frac{4.91\,\text{m/s}^2}{(26.8\,\text{m/s})^2}R^2$$

Solve for R to obtain:

$$R = \boxed{17.1\,\text{m}}$$

(b) Using trigonometry, solve for v_{0x} and v_{0y}:

$$v_{0x} = v_0 \cos\theta_0 = (26.8\,\text{m/s})\cos 45°$$
$$= 19.0\,\text{m/s}$$
and
$$v_{0y} = v_0 \sin\theta_0 = (26.8\,\text{m/s})\sin 45°$$
$$= 19.0\,\text{m/s}$$

Substitute in equations (1) and (2) to obtain:

$$x = (19.0\,\text{m/s})t$$
and

$$y = 2\,\mathrm{m} + (19.0\,\mathrm{m/s})t + \tfrac{1}{2}\left(-9.81\,\mathrm{m/s^2}\right)t^2$$

Eliminate t between these equations to obtain:

$$y = 2\,\mathrm{m} + x - \frac{4.905\,\mathrm{m/s^2}}{(19.0\,\mathrm{m/s})^2}x^2$$

At impact, $y = 0$ and $x = R$. Hence:

$$0 = 2\,\mathrm{m} + R - \frac{4.905\,\mathrm{m/s^2}}{(19.0\,\mathrm{m/s})^2}R^2$$

or

$$R^2 - (73.60\,\mathrm{m})R - 147.2\,\mathrm{m^2} = 0$$

Solve for R (you can use the "solver" or "graph" functions of your calculator) to obtain:

$$R = \boxed{75.6\,\mathrm{m}}$$

(c) Solve for v_{0x} and v_{0y}:

$$v_{0x} = v_0 = 26.8\,\mathrm{m/s}$$

and

$$v_{0y} = 0$$

Substitute in equations (1) and (2) to obtain:

$$x = (26.8\,\mathrm{m/s})t$$

and

$$y = 14\,\mathrm{m} + \tfrac{1}{2}\left(-9.81\,\mathrm{m/s^2}\right)t^2$$

Eliminate t between these equations to obtain:

$$y = 14\,\mathrm{m} - \frac{4.905\,\mathrm{m/s^2}}{(26.8\,\mathrm{m/s})^2}x^2$$

At impact, $y = 0$ and $x = R$:

$$0 = 14\,\mathrm{m} - \frac{4.905\,\mathrm{m/s^2}}{(26.8\,\mathrm{m/s})^2}R^2$$

Solve for R to obtain:

$$R = \boxed{45.3\,\mathrm{m}}$$

(d) Using trigonometry, solve for v_{0x} and v_{0y}:

$$v_{0x} = v_{0y} = 19.0\,\mathrm{m/s}$$

Substitute in equations (1) and (2) to obtain:

$$x = (19.0\,\mathrm{m/s})t$$

and

$$y = 14\,\mathrm{m} + (19.0\,\mathrm{m/s})t + \tfrac{1}{2}\left(-9.81\,\mathrm{m/s^2}\right)t^2$$

Eliminate t between these equations to obtain:

$$y = 14\,\mathrm{m} + x - \frac{4.905\,\mathrm{m/s^2}}{(19.0\,\mathrm{m/s})^2}x^2$$

At impact, $y = 0$ and $x = R$:

$$0 = 14\,\mathrm{m} + R - \frac{4.905\,\mathrm{m/s^2}}{(19.0\,\mathrm{m/s})^2} R^2$$

Solve for R (you can use the "solver" or "graph" function of your calculator) to obtain:

$$R = \boxed{85.6\,\mathrm{m}}$$

Vectors, Vector Addition, and Coordinate Systems

*39 •   A bear walks northeast for 12 m and then east for 12 m. Show each displacement graphically and find the resultant displacement vector graphically, as in Example 3-2(*a*).

Picture the Problem The resultant displacement is the vector sum of the individual displacements.

The two displacements of the bear and its resultant displacement are shown to the right:

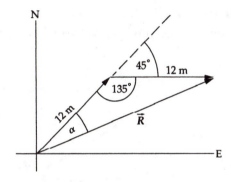

Using the law of cosines, solve for the resultant displacement:

$$R^2 = (12\,\mathrm{m})^2 + (12\,\mathrm{m})^2$$
$$- 2(12\,\mathrm{m})(12\,\mathrm{m})\cos 135°$$

and

$$R = \boxed{22.2\,\mathrm{m}}$$

Using the law of sines, solve for α:

$$\frac{\sin \alpha}{12\,\mathrm{m}} = \frac{\sin 135°}{22.2\,\mathrm{m}}$$

$\therefore \alpha = 22.5°$ and the angle with the horizontal is $45° - 22.5° = \boxed{22.5°}$

*41 •   For the two vectors $\vec{A}$ and $\vec{B}$ of Figure 3-45, find the following graphically as in Example 3-2(a): (*a*) $\vec{A}+\vec{B}$, (*b*) $\vec{A}-\vec{B}$, (*c*) $2\vec{A}+\vec{B}$, (*d*) $\vec{B}-\vec{A}$, (*e*) $2\vec{B}-\vec{A}$.

Figure 3-45 Problem 41

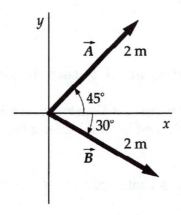

Picture the Problem Use the standard rules for vector addition. Remember that changing the sign of a vector reverses its direction.

(*a*)

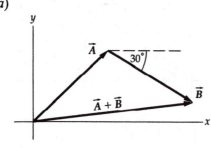

(*b*)

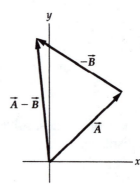

(*c*)

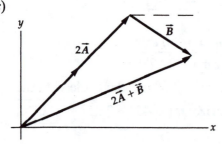

(*d*)

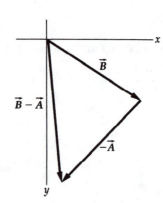

(e)

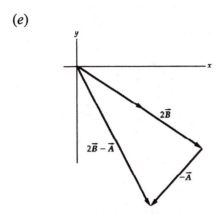

*46 •   Vector $\vec{A}$ has a magnitude of 8 m at an angle of 37° with the positive x-axis; vector $\vec{B} = (3\,\text{m})\hat{i} - (5\,\text{m})\hat{j}$; vector $\vec{C} = (-6\,\text{m})\hat{i} + (3\,\text{m})\hat{j}$. Find the following vectors: (a) $\vec{D} = \vec{A} + \vec{C}$, (b) $\vec{E} = \vec{B} - \vec{A}$, (c) $\vec{F} = \vec{A} - 2\vec{B} + 3\vec{C}$, (d) A vector $\vec{G}$ such that $\vec{G} - \vec{B} = \vec{A} + 2\vec{C} + 3\vec{G}$.

Picture the Problem Vectors can be added and subtracted by adding and subtracting their components.

Write $\vec{A}$ in component form:	$A_x = (8\text{ m})\cos 37° = 6.4\text{ m}$ $A_y = (8\text{ m})\sin 37° = 4.8\text{ m}$ $\therefore \vec{A} = (6.4\,\text{m})\hat{i} + (4.8\,\text{m})\hat{j}$
(a), (b), (c) Add (or subtract) x and y components:	$\vec{D} = \boxed{(0.4\text{m})\hat{i} + (7.8\text{m})\hat{j}}$ $\vec{E} = \boxed{(-3.4\text{m})\hat{i} - (9.8\text{m})\hat{j}}$ $\vec{F} = \boxed{(-17.6\text{m})\hat{i} + (23.8\text{m})\hat{j}}$
(d) Solve for $\vec{G}$ and add components to obtain:	$\vec{G} = -\dfrac{1}{2}\left(\vec{A} + \vec{B} + 2\vec{C}\right)$ $= \boxed{(1.3\text{m})\hat{i} - (2.9\text{m})\hat{j}}$

*51 ••   The faces of a cube with 3-m long edges are parallel to the coordinate planes. The cube has one corner at the origin. A fly begins at the origin and walks along three edges until it is at the far corner. Write the displacement vector of the fly using the unit vectors $\hat{i}$, $\hat{j}$, and $\hat{k}$, and find the magnitude of this displacement.

Picture the Problem While there are
several walking routes the fly could
take to get from the origin to point C,
its displacement will be the same for all
of them. One possible route is shown in
the figure.

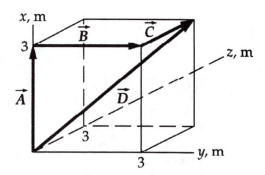

Express the fly's displacement
$\vec{D}$ during its trip from the origin to
point C and find its magnitude:

$$\vec{D} = \vec{A} + \vec{B} + \vec{C}$$
$$= (3\,\text{m})\hat{i} + (3\,\text{m})\hat{j} + (3\,\text{m})\hat{k}$$

and

$$D = \sqrt{(3\,\text{m})^2 + (3\,\text{m})^2 + (3\,\text{m})^2}$$
$$= \boxed{5.20\,\text{m}}$$

*52 •   A ship at sea receives radio signals from two transmitters A and B ,
which are 100 km apart, one due south of the other. The direction finder shows
that transmitter A is $\theta = 30°$ south of east, while transmitter B is due east.
Calculate the distance between the ship and transmitter B.

Picture the Problem The diagram shows the locations of the transmitters relative
to the ship and defines the distances separating the transmitters from each other
and from the ship. We can find the distance between the ship and transmitter B
using trigonometry.

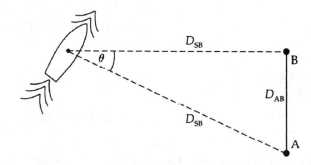

Relate the distance between A and B to
the distance from the ship to A and the
angle θ.

$$\tan\theta = \frac{D_{AB}}{D_{SB}}$$

Solve for and evaluate the distance
from the ship to transmitter B:

$$D_{SB} = \frac{D_{AB}}{\tan\theta} = \frac{100\,\text{km}}{\tan 30°} = \boxed{173\,\text{km}}$$

Velocity and Acceleration Vectors

*55 •   A particle moving at a velocity of 4.0 m/s in the positive x direction is given an acceleration of 3.0 m/s^2 in the positive y direction for 2.0 s. The final speed of the particle is (a) –2.0 m/s, (b) 7.2 m/s, (c) 6.0 m/s, (d) 10 m/s, (e) None of the above.

Picture the Problem The magnitude of the velocity vector at the end of the 2 s of acceleration will give us its speed at that instant. This is a constant-acceleration problem.

Find the final velocity vector of the particle:

$$\vec{v} = v_x\hat{i} + v_y\hat{j} = v_{x0}\hat{i} + a_y t\hat{j}$$
$$= (4.0\,\text{m/s})\hat{i} + (3.0\,\text{m/s}^2)(2.0\,\text{s})\hat{j}$$
$$= (4.0\,\text{m/s})\hat{i} + (6.0\,\text{m/s})\hat{j}$$

Find the magnitude of $\vec{v}$:

$$v = \sqrt{(4.0\,\text{m/s})^2 + (6.0\,\text{m/s})^2} = 7.21\,\text{m/s}$$

and $\boxed{(b)\text{ is correct.}}$

*58 ••   A particle moves in the xy plane with constant acceleration. At time zero, the particle is at $x = 4$ m, $y = 3$ m and has velocity $\vec{v} = (2\,\text{m/s})\hat{i} + (-9\,\text{m/s})\hat{j}$. The acceleration is given by $\vec{a} = (4\,\text{m/s}^2)\hat{i} + (3\,\text{m/s}^2)\hat{j}$. (a) Find the velocity at $t = 2$ s. (b) Find the position at $t = 4$ s. Give the magnitude and direction of the position vector.

Picture the Problem The acceleration is constant so we can use the constant acceleration equations in vector form to find the velocity at $t = 2$ s and the position vector at $t = 4$ s.

(a) The velocity of the particle, as a function of time, is given by:

$$\vec{v} = \vec{v}_0 + \vec{a}t$$

Substitute to find the velocity at $t = 2$ s:

$$\vec{v} = (2\,\text{m/s})\hat{i} + (-9\,\text{m/s})\hat{j}$$
$$+ \left[(4\,\text{m/s}^2)\hat{i} + (3\,\text{m/s}^2)\hat{j}\right](2\text{s})$$
$$= \boxed{(10\,\text{m/s})\hat{i} + (-3\,\text{m/s})\hat{j}}$$

(b) Express the position vector as a function of time:

$$\vec{r} = \vec{r}_0 + \vec{v}_0 t + \tfrac{1}{2}\vec{a}t^2$$

Substitute and simplify:

$$\vec{r} = (4\,\text{m})\hat{i} + (3\,\text{m})\hat{j}$$
$$+ \left[(2\,\text{m/s})\hat{i} + (\text{-9 m/s})\hat{j}\right](4\,\text{s})$$
$$+ \tfrac{1}{2}\left[(4\,\text{m/s}^2)\hat{i} + (3\,\text{m/s}^2)\hat{j}\right](4\,\text{s})^2$$
$$= \boxed{(44\,\text{m})\hat{i} + (-9\,\text{m})\hat{j}}$$

Find the magnitude and direction of $\vec{r}$ at $t = 4$ s:

$$r(4\,\text{s}) = \sqrt{(44\,\text{m})^2 + (-9\,\text{m})^2} = \boxed{44.9\,\text{m}}$$

and, because $\vec{r}$ is in the 4th quadrant,

$$\theta = \tan^{-1}\left(\frac{-9\,\text{m}}{44\,\text{m}}\right) = \boxed{-11.6°}$$

*62 •••   Mary and Robert decide to rendezvous on Lake Michigan. Mary departs in her boat from Petoskey at 9:00 a.m. and travels due north at 8 mi/h. Robert leaves from his home on the shore of Beaver Island, 26 miles 30° west of north of Petoskey, at 10:00 a.m. and travels at a constant speed of 6 mi/h. In what direction should Robert be heading to intercept Mary, and where and when will they meet?

Picture the Problem Choose a coordinate system with the origin at Petoskey, the positive x direction to the east, and the positive y direction to the north. Let $t = 0$ at 9:00 a.m. and θ be the angle between the velocity vector of Robert's boat and the easterly direction. Let "M" and "R" denote Mary and Robert, respectively.

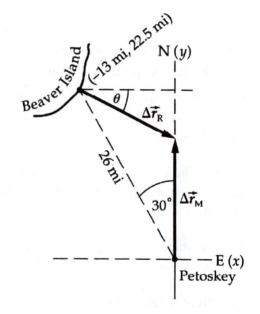

Express Mary's displacement from Petoskey:

$$\Delta \vec{r}_\text{M} = v_\text{M} t \hat{j} = (8t)\hat{j}$$

where $\Delta \vec{r}_\text{M}$ is in miles if t is in hours.

Note that Robert's initial position coordinates (x_i, y_i) are:

$(x_i, y_i) = (-13\,\text{mi}, 22.5\,\text{mi})$

Express Robert's displacement from Beaver Island:

$$\Delta \vec{r}_R = [x_i + (v_R \cos\theta)(t-1)])\hat{i}$$
$$+ [y_i + (v_R \sin\theta)(t-1)]\hat{j}$$
$$= [-13 + \{6(t-1)\cos\theta\}]\hat{i}$$
$$+ [22.5 + \{6(t-1)\sin\theta\}]\hat{j}$$

where the units are as above.

When Mary and Robert rendezvous, their coordinates will be the same. Equating their north and east coordinates yields:

East: $-13 + (6t\cos\theta) - (6\cos\theta) = 0$

North: $22.5 + (6t\sin\theta) - (6\sin\theta) = 8t$

Eliminate t between the two equations to obtain:

$(78\tan\theta + 87)\cos\theta = 104$

This transcendental equation can be solved by writing it as

$$f(\theta) = (78\tan\theta + 87)\cos\theta - 104$$

and then plotting its graph. The graph shown to the right was plotted using a spreadsheet program and the root at 0.258 rad (14.8°) was found using a calculator's "trace" function.

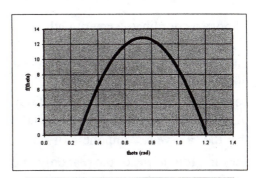

Robert should head 14.8° south of east.

Use either the north or east equation to solve for t:

$t = $ | 3.24 h = 3 h 15 min

Finally, find the distance traveled due north by Mary:

$r_M = v_M t$
$= (8 \text{ mi/h}) (3.24 \text{ h})$
$= $ | 25.9 mi, due north of Petoskey

Remarks: Two alternatives to solving the transcendental equation using a calculator's "trace" function are: (a) to search the spreadsheet program used to generate data for the function $f(\theta) = (78\tan\theta + 87)\cos\theta - 104$ for values of θ that satisfy the condition $f(\theta) = 0$, or (b) a trial-and-error sequence of substitutions for θ ... using the result of each substitution (e.g., a change in sign) to motivate the next substitution ... until a root is found.

Relative Velocity

*65 •• A small plane departs from point A heading for an airport 520 km due north at point B. The airspeed of the plane is 240 km/h and there is a steady wind of 50 km/h blowing northwest to southeast. Determine the proper heading for the plane and the time of flight.

Picture the Problem Let the velocity of the plane relative to the ground be represented by $\vec{v}_{PG}$; the velocity of the plane relative to the air by $\vec{v}_{PA}$, and the velocity of the air relative to the ground by $\vec{v}_{AG}$. Then

$$\vec{v}_{PG} = \vec{v}_{PA} + \vec{v}_{AG} \ (1)$$

Choose a coordinate system with the origin at point A, the positive x direction to the east, and the positive y direction to the north. θ is the angle between north and the direction of the plane's heading. The pilot must head so that the east-west component of $\vec{v}_{PG}$ is zero in order to make the plane fly due north.

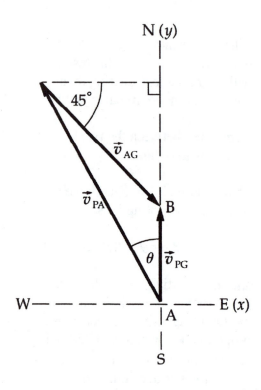

Use the diagram to express the condition relating the eastward component of $\vec{v}_{AG}$ and the westward component of $\vec{v}_{PA}$. This must be satisfied if the plane is to stay on its northerly course. [Note: this is equivalent to equating the x-components of equation (1).]

$$(50 \text{ km/h}) \cos 45° = (240 \text{ km/h}) \sin\theta$$

Now solve for θ to obtain:

$$\theta = \sin^{-1}\left[\frac{(50\,\text{km/h})\cos 45°}{240\,\text{km/h}}\right] = \boxed{8.47°}$$

Add the north components of $\vec{v}_{PA}$ and $\vec{v}_{AG}$ to find the velocity of the plane relative to the ground:

$v_{PG} + v_{AG}\sin 45° = v_{PA}\cos 8.47°$
and
$v_{PG} = (240 \text{ km/h})\cos 8.47°$
$\qquad\quad - (50 \text{ km/h})\sin 45°$
$\qquad = 202 \text{ km/h}$

Finally, find the time of flight:

$$t_{\text{flight}} = \frac{\text{distance travelled}}{v_{PG}}$$

$$= \frac{520\,\text{km}}{202\,\text{km/h}} = \boxed{2.57\,\text{h}}$$

*69 ••   Car A is traveling east at 20 m/s toward an intersection. As car A crosses the intersection, car B starts from rest 40 m north of the intersection and moves south with a constant acceleration of 2 m/s^2. Six seconds after A crosses the intersection find (*a*) the position of B relative to A, (*b*) the velocity of B relative to A, (*c*) the acceleration of B relative to A.

Picture the Problem The position of B relative to A is the vector from A to B; i.e.,

$$\vec{r}_{AB} = \vec{r}_B - \vec{r}_A$$

The velocity of B relative to A is

$$\vec{v}_{AB} = d\vec{r}_{AB}/dt$$

and the acceleration of B relative to A is

$$\vec{a}_{AB} = d\vec{v}_{AB}/dt$$

Choose a coordinate system with the origin at the intersection, the positive *x* direction to the east, and the positive *y* direction to the north.

(*a*) Find $\vec{r}_B$, $\vec{r}_A$, and $\vec{r}_{AB}$:

$$\vec{r}_B = \left[40\text{m} - \tfrac{1}{2}\left(2\text{m/s}^2\right)t^2\right]\hat{j}$$

$$\vec{r}_A = \left[(20\text{m/s})t\right]\hat{i}$$

and

$$\vec{r}_{AB} = \vec{r}_B - \vec{r}_A$$

$$= \left[(-20\text{m/s})t\right]\hat{i}$$

$$+ \left[40\text{m} - \tfrac{1}{2}\left(2\text{m/s}^2\right)t^2\right]\hat{j}$$

Evaluate $\vec{r}_{AB}$ at $t = 6$ s:

$$\vec{r}_{AB}(6\text{s}) = \boxed{(120\,\text{m})\,\hat{i} + (4\,\text{m})\,\hat{j}}$$

(*b*) Find $\vec{v}_{AB} = d\vec{r}_{AB}/dt$:

$$\vec{v}_{AB} = \frac{d\vec{r}_{AB}}{dt} = \frac{d}{dt}\left[\{(-20\,\text{m/s})t\}\vec{i}\right.$$

$$+ \left\{40\,\text{m} - \tfrac{1}{2}\left(2\,\text{m/s}^2\right)t^2\right\}\hat{j}\right]$$

$$= (-20\,\text{m/s})\,\hat{i} + (-2\,\text{m/s}^2)t\,\hat{j}$$

Evaluate $\vec{v}_{AB}$ at $t = 6$ s:

$$\vec{v}_{AB}(6\,\mathrm{s}) = \boxed{(-20\,\mathrm{m/s})\hat{i} - (12\,\mathrm{m/s})\hat{j}}$$

(c) Find $\vec{a}_{AB} = d\vec{v}_{AB}/dt$:

$$\vec{a}_{AB} = \frac{d}{dt}\left[(-20\,\mathrm{m/s})\,\hat{i} + (-2\,\mathrm{m/s}^2)t\,\hat{j}\right]$$

$$= \boxed{(-2\,\mathrm{m/s}^2)\hat{j}}$$

Note that $\vec{a}_{AB}$ is independent of time.

*70 •••   A tennis racket is held horizontally and a tennis ball is held above the racket. When the ball is dropped from rest, and bounces off of the strings of the racket, the ball always rebounds to 64% of its initial height. (a) Express the speed of the tennis ball just after it bounces as some fraction of the speed of the ball just before the bounce. (b) The tennis ball is now thrown up into the air and served using the same racket. Assuming the ball's preimpact speed is zero, and that the racket's speed through impact is 25 m/s, with what speed does the tennis ball come off the racket strings? *Hint: Using the results of part (a), solve for the post-impact speed of the ball in the reference frame of the racket, and then calculate the ball's speed in the reference frame of the earth.* (c) From some well-established laws of physics, we never see a ball bounce higher than the point from which it was released. From this, can you give an upper bound on the speed of a served tennis ball in relation to the speed of the racket, no matter how well the racket is designed? (We will see later that these results can be explained in a different context: the idea of *conservation of momentum.*)

Picture the Problem Let h and h' represent the heights from which the ball is dropped and to which it rebounds, respectively. Let v and v' represent the speeds with which the ball strikes the racket and rebounds from it. We can use a constant acceleration equation to relate the pre- and post-collision speeds of the ball to its drop and rebound heights.

(a) Using a constant-acceleration equation, relate the impact speed of the ball to the distance it has fallen:

$$v^2 = v_0^2 + 2gh$$
or, because $v_0 = 0$,
$$v = \sqrt{2gh}$$

Relate the rebound speed of the ball to the height to which it rebounds:

$$v^2 = v'^2 - 2gh'$$
or because $v = 0$,
$$v' = \sqrt{2gh'}$$

Divide the second of these equations by the first to obtain:

$$\frac{v'}{v} = \frac{\sqrt{2gh'}}{\sqrt{2gh}} = \sqrt{\frac{h'}{h}}$$

Substitute for h' and evaluate the ratio of the speeds:

$$\frac{v'}{v} = \sqrt{\frac{0.64h}{h}} = 0.8 \Rightarrow v' = \boxed{0.8v}$$

(b) Call the speed of the racket V. In a reference frame where the racket is unmoving, the ball initially has speed V, moving *toward* the racket. After it "bounces" from the racket, it will have speed $0.8\,V$, moving *away* from the racket.

In the reference frame where the racket is moving and the ball initially unmoving, we need to add the speed of the racket to the speed of the ball in the racket's rest frame. Therefore, the ball's speed is:

$v' = V + 0.8V = 1.8V = 45\,\text{m/s}$

$\approx \boxed{100\,\text{mi/h}}$

This speed is close to that of a tennis pro's serve. Note that this result tells us that the ball is moving significantly faster than the racket.

(c) | From the result in part (b), the ball can never move more than twice as fast as the racket.

Projectile Motion and Projectile Range

*78 ••   A cannonball is fired with initial speed v_0 at an angle 30° above the horizontal from a height of 40 m above the ground. The projectile strikes the ground with a speed of $1.2\,v_0$. Find v_0.

Picture the Problem Choose the coordinate system shown to the right. Because, in the absence of air resistance, the horizontal and vertical speeds are independent of each other, we can use constant-acceleration equations to relate the impact speed of the projectile to its components.

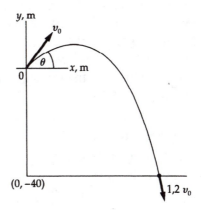

The horizontal and vertical velocity components are:

$v_{0x} = v_x = v_0\cos\theta$
and
$v_{0y} = v_0\sin\theta$

Using a constant-acceleration equation, relate the vertical component of the velocity to the vertical displacement of the projectile:

$v_y^2 = v_{0y}^2 + 2a_y\Delta y$
or, because $a_y = -g$ and $\Delta y = -h$,
$v_y^2 = (v_0\sin\theta)^2 + 2gh$

Express the relationship between the magnitude of a velocity vector and its components, substitute for the components, and simplify to obtain:

$$v^2 = v_x^2 + v_y^2 = (v_0 \cos\theta)^2 + v_y^2$$
$$= v_0^2(\sin^2\theta + \cos^2\theta) + 2gh$$
$$= v_0^2 + 2gh$$

Substitute for v:

$$(1.2v_0)^2 = v_0^2 + 2gh$$

Set $v = 1.2\,v_0$, $h = 40$ m and solve for v_0:

$$v_0 = \boxed{42.2\,\text{m/s}}$$

Remarks: Note that v is independent of θ. This will be more obvious once conservation of energy has been studied.

*82 •• Consider a ball that is thrown with initial speed v_0 at an angle θ above the horizontal. If we consider its speed v at some height h above the ground, show that $v(h)$ is independent of θ.

Picture the Problem In the absence of friction, the acceleration of the ball is constant and we can use the constant-acceleration equations to describe its motion. The figure shows the launch conditions and an appropriate coordinate system. The speeds v, v_x, and v_y are related through the Pythagorean Theorem.

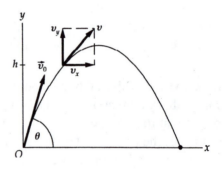

The squares of the vertical and horizontal components of the object's velocity are:

$$v_y^2 = v_0^2 \sin^2\theta - 2gh$$
and
$$v_x^2 = v_0^2 \cos^2\theta$$

The relationship between these variables is:

$$v^2 = v_x^2 + v_y^2$$

Substitute and simplify to obtain:

$$v^2 = \boxed{v_0^2 - 2gh}$$

Note that v is independent of θ ... as was to be shown.

*85 •• Wiley Coyote (*Carnivorous hungribilous*) is chasing the Roadrunner (*Speedibus cantcatchmi*) yet again.   While running down the road, they come to a deep gorge, 15 m straight across and 100 m deep. The Roadrunner launches itself across the gorge at a launch angle of 15° above the horizontal, and lands with 1.5 m to spare. (*a*) What was the Roadrunner's launch speed? Ignore air resistance.

(b) Wiley Coyote launches himself across the gorge with the same initial speed, but at a different launch angle. To his horror, he is short the other lip by 0.5 m. What was his launch angle? (Assume that it was lower than 15°.)

Picture the Problem In the absence of air resistance, the accelerations of both Wiley Coyote and the Roadrunner are constant and we can use constant acceleration equations to express their coordinates at any time during their leaps across the gorge. By eliminating the parameter t between these equations, we can obtain an expression that relates their y coordinates to their x coordinates and that we can solve for their launch angles.

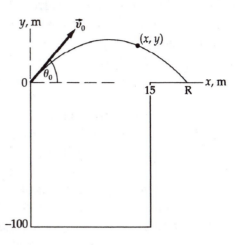

(a) Using constant-acceleration equations, express the x coordinate of the Roadrunner while it is in flight across the gorge:

$x = x_0 + v_{0x}t + \frac{1}{2}a_x t^2$
or, because $x_0 = 0$, $a_x = 0$ and $v_{0x} = v_0 \cos\theta_0$,
$x = (v_0 \cos\theta_0)t$

Using constant-acceleration equations, express the y coordinate of the Roadrunner while it is in flight across the gorge:

$y = y_0 + v_{0y}t + \frac{1}{2}a_y t^2$
or, because $y_0 = 0$, $a_y = -g$ and $v_{0y} = v_0 \sin\theta_0$,
$y = (v_0 \sin\theta_0)t - \frac{1}{2}gt^2$

Eliminate the parameter t to obtain:

$$y = (\tan\theta_0)x - \frac{g}{2v_0^2 \cos^2\theta_0}x^2 \qquad (1)$$

Letting R represent the Roadrunner's range and using the trigonometric identity $\sin 2\theta = 2\sin\theta\cos\theta$, solve for and evaluate its launch speed:

$$v_0 = \sqrt{\frac{Rg}{\sin 2\theta_0}} = \sqrt{\frac{(15\,\text{m})(9.81\,\text{m/s}^2)}{\sin 30°}}$$

$$= \boxed{17.2\,\text{m/s}}$$

(b) Letting R represent Wiley's range, solve equation (1) for his launch angle:

$$\theta_0 = \frac{1}{2}\sin^{-1}\left[\frac{Rg}{v_0^2}\right]$$

Substitute numerical values and evaluate θ_0:

$$\theta_0 = \frac{1}{2}\sin^{-1}\left[\frac{(14.5\,\text{m})(9.81\,\text{m/s}^2)}{(17.2\,\text{m/s})^2}\right]$$

$$= \boxed{14.4°}$$

*92 ••   The speed of an arrow fired from a compound bow is about 45 m/s.
(*a*) A Tartar archer sits astride his horse and launches an arrow into the air,
elevating the bow at an angle of 10° above the horizontal. If the bow is 2.25 m
above the ground, what is the arrow's range? Assume that the ground is level, and
ignore air resistance. (*b*) Now assume that his horse is at full gallop, moving in
the same direction as he will fire the arrow, and that he elevates the bow in the
same way as in part (*a*) and fires.   If the horse's speed is 12 m/s, what is the
arrow's range now?

Picture the Problem Choose a
coordinate system in which the origin is
at ground level. Let the positive *x*
direction be to the right and the positive
y direction be upward. We can apply
constant-acceleration equations to
obtain parametric equations in time that
relate the range to the initial horizontal
speed and the height *h* to the initial
upward speed. Eliminating the
parameter will leave us with a quadratic
equation in *R,* the solution to which
will give us the range of the arrow. In
(*b*), we'll find the launch speed and
angle as viewed by an observer who is
at rest on the ground and then use these
results to find the arrow's range when
the horse is moving at 12 m/s.

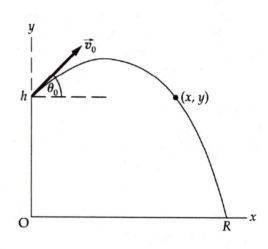

(*a*) Use constant-acceleration
equations to express the
horizontal and vertical
coordinates of the arrow's
motion:

$$R = \Delta x = x - x_0 = v_{0x}t$$
and
$$y = h + v_{0y}t + \tfrac{1}{2}(-g)t^2$$
where
$$v_{0x} = v_0\cos\theta \text{ and } v_{0y} = v_0\sin\theta$$

Solve the *x*-component equation
for time:

$$t = \frac{R}{v_{0x}} = \frac{R}{v_0\cos\theta}$$

Eliminate time from the
y-component equation:

$$y = h + v_{0y}\frac{R}{v_{0x}} - \frac{1}{2}g\left(\frac{R}{v_{0x}}\right)^2$$
and, at (*R*, 0),
$$0 = h + (\tan\theta)R - \frac{g}{2v_0^2\cos^2\theta}R^2$$

Solve for the range to obtain:

$$R = \frac{v_0^2}{2g} \sin 2\theta \left(1 + \sqrt{1 + \frac{2gh}{v_0^2 \sin^2 \theta}} \right)$$

Substitute numerical values and evaluate R:

$$R = \frac{(45\,\text{m/s})^2}{2(9.81\,\text{m/s}^2)} \sin 20° \left(1 + \sqrt{1 + \frac{2(9.81\,\text{m/s}^2)(2.25\,\text{m})}{(45\,\text{m/s})^2 (\sin^2 10°)}} \right) = \boxed{81.6\,\text{m}}$$

(b) Express the speed of the arrow in the horizontal direction:

$$v_x = v_{\text{arrow}} + v_{\text{archer}}$$
$$= (45\,\text{m/s})\cos 10° + 12\,\text{m/s}$$
$$= 56.3\,\text{m/s}$$

Express the vertical speed of the arrow:

$$v_y = (45\,\text{m/s})\sin 10° = 7.81\,\text{m/s}$$

Express the angle of elevation from the perspective of someone on the ground:

$$\theta = \tan^{-1} \frac{v_y}{v_x} = \tan^{-1} \left(\frac{7.81\,\text{m/s}}{56.3\,\text{m/s}} \right) = 7.90°$$

Express the arrow's speed relative to the ground:

$$v_0 = \sqrt{v_x^2 + v_y^2}$$
$$= \sqrt{(56.3\,\text{m/s})^2 + (7.81\,\text{m/s})^2}$$
$$= 56.8\,\text{m/s}$$

Substitute numerical values and evaluate R:

$$R = \frac{(56.8\,\text{m/s})^2}{2(9.81\,\text{m/s}^2)} \sin 15.8° \left(1 + \sqrt{1 + \frac{2(9.81\,\text{m/s}^2)(2.25\,\text{m})}{(56.8\,\text{m/s})^2 (\sin^2 7.9°)}} \right) = \boxed{104\,\text{m}}$$

Remarks: An alternative solution for part (b) is to solve for the range in the reference frame of the archer and then add to it the distance the frame travels, relative to the earth, during the time of flight.

*99 •• A projectile is launched over level ground at an elevation angle θ. An observer standing at the launch site sights the projectile at the point of its highest elevation, and measures the angle ϕ shown in Figure 3-51. Show that $\tan \phi = \frac{1}{2} \tan \theta$.

Figure 3-51 Problem 99

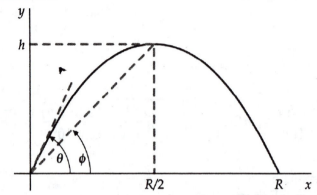

Picture the Problem We can use trigonometry to relate the maximum height of the projectile to its range and the sighting angle at maximum elevation and the range equation to express the range as a function of the launch speed and angle. We can use a constant-acceleration equation to express the maximum height reached by the projectile in terms of its launch angle and speed. Combining these relationships will allow us to conclude that $\tan\phi = \frac{1}{2}\tan\theta$.

Referring to the figure, relate the maximum height of the projectile to its range and the sighting angle ϕ:	$\tan\phi = \dfrac{h}{R/2}$
Express the range of the rocket and use the trigonometric identity $\sin 2\theta = 2\sin\theta\cos\theta$ to rewrite the expression as:	$R = \dfrac{v^2}{g}\sin(2\theta) = 2\dfrac{v^2}{g}\sin\theta\cos\theta$
Using a constant-acceleration equation, relate the maximum height of a projectile to the vertical component of its launch speed:	$v_y^2 = v_{0y}^2 - 2gh$ or, because $v_y = 0$ and $v_{0y} = v_0\sin\theta$, $v_0^2\sin^2\theta = 2gh$
Solve for the maximum height h:	$h = \dfrac{v^2}{2g}\sin^2\theta$
Substitute for R and h and simplify to obtain:	$\tan\phi = \dfrac{2\dfrac{v^2}{2g}\sin^2\theta}{2\dfrac{v^2}{g}\sin\theta\cos\theta} = \boxed{\tfrac{1}{2}\tan\theta}$

*101 •• A stone is thrown horizontally from the top of an incline that makes an angle θ with the horizontal. If the stone's initial speed is v_0, how far down the incline will it land?

Picture the Problem In the absence of
air resistance, the acceleration of the
stone is constant and the horizontal and
vertical motions are independent of
each other. Choose a coordinate
system with the origin at the throwing
location and the axes oriented as shown
in the figure and use constant-
acceleration equations to express the x
and y coordinates of the stone while it
is in flight.

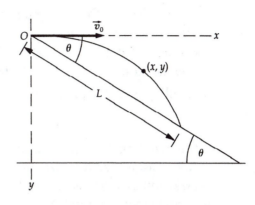

Using a constant-acceleration
equation, express the x coordinate
of the stone in flight:

$x = x_0 + v_{0x}t + \frac{1}{2}a_x t^2$

or, because $x_0 = 0$, $v_{0x} = v_0$ and $a_x = 0$,

$x = v_0 t$

Using a constant-acceleration
equation, express the y coordinate
of the stone in flight:

$y = y_0 + v_{0y}t + \frac{1}{2}a_y t^2$

or, because $y_0 = 0$, $v_{0y} = 0$ and $a_y = g$,

$y = \frac{1}{2}gt^2$

Referring to the diagram, express
the relationship between θ, y and
x at impact:

$\tan\theta = \dfrac{y}{x}$

Substitute for x and y and solve
for the time to impact:

$\tan\theta = \dfrac{gt^2}{2v_0 t} = \dfrac{g}{2v_0}t$

Solve for t to obtain:

$t = \dfrac{2v_0}{g}\tan\theta$

Referring to the diagram, express
the relationship between θ, L, y,
and x at impact:

$x = L\cos\theta = \dfrac{y}{\tan\theta}$

Substitute for y to obtain:

$\dfrac{gt^2}{2g} = L\cos\theta$

Substitute for t and solve for L to
obtain:

$L = \boxed{\dfrac{2v_0^2 \tan\theta}{g\cos\theta}}$

Hitting Targets and Related Problems

*106 •• The distance from the pitcher's mound to home plate is 18.4 m. The mound is 0.2 m above the level of the field. A pitcher throws a fastball with an initial speed of 37.5 m/s. At the moment the ball leaves the pitcher's hand, it is 2.3 m above the mound. What should the angle between $\vec{v}_0$ and the horizontal be so that the ball crosses the plate 0.7 m above ground? (Neglect interaction with air.)

Picture the Problem The acceleration of the ball is constant (zero horizontally and $-g$ vertically) and the vertical and horizontal components are independent of each other. Choose the coordinate system shown in the figure and assume that v and t are unchanged by throwing the ball slightly downward.

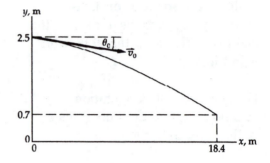

Express the horizontal displacement of the ball as a function of time:

$$\Delta x = v_{0x}\Delta t + \tfrac{1}{2}a_x(\Delta t)^2$$
or, because $a_x = 0$,
$$\Delta x = v_{0x}\Delta t$$

Solve for the time of flight if the ball were thrown horizontally:

$$\Delta t = \frac{\Delta x}{v_{0x}} = \frac{18.4\,\text{m}}{37.5\,\text{m/s}} = 0.491\,\text{s}$$

Using a constant-acceleration equation, express the distance the ball would drop (vertical displacement) if it were thrown horizontally:

$$\Delta y = v_{0y}\Delta t + \tfrac{1}{2}a_y(\Delta t)^2$$
or, because $v_{0y} = 0$ and $a_y = -g$,
$$\Delta y = -\tfrac{1}{2}g(\Delta t)^2$$

Substitute numerical values and evaluate Δy:

$$\Delta y = -\tfrac{1}{2}(9.81\,\text{m/s}^2)(0.491\,\text{s})^2 = -1.18\,\text{m}$$

The ball must drop an additional 0.62 m before it gets to home plate:

$$y = (2.5 - 1.18)\,\text{m}$$
$$= 1.32\,\text{m above ground}$$

Calculate the initial downward speed the ball must have to drop 0.62 m in 0.491 s:

$$v_y = \frac{-0.62\,\text{m}}{0.491\,\text{s}} = -1.26\,\text{m}$$

Find the angle with horizontal:

$$\theta = \tan^{-1}\frac{v_y}{v_x} = \tan^{-1}\left(\frac{-1.26\,\text{m/s}}{37.5\,\text{m/s}}\right)$$

$$= \boxed{-1.92°}$$

Remarks: One can readily show that $\sqrt{v_x^2 + v_y^2} = 37.5$ m/s to within 1%; so the assumption that v and t are unchanged by throwing the ball downward at an angle of 1.93° is justified.

General Problems

*111 •   A plane is inclined at an angle of 30° from the horizontal. Choose the x axis pointing down the slope of the plane and the y axis perpendicular to the plane. Find the x and y components of the acceleration of gravity, which has the magnitude 9.81 m/s² and points vertically down.

Picture the Problem A vector quantity can be resolved into its components relative to any coordinate system. In this example, the axes are orthogonal and the components of the vector can be found using trigonometric functions.

The x and y components of $\vec{g}$ are related to g through the sine and cosine functions.

$$g_x = g\sin 30° = \boxed{4.91\,\text{m/s}^2}$$
and
$$g_y = g\cos 30° = \boxed{8.50\,\text{m/s}^2}$$

*117 ••   A small steel ball is projected horizontally off the top landing of a long rectangular staircase. The initial speed of the ball is 3 m/s. Each step is 0.18 m high and 0.3 m wide. Which step does the ball strike first?

Picture the Problem In the absence of air resistance, the steel ball will experience constant acceleration. Choose a coordinate system with its origin at the initial position of the ball, the x direction to the right, and the y direction downward. In this coordinate system $y_0 = 0$ and $a = g$. Letting (x, y) be a point on the path of the ball, we can use constant-acceleration equations to express both x and y as functions of time and, using the geometry of the staircase, find an expression for the time of flight of the ball. Knowing its time of flight, we can find its range and identify the step it strikes first.

The angle of the steps, with respect to the horizontal, is:

$$\theta = \tan^{-1}\left(\frac{0.18\,\text{m}}{0.3\,\text{m}}\right) = 31.0°$$

Using a constant-acceleration equation, express the x coordinate of the steel ball in its flight:

$$x = x_0 + v_0 t + \tfrac{1}{2}a_x t^2$$
or, because $x_0 = 0$ and $a_y = 0$,
$$x = v_0 t$$

Using a constant-acceleration equation, express the y coordinate of the steel ball in its flight:

$$y = y_0 + v_{0y}t + \tfrac{1}{2}a_y t^2$$
or, because $y_0 = 0$, $v_{0y} = 0$, and $a_y = g$,
$$y = \tfrac{1}{2}gt^2$$

The equation of the dashed line in the figure is:	$$\frac{y}{x} = \tan\theta = \frac{gt}{2v_0}$$
Solve for the flight time:	$$t = \frac{2v_0}{g}\tan\theta$$
Find the x coordinate of the landing position:	$$x = \frac{y}{\tan\theta} = \frac{2v_0^2}{g}\tan\theta$$
Substitute the angle determined in the first step:	$$x = \frac{2(3\,\text{m/s})^2}{9.81\,\text{m/s}^2}\tan31° = 1.10\,\text{m}$$
Find the first step with $x > 1.10$ m:	The first step with $x > 1.10$ m is the 4th step.

*121 •• Galileo showed that, if air resistance is neglected, the ranges for projectiles whose angles of projection exceed or fall short of 45° by the same amount are equal. Prove Galileo's result.

Picture the Problem In the absence of air resistance, the acceleration of the projectile is constant and the equation of a projectile for equal initial and final elevations, which was derived from the constant-acceleration equations, is applicable. We can use the equation giving the range of a projectile for equal initial and final elevations to evaluate the ranges of launches that exceed or fall short of 45° by the same amount.

Express the range of the projectile as a function of its initial speed and angle of launch:	$$R = \frac{v_0^2}{g}\sin 2\theta_0$$
Let $\theta_0 = 45° \pm \theta$.	$$R = \frac{v_0^2}{g}\sin(90° \pm 2\theta)$$ $$= \frac{v_0^2}{g}\cos(\pm 2\theta)$$
Because $\cos(-\theta) = \cos(+\theta)$ (the cosine function is an *even* function):	$$R(45° + \theta) = R(45° - \theta)$$

Chapter 4
Newton's Laws

Conceptual Problems

*1 •• How can you tell if a particular reference frame is an inertial reference frame?

Determine the Concept A reference frame in which the law of inertia holds is called an inertial reference frame.

If an object with no net force acting on it is at rest or is moving with a constant speed in a straight line (i.e., with constant velocity) relative to the reference frame, then the reference frame is an inertial reference frame. Consider sitting at rest in an accelerating train or plane. The train or plane is not an inertial reference frame even though you are at rest relative to it. In an inertial frame, a dropped ball lands at your feet. You are in a noninertial frame when the driver of the car in which you are riding steps on the gas and you are pushed back into your seat.

*4 • If only a single nonzero force acts on an object, must the object have an acceleration relative to an inertial reference frame? Can it ever have zero velocity?

Determine the Concept An object accelerates when a *net* force acts on it. The fact that an object is accelerating tells us nothing about its velocity other than that it is always changing.

Yes, the object must have an acceleration relative to the inertial frame of reference. According to Newton's 1st and 2nd laws, an object must accelerate, relative to any inertial reference frame, in the direction of the net force. If there is "only a single nonzero force," then this force is the net force.

Yes, the object's velocity may be momentarily zero. During the period in which the force is acting, the object may be momentarily at rest, but its velocity cannot remain zero because it must continue to accelerate. Thus, its velocity is always changing.

*8 •   How would an astronaut in apparent weightlessness be aware of her mass?

Determine the Concept If there is a force on her in addition to the gravitational force, she will experience an additional acceleration relative to her space vehicle that is proportional to the net force required producing that acceleration and inversely proportional to her mass.

She could do an experiment in which she uses her legs to push off from the wall of her space vehicle and measures her acceleration and the force exerted by the wall. She could calculate her mass from the ratio of the force exerted by the wall to the acceleration it produced.

*9 •   Under what circumstances would your apparent weight be greater than your true weight?

Determine the Concept One's apparent weight is the reading of a scale in one's reference frame.

Imagine yourself standing on a scale that, in turn, is on a platform accelerating upward with an acceleration a. The free-body diagram shows the force the gravitational field exerts on you, $m\vec{g}$, and the force the scale exerts on you, $\vec{w}_{app}$, The scale reading (the force the scale exerts on you) is your apparent weight.

Choose the coordinate system shown in the free-body diagram and apply $\sum \vec{F} = m\vec{a}$ to the scale:

$$\sum F_y = w_{app} - mg = ma_y$$

or

$$w_{app} = mg + ma_y$$

Your apparent weight would be greater than your true weight when observed from a reference frame that is accelerating upward. That is, when the surface on which you are standing has an acceleration a such that a_y is positive: $\boxed{a_y > 0}$

*12 •   True or false. (*a*) If two external forces that are both equal in magnitude and opposite in direction act on the same object, the two forces can never be an action-reaction force pair. (*b*) Action equals reaction only if the objects are not accelerating.

(*a*) True. By definition, action-reaction force pairs cannot act on the same object.

(b) False. Action equals reaction independent of any motion of the two objects.

*17 • A 2.5-kg object hangs at rest from a string attached to the ceiling. (a) Draw a diagram showing all the forces on the object and indicate the reaction force to each. (b) Do the same for each force acting on the string.

Determine the Concept The force diagrams will need to include the ceiling, string, object, and earth if we are to show all of the reaction forces as well as the forces acting on the object.

(a) The forces acting on the 2.5-kg object are its weight $\vec{W}$, and the tension $\vec{T}_1$, in the string. The reaction forces are $\vec{W}'$ acting on the earth and $\vec{T}_1'$ acting on the string.

(b) The forces acting on the string are its weight, the weight of the object, and $\vec{F}$, the force exerted by the ceiling. The reaction forces are $\vec{T}_1$ acting on the string and $\vec{F}'$ acting on the ceiling.

*21 • The net force on a moving object is suddenly reduced to zero and remains zero. As a consequence, the object (a) stops abruptly, (b) stops during a short time interval, (c) changes direction, (d) continues at constant velocity, (e) changes velocity in an unknown manner.

Determine the Concept In considering these statements, one needs to decide whether they are consistent with Newton's laws of motion. In the absence of a *net* force, an object moves with constant velocity. $\boxed{(d)\text{ is correct.}}$

Estimation and Approximation

*25 ••• Making any necessary assumptions, find the normal force and the tangential force exerted by the road on the wheels of your bicycle (a) as you climb an 8% grade at constant speed, and (b) as you descend the 8% grade at constant speed. (An 8% grade means that the angle of inclination θ is given by $\tan\theta = 0.08$.)

Picture the Problem The free-body diagram shows the forces acting on you and your bicycle as you are either ascending or descending the grade. The magnitude of the normal force acting on you and your bicycle is equal to the component of your weight in the y direction and the magnitude of the tangential force is the x component of your weight. Assume a combined mass (you plus your bicycle) of 80 kg.

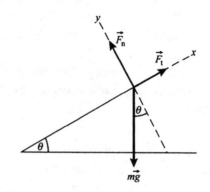

(a) Apply $\sum F_y = ma_y$ to you and your bicycle and solve for F_n:

$F_n - mg\cos\theta = 0$, because there is no acceleration in the y direction.

$\therefore F_n = mg\cos\theta$

Determine θ from the information concerning the grade:

$\tan\theta = 0.08$
and
$\theta = \tan^{-1}(0.08) = 4.57°$

Substitute to determine F_n:

$F_n = (80\text{ kg})(9.81\text{ m/s}^2)\cos4.57°$
$\quad = \boxed{782\text{ N}}$

Apply $\sum F_x = ma_x$ to you and your bicycle and solve for F_t, the tangential force exerted by the road on the wheels:

$F_t - mg\sin\theta = 0$, because there is no acceleration in the x direction.

Evaluate F_t:

$F_t = (80\text{ kg})(9.81\text{ m/s}^2)\sin4.57°$
$\quad = \boxed{62.6\text{ N}}$

(b)

Because there is no acceleration, the forces are the same going up and going down the incline.

Newton's First and Second Laws: Mass, Inertia, and Force

*29 ••   A bullet of mass 1.8×10^{-3} kg moving at 500 m/s impacts a large fixed block of wood and travels 6 cm before coming to rest. Assuming that the acceleration of the bullet is constant, find the force exerted by the wood on the bullet.

Picture the Problem Because the deceleration of the bullet is constant, we can use a constant-acceleration equation to determine its acceleration and Newton's 2nd law of motion to find the average resistive force that brings it to a stop.

Apply $\sum \vec{F} = m\vec{a}$ to express the force exerted on the bullet by the wood:

$$F_{wood} = ma$$

Using a constant-acceleration equation, express the final velocity of the bullet in terms of its acceleration and solve for the acceleration:

$$v^2 = v_0^2 + 2a\Delta x$$
and
$$a = \frac{v^2 - v_0^2}{2\Delta x} = \frac{-v_0^2}{2\Delta x}$$

Substitute to obtain:

$$F_{wood} = -\frac{mv_0^2}{2\Delta x}$$
$$= -\frac{\left(1.8\times10^{-3}\,\text{kg}\right)\left(500\,\text{m/s}\right)^2}{2(0.06\,\text{m})}$$
$$= \boxed{-3.75\,\text{kN}}$$

where the negative sign means that the direction of the force is opposite the velocity.

*30 •• A cart on a horizontal, linear track has a fan attached to it. The cart is positioned at one end of the track, and the fan is turned on. Starting from rest, the cart takes 4.55 s to travel a distance of 1.5 m. The mass of the cart plus fan is 355 g. Assume that the cart travels with constant acceleration. (*a*) What is the net force exerted on the cart? (*b*) Weights are added to the cart until its mass is 722 g, and the experiment is repeated. How long does it take for the cart to travel 1.5 m now? Ignore the effects of friction.

Picture the Problem The pictorial representation summarizes what we know about the motion. We can find the acceleration of the cart by using a constant-acceleration equation.

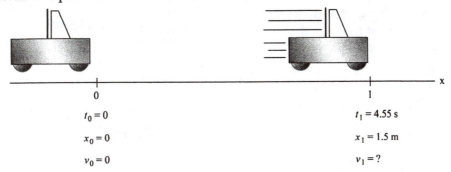

The free-body diagram shows the forces acting on the cart as it accelerates along the air track. We can determine the net force acting on the cart using Newton's 2nd law and our knowledge of its acceleration.

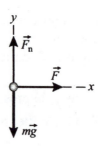

(a) Apply $\sum F_x = ma_x$ to the cart to obtain an expression for the net force F:

$$F = ma$$

Using a constant-acceleration equation, relate the displacement of the cart to its acceleration, initial speed, and travel time:

$$\Delta x = v_0 \Delta t + \tfrac{1}{2} a (\Delta t)^2$$
or, because $v_0 = 0$,
$$\Delta x = \tfrac{1}{2} a (\Delta t)^2$$

Solve for a:

$$a = \frac{2\Delta x}{(\Delta t)^2}$$

Substitute for a in the force equation to obtain:

$$F = m\frac{2\Delta x}{(\Delta t)^2} = \frac{2m\Delta x}{(\Delta t)^2}$$

Substitute numerical values and evaluate F:

$$F = \frac{2(0.355\,\text{kg})(1.5\,\text{m})}{(4.55\,\text{s})^2} = \boxed{0.0514\,\text{N}}$$

(b) Using a constant-acceleration equation, relate the displacement of the cart to its acceleration, initial speed, and travel time:

$$\Delta x = v_0 \Delta t + \tfrac{1}{2} a' (\Delta t)^2$$
or, because $v_0 = 0$,
$$\Delta x = \tfrac{1}{2} a' (\Delta t)^2$$

Solve for Δt:

$$\Delta t = \sqrt{\frac{2\Delta x}{a'}}$$

If we assume that air resistance is negligible, the net force on the cart is still 0.0514 N and its acceleration is:

$$a' = \frac{0.0514\,\text{N}}{0.722\,\text{kg}} = 0.0713\,\text{m/s}^2$$

Substitute numerical values and evaluate Δt:

$$\Delta t = \sqrt{\frac{2(1.5\,\text{m})}{0.0713\,\text{m/s}^2}} = \boxed{6.49\,\text{s}}$$

*34 •   Al and Bert stand in the middle of a large frozen lake. Al pushes on Bert with a force of 20 N for a period of 1.5 s. Bert's mass is 100 kg. Assume that both are at rest before Al pushes Bert. (*a*) What is the speed that Bert reaches as he is pushed away from Al? Treat the ice as frictionless. (*b*) What speed does Al reach if his mass is 80 kg?

Picture the Problem The speed of either Al or Bert can be obtained from their accelerations; in turn, they can be obtained from Newtons 2nd law applied to each person. The free-body diagrams to the right show the forces acting on Al and Bert. The forces that Al and Bert exert on each other are action-and-reaction forces.

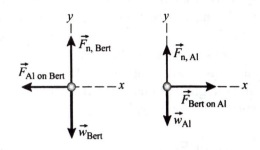

(*a*) Apply $\sum F_x = ma_x$ to Bert and solve for his acceleration:

$$-F_{\text{Al on Bert}} = m_{\text{Bert}}a_{\text{Bert}}$$

$$a_{\text{Bert}} = \frac{-F_{\text{Al on Bert}}}{m_{\text{Bert}}} = \frac{-20\,\text{N}}{100\,\text{kg}}$$

$$= -0.200\,\text{m/s}^2$$

Using a constant-acceleration equation, relate Bert's speed to his initial speed, speed after 1.5 s, and acceleration and solve for his speed at the end of 1.5 s:

$$v = v_0 + a\Delta t$$
$$= 0 + (-0.200\,\text{m/s}^2)\,(1.5\,\text{s})$$
$$= \boxed{-0.300\,\text{m/s}}$$

(*b*) From Newton's 3rd law, an equal but oppositely directed force acts on Al while he pushes Bert. Because the ice is frictionless, Al speeds off in the opposite direction. Apply Newton's 2nd law to the forces acting on Al and solve for his acceleration:

$$\sum F_{x,\text{Al}} = F_{\text{Bert on Al}} = m_{\text{Al}}a_{\text{Al}}$$
and
$$a_{\text{Al}} = \frac{F_{\text{Bert on Al}}}{m_{\text{Al}}} = \frac{20\,\text{N}}{80\,\text{kg}}$$
$$= 0.250\,\text{m/s}^2$$

Using a constant-acceleration equation, relate Al's speed to his initial speed, speed after 1.5 s, and acceleration; solve for his speed at the end of 1.5 s:

$$v = v_0 + a\Delta t$$
$$= 0 + (0.250\,\text{m/s}^2)\,(1.5\,\text{s})$$
$$= \boxed{0.375\,\text{m/s}}$$

Mass and Weight

*38 •   On the moon, the acceleration due to gravity is only about 1/6 of that on earth. An astronaut whose weight on earth is 600 N travels to the lunar surface.

His mass as measured on the moon will be (*a*) 600 kg, (*b*) 100 kg, (*c*) 61.2 kg, (*d*) 9.81 kg, (*e*) 360 kg.

Picture the Problem The mass of the astronaut is independent of gravitational fields and will be the same on the moon or, for that matter, out in deep space.

Express the mass of the astronaut in terms of his weight on earth and the gravitational field at the surface of the earth:	$m = \dfrac{W_{earth}}{g_{earth}} = \dfrac{600\,\text{N}}{9.81\,\text{N/kg}} = 61.2\,\text{kg}$ and $\boxed{(c) \text{ is correct.}}$

Contact Forces

*41 •   A vertical spring of force constant 600 N/m has one end attached to the ceiling and the other to a 12-kg block resting on a horizontal surface so that the spring exerts an upward force on the block. The spring stretches by 10 cm. (*a*) What force does the spring exert on the block? (*b*) What is the force that the surface exerts on the block?

Picture the Problem Draw a free-body diagram showing the forces acting on the block. $\vec{F}_k$ is the force exerted by the spring, $\vec{W} = m\vec{g}$ is the weight of the block, and $\vec{F}_n$ is the normal force exerted by the horizontal surface. Because the block is resting on a surface, $F_k + F_n = W$.

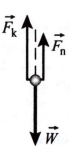

(*a*) Calculate the force exerted by the spring on the block:	$F_k = kx = (600\ \text{N/m})(0.1\ \text{m})$ $= \boxed{60.0\,\text{N}}$
(*b*) Choosing the upward direction to be positive, sum the forces acting on the block and solve for F_n:	$\sum \vec{F} = 0 \Rightarrow F_k + F_n - W = 0$ and $F_n = W - F_k$
Substitute numerical values and evaluate F_n:	$F_n = (12\ \text{kg})(9.81\ \text{N/kg}) - 60\ \text{N}$ $= \boxed{57.7\,\text{N}}$

Free-Body Diagrams: Static Equilibrium

*45 •• In figure 4-33a, a 0.500-kg block is suspended from a 1.25-m-long string. The ends of the string are attached to the ceiling at points separated by 1.00 m. (a) What angle does the string make with the ceiling? (b) What is the tension in the string? (c) The 0.500-kg block is removed and two 0.250-kg blocks are attached to the string such that the lengths of the three string segments are equal (Figure 4-33b). What is the tension in each segment of the string?

Figure 4-33 Problem 45

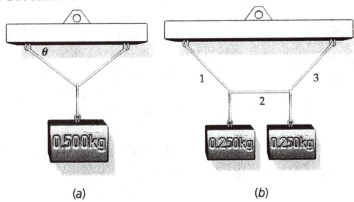

(a)   (b)

Picture the Problem The free-body diagrams for parts (a), (b), and (c) are shown below. In both cases, the block is in equilibrium under the influence of the forces and we can use Newton's 2nd law of motion and geometry and trigonometry to obtain relationships between θ and the tensions.

(a) and (b)

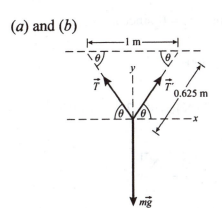

(c)

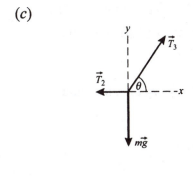

(a) Referring to the free-body diagram for part (a), use trigonometry to determine θ:

$$\theta = \cos^{-1}\frac{0.5\,\text{m}}{0.625\,\text{m}} = \boxed{36.9°}$$

(b) Noting that $T = T'$, apply $\sum F_y = ma_y$ to the 0.500-kg block and solve for the tension T:

$$2T\sin\theta - mg = 0 \text{ since } a = 0$$
and
$$T = \frac{mg}{2\sin\theta}$$

Substitute numerical values and evaluate T:

$$T = \frac{(0.5\,\text{kg})(9.81\,\text{m/s}^2)}{2\sin 36.9°} = \boxed{4.08\,\text{N}}$$

(c) The length of each segment is:

$$\frac{1.25\,\text{m}}{3} = 0.417\,\text{m}$$

Find the distance d:

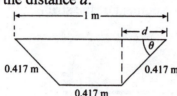

$$d = \frac{1\,\text{m} - 0.417\,\text{m}}{2}$$

$$= 0.2915\,\text{m}$$

Express θ in terms of d and solve for its value:

$$\theta = \cos^{-1}\left(\frac{d}{0.417\,\text{m}}\right) = \cos^{-1}\left(\frac{0.2915\,\text{m}}{0.417\,\text{m}}\right)$$

$$= 45.7°$$

Apply $\sum F_y = ma_y$ to the 0.250-kg block and solve for the tension T_3:

$T_3 \sin\theta - mg = 0$ since $a = 0$.
and

$$T_3 = \frac{mg}{\sin\theta}$$

Substitute numerical values and evaluate T_3:

$$T_3 = \frac{(0.25\,\text{kg})(9.81\,\text{m/s}^2)}{\sin 45.7°} = \boxed{3.43\,\text{N}}$$

Apply $\sum F_x = ma_x$ to the 0.250-kg block and solve for the tension T_2:

$T_3 \cos\theta - T_2 = 0$ since $a = 0$.
and
$T_2 = T_3 \cos\theta$

Substitute numerical values and evaluate T_2:

$$T_2 = (3.43\,\text{N})\cos 45.7° = \boxed{2.40\,\text{N}}$$

By symmetry:

$$T_1 = T_3 = \boxed{3.43\,\text{N}}$$

*48 •   A vertical force $\vec{T}$ is exerted on a 5-kg object near the surface of the earth, as shown in Figure 4-36. Find the acceleration of the object if (a) $T = 5$ N, (b) $T = 10$ N, and (c) $T = 100$ N.

Figure 4-36 Problem 48

Picture the Problem The acceleration of the object equals the net force, $\vec{T} - m\vec{g}$, divided by the mass. Choose a coordinate system in which upward is the positive y direction. Apply Newton's 2nd law to the forces acting on this body to find the acceleration of the object as a function of T.

(a) Apply $\sum F_y = ma_y$ to the object:	$T - w = T - mg = ma_y$
Solve this equation for a as a function of T:	$a_y = \dfrac{T}{m} - g$
Substitute numerical values and evaluate a_y:	$a_y = \dfrac{5\,\text{N}}{5\,\text{kg}} - 9.81\,\text{m/s}^2 = \boxed{-8.81\,\text{m/s}^2}$
(b) Proceed as in (a) with $T = 10$ N:	$a = \boxed{-7.81\,\text{m/s}^2}$
(c) Proceed as in (a) with $T = 100$ N:	$a = \boxed{10.2\,\text{m/s}^2}$

*50 •••   Balloon arches are often seen at festivals or celebrations; they are made by attaching helium-filled balloons to a rope that is fixed to the ground at each end. The lift from the balloons raises the structure into the arch shape. Figure 4-38*a* shows the geometry of such a structure: N balloons are attached at equally spaced intervals along a massless rope of length L, which is attached to two supports. Each balloon provides a lift force F. The horizontal and vertical coordinates of the point on the rope where the ith balloon is attached are x_i and y_i, and T_i is the tension in the ith segment (with segment 0 being the segment between the point of attachment and the first balloon, and segment N being the segment between the last balloon and the other point of attachment).

Figure 4-38 Problem 50

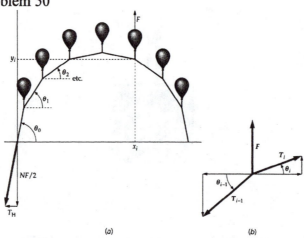

(*a*)   (*b*)

(*a*) Figure 4-38*b* shows a free-body diagram for the *i*th balloon. From this diagram, show that the horizontal component of the force T_i (call it T_H) is the same for all the balloons, and that by considering the vertical component of the force, one can derive the following equation relating the tension in the *i*th and (*i*-1)th segments:

$$T_{i-1} \sin(\theta_{i-1}) - T_i \sin(\theta_i) = F$$

(*b*) Show that $\tan(\theta_0) = -\tan(\theta_{N+1}) = NF/2T_H$.
(*c*) From the diagram and the two expressions above, show that:

$$\tan(\theta_i) = (N - 2i)F/2T_H$$

and that

$$x_i = \frac{L}{N+1} \sum_{j=0}^{i-1} \cos\theta_j \qquad\qquad y_i = \frac{L}{N+1} \sum_{j=0}^{i-1} \sin\theta_j$$

(*d*) Write a spreadsheet program to make a graph of the shape of a balloon arch with the following parameters: $N = 10$ balloons giving a lift force $F = 1$ N each attached to a rope length $L = 10$ m, with a horizontal component of tension $T_H = 10$ N. How far apart are the two points of attachment? How high is the arch at its highest point?

(*e*) Note that we haven't specified the spacing between the supports - it is determined by the other parameters. Vary T_H while keeping the other parameters the same until you create an arch that has a spacing of 8 m between the supports. What is T_H then? As you increase T_H, the arch should get flatter and more spread out. Does your spreadsheet model show this?

Picture the Problem In part (*a*) we can apply Newton's 2nd law to obtain the given expression for F. In (*b*) we can use a symmetry argument to find an expression for $\tan \theta_0$. In (*c*) we can use our results obtained in (*a*) and (*b*) to express x_i and y_i.

(a) Apply $\sum F_y = 0$ to the balloon:

$$F + T_i \sin\theta_i - T_{i-1}\sin\theta_{i-1} = 0$$

Solve for F to obtain:

$$F = \boxed{T_{i-1}\sin\theta_{i-1} - T_i\sin\theta_i}$$

(b) By symmetry, each support must balance half of the force acting on the entire arch. Therefore, the vertical component of the force on the support must be $NF/2$. The horizontal component of the tension must be T_H. Express $\tan\theta_0$ in terms of $NF/2$ and T_H:

$$\tan\theta_0 = \frac{NF/2}{T_H} = \frac{NF}{2T_H}$$

By symmetry, $\theta_{N+1} = -\theta_0$. Therefore, because the tangent function is odd:

$$\tan\theta_0 = \boxed{-\tan\theta_{N+1} = \frac{NF}{2T_H}}$$

(c) Using $T_H = T_i\cos\theta_i = T_{i-1}\cos\theta_{i-1}$, divide both sides of our result in (a) by T_H and simplify to obtain:

$$\frac{F}{T_H} = \frac{T_{i-1}\sin\theta_{i-1}}{T_{i-1}\cos\theta_{i-1}} - \frac{T_i\sin\theta_i}{T_i\cos\theta_i}$$
$$= \tan\theta_{i-1} - \tan\theta_i$$

Using this result, express $\tan\theta_1$:

$$\tan\theta_1 = \tan\theta_0 - \frac{F}{T_H}$$

Substitute for $\tan\theta_0$ from (a):

$$\tan\theta_1 = \frac{NF}{2T_H} - \frac{F}{T_H} = (N-2)\frac{F}{2T_H}$$

Generalize this result to obtain:

$$\tan\theta_i = \boxed{(N-2i)\frac{F}{2T_H}}$$

Express the length of rope between two balloons:

$$\ell_{\text{between balloons}} = \frac{L}{N+1}$$

Express the horizontal coordinate of the point on the rope where the ith balloon is attached, x_i, in terms of x_{i-1} and the length of rope between two balloons:

$$x_i = x_{i-1} + \frac{L}{N+1}\cos\theta_{i-1}$$

Sum over all the coordinates to obtain:

$$x_i = \frac{L}{N+1}\sum_{j=0}^{i-1}\cos\theta_j$$

Proceed similarly for the vertical coordinates to obtain:

$$y_i = \frac{L}{N+1}\sum_{j=0}^{i-1}\sin\theta_j$$

(*d*) A spreadsheet program is shown below. The formulas used to calculate the quantities in the columns are as follows:

Cell	Content/Formula	Algebraic Form
C9	($B$2–2*B9)/(2*$B$4)	$(N-2i)\dfrac{F}{2T_H}$
D9	SIN(ATAN(C9))	$\sin\left(\tan^{-1}\theta_i\right)$
E9	COS(ATAN(C9))	$\cos\left(\tan^{-1}\theta_i\right)$
F10	F9+$B$1/($B$2+1)*E9	$x_{i-1}+\dfrac{L}{N+1}\cos\theta_{i-1}$
G10	G9+$B$1/($B$2+1)*D9	$y_{i-1}+\dfrac{L}{N+1}\cos\theta_{i-1}$

	A	B	C	D	E	F	G
1	L =	10	m				
2	N =	10					
3	F =	1	N				
4	TH=	3.72	N				
5							
6							
7							
8		I	tan(thetai)	sin(thetai)	cos(thetai)	xi	yi
9		0	1.344	0.802	0.597	0.000	0.000
10		1	1.075	0.732	0.681	0.543	0.729
11		2	0.806	0.628	0.778	1.162	1.395
12		3	0.538	0.474	0.881	1.869	1.966
13		4	0.269	0.260	0.966	2.670	2.396
14		5	0.000	0.000	1.000	3.548	2.632
15		6	–0.269	–0.260	0.966	4.457	2.632
16		7	–0.538	–0.474	0.881	5.335	2.396
17		8	–0.806	–0.628	0.778	6.136	1.966
18		9	–1.075	–0.732	0.681	6.843	1.395
19		10	–1.344	–0.802	0.597	7.462	0.729

(*e*) A horizontal component of tension 3.72 N gives a spacing of 8 m. At this spacing, the arch is 2.63 m high, tall enough for someone to walk through.

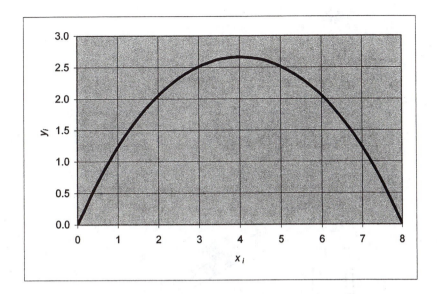

Free-Body Diagrams: Inclined Planes and the Normal Force

*54 • A large box whose mass is 20 kg rests on a frictionless floor. A mover pushes on the box with a force of 250 N at an angle 35° below the horizontal. What is the acceleration of the box across the floor?

Picture the Problem The free-body diagram shows the forces acting on the box as the man pushes it across the frictionless floor. We can apply Newton's 2nd law of motion to the box to find its acceleration.

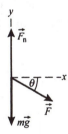

Apply $\sum F_x = ma_x$ to the box:

$$F\cos\theta = ma_x$$

Solve for a_x:

$$a_x = \frac{F\cos\theta}{m}$$

Substitute numerical values and evaluate a_x:

$$a_x = \frac{(250\,\text{N})\cos35°}{20\,\text{kg}} = \boxed{10.2\,\text{m/s}^2}$$

*58 • In Figure 4-43, the objects are attached to spring balances calibrated in newtons. Give the reading of the balance(s) in each case, assuming that the strings are massless and the incline is frictionless.

Figure 4-43 Problem 58

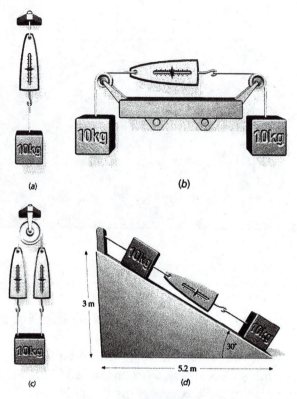

(a)

(b)

(c)

(d)

Picture the Problem The balance(s) indicate the tension in the string(s). Draw free-body diagrams for each of these systems and apply the condition(s) for equilibrium.

(*a*)

$$\sum F_y = T - mg = 0$$

and

$$T = mg = (10\,\text{kg})(9.81\,\text{m/s}^2) = \boxed{98.1\,\text{N}}$$

(*b*)

$$\sum F_x = T - T' = 0$$

or, because $T' = mg$,

$$T = T' = mg$$

$$= (10\,\text{kg})(9.81\,\text{m/s}^2) = \boxed{98.1\,\text{N}}$$

(*c*)

$$\sum F_y = 2T - mg = 0$$

and

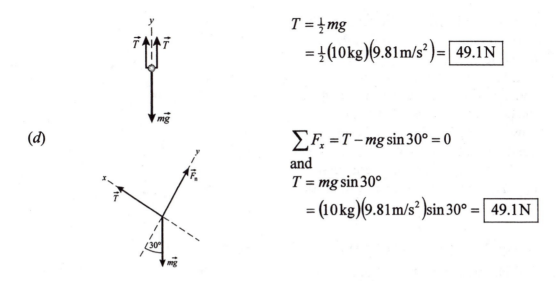

$$T = \tfrac{1}{2}mg$$
$$= \tfrac{1}{2}(10\,\text{kg})(9.81\,\text{m/s}^2) = \boxed{49.1\,\text{N}}$$

(d)

$$\sum F_x = T - mg\sin 30° = 0$$
and
$$T = mg\sin 30°$$
$$= (10\,\text{kg})(9.81\,\text{m/s}^2)\sin 30° = \boxed{49.1\,\text{N}}$$

Remarks: Note that (a) and (b) give the same answers ... a rather surprising result until one has learned to draw FBDs and apply the conditions for translational equilibrium.

*61 ••   A 65-kg student weighs himself by standing on a scale mounted on a skateboard that is rolling down an incline, as shown in Figure 4-45. Assume there is no friction so that the force exerted by the incline on the skateboard is normal to the incline. What is the reading on the scale if $\theta = 30°$?

Figure 4-45 Problem 61

Picture the Problem The scale reading (the boy's apparent weight) is the force the scale exerts on the boy. Draw a free-body diagram for the boy, choosing a coordinate system in which the positive x-axis is parallel to and down the inclined plane and the positive y-axis is in the direction of the normal force the incline exerts on the boy. Apply Newton's 2nd law of motion in the y direction.

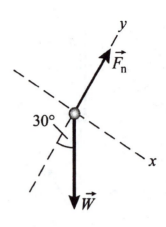

Apply $\sum F_y = ma_y$ to the boy to find F_n. Remember that there is no acceleration in the y direction:

$$F_n - W\cos 30° = 0$$

Substitute for W to obtain:

$$F_n - mg\cos 30° = 0$$

Solve for F_n:

$$F_n = mg\cos 30°$$

Substitute numerical values and evaluate F_n:

$$F_n = (65\,\text{kg})(9.81\,\text{m/s}^2)\cos 30°$$
$$= \boxed{552\,\text{N}}$$

Free-Body Diagrams: Elevators

*65 •   A person of weight w is in an elevator going up when the cable suddenly breaks. What is the person's apparent weight immediately after the cable breaks? (a) w, (b) Greater than w, (c) Less than w, (d) $9.8w$, (e) Zero.

Picture the Problem The sketch to the right shows a person standing on a scale in the elevator immediately after the cable breaks. To its right is the free-body diagram showing the forces acting on the person. The force exerted by the scale on the person, $\vec{w}_{app}$, is the person's apparent weight.

From the free-body diagram we can see that $\vec{w}_{app} + m\vec{g} = m\vec{a}$ where $\vec{g}$ is the local gravitational field and $\vec{a}$ is the acceleration of the reference frame (elevator). When the elevator goes into free fall ($\vec{a} = \vec{g}$), our equation becomes

$\vec{w}_{app} + m\vec{g} = m\vec{a} = m\vec{g}$ This tells us that $\vec{w}_{app} = 0.$ $\boxed{(e)\text{ is correct.}}$

Free-Body Diagrams: Ropes, Tension, and Newton's Third Law

*70 •• Two blocks are in contact on a frictionless horizontal surface. The blocks are accelerated by a horizontal force $\vec{F}$ applied to one of them (Figure 4-50). Find the acceleration and the contact force for (*a*) general values of *F*, m_1, and m_2, and (*b*) $F = 3.2$ N, $m_1 = 2$ kg, and $m_2 = 6$ kg.

Figure 4-50 Problem 70

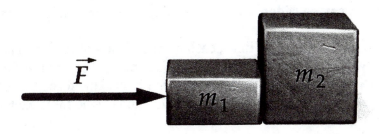

Picture the Problem Choose a coordinate system in which the positive *x* direction is to the right and the positive *y* direction is upward. Let $\vec{F}_{2,1}$ be the contact force exerted by m_2 on m_1 and $\vec{F}_{1,2}$ be the force exerted by m_1 on m_2. These forces are equal and opposite so $\vec{F}_{2,1} = -\vec{F}_{1,2}$. The free-body diagrams for the blocks are shown to the right. Apply Newton's 2nd law to each block separately and use the fact that their accelerations are equal.

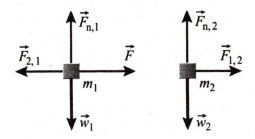

(*a*) Apply $\sum F_x = ma_x$ to the first block:

$$F - F_{2,1} = m_1 a_1 = m_1 a$$

Apply $\sum F_x = ma_x$ to the second block:

$$F_{1,2} = m_2 a_2 = m_2 a \qquad (1)$$

Add these equations to eliminate $F_{2,1}$ and $F_{1,2}$ and solve for $a = a_1 = a_2$:

$$a = \boxed{\frac{F}{m_1 + m_2}}$$

Substitute your value for *a* into equation (1) and solve for $F_{1,2}$:

$$F_{1,2} = \boxed{\frac{F m_2}{m_1 + m_2}}$$

(b) Substitute numerical values in the equations derived in part (a) and evaluate a and $F_{1,2}$:

$$a = \frac{3.2\,\text{N}}{2\,\text{kg} + 6\,\text{kg}} = \boxed{0.400\,\text{m/s}^2}$$

and

$$F_{1,2} = \frac{(3.2\,\text{N})(6\,\text{kg})}{2\,\text{kg} + 6\,\text{kg}} = \boxed{2.40\,\text{N}}$$

Remarks: Note that our results for the acceleration are the same as if the force F had acted on a single object whose mass is equal to the sum of the masses of the two blocks. In fact, because the two blocks have the same acceleration, we can consider them to be a single system with mass $m_1 + m_2$.

*74 ••   A chain consists of 5 links, each having a mass of 0.1 kg. The chain is lifted vertically with an upward acceleration of 2.5 m/s². The chain is held at the top link; no point of the chain touches the floor. Find (a) the force F exerted on the top of the chain, (b) the net force on each link, and (c) the force each link exerts on the link below it.

Picture the Problem Choose a coordinate system with the positive y direction upward and denote the top link with the numeral 1, the second with the numeral 2, etc.. The free-body diagrams show the forces acting on links 1 and 2. We can apply Newton's 2nd law to each link to obtain a system of simultaneous equations that we can solve for the force each link exerts on the link below it. Note that the net force on each link is the product of its mass and acceleration.

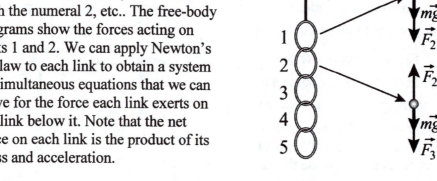

(a) Apply $\sum F_y = ma_y$ to the top link and solve for F:

$$F - 5mg = 5ma$$

and

$$F = 5m(g + a)$$

Substitute numerical values and evaluate F:

$$F = 5(0.1\,\text{kg})(9.81\,\text{m/s}^2 + 2.5\,\text{m/s}^2)$$
$$= \boxed{6.16\,\text{N}}$$

(b) Apply $\sum F_y = ma_y$ to a single link:

$$F_{1\,\text{link}} = m_{1\,\text{link}}\,a = (0.1\,\text{kg})(2.5\,\text{m/s}^2)$$
$$= \boxed{0.250\,\text{N}}$$

(c) Apply $\sum F_y = ma_y$ to the 1st through 5th links to obtain:

$$F - F_2 - mg = ma, \qquad (1)$$
$$F_2 - F_3 - mg = ma, \qquad (2)$$

$$F_3 - F_4 - mg = ma, \qquad (3)$$
$$F_4 - F_5 - mg = ma, \text{ and} \qquad (4)$$
$$F_5 - mg = ma \qquad (5)$$

Add equations (2) through (5) to obtain:

$$F_2 - 4mg = 4ma$$

Solve for F_2 to obtain:

$$F_2 = 4mg + 4ma = 4m(a + g)$$

Substitute numerical values and evaluate F_2:

$$F_2 = 4(0.1\,\text{kg})(9.81\,\text{m/s}^2 + 2.5\,\text{m/s}^2)$$
$$= \boxed{4.92\,\text{N}}$$

Substitute for F_2 to find F_3, and then substitute for F_3 to find F_4:

$$F_3 = \boxed{3.69\,\text{N}} \text{ and } F_4 = \boxed{2.46\,\text{N}}$$

Solve equation (5) for F_5:

$$F_5 = m(g + a)$$

Substitute numerical values and evaluate F_5:

$$F_5 = (0.1\,\text{kg})(9.81\,\text{m/s}^2 + 2.5\,\text{m/s}^2)$$
$$= \boxed{1.23\,\text{N}}$$

*81 ••   A 60-kg housepainter stands on a 15-kg aluminum platform. The platform is attached to a rope that passes through an overhead pulley, which allows the painter to raise herself and the platform (Figure 4-55). (*a*) To accelerate herself and the platform at a rate of 0.8 m/s², with what force *F* must she pull on the rope? (*b*) When her speed reaches 1 m/s, she pulls in such a way that she and the platform go up at a constant speed. What force is she exerting on the rope? (Ignore the mass of the rope.)

Figure 4-55 Problem 81

Picture the Problem Choose a coordinate system in which upward is the positive y direction and draw the free-body diagram for the frame-plus-painter. Noting that $\vec{F} = -\vec{T}$, apply Newton's 2nd law of motion.

(a) Letting $m_{tot} = m_{frame} + m_{painter}$, apply $\sum F_y = ma_y$ to the frame-plus-painter and solve T:

$$2T - m_{tot}g = m_{tot}a$$
and
$$T = \frac{m_{tot}(a + g)}{2}$$

Substitute numerical values and evaluate T:

$$T = \frac{(75\,\text{kg})(0.8\,\text{m/s}^2 + 9.81\,\text{m/s}^2)}{2}$$
$$= 398\,\text{N}$$

Because $F = T$:

$$F = \boxed{398\,\text{N}}$$

(b) Apply $\sum F_y = ma_y$ with $a = 0$ to obtain:

$$2T - m_{tot}g = 0$$

Solve for T:

$$T = \tfrac{1}{2}m_{tot}g$$

Substitute numerical values and evaluate T:

$$T = \tfrac{1}{2}(75\,\text{kg})(9.81\,\text{m/s}^2) = \boxed{368\,\text{N}}$$

Free-Body Diagrams: The Atwood's Machine

Figure 4-58 Problems 84-87

*84 ••   The apparatus in Figure 4-58 is called an *Atwood's machine* and is used to measure the free-fall acceleration g by measuring the acceleration of the two blocks. Assuming a massless, frictionless pulley and a massless string, show that the magnitude of the acceleration of either body and the tension in the string are

$$a = \frac{m_1 - m_2}{m_1 + m_2} g \text{ and } T = \frac{2m_1 m_2 g}{m_1 + m_2}$$

Picture the Problem Assume that $m_1 > m_2$. Choose a coordinate system in which the positive y direction is downward for the block whose mass is m_1 and upward for the block whose mass is m_2 and draw free-body diagrams for each block. Apply Newton's 2nd law of motion to both blocks and solve the resulting equations simultaneously.

Draw a FBD for the block whose mass is m_2:

Apply $\sum F_y = ma_y$ to this block: $T - m_2 g = m_2 a_2$

Draw a FBD for the block whose mass is m_1:

Apply $\sum F_y = ma_y$ to this block: $m_1 g - T = m_1 a_1$

Because the blocks are connected by a taut string, let a represent their common acceleration: $a = a_1 = a_2$

Add the two force equations to eliminate T and solve for a:

$m_1 g - m_2 g = m_1 a + m_2 a$
and

$$a = \frac{m_1 - m_2}{m_1 + m_2} g$$

Substitute for a in either of the force equations and solve for T:

$$T = \frac{2m_1 m_2 g}{m_1 + m_2}$$

*89 ••   You are given an Atwood's machine and a set of weights whose total mass is M. You are told to attach some of the weights to one side of the machine, and the rest to the other side. If m_1 represents the mass attached to the left side and m_2 is the mass attached to the right side, the tension in the rope is given by the expression:

$$T = \frac{2m_1 m_2}{m_1 + m_2} g$$

as was shown in Problem 85. Show that the tension will be greatest when $m_1 = m_2 = M/2$.

Picture the Problem We can reason to this conclusion as follows: in the two extreme cases when the mass on one side or the other is zero, the tension is zero as well, because the mass is in free-fall. By symmetry, the maximum tension must occur when the masses on each side are equal. An alternative approach that is shown below is to treat the problem as an extreme-value problem.

Express m_2 in terms of M and m_1:

$$m_2 = M - m_1$$

Substitute in the equation from Problem 84 and simplify to obtain:

$$T = \frac{2gm_1(M - m_1)}{m_1 + M - m_1} = 2g\left(m_1 - \frac{m_1^2}{M}\right)$$

Differentiate this expression with respect to m_1 and set the derivative equal to zero for extreme values:

$$\frac{dT}{dm_1} = 2g\left(1 - \frac{2m_1}{M}\right) = 0 \text{ for extreme values}$$

Solve for m_1 to obtain:

$$m_1 = \tfrac{1}{2}M$$

Show that $m_1 = M/2$ is a maximum value by evaluating the second derivative of T with respect to m_1 at $m_1 = M/2$:

$$\frac{d^2 T}{dm_1^2} = -\frac{4g}{M} < 0 \text{ independently of } m_1$$

and we have shown that

T is a maximum when
$m_1 = m_2 = \tfrac{1}{2}M$.

Remarks: An alternative solution is to use a graphing calculator to show that T as a function of m_1 is concave downward and has its maximum value when $m_1 = m_2 = M/2$.

General Problems

*92 ••   A simple accelerometer can be made by suspending a small object from a string attached to a fixed point on an accelerating object. Suppose such an accelerometer is attached to the ceiling of an automobile traveling on a large flat surface. When there is acceleration, the object will deflect and the string will make some angle with the vertical. (*a*) How is the direction in which the suspended object is deflected related to the direction of the acceleration? (*b*) Show that the acceleration a is related to the angle θ that the string makes by $a = g \tan \theta$. (c) Suppose the automobile brakes to rest from 50 km/h in a distance of 60 m. What angle will the accelerometer make? Will the object swing forward or backward?

Picture the Problem The free-body diagram shown to the right shows the forces acting on an object suspended from the ceiling of a car that is accelerating to the right. Choose the coordinate system shown and use Newton's laws of motion and constant-acceleration equations in the determination of the influence of the forces on the behavior of the suspended object.

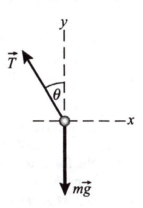

The second free-body diagram shows the forces acting on an object suspended from the ceiling of a car that is braking while it moves to the right.

(a) | In accordance with Newton's law of inertia, the object's displacement will be in the direction opposite that of the acceleration.

(b) Resolve the tension, T, into its components and apply $\sum \vec{F} = m\vec{a}$ to the object:

$\Sigma F_x = T \sin\theta = ma$
and
$\Sigma F_y = T \cos\theta - mg = 0$

Take the ratio of these two equations to eliminate T and m:

$$\frac{T \sin\theta}{T \cos\theta} = \frac{ma}{mg}$$

or

$$\tan\theta = \frac{a}{g} \Rightarrow \boxed{a = g \tan\theta}$$

> (c) | Because the acceleration is opposite the direction the car is moving, the accelerometer will swing forward.

Using a constant-acceleration equation, express the velocity of the car in terms of its acceleration and solve for the acceleration:

$$v^2 = v_0^2 + 2a\Delta x$$
or, because $v = 0$,
$$0 = v_0^2 + 2a\Delta x$$

Solve for and evaluate a:

$$a = \frac{-v_0^2}{2\Delta x}$$

Substitute numerical values and evaluate a:

$$a = \frac{-(50\,\text{km/h})^2}{2(60\,\text{m})} = \boxed{-1.61\,\text{m/s}^2}$$

Solve the equation derived in (b) for θ:

$$\theta = \tan^{-1}\frac{a}{g}$$

Substitute numerical values and evaluate θ:

$$\theta = \tan^{-1}\frac{1.61\,\text{m/s}^2}{9.81\,\text{m/s}^2} = \boxed{9.32°}$$

*95 ••• A man pushes a 24-kg box across a frictionless floor. The box begins moving from rest. He initially pushes on the box gently, but gradually increases his force so that the force he exerts on the box varies in time as $F = (8.00\,\text{N/s})\,t$. After 3 s, he stops pushing the box. The force is always exerted in the same direction. (a) What is the velocity of the box after 3 s? (b) How far has the man pushed the box in 3 s? (c) What is the average velocity of the box between 0 s and 3 s? (d) What is the average force that the man exerts on the box while he is pushing it?

Picture the Problem The free-body diagram shows the forces acting on the box as the man pushes it across a frictionless floor. Because the force is time-dependent, the acceleration will be, too. We can obtain the acceleration as a function of time from the application of Newton's 2nd law and then find the velocity of the box as a function of time by integration. Finally, we can derive an expression for the displacement of the box as a function of time by integration of the velocity function.

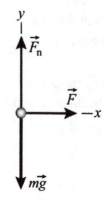

(a) The velocity is related to the acceleration according to:

$$\frac{dv}{dt} = a(t) \qquad (1)$$

Apply $\sum F_x = ma_x$ to the box and solve for its acceleration:

$$F = ma$$
and
$$a = \frac{F}{m} = \frac{(8\,\text{N/s})t}{24\,\text{kg}} = \left(\tfrac{1}{3}\,\text{m/s}^3\right)t$$

Because the box's acceleration is a function of time, separate variables in equation (1) and integrate to find v as a function of time:

$$v(t) = \int_0^t a(t')dt' = \left(\tfrac{1}{3}\,\text{m/s}^3\right)\int_0^t t'\,dt'$$

$$= \left(\tfrac{1}{3}\,\text{m/s}^3\right)\frac{t^2}{2} = \left(\tfrac{1}{6}\,\text{m/s}^3\right)t^2$$

Evaluate v at $t = 3$ s:

$$v(3\,\text{s}) = \left(\tfrac{1}{6}\,\text{m/s}^3\right)(3\,\text{s})^2 = \boxed{1.50\,\text{m/s}}$$

(b) Integrate $v = dx/dt$ between 0 and 3 s to find the displacement of the box during this time:

$$\Delta x = \int_0^{3\,\text{s}} v(t')dt' = \left(\tfrac{1}{6}\,\text{m/s}^3\right)\int_0^{3\,\text{s}} t'^2\,dt'$$

$$= \left[\left(\tfrac{1}{6}\,\text{m/s}^3\right)\frac{t'^3}{3}\right]_0^{3\,\text{s}} = \boxed{1.50\,\text{m}}$$

(c) The average velocity is given by:

$$v_{\text{ave}} = \frac{\Delta x}{\Delta t} = \frac{1.5\,\text{m}}{3\,\text{s}} = \boxed{0.500\,\text{m/s}}$$

(d) Use Newton's 2nd law to express the average force exerted on the box by the man:

$$F_{\text{av}} = ma_{\text{av}} = m\frac{\Delta v}{\Delta t}$$

$$= (24\,\text{kg})\frac{1.5\,\text{m/s} - 0\,\text{m/s}}{3\,\text{s}}$$

$$= \boxed{12.0\,\text{N}}$$

*98 ••   A 2-kg block rests on a frictionless wedge that has an inclination of 60° and an acceleration a to the right such that the mass remains stationary relative to the wedge (Figure 4-62). (a) Find a. (b) What would happen if the wedge were given a greater acceleration?

Figure 4-62 Problem 98

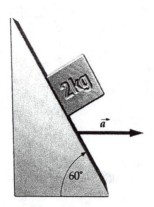

Picture the Problem The free-body diagram shows the forces acting on the block. Choose the coordinate system shown on the diagram. Because the surface of the wedge is frictionless, the force it exerts on the block must be normal to its surface.

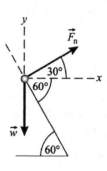

(a) Apply $\sum F_y = ma_y$ to the block to obtain:

$F_n \sin 30° - w = ma_y$

or, because $a_y = 0$ and $w = mg$,

$F_n \sin 30° - mg = 0$

or

$F_n \sin 30° = mg$ \hspace{2cm} (1)

Apply $\sum F_x = ma_x$ to the block:

$F_n \cos 30° = ma_x$ \hspace{2cm} (2)

Divide equation (2) by equation (1) to obtain:

$\dfrac{a_x}{g} = \cot 30°$

Solve for and evaluate a_x:

$a_x = g \cot 30° = (9.81\,\text{m/s}^2)\cot 30°$

$= \boxed{17.0\,\text{m/s}^2}$

(b) An acceleration of the wedge greater than $g \cot 30°$ would require that the normal force exerted on the body by the wedge be greater than that given in part (a); i.e., $F_n > mg/\sin 30°$.

> Under this condition, there would be a net force in the y direction and the block would accelerate up the wedge.

*102 ••• The pulley in an Atwood's machine is given an upward acceleration a, as shown in Figure 4-65. Find the acceleration of each mass and the tension in the string that connects them. *Hint: A constant upward acceleration has the same effect as an increase in the acceleration due to gravity.*

Figure 4-65 Problem 102

Picture the Problem Because a constant upward acceleration has the same effect as an increase in the acceleration due to gravity, we can use the result of Problem 89 (for the tension) with a replaced by $a + g$. The application of Newton's 2nd law to the object whose mass is m_2 will connect the acceleration of this body to tension from Problem 84.

In Problem 84 it is given that, when the support pulley is not accelerating, the tension in the rope and the acceleration of the masses are related according to:

$$T = \frac{2m_1 m_2}{m_1 + m_2} g$$

Replace a with $a + g$:

$$T = \frac{2m_1 m_2}{m_1 + m_2} (a + g)$$

Apply $\sum F_y = ma_y$ to the object whose mass is m_2 and solve for a_2:

$T - m_2 g = m_2 a_2$
and
$$a_2 = \frac{T - m_2 g}{m_2}$$

Substitute for T and simplify to obtain:

$$a_2 = \boxed{\frac{(m_1 - m_2)g + 2m_1 a}{m_1 + m_2}}$$

The expression for a_1 is the same as for a_2 with all subscripts interchanged (note that a positive value for a_1 represents acceleration upward):

$$a_1 = \boxed{\frac{(m_2 - m_1)g + 2m_2a}{m_1 + m_2}}$$

Chapter 5
Applications of Newton's Laws

Conceptual Problems

*2 • Any object resting on the floor of a truck will slide if the truck's acceleration is too great. How does the critical acceleration at which a light object slips compare with that at which a much heavier object slips?

Determine the Concept The forces acting on an object are the normal force exerted by the floor of the truck, the weight of the object, and the friction force; also exerted by the floor of the truck. Of these forces, the only one that acts in the direction of the acceleration (chosen to be to the right in the free-body diagram) is the friction force. Apply Newton's 2nd law to the object to determine how the critical acceleration depends on its weight.

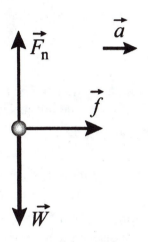

Taking the positive x direction to be to the right, apply $\Sigma F_x = ma_x$ and solve for a_x:

$$f = \mu_s w = \mu_s mg = ma_x$$
and
$$a_x = \mu_s g$$

> Because a_x is independent of m and w, the critical accelerations are the same.

*4 • A block of mass m is at rest on a plane inclined at angle of 30° with the horizontal, as shown in Figure 5-34. Which of the following statements about the force of static friction is necessarily true? $(a) f_s > mg$. $(b) f_s > mg \cos 30°$. $(c) f_s = mg \cos 30°$. $(d) f_s = mg \sin 30°$. (e) None of these statements are true.

Figure 5-34 Problem 4

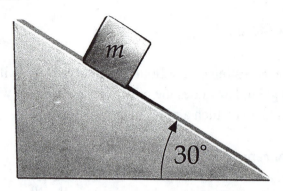

Determine the Concept The block is in equilibrium under the influence of $\vec{F}_n$, $m\vec{g}$, and $\vec{f}_s$; i.e.,

$$\vec{F}_n + m\vec{g} + \vec{f}_s = 0$$

We can apply Newton's 2nd law in the x direction to determine the relationship between f_s and mg.

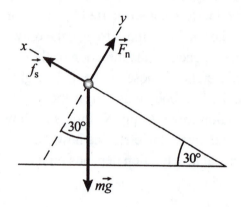

Apply $\sum F_x = 0$ to the block:

$$f_s - mg\sin\theta = 0$$

Solve for f_s:

$$f_s = mg\sin\theta$$
and $\boxed{(d)}$ is correct.

*6 •• Show with a force diagram how a motorcycle can travel in a circle on the inside vertical wall of a hollow cylinder. Assume reasonable parameters (coefficient of friction, radius of the circle, mass of the motorcycle, or whatever is required), and calculate the minimum speed needed.

Picture the Problem The normal reaction force F_n provides the centripetal force and the force of static friction, $\mu_s F_n$, keeps the cycle from sliding down the wall. We can apply Newton's 2nd law and the definition of $f_{s,max}$ to derive an expression for v_{min}:

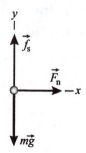

Apply $\sum \vec{F} = m\vec{a}$ to the motorcycle:

$$\sum F_x = F_n = m\frac{v^2}{R}$$

and

$$\sum F_y = f_s - mg = 0$$

For the minimum speed:

$$f_s = f_{s,max} = \mu_s F_n$$

Substitute for f_s, eliminate F_n between the force equations, and solve for v_{min}:

$$v_{min} = \sqrt{\frac{Rg}{\mu_s}}$$

Assume that $R = 6$ m and $\mu_s = 0.8$ and solve for v_{min}:

$$v_{min} = \sqrt{\frac{(6\,m)(9.81\,m/s^2)}{0.8}}$$

$$= \boxed{8.58\,m/s = 30.9\,km/h}$$

*10 •   You place a lightweight piece of iron on a table and a small kitchen magnet above the iron at a distance of 1 cm. You find that the magnet cannot lift the iron, even though there is obviously a force between the iron and the magnet. Next, you again hold the piece of iron and the magnet 1 cm apart with the magnet above the iron, but this time you drop them from arm's length. As they fall, the magnet and the piece of iron are pulled together before hitting the floor. (*a*) Draw free-body diagrams illustrating all of the forces on the magnet and the iron for each demonstration. (*b*) Explain why the magnet and iron are pulled together when they are dropped even though the magnet cannot pull up the piece of iron when it is sitting on the table.

Determine the Concept We can analyze these demonstrations by drawing force diagrams for each situation. In both diagrams, h denotes "hand", g denotes "gravitational", m denotes "magnetic", and n denotes "normal".

(*a*) Demonstration 1:   Demonstration 2:

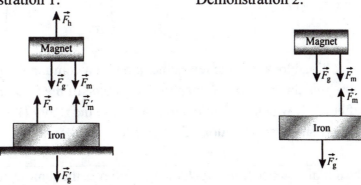

(*b*) Because the magnet doesn't lift the iron in the first demonstration, the force exerted on the iron must be less than its (the iron's) weight. This is still true when the two are falling, but the motion of the iron is not restrained by the table, and the motion of the magnet is not restrained by the hand. Looking at the second diagram, the net force pulling the magnet down is greater than its weight, implying that its acceleration is greater than g. The opposite is true for the iron: the magnetic force acts upwards, slowing it down, so its acceleration will be less than g. Because of this, the magnet will catch up to the iron piece as they fall.

*11 ••• The following question is an excellent "braintwister" invented by Boris Korsunsky[†]. Two identical blocks are attached together by a massless string running over a pulley as shown in Figure 5-35. The rope initially runs over the pulley at its (the rope's) midpoint, and the surface which block 1 rests on is frictionless. Blocks 1 and 2 are initially at rest when block 2 is released with the string taut and horizontal. Will block 1 hit the pulley before or after block 2 hits the wall? (Assume the initial distance from block 1 to the pulley is the same as the initial distance of block 2 to the wall.) There is a very simple solution.

Figure 5-35 Problem 11

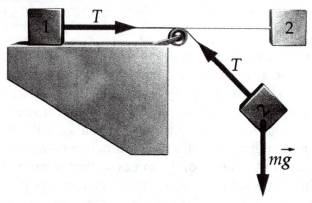

[†] "Braintwisters for Physics Students", Boris Korsunsky, *The Physics Teacher*, 33, p. 550 (1995).

Picture the Problem The free-body diagrams show the forces acting on the two objects some time after block 2 is dropped. While $\vec{T}_1 \neq \vec{T}_2$, $T_1 = T_2$.

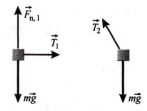

The only force pulling block 2 to the left is the horizontal component of the tension. Because this force is smaller than the magnitude of the tension, the acceleration of block 1, which is identical to block 2, to the right ($T_1 = T_2$) will always be greater than the acceleration of block 2 to the left.

Because the initial distance from block 1 to the pulley is the same as the initial distance of block 2 to the wall, block 1 will hit the pulley before block 2 hits the wall.

*15 • The mass of the moon is about 1% of that of the earth. The centripetal force that keeps the moon in its orbit around the earth (*a*) is much smaller than the gravitational force exerted by the moon on the earth, (*b*) depends on the phase of the moon, (*c*) is much greater than the gravitational force exerted by the moon on the earth, (*d*) is the gravitational force exerted by the moon on the earth, (*e*) cannot be answered; we haven't studied Newton's law of gravity yet.

Determine the Concept The centripetal force that keeps the moon in its orbit around the earth is provided by the gravitational force the earth exerts on the moon. As described by Newton's 3rd law, this force is equal in magnitude to the force the moon exerts on the earth. $\boxed{(d)\text{ is correct.}}$

Estimation and Approximation

*18 • To determine the aerodynamic drag on a car, the "coast-down" method is often used. The car is driven on a long, flat road at some convenient speed (60 mph is typical), shifted into neutral, and allowed to coast to a stop. The time which it takes for the speed to drop by successive 5-mph intervals is measured and used to compute the net force slowing down the car. (*a*) It was found that a Toyota Tercel with mass 1020 kg coasted down from 60 to 55 mph in 3.92 s. What is the average force slowing down the car? (*b*) If the coefficient of rolling friction for the car is 0.02, what is the force of rolling friction which is acting to slow down the car? If we assume that the only two forces acting on the car are rolling friction and aerodynamic drag, what is the average drag force acting on the car? (*c*) The drag force will have the form $\frac{1}{2}C\rho Av^2$, where A is the cross-sectional area of the car facing the wind, v is the car's speed, ρ is the density of air, and C is a dimensionless constant of order 1. If the cross-sectional area of the car is 1.91 m^2, determine C from the data given above. (The density of air is 1.21 kg/m^3; use the average speed of the car in this computation.)

Picture the Problem The free-body diagram shows the forces on the Tercel as it slows from 60 to 55 mph. We can use Newton's 2nd law to calculate the average force from the rate at which the car's speed decreases and the rolling force from its definition. The drag force can be inferred from the average and rolling friction forces and the drag coefficient from the defining equation for the drag force.

(*a*) Apply $\sum F_x = ma_x$ to the car to relate the average force acting

$$F_{av} = ma_{av} = m\frac{\Delta v}{\Delta t}$$

on it to its average velocity:

Substitute numerical values and evaluate F_{av}:

$$F_{av} = (1020\,\text{kg})\frac{5\dfrac{\text{mi}}{\text{h}}\times 1.609\dfrac{\text{km}}{\text{mi}}\times\dfrac{1\text{h}}{3600\,\text{s}}\times\dfrac{1000\,\text{m}}{\text{km}}}{3.92\,\text{s}} = \boxed{581\,\text{N}}$$

(b) Using its definition, express and evaluate the force of rolling friction:

$$f_{\text{rolling}} = \mu_{\text{rolling}} F_n = \mu_{\text{rolling}}\, mg$$
$$= (0.02)(1020\,\text{kg})(9.81\,\text{m/s}^2)$$
$$= \boxed{200\,\text{N}}$$

Assuming that only two forces are acting on the car in the direction of its motion, express their relationship and solve for and evaluate the drag force:

$$F_{av} = F_{\text{drag}} + F_{\text{rolling}}$$
and
$$F_{\text{drag}} = F_{av} - F_{\text{rolling}}$$
$$= 581\,\text{N} - 200\,\text{N} = \boxed{381\,\text{N}}$$

(c) Convert 57.5 mi/h to m/s:

$$57.5\frac{\text{mi}}{\text{h}} = 57.5\frac{\text{mi}}{\text{h}}\times\frac{1.609\,\text{km}}{\text{mi}}$$
$$\times\frac{1\text{h}}{3600\,\text{s}}\times\frac{10^3\,\text{m}}{\text{km}}$$
$$= 25.7\,\text{m/s}$$

Using the definition of the drag force and its calculated value from (b) and the average speed of the car during this 5 mph interval, solve for C:

$$F_{\text{drag}} = \tfrac{1}{2}C\rho A v^2 \Rightarrow C = \frac{2F_{\text{drag}}}{\rho A v^2}$$

Substitute numerical values and evaluate C:

$$C = \frac{2(381\,\text{N})}{(1.21\,\text{kg/m}^3)(1.91\,\text{m}^2)(25.7\,\text{m/s})^2}$$
$$= \boxed{0.499}$$

Friction

*21 •   A block of mass m slides at constant speed down a plane inclined at an angle θ with the horizontal. It follows that (a) $\mu_k = mg\sin\theta$. (b) $\mu_k = \tan\theta$. (c) $\mu_k = 1 - \cos\theta$. (d) $\mu_k = \cos\theta - \sin\theta$.

Picture the Problem The block is in equilibrium under the influence of $\vec{F}_n$, $m\vec{g}$, and $\vec{f}_k$; i.e.,

$$\vec{F}_n + m\vec{g} + \vec{f}_k = 0$$

We can apply Newton's 2nd law to determine the relationship between f_k, θ, and mg.

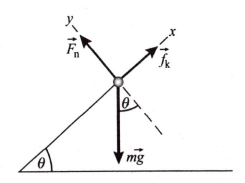

Using its definition, express the coefficient of kinetic friction:	$\mu_k = \dfrac{f_k}{F_n}$ (1)
Apply $\sum F_x = ma_x$ to the block:	$f_k - mg\sin\theta = ma_x = 0$ because $a_x = 0$
Solve for f_k:	$f_k = mg\sin\theta$
Apply $\sum F_y = ma_y$ to the block:	$F_n - mg\cos\theta = ma_y = 0$ because $a_y = 0$
Solve for F_n:	$F_n = mg\cos\theta$
Substitute in equation (1) to obtain:	$\mu_k = \dfrac{mg\sin\theta}{mg\cos\theta} = \tan\theta$
	and $\boxed{(b) \text{ is correct.}}$

*23 • A 20-N block rests on a horizontal surface. The coefficients of static and kinetic friction between the surface and the block are $\mu_s = 0.8$ and $\mu_k = 0.6$. A horizontal string is attached to the block and a constant tension T is maintained in the string. What is the force of friction acting on the block if (a) $T = 15$ N, or (b) $T = 20$ N.

Picture the Problem Whether the friction force is that due to static friction or kinetic friction depends on whether the applied tension is greater than the maximum static friction force. We can apply the definition of the maximum static friction to decide whether $f_{s,max}$ or T is greater.

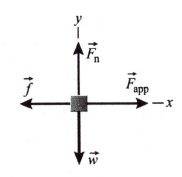

Calculate the maximum static friction force:

$f_{s,max} = \mu_s F_n = \mu_s w = (0.8)(20 \text{ N}) = 16 \text{ N}$

(a) Because $f_{s,max} > T$:

$$f = f_s = T = \boxed{15.0\,\text{N}}$$

(b) Because $T > f_{s,max}$:

$$f = f_k = \mu_k w = (0.6)(20\,\text{N}) = \boxed{12.0\,\text{N}}$$

*28 •   The force that accelerates a car along a flat road is the frictional force exerted by the road on the car's tires. (a) Explain why the acceleration can be greater when the wheels do not slip. (b) If a car is to accelerate from 0 to 90 km/h in 12 s at constant acceleration, what is the minimum coefficient of friction needed between the road and tires? Assume that half the weight of the car is supported by the drive wheels.

Picture the Problem The free-body diagram shows the forces acting on the drive wheels–the ones we're assuming support half the weight of the car. We can use the definition of acceleration and apply Newton's 2nd law to the horizontal and vertical components of the forces to determine the minimum coefficient of friction between the road and the tires.

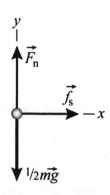

(a) $\boxed{\text{Because } \mu_s > \mu_k,\ f \text{ will be greater if the wheels do not slip.}}$

(b) Apply $\sum F_x = ma_x$ to the car:

$$f_s = \mu_s F_n = ma_x \qquad (1)$$

Apply $\sum F_y = ma_y$ to the car and solve for F_n:

$$F_n - \tfrac{1}{2}mg = ma_y$$
Because $a_y = 0$,
$$F_n - \tfrac{1}{2}mg = 0 \Rightarrow F_n = \tfrac{1}{2}mg$$

Find the acceleration of the car:

$$a_x = \frac{\Delta v}{\Delta t} = \frac{(90\,\text{km/h})(1000\,\text{m/km})}{12\,\text{s}}$$
$$= 2.08\,\text{m/s}^2$$

Solve equation (1) for μ_s:

$$\mu_s = \frac{ma_x}{\tfrac{1}{2}mg} = \frac{2a_x}{g}$$

Substitute numerical values and evaluate a_x:

$$\mu_s = \frac{2(2.08\,\text{m/s}^2)}{9.81\,\text{m/s}^2} = \boxed{0.424}$$

*32 •   A 50-kg box that is resting on a level floor must be moved. The coefficient of static friction between the box and the floor is 0.6. One way to move the box is to push down on it at an angle θ with the horizontal. Another method is to pull up on the box at an angle θ with the horizontal. (a) Explain why one method is better than the other. (b) Calculate the force necessary to move the box by each method if $\theta = 30°$ and compare the answers with the result when $\theta = 0°$.

Picture the Problem The free-body diagrams for the two methods are shown to the right. Method 1 results in the box being pushed into the floor, increasing the normal force and the static friction force, whereas method 2 partially lifts the box, reducing the normal force and the static friction force. We can apply Newton's 2nd law to obtain expressions that relate the maximum static friction force to the applied force $\vec{F}$.

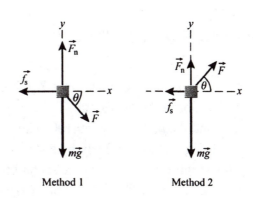

Method 1   Method 2

(a) | Method 2 is preferable as it reduces F_n and, therefore, f_s. |

(b) Apply $\sum F_x = ma_x$ to the box:

$$F\cos\theta - f_s = F\cos\theta - \mu_s F_n = 0$$

Method 1: Apply $\sum F_y = ma_y$ to the block and solve for F_n:

$$F_n - mg - F\sin\theta = 0$$
$$\therefore F_n = mg + F\sin\theta$$

Relate $f_{s,max}$ to F_n:

$$f_{s,max} = \mu_s F_n = \mu_s(mg + F\sin\theta) \quad (1)$$

Method 2: Apply $\sum F_y = ma_y$ to the forces in the y direction and solve for F_n:

$$F_n - mg + F\sin\theta = 0$$
$$\therefore F_n = mg - F\sin\theta$$

Relate $f_{s,max}$ to F_n:

$$f_{s,max} = \mu_s F_n = \mu_s(mg - F\sin\theta) \quad (2)$$

Express the condition that must be satisfied to move the box by either method:

$$f_{s,max} = F\cos\theta \quad (3)$$

Method 1: Substitute (1) in (3) and solve for F:

$$F_1 = \frac{\mu_s mg}{\cos\theta - \mu_s \sin\theta} \qquad (4)$$

Method 2: Substitute (2) in (3) and solve for F:

$$F_2 = \frac{\mu_s mg}{\cos\theta + \mu_s \sin\theta} \qquad (5)$$

Evaluate (4) and (5) with $\theta = 30°$:

$$F_1(30°) = \boxed{520\,\text{N}}$$
$$F_2(30°) = \boxed{252\,\text{N}}$$

Evaluate (4) and (5) with $\theta = 0°$:

$$F_1(0°) = F_2(0°) = \mu_s mg = \boxed{294\,\text{N}}$$

*35 ••   A block of mass $m_1 = 250$ g is at rest on a plane that makes an angle $\theta = 30°$ above the horizontal. The coefficient of kinetic friction between the block and the plane is $\mu_k = 0.100$. The block is attached to a second block of mass $m_2 = 200$ g that hangs freely by a string that passes over a frictionless, massless pulley (Figure 5-41). When the second block has fallen 30.0 cm, its speed is (*a*) 83 cm/s, (*b*) 48 cm/s, (*c*) 160 cm/s, (*d*) 59 cm/s, (*e*) 72 cm/s.

Figure 5-41 Problems 35-37

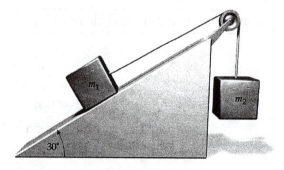

Picture the Problem We can find the speed of the system when it has moved a given distance by using a constant acceleration equation. Under the influence of the forces shown in the free-body diagrams, the blocks will have a common acceleration a. The application of Newton's 2nd law to each block, followed by the elimination of the tension T and the use of the definition of f_k, will allow us to determine the acceleration of the system.

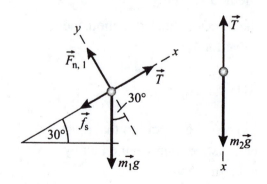

Using a constant-acceleration equation, relate the speed of the system to its acceleration and displacement; solve for its speed:

$$v^2 = v_0^2 + 2a\Delta x$$
and, because $v_0 = 0$,
$$v = \sqrt{2a\Delta x}$$

Apply $\vec{F}_{net} = m\vec{a}$ to the block whose mass is m_1:

$$\Sigma F_x = T - f_k - m_1 g \sin 30° = m_1 a \quad (1)$$
and
$$\Sigma F_y = F_{n,1} - m_1 g \cos 30° = 0 \quad (2)$$

Using $f_k = \mu_k F_n$, substitute (2) in (1) to obtain:

$$T - \mu_k m_1 g \cos 30° - m_1 g \sin 30° = m_1 a$$

Apply $\sum F_x = ma_x$ to the block whose mass is m_2:

$$m_2 g - T = m_2 a$$

Add the last two equations to eliminate T and solve for a to obtain:

$$a = \frac{(m_2 - \mu_k m_1 \cos 30° - m_1 \sin 30°)g}{m_1 + m_2}$$
$$= 1.16\,\text{m/s}^2$$

Substitute and evaluate a:

$$v = \sqrt{2(1.16\,\text{m/s}^2)(0.3\,\text{m})} = 0.835\,\text{m/s}$$
and $\boxed{(a) \text{ is correct.}}$

*38 ••   The coefficient of static friction between the bed of a truck and a box resting on it is 0.30. The truck is traveling at 80 km/h along a horizontal road. What is the least distance in which the truck can stop if the box is not to slide?

Picture the Problem The truck will stop in the shortest possible distance when its acceleration is a maximum. The maximum acceleration is, in turn, determined by the maximum value of the static friction force. The free-body diagram shows the forces acting on the box as the truck brakes to a stop. Assume that the truck is moving in the positive x direction and apply Newton's 2nd law and the definition of $f_{s,max}$ to find the shortest stopping distance.

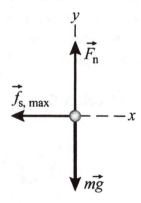

Using a constant-acceleration equation, relate the truck's stopping distance to its acceleration and initial velocity; solve for the stopping distance:

$$v^2 = v_0^2 + 2a\Delta x$$
or, since $v = 0$,

$$\Delta x_{min} = \sqrt{\frac{-v_0^2}{2a_{max}}}$$

Apply $\vec{F}_{net} = m\vec{a}$ to the block:

$$\Sigma F_x = -f_{s,max} = ma_{max} \qquad (1)$$
and
$$\Sigma F_y = F_n - mg = 0 \qquad (2)$$

Using the definition of $f_{s,max}$, solve equations (1) and (2) simultaneously for a:

$$f_{s,max} = \mu_s F_n$$
and
$$a_{max} = -\mu_s g$$

Substitute numerical values and evaluate a_{max}:

$$a_{max} = -(0.3)(9.81\,\text{m/s}^2) = -2.943\,\text{m/s}^2$$

Substitute numerical values and evaluate Δx_{min}:

$$\Delta x_{min} = \sqrt{\frac{-(80\,\text{km/h})^2(1000\,\text{km/m})^2(1\text{h}/3600\text{s})^2}{2(-2.943\,\text{m/s}^2)}} = \boxed{9.16\,\text{m}}$$

*42 ••   Lou bets an innocent stranger that he can place a 2-kg box against the side of a cart, as in Figure 5-42, and that the box will not fall to the ground, even though Lou will use no hooks, ropes, fasteners, magnets, glue, or adhesives of any kind. When the stranger accepts the bet, Lou begins to push the cart in the direction shown. The coefficient of static friction between the box and the cart is 0.6. (*a*) Find the minimum acceleration for which Lou will win the bet. (*b*) What is the magnitude of the frictional force in this case? (*c*) Find the force of friction on the box if *a* is twice the minimum needed for the box not to fall. (*d*) Show that, for a box of any mass, the box will not fall if the acceleration is $a \geq g/\mu_s$, where μ_s is the coefficient of static friction.

Figure 5-42 Problem 42

Picture the Problem To hold the box in place, the acceleration of the cart and box must be great enough so that the static friction force acting on the box will equal the weight of the box. We can use Newton's 2nd law to determine the minimum acceleration required.

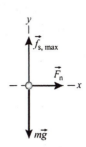

(a) Apply $\sum \vec{F} = m\vec{a}$ to the box:

$$\Sigma F_x = F_n = ma_{min} \qquad (1)$$
and
$$\Sigma F_y = f_{s,max} - mg = 0 \qquad (2)$$

Substitute μF_n for $f_{s,max}$ in equation (2), eliminate F_n between the two equations and solve for and evaluate a_{min}:

$$\mu F_n - mg = 0, \quad \mu(ma_{min}) - mg = 0,$$
and
$$a_{min} = \frac{g}{\mu_s} = \frac{9.81\,\text{m/s}^2}{0.6} = \boxed{16.4\,\text{m/s}^2}$$

(b) Solve equation (2) for $f_{s,max}$, and substitute numerical values and evaluate $f_{s,max}$:

$$f_{s,max} = mg = (2\,\text{kg})(9.81\,\text{m/s}^2)$$
$$= \boxed{19.6\,\text{N}}$$

(c) If a is twice that required to hold the box in place, f_s will still have its maximum value given by:

$$f_{s,max} = \boxed{19.6\,\text{N}}$$

(d) $\boxed{\text{Because } g/\mu_s \text{ is } a_{min}, \text{ the box will not fall if } a \geq g/\mu_s.}$

*44 •• As in Problem 43, two blocks of masses m_1 and m_2 are sliding down an incline as shown in Figure 5-43. They are connected by a massless *rod* this time; the rod behaves in exactly the same way as a string, except that the force can be compressive as well as tensile. The coefficient of kinetic friction for block 1 is μ_1 and the coefficient of kinetic friction for block 2 is μ_2. (a) Determine the acceleration of the two blocks. (b) Determine the force that the rod exerts on the two blocks. Show that the force is 0 when $\mu_1 = \mu_2$, and give a simple, nonmathematical argument why this is true.

Figure 5-43 Problem 44

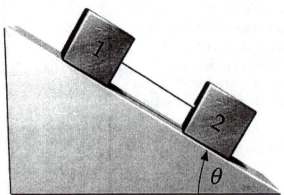

Picture the Problem The free-body diagram shows the forces acting on the two blocks as they slide down the incline. Down the incline has been chosen as the positive x direction. T is the force transmitted by the stick; it can be either tensile ($T > 0$) or compressive ($T < 0$). By applying Newton's 2nd law to these blocks, we can obtain equations in T and a from which we can eliminate either by solving them simultaneously. Once we have expressed T, the role of the stick will become apparent.

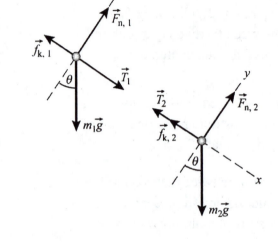

(a) Apply $\sum \vec{F} = m\vec{a}$ to block 1:

$$\sum F_x = T_1 + m_1 g \sin\theta - f_{k,1} = m_1 a$$

and

$$\sum F_y = F_{n,1} - m_1 g \cos\theta = 0$$

Apply $\sum \vec{F} = m\vec{a}$ to block 2:

$$\sum F_x = m_2 g \sin\theta - T_2 - f_{k,2} = m_2 a$$

and

$$\sum F_y = F_{n,2} - m_2 g \cos\theta = 0$$

Letting $T_1 = T_2 = T$, use the definition of the kinetic friction force to eliminate $f_{k,1}$ and $F_{n,1}$ between the equations for block 1 and $f_{k,2}$ and $F_{n,1}$ between the equations for block 2 to obtain:

$$m_1 a = m_1 g \sin\theta + T - \mu_1 m_1 g \cos\theta \qquad (1)$$

and

$$m_2 a = m_2 g \sin\theta - T - \mu_2 m_2 g \cos\theta \qquad (2)$$

Add equations (1) and (2) to eliminate T and solve for a:

$$a = \left| g\left(\sin\theta - \frac{\mu_1 m_1 + \mu_2 m_2}{m_1 + m_2}\cos\theta \right) \right|$$

(b) Rewrite equations (1) and (2) by dividing both sides of (1) by m_1 and both sides of (2) by m_2 to obtain:

$$a = g\sin\theta + \frac{T}{m_1} - \mu_1 g\cos\theta \qquad (3)$$

and

$$a = g\sin\theta - \frac{T}{m_2} - \mu_2 g\cos\theta \qquad (4)$$

Subtracting (4) from (3) and rearranging yields:

$$T = \left| \left(\frac{m_1 m_2}{m_1 - m_2} \right)(\mu_1 - \mu_2)g\cos\theta \right|$$

If $\mu_1 = \mu_2$, $T = 0$ and the blocks move down the incline with the same acceleration $g(\sin\theta - \mu\cos\theta)$. Inserting a stick between them can't change this, so the stick must exert no force on either block.

*47 ••   A block of mass m rests on a horizontal table (Figure 5-44). The block is pulled by a massless rope with a force $\vec{F}$ at an angle θ. The coefficient of static friction is 0.6. The minimum value of the force needed to move the block depends on the angle θ. (a) Discuss qualitatively how you would expect this force to depend on θ. (b) Compute the force for the angles $\theta = 0°, 10°, 20°, 30°, 40°, 50°$, and $60°$, and make a plot of F versus θ for $mg = 400$ N. From your plot, at what angle is it most efficient to apply the force to move the block?

Figure 5-44 Problem 47

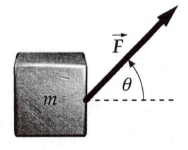

Picture the Problem The vertical component of $\vec{F}$ reduces the normal force; hence, the static friction force between the surface and the block. The horizontal component is responsible for any tendency to move and equals the static friction force until it exceeds its maximum value. We can apply Newton's 2nd law to the box, under equilibrium conditions, to relate F to θ.

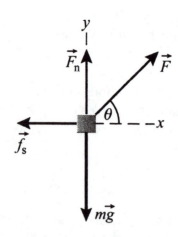

(*a*) The static-frictional force opposes the motion of the object, and the maximum value of the static-frictional force is proportional to the normal force F_N. The normal force is equal to the weight minus the vertical component F_V of the force F. Keeping the magnitude F constant while increasing θ from zero results in a decrease in F_V and thus a corresponding decrease in the maximum static-frictional force f_{max}. The object will begin to move if the horizontal component F_H of the force F exceeds f_{max}. An increase in θ results in a decrease in F_H. As θ increases from 0, the decrease in F_N is larger than the decrease in F_H, so the object is more and more likely to slip. However, as θ approaches 90°, F_H approaches zero and no movement will be initiated. If F is large enough and if θ increases from 0, then at some value of θ the block will start to move.

(*b*) Apply $\sum \vec{F} = m\vec{a}$ to the block:

$$\Sigma F_x = F\cos\theta - f_s = 0 \qquad (1)$$

and

$$\Sigma F_y = F_n + F\sin\theta - mg = 0 \qquad (2)$$

Assuming that $f_s = f_{s,max}$, eliminate f_s and F_n between equations (1) and (2) and solve for F:

$$F = \frac{\mu_s mg}{\cos\theta + \mu_s \sin\theta}$$

Use this function with $mg = 240$ N to generate the table shown below:

θ	(deg)	0	10	20	30	40	50	60
F	(N)	240	220	210	206	208	218	235

A spreadsheet-generated graph of $F(\theta)$ follows.

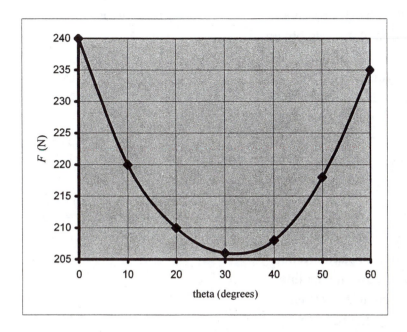

From the graph, we can see that the minimum value for F occurs when $\theta \approx 32°$.

Remarks: An alternative to manually plotting F as a function of θ or using a spreadsheet program is to use a graphing calculator to enter and graph the function.

*52 ••   The coefficient of static friction between a rubber tire and the road surface is 0.85. What is the maximum acceleration of a 1000-kg four-wheel-drive truck if the road makes an angle of 12° with the horizontal and the truck is (*a*) climbing, and (*b*) descending?

Picture the Problem The accelerations of the truck can be found by applying Newton's 2nd law of motion. The free-body diagram for the truck climbing the incline with maximum acceleration is shown to the right.

(*a*) Apply $\sum \vec{F} = m\vec{a}$ to the truck when it is climbing the incline:

$$\Sigma F_x = f_{s,max} - mg\sin 12° = ma \qquad (1)$$
and

$$\Sigma F_y = F_n - mg\cos 12° = 0 \qquad (2)$$

Solve equation (2) for F_n and use the definition of $f_{s,max}$ to obtain:

$$f_{s,max} = \mu_s mg\cos 12° \qquad (3)$$

Substitute equation (3) into equation (1) and solve for a:

$$a = g(\mu_s \cos 12° - \sin 12°)$$

Substitute numerical values and evaluate a:

$$a = (9.81\,\text{m/s}^2)[(0.85)\cos 12° - \sin 12°]$$
$$= \boxed{6.12\,\text{m/s}^2}$$

(b) When the truck is descending the incline with maximum acceleration, the static friction force points down the incline; i.e., its direction is reversed on the free-body diagram. Apply $\sum F_x = ma_x$ to the truck under these conditions:

$$-f_{s,max} - mg\sin 12° = ma \qquad (4)$$

Substitute equation (3) into equation (4) and solve for a:

$$a = -g(\mu_s \cos 12° + \sin 12°)$$

Substitute numerical values and evaluate a:

$$a = (-9.81\,\text{m/s}^2)[(0.85)\cos 12° + \sin 12°]$$
$$= \boxed{-10.2\,\text{m/s}^2}$$

*57 ••   On planet Vulcan, an introductory physics class performs several experiments involving friction. In one of these experiments the acceleration of a block is measured both when it is sliding up an incline and when it is sliding down the same incline. You copy the following data and diagram (Figure 5-52) out of one of the lab notebooks, but can't find any translations into metric units. (Negative sign indicates that the acceleration is pointing down the incline.)

Acceleration of block
    Going up inclined plane   −1.73 glapp/plip² *
    Going down plane   −1.42 glapp/plip²

Figure 5-52 Problem 57

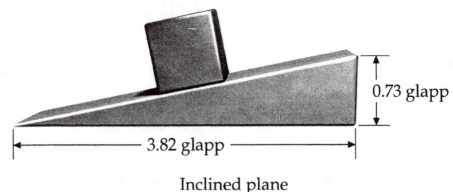

0.73 glapp

3.82 glapp

Inclined plane

From these data, determine the acceleration of gravity on Vulcan (in glapps/plip^2) and the coefficient of kinetic friction between the block and the incline.

Picture the Problem The free-body diagram shows the forces acting on the block as it is moving up the incline. By applying Newton's 2nd law, we can obtain expressions for the accelerations of the block up and down the incline. Adding and subtracting these equations, together with the data found in the notebook, will lead to values for g_V and μ_k.

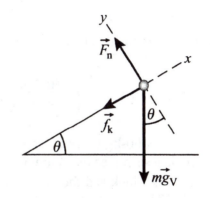

Apply $\sum_i \vec{F}_i = m\vec{a}$ to the block when it is moving up the incline:	$\sum F_x = -f_k - mg_V \sin\theta = ma_{up}$ and $\sum F_y = F_n - mg_V \cos\theta = 0$
Using the definition of f_k, eliminate F_n between the two equations to obtain:	$a_{up} = -\mu_k g_V \cos\theta - g_V \sin\theta \qquad (1)$
When the block is moving down the incline, f_k is in the positive x direction, and its acceleration is:	$a_{down} = \mu_k g_V \cos\theta - g_V \sin\theta \qquad (2)$
Add equations (1) and (2) to obtain:	$a_{up} + a_{down} = -2g_V \sin\theta \qquad (3)$
Solve equation (3) for g_V:	$g_V = \dfrac{a_{up} + a_{down}}{-2\sin\theta} \qquad (4)$

Determine θ from the figure:

$$\theta = \tan^{-1}\left[\frac{0.73\,\text{glapp}}{3.82\,\text{glapp}}\right] = 10.8°$$

Substitute the data from the notebook in equation (4) to obtain:

$$g_V = \frac{1.73\,\text{glapp/plipp}^2 + 1.42\,\text{glapp/plipp}^2}{-2\sin10.8°}$$

$$= \boxed{-8.41\,\text{glapp/plipp}^2}$$

Subtract equation (1) from equation (2) to obtain:

$$a_{\text{down}} - a_{\text{up}} = 2\mu_k g_V \cos\theta$$

Solve for μ_k:

$$\mu_k = \frac{a_{\text{down}} - a_{\text{up}}}{2g_V \cos\theta}$$

Substitute numerical values and evaluate μ_k:

$$\mu_k = \frac{-1.42\,\text{glapp/plipp}^2 - 1.73\,\text{glapp/plipp}^2}{2\left(-8.41\,\text{glapp/plipp}^2\right)\cos10.8°} = \boxed{0.191}$$

*58 ••   A 100-kg block on an inclined plane is attached to another block of mass m via a string, as in Figure 5-53. The coefficients of static and kinetic friction of the block and the incline are $\mu_s = 0.4$ and $\mu_k = 0.2$. The angle of the incline is 18° above the horizontal. (a) Determine the range of values for m, the mass of the hanging block, for which the block on the incline will not move unless disturbed, but if nudged, will slide *down* the incline. (b) Determine a range of values for m for which the block on the incline will not move unless nudged, but if nudged will slide *up* the incline.

Figure 5-53 Problem 58

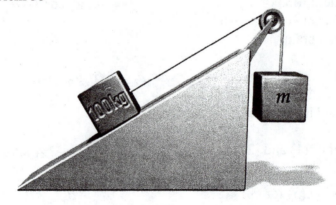

Picture the Problem The free-body diagram shows the block sliding down the incline under the influence of a friction force, its weight, and the normal force exerted on it by the inclined surface. We can find the range of values for m for the two situations described in the problem statement by applying Newton's 2nd law of motion to, first, the conditions under which the block will not move or slide if pushed, and secondly, if pushed, the block will move up the incline.

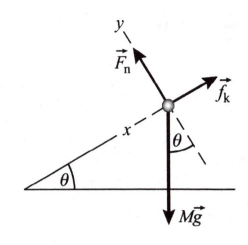

(*a*) Assume that the block is sliding down the incline with a constant velocity and with no hanging weight ($m = 0$) and apply $\sum \vec{F} = m\vec{a}$ to the block:

$$\sum F_x = -f_k + Mg\sin\theta = 0$$
and
$$\sum F_y = F_n - Mg\cos\theta = 0$$

Using $f_k = \mu_k F_n$, eliminate F_n between the two equations and solve for the net force acting on the block:

$$F_{net} = -\mu_k Mg\cos\theta + Mg\sin\theta$$

If the block is moving, this net force must be nonnegative and:

$$\left(-\mu_k\cos\theta + \sin\theta\right)Mg \geq 0$$

This condition requires that:

$$\mu_k \leq \tan\theta = \tan 18° = 0.325$$

Because $\mu_k = 0.2$, this condition is satisfied and:

$$m_{min} = 0$$

To find the maximum value, note that the maximum possible value for the tension in the rope is mg. For the block to move down the incline, the component of the block's weight parallel to the incline minus the frictional force must be greater than or equal to the tension in the rope:

$$Mg\sin\theta - \mu_k Mg\cos\theta \geq mg$$

Solve for m_{max}:

$$m_{max} \leq M\left(\sin\theta - \mu_k\cos\theta\right)$$

Substitute numerical values and evaluate m_{max}:	$m_{max} \leq (100\,kg)[\sin 18° - (0.2)\cos 18°]$ $= 11.9\,kg$
The range of values for m is:	$\boxed{0 \leq m \leq 11.9\,kg}$
(b) If the block is being dragged up the incline, the frictional force will point down the incline, and:	$Mg\sin\theta + \mu_k Mg\cos\theta < mg$
Solve for and evaluate m_{min}:	$m_{min} > M(\sin\theta + \mu_k\cos\theta)$ $= (100\ kg)[\sin 18° + (0.2)\cos 18°]$ $= 49.9\,kg$
If the block is not to move unless pushed:	$Mg\sin\theta + \mu_s Mg\cos\theta > mg$
Solve for and evaluate m_{max}:	$m_{max} < M(\sin\theta + \mu_s\cos\theta)$ $= (100\ kg)\,[\sin 18° + (0.4)\cos 18°]$ $= 68.9\,kg$
The range of values for m is:	$\boxed{49.9\,kg \leq m \leq 68.9\,kg}$

*62 ••• A block of wood with mass 10 kg is pushed, starting from rest, with a constant horizontal force of 70 N across a wooden floor. Assuming that the coefficient of kinetic friction varies with particle speed as $\mu_k = 0.11/(1 + 2.3 \times 10^{-4}\,v^2)^2$ (see Problem 61), write a spreadsheet program using Euler's method to calculate and graph the speed of the block and its displacement as a function of time from 0 to 10 s. Compare this to the case where the coefficient of kinetic friction is equal to 0.11, independent of v.

Picture the Problem The kinetic friction force f_k is the product of the coefficient of sliding friction μ_k and the normal force F_n the surface exerts on the sliding object. By applying Newton's 2nd law in the vertical direction, we can see that, on a horizontal surface, the normal force is the weight of the sliding object. We can apply Newton's 2nd law in the horizontal (x) direction to relate the block's acceleration to the net force acting on it. In the spreadsheet program, we'll find the acceleration of the block from this net force (which is velocity dependent), calculate the increase in the block's speed from its acceleration and the elapsed time and add this increase to its speed at end of the previous time interval, determine how far it has moved in this time interval, and add this distance to its previous position to find its current position. We'll also calculate the position of the block x_2, under the assumption that $\mu_k = 0.11$, using a constant-acceleration equation.

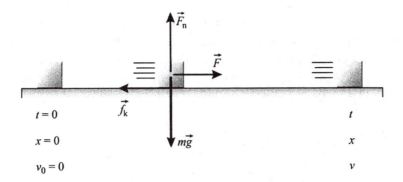

The spreadsheet solution follows. The formulas used to calculate the quantities in the columns are as follows:

Cell	Formula/Content	Algebraic Form
C9	C8+$B$6	$t + \Delta t$
D9	D8+F9*$B$6	$v + a\Delta t$
E9	$B$5–($B$3)*($B$2)*$B$5/(1+$B$4*D9^2)^2	$F - \dfrac{\mu_k mg}{\left(1 + 2.34 \times 10^{-4} v^2\right)^2}$
F9	E10/$B$5	F_{net} / m
G9	G9+D10*$B$6	$x + v\Delta t$
K9	0.5*5.922*I10^2	$\frac{1}{2}at^2$
L9	J10–K10	$x - x_2$

	A	B	C	D	E	F	G	H	I	J
1	g=	9.81	m/s^2							
2	Coeff1=	0.11								
3	Coeff2=	2.30E-4								
4	Mass=	10	kg							
5	Applied Force=	70	N							
6	Time step=	0.05	s			t	x	x2	x–x2
7										
8										
9	t	v	Net force	a	x			mu=variable	mu=constant	
10	0.00	0.00			0.00		0.00	0.00	0.00	0.00
11	0.05	0.30	59.22	5.92	0.01		0.05	0.01	0.01	0.01
12	0.10	0.59	59.22	5.92	0.04		0.10	0.04	0.03	0.01
13	0.15	0.89	59.22	5.92	0.09		0.15	0.09	0.07	0.02
14	0.20	1.18	59.22	5.92	0.15		0.20	0.15	0.12	0.03
15	0.25	1.48	59.23	5.92	0.22		0.25	0.22	0.19	0.04
205	9.75	61.06	66.84	6.68	292.37		9.75	292.37	281.48	10.89
206	9.80	61.40	66.88	6.69	295.44		9.80	295.44	284.37	11.07
207	9.85	61.73	66.91	6.69	298.53		9.85	298.53	287.28	11.25
208	9.90	62.07	66.94	6.69	301.63		9.90	301.63	290.21	11.42
209	9.95	62.40	66.97	6.70	304.75		9.95	304.75	293.15	11.61
210	10.00	62.74	67.00	6.70	307.89		10.00	307.89	296.10	11.79

The displacement of the block as a function of time, for a constant coefficient of friction ($\mu_k = 0.11$) is shown as a solid line on the graph and for a variable coefficient of friction, is shown as a dotted line. Because the coefficient of friction decreases with increasing particle speed, the particle travels slightly farther when the coefficient of friction is variable.

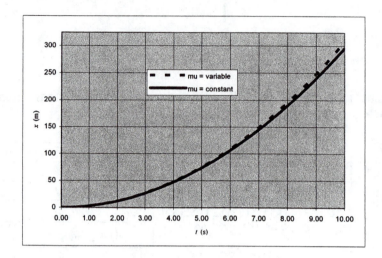

The velocity of the block, with variable coefficient of kinetic friction, is shown below.

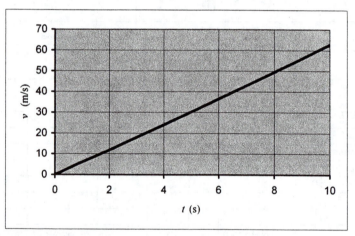

*64 •• The following data show the acceleration of a block down an inclined plane as a function of the angle of incline $\theta^{\ddagger}$:

θ (degrees)	Acceleration (m/s^2)
25	1.6909
27	2.1043
29	2.4064
31	2.8883
33	3.1750
35	3.4886
37	3.7812
39	4.1406

41	4.3257
43	4.7178
45	5.056

†Data taken from "Science Friction Adventure- Part II," Dennis W. Phillips, *The Physics Teacher*, **41**, 553, (Nov. 1990).
(*a*) Show that, for a block sliding down an incline, graphing $a/\cos\theta$ versus $\tan\theta$ should give a straight line with slope g and y-intercept $-\mu_k g$. (*b*) Using a spreadsheet program, graph these data and fit a straight line to them to determine μ_k and g. What is the percentage error in g from the commonly accepted value of 9.81 m/s^2?

Picture the Problem The free-body diagram shows the forces acting on the block as it slides down an incline. We can apply Newton's 2nd law to these forces to obtain the acceleration of the block and then manipulate this expression algebraically to show that a graph of $a/\cos\theta$ versus $\tan\theta$ will be linear with a slope equal to the acceleration due to gravity and an intercept whose absolute value is the coefficient of kinetic friction.

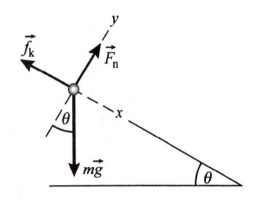

(*a*) Apply $\sum \vec{F} = m\vec{a}$ to the block as it slides down the incline:

$$\sum F_x = mg\sin\theta - f_k = ma$$
and
$$\sum F_y = F_n - mg\cos\theta = 0$$

Substitute $\mu_k F_n$ for f_k and eliminate F_n between the two equations to obtain:

$$a = g(\sin\theta - \mu_k \cos\theta)$$

Divide both sides of this equation by $\cos\theta$ to obtain:

$$\frac{a}{\cos\theta} = g\tan\theta - g\mu_k$$

Note that this equation is of the form $y = mx + b$:

Thus, if we graph $a/\cos\theta$ versus $\tan\theta$, we should get a straight line with slope g and y-intercept $-g\mu_k$.

(*b*) A spreadsheet solution follows. The formulas used to calculate the quantities in the columns are as follows:

Cell	Formula/Content	Algebraic Form
C7	θ	
D7	a	
E7	TAN(C7*PI()/180)	$\tan\left(\theta \times \dfrac{\pi}{180}\right)$
F7	D7/COS(C7*PI()/180)	$\dfrac{a}{\cos\left(\theta \times \dfrac{\pi}{180}\right)}$

	C	D	E	F
6	theta	a	tan(theta)	a/cos(theta)
7	25	1.691	0.466	1.866
8	27	2.104	0.510	2.362
9	29	2.406	0.554	2.751
10	31	2.888	0.601	3.370
11	33	3.175	0.649	3.786
12	35	3.489	0.700	4.259
13	37	3.781	0.754	4.735
14	39	4.149	0.810	5.338
15	41	4.326	0.869	5.732
16	43	4.718	0.933	6.451
17	45	5.106	1.000	7.220

A graph of $a/\cos\theta$ versus $\tan\theta$ follows. From the curve fit (Excel's Trendline was used), $g = 9.77$ m/s^2 and $\mu_k = \dfrac{2.62\,\text{m/s}^2}{9.77\,\text{m/s}^2} = 0.268$.

The percentage error in g from the commonly accepted value of 9.81 m/s^2 is

$$100\left(\frac{9.81\,\text{m/s}^2 - 9.77\,\text{m/s}^2}{9.81\,\text{m/s}^2}\right) = \boxed{0.408\%}$$

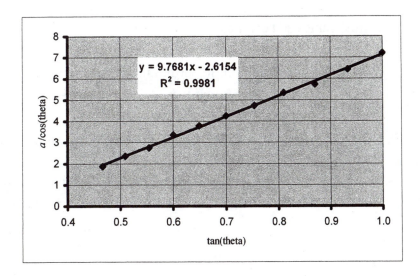

Motion Along a Curved Path

*68 ••   A 50-kg pilot comes out of a vertical dive in a circular arc such that at the bottom of the arc her upward acceleration is 8.5g. (*a*) What is the magnitude of the force exerted by the airplane seat on the pilot at the bottom of the arc? (*b*) If the speed of the plane is 345 km/h, what is the radius of the circular arc?

Picture the Problem The sketch shows the forces acting on the pilot when her plane is at the lowest point of its dive. $\vec{F}_n$ is the force the airplane seat exerts on her. We'll apply Newton's 2nd law for circular motion to determine F_n and the radius of the circular path followed by the airplane.

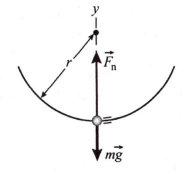

(*a*) Apply $\sum F_y = ma_y$ to the pilot:

$$F_n - mg = ma_c$$

Solve for F_n:

$$F_n = mg + ma_c = m(g + a_c)$$
$$= m(g + 8.5g) = 9.5mg$$

Substitute numerical values and evaluate F_n:

$$F_n = (9.5)(50\,\text{kg})(9.81\,\text{m/s}^2) = \boxed{4.66\,\text{kN}}$$

(*b*) Relate her acceleration to her velocity and the radius of the circular arc and solve for the radius:

$$a_c = \frac{v^2}{r} \Rightarrow r = \frac{v^2}{a_c}$$

Substitute numerical values and evaluate r:

$$r = \frac{[(345\,\text{km/h})(1\,\text{h}/3600\,\text{s})(1000\,\text{m/km})]^2}{(8.5)(9.81\,\text{m/s}^2)} = \boxed{110\,\text{m}}$$

*71 ••   A block of mass m_1 is attached to a cord of length L_1, which is fixed at one end. The block moves in a horizontal circle on a frictionless table. A second block of mass m_2 is attached to the first by a cord of length L_2 and also moves in a circle, as shown in Figure 5-56. If the period of the motion is T, find the tension in each cord.

Figure 5-56 Problem 71

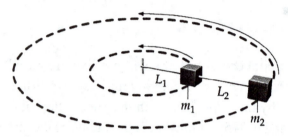

Picture the Problem The free-body diagrams show the forces acting on each block. We can use Newton's 2nd law to relate these forces to each other and to the masses and accelerations of the blocks.

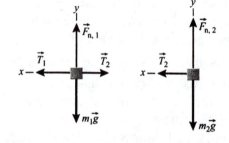

Apply $\sum F_x = ma_x$ to the block whose mass is m_1:

$$T_1 - T_2 = m_1 \frac{v_1^2}{L_1}$$

Apply $\sum F_x = ma_x$ to the block whose mass is m_2:

$$T_2 = m_2 \frac{v_2^2}{L_1 + L_2}$$

Relate the speeds of each block to their common period and their distance from the center of the circle:

$$v_1 = \frac{2\pi L_1}{T} \quad \text{and} \quad v_2 = \frac{2\pi(L_1 + L_2)}{T}$$

Solve the first force equation for T_2, substitute for v_2, and simplify

$$T_2 = \boxed{[m_2(L_1 + L_2)]\left(\frac{2\pi}{T}\right)^2}$$

to obtain:

Substitute for T_2 and v_1 in the first force equation to obtain:

$$T_1 = \left[m_2(L_1 + L_2) + m_1 L_1 \right] \left(\frac{2\pi}{T} \right)^2$$

*72 •• A particle moves with constant speed in a circle of radius 4 cm. It takes 8 s to complete each revolution. Draw the path of the particle to scale, and indicate the particle's position at 1-s intervals. Draw displacement vectors for each interval. These vectors also indicate the directions for the average-velocity vectors for each interval. Find graphically the change in the average velocity $|\vec{v}|$ for two consecutive 1-s intervals. Compare $|\vec{v}|/\Delta t$, measured in this way, with the instantaneous acceleration computed from $a = v^2/r$.

Picture the Problem The path of the particle and its position at 1-s intervals are shown. The displacement vectors are also shown. The velocity vectors for the average velocities in the first and second intervals are along $\vec{r}_{01}$ and $\vec{r}_{12}$, respectively, and are shown in the lower diagram. $\Delta\vec{v}$ points toward the center of the circle.

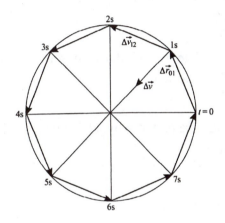

Use the diagram to the right to find Δr:

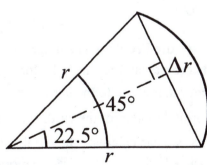

$\Delta r = 2r\sin 22.5° = 2(4\ cm)\sin 22.5°$
$\quad\ = 3.06\ cm$

Find the average velocity of the particle along the chords:

$$v_{av} = \frac{\Delta r}{\Delta t} = \frac{3.06\ cm}{1s} = 3.06\ cm/s$$

Using the lower diagram and the fact that the angle between $\vec{v}_1$ and $\vec{v}_2$ is 45°, express Δv in terms of v_1 (= v_2):	$\Delta v = 2v_1\sin 22.5°$
Evaluate Δv using v_{av} as v_1:	$\Delta v = 2(3.06 \text{ cm/s})\sin 22.5° = 2.34 \text{ cm/s}$
Now we can determine $a = \Delta v/\Delta t$:	$a = \dfrac{2.34 \text{ cm/s}}{1 \text{ s}} = \boxed{2.34 \text{ cm/s}^2}$
Find the speed v (= $v_1 = v_2$ ...) of the particle along its circular path:	$v = \dfrac{2\pi r}{T} = \dfrac{2\pi(4 \text{ cm})}{8 \text{ s}} = 3.14 \text{ cm/s}$
Calculate the radial acceleration of the particle:	$a_c = \dfrac{v^2}{r} = \dfrac{(3.14 \text{ cm/s})^2}{4 \text{ cm}} = \boxed{2.46 \text{ cm/s}^2}$
Compare a_c and a by taking their ratio:	$\dfrac{a_c}{a} = \dfrac{2.46 \text{ cm/s}^2}{2.34 \text{ cm/s}^2} = 1.05$ or $\boxed{a_c = 1.05a}$

*77 ••   An object on the equator has both an acceleration toward the center of the earth because of the earth's rotation, and an acceleration toward the sun because of the earth's motion along its orbit. Calculate the magnitudes of both accelerations, and express them as fractions of the free-fall acceleration due to gravity g.

Picture the Problem The diagram includes a pictorial representation of the earth in its orbit about the sun and a force diagram showing the force on an object at the equator that is due to the earth's rotation, $\vec{F}_R$, and the force on the object due to the orbital motion of the earth about the sun, $\vec{F}_o$. Because these are centripetal forces, we can calculate the accelerations they require from the speeds and radii associated with the two circular motions.

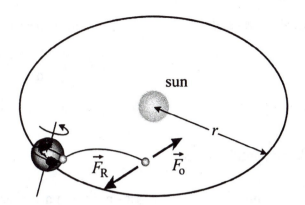

Express the radial acceleration due to the rotation of the earth:

$$a_R = \frac{v_R^2}{R}$$

Express the speed of the object on the equator in terms of the radius of the earth R and the period of the earth's rotation T_R:

$$v_R = \frac{2\pi R}{T_R}$$

Substitute for v_R in the expression for a_R to obtain:

$$a_R = \frac{4\pi^2 R}{T_R^2} = \frac{4\pi^2 (6370\,\text{km})(1000\,\text{m/km})}{\left[(24\,\text{h})\left(\dfrac{3600\,\text{s}}{1\,\text{h}}\right)\right]^2}$$

$$= 3.37\times10^{-2}\,\text{m/s}^2$$

$$= \boxed{3.44\times10^{-3}\,g}$$

Express the radial acceleration due to the orbital motion of the earth:

$$a_o = \frac{v_o^2}{r}$$

Express the speed of the object on the equator in terms of the earth-sun distance r and the period of the earth's motion about the sun T_o:

$$v_o = \frac{2\pi r}{T_o}$$

Substitute for v_o in the expression for a_o to obtain:

$$a_o = \frac{4\pi^2 r}{T_o^2}$$

Substitute numerical values and evaluate a_c:

$$a_o = \frac{4\pi^2\left(1.5\times10^{11}\,\text{m}\right)}{\left[\left(365\,\text{d}\right)\left(\frac{24\,\text{h}}{1\,\text{d}}\right)\left(\frac{3600\,\text{s}}{1\,\text{h}}\right)\right]^2}$$

$$= 5.95\times10^{-3}\,\text{m/s}^2 = \boxed{6.07\times10^{-4}\,g}$$

Concepts of Centripetal Force

*82 •   A ride at an amusement park carries people in a vertical circle at constant speed such that the normal forces exerted by the seats are always inward–toward the center of the circle. At the top, the normal force exerted by a seat equals the person's weight, mg. At the bottom of the loop, the force exerted by the seat will be (a) 0, (b) mg, (c) $2mg$, (d) $3mg$, (e) greater than mg, but it cannot be calculated from the information given.

Picture the Problem The diagram depicts a seat at its highest and lowest points. Let "t" denote the top of the loop and "b" the bottom of the loop. Applying Newton's 2nd law to the seat at the top of the loop will establish the value of mv^2/r; this can then be used at the bottom of the loop to determine $F_{n,b}$.

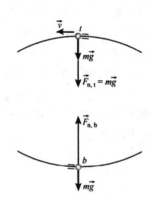

Apply $\sum F_r = ma_r$ to the seat at the top of the loop:

$$mg + F_{n,t} = 2mg = ma_r = mv^2/r$$

Apply $\sum F_r = ma_r$ to the seat at the bottom of the loop:

$$F_{n,b} - mg = mv^2/r$$

Solve for $F_{n,b}$ and substitute for mv^2/r to obtain:

$F_{n,b} = 3mg$ and $\boxed{(d)\text{ is correct.}}$

*85 •••   Suppose you ride a bicycle on a horizontal surface in a circle with a radius of 20 m. The resultant force exerted by the road on the bicycle (normal force plus frictional force) makes an angle of 15° with the vertical. (a) What is your speed? (b) If the frictional force is half its maximum value, what is the coefficient of static friction?

Picture the Problem The forces acting on the bicycle are shown in the force diagram. The static friction force is the centripetal force exerted by the surface on the bicycle that allows it to move in a circular path. $\vec{F}_n + \vec{f}_s$ makes an angle θ with the vertical direction. The application of Newton's 2nd law will allow us to relate this angle to the speed of the bicycle and the coefficient of static friction.

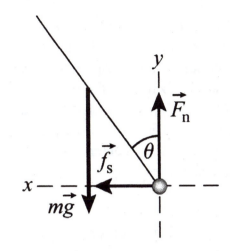

(a) Apply $\sum \vec{F} = m\vec{a}$ to the bicycle:

$$\sum F_x = f_s = \frac{mv^2}{r}$$

and

$$\sum F_y = F_n - mg = 0$$

Relate F_n and f_s to θ:

$$\tan\theta = \frac{f_s}{F_n} = \frac{\dfrac{mv^2}{r}}{mg} = \frac{v^2}{rg}$$

Solve for v:

$$v = \sqrt{rg\tan\theta}$$

Substitute numerical values and evaluate v:

$$v = \sqrt{(20\,\text{m})(9.81\,\text{m/s}^2)\tan 15°}$$

$$= \boxed{7.25\,\text{m/s}}$$

(b) Relate f_s to μ_s and F_n:

$$f_s = \tfrac{1}{2} f_{s,\text{max}} = \tfrac{1}{2}\mu_s mg$$

Solve for μ_s and substitute for f_s to obtain:

$$\mu_s = \frac{2f_s}{mg} = \frac{2v^2}{rg}$$

Substitute numerical values and evaluate μ_s

$$\mu_s = \frac{2(7.25\,\text{m/s})^2}{(20\,\text{m})(9.81\,\text{m/s}^2)} = \boxed{0.536}$$

*88 ••   A curve of radius 150 m is banked at an angle of 10°. An 800-kg car negotiates the curve at 85 km/h without skidding. Find (a) the normal force on the tires exerted by the pavement, (b) the frictional force exerted by the pavement on the tires, (c) the minimum coefficient of static friction between the pavement and tires.

Picture the Problem Both the normal force and the static friction force contribute to the centripetal force in the situation described in this problem. We can apply Newton's 2nd law to relate f_s and F_n and then solve these equations simultaneously to determine each of these quantities.

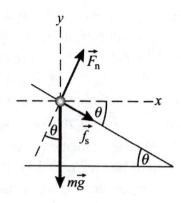

(a) Apply $\sum \vec{F} = m\vec{a}$ to the car:

$$\sum F_x = F_n \sin\theta + f_s \cos\theta = m\frac{v^2}{r}$$

and

$$\sum F_y = F_n \cos\theta - f_s \sin\theta - mg = 0$$

Multiply the x equation by $\sin\theta$ and the y equation by $\cos\theta$ to obtain:

$$f_s \sin\theta\cos\theta + F_n \sin^2\theta = m\frac{v^2}{r}\sin\theta$$

and

$$F_n \cos^2\theta - f_s \sin\theta\cos\theta - mg\cos\theta = 0$$

Add these equations to eliminate f_s:

$$F_n - mg\cos\theta = m\frac{v^2}{r}\sin\theta$$

Solve for F_n:

$$F_n = mg\cos\theta + m\frac{v^2}{r}\sin\theta$$

$$= m\left(g\cos\theta + \frac{v^2}{r}\sin\theta \right)$$

Substitute numerical values and evaluate F_n:

$$F_n = (800\,\text{kg})\left[(9.81\,\text{m/s}^2)\cos10° + \frac{(85\,\text{km/h})^2(1000\,\text{m/km})^2(1\,\text{h/3600 s})^2}{150\,\text{m}}\sin10° \right]$$

$$= \boxed{8.25\,\text{kN}}$$

(b) Solve the y equation for f_s:

$$f_s = \frac{F_n \cos\theta - mg}{\sin\theta}$$

Substitute numerical values and evaluate f_s:

$$f_s = \frac{(8.25\,\text{kN})\cos 10° - (800\,\text{kg})(9.81\,\text{m/s}^2)}{\sin 10°} = \boxed{1.59\,\text{kN}}$$

(c) Express $\mu_{s,\min}$ in terms of f_s and F_n:

$$\mu_{s,\min} = \frac{f_s}{F_n}$$

Substitute numerical values and evaluate $\mu_{s,\min}$:

$$\mu_{s,\min} = \frac{1.59\,\text{kN}}{8.25\,\text{kN}} = \boxed{0.193}$$

*90 •••   A civil engineer is asked to design a curved section of roadway that meets the following conditions: With ice on the road, when the coefficient of static friction between the road and rubber is 0.08, a car at rest must not slide into the ditch and a car traveling less than 60 km/h must not skid to the outside of the curve. What is the minimum radius of curvature of the curve and at what angle should the road be banked?

Picture the Problem The free-body diagram to the left is for the car at rest. The static friction force up the incline balances the downward component of the car's weight and prevents it from sliding. In the free-body diagram to the right, the static friction force points in the opposite direction as the tendency of the moving car is to slide toward the outside of the curve.

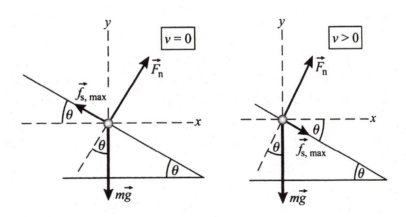

Apply $\sum \vec{F} = m\vec{a}$ to the car that is at rest:

$$\sum F_y = F_n \cos\theta + f_s \sin\theta - mg = 0 \quad (1)$$

and

$$\sum F_x = F_n \sin\theta - f_s \cos\theta = 0 \quad (2)$$

Substitute $f_s = f_{s,\max} = \mu_s F_n$ in equation (2) and solve for and evaluate the maximum allowable value of θ:

$$\theta = \tan^{-1}\mu_s = \tan^{-1}0.08 = \boxed{4.57°}$$

Apply $\sum \vec{F} = m\vec{a}$ to the car that is moving with speed v:

$$\sum F_y = F_n \cos\theta - f_s \sin\theta - mg = 0 \quad (3)$$

$$\sum F_x = F_n \sin\theta + f_s \cos\theta = m\frac{v^2}{r} \quad (4)$$

Substitute $f_s = \mu_s F_n$ in equations (3) and (4) and simplify to obtain:

$$F_n(\cos\theta - \mu_s \sin\theta) = mg \quad (5)$$

$$F_n(\mu_s \cos\theta + \sin\theta) = m\frac{v^2}{r} \quad (6)$$

Substitute numerical values into (5) and (6) to obtain:

$0.9904 F_n = mg$

and

$$0.1595 F_n = m\frac{v^2}{r}$$

Eliminate F_n and solve for r:

$$r = \frac{v^2}{0.1610g}$$

Substitute numerical values and evaluate r:

$$r = \frac{(60\,\text{km/h} \times 1\,\text{h}/3600\,\text{s} \times 1000\,\text{m/km})^2}{0.1610(9.81\,\text{m/s}^2)}$$

$$= \boxed{176\,\text{m}}$$

Drag Forces

*94 •   A sky diver of mass 60 kg can slow herself to a constant speed of 90 km/h by adjusting her form. (*a*) What is the magnitude of the upward drag force on the sky diver? (*b*) If the drag force is equal to bv^2, what is the value of b?

Picture the Problem Let the upward direction be the positive y direction and apply Newton's 2nd law to the sky diver.

(*a*) Apply $\sum F_y = ma_y$ to the sky diver:

$F_d - mg = ma_y$

or, because $a_y = 0$,

$F_d = mg \quad (1)$

Substitute numerical values and evaluate F_d:

$F_d = (60\,\text{kg})(9.81\,\text{m/s}^2) = \boxed{589\,\text{N}}$

(*b*) Substitute $F_d = bv_t^2$ in equation (1) to obtain:

$bv_t^2 = mg$

Solve for b:

$$b = \frac{mg}{v_t^2} = \frac{F_d}{v_t^2}$$

Substitute numerical values and evaluate b:

$$b = \frac{589\,\text{N}}{(25\,\text{m/s})^2} = \boxed{0.942\,\text{kg/m}}$$

*97 •••   A sample of air containing pollution particles of the size and density given in Problem 96 is captured in a test tube 8.0 cm long. The test tube is then placed in a centrifuge with the midpoint of the test tube 12 cm from the center of the centrifuge. The centrifuge spins at 800 revolutions per minute. Estimate the time required for nearly all of the pollution particles to settle at the end of the test tube and compare this to the time required for a pollution particle to fall 8.0 cm under the action of gravity and subject to the viscous drag of air.

Picture the Problem The motion of the centrifuge will cause the pollution particles to migrate to the end of the test tube. We can apply Newton's 2nd law and Stokes' law to derive an expression for the terminal speed of the sedimentation particles. We can then use this terminal speed to calculate the sedimentation time. We'll use the 12 cm distance from the center of the centrifuge as the average radius of the pollution particles as they settle in the test tube. Let R represent the radius of a particle and r the radius of the particle's circular path in the centrifuge.

Express the sedimentation time in terms of the sedimentation speed v_t:

$$\Delta t_{\text{sediment}} = \frac{\Delta x}{v_t}$$

Apply $\sum F_{\text{radial}} = ma_{\text{radial}}$ to a pollution particle:

$$6\pi\eta R v_t = ma_c$$

Express the mass of the particle in terms of its radius R and density ρ:

$$m = \rho V = \tfrac{4}{3}\pi R^3 \rho$$

Express the acceleration of the pollution particles due to the motion of the centrifuge in terms of their orbital radius r and period T:

$$a_c = \frac{v^2}{r} = \frac{\left(\dfrac{2\pi r}{T}\right)^2}{r} = \frac{4\pi^2 r}{T^2}$$

Substitute for m and a_c and simplify to obtain:

$$6\pi\eta R v_t = \tfrac{4}{3}\pi R^3 \rho\left(\frac{4\pi^2 r}{T^2}\right) = \frac{16\pi^3 \rho r R^3}{3T^2}$$

Solve for v_t:

$$v_t = \frac{8\pi^2 \rho r R^2}{9\eta T^2}$$

Find the period T of the motion from the number of revolutions the centrifuge makes in 1 second:

$$T = \frac{1}{800\,\text{rev/min}} = 1.25 \times 10^{-3}\,\text{min/rev}$$

$$= 1.25 \times 10^{-3}\,\text{min/rev} \times 60\,\text{s/min}$$

$$= 75.0 \times 10^{-3}\,\text{s/rev}$$

Substitute numerical values and evaluate v_t:

$$v_t = \frac{8\pi^2 (2000\,\text{kg/m}^3)(0.12\,\text{m})(10^{-5}\,\text{m})^2}{9(1.8 \times 10^{-5}\,\text{N} \cdot \text{s/m}^2)(75 \times 10^{-3}\,\text{s})^2}$$

$$= 2.08\,\text{m/s}$$

Find the time it takes the particles to move 8 cm as they settle in the test tube:

$$\Delta t_{\text{sediment}} = \frac{\Delta x}{v} = \frac{8\,\text{cm}}{208\,\text{cm/s}}$$

$$= \boxed{38.5\,\text{ms}}$$

In Problem 96 it was shown that the rate of fall of the particles in air is 2.42 cm/s. Find the time required to fall 8 cm in air under the influence of gravity:

$$\Delta t_{\text{air}} = \frac{\Delta x}{v} = \frac{8\,\text{cm}}{2.42\,\text{cm/s}}$$

$$= \boxed{3.31\,\text{s}}$$

Find the ratio of the two times:

$$\Delta t_{\text{air}}/\Delta t_{\text{sediment}} \approx \boxed{100}$$

Euler's Method

*99 ••   You throw a baseball straight up with an initial speed of 150 km/h. Its terminal speed when falling is also 150 km/h. Use Euler's method to estimate its height 3.5 s after release. What is the maximum height it reaches? How long after release does it reach its maximum height? How much later does it return to the ground? Is the time the ball spends while on the way up less than, the same as, or greater than the time it spends on the way down?

Picture the Problem The free-body diagram shows the forces acting on the baseball after it has left your hand. In order to use Euler's method we'll need to determine how the acceleration of the ball varies with its speed. We can do this by applying Newton's 2nd law to the baseball. We can then use $v_{n+1} = v_n + a_n \Delta t$ and $x_{n+1} = x_n + v_n \Delta t$ to find the speed and position of the ball.

Apply $\sum F_y = ma_y$ to the baseball:

$$-bv|v| - mg = m\frac{dv}{dt}$$

where $|v| = v$ for the upward part of the flight of the ball and $|v| = -v$ for the downward part of the flight.

Solve for dv/dt:

$$\frac{dv}{dt} = -g - \frac{b}{m}v|v|$$

Under terminal speed conditions ($|v| = -v_t$):

$$0 = -g + \frac{b}{m}v_t^2$$

and

$$\frac{b}{m} = \frac{g}{v_t^2}$$

Substitute to obtain:

$$\frac{dv}{dt} = -g - \frac{g}{v_t^2}v|v| = -g\left(1 + \frac{v|v|}{v_t^2}\right)$$

Letting a_n be the acceleration of the ball at time t_n, express its position and speed when $t = t_n + 1$:

$$y_{n+1} = y_n + \tfrac{1}{2}(v_n + v_{n-1})\Delta t$$
and
$$v_{n+1} = v_n + a_n\Delta t$$
where

$$a_n = -g\left(1 + \frac{v_n|v_n|}{v_t^2}\right)$$

and Δt is an arbitrarily small interval of time.

A spreadsheet solution follows. The formulas used to calculate the quantities in the columns are as follows:

Cell	Formula/Content	Algebraic Form
D11	D10+$B$6	$t + \Delta t$
E10	41.7	v_0
E11	E10–$B$4* (1+E10*ABS(E10)/($B$5^2))*$B$6	$v_{n+1} = v_n + a_n \Delta t$
F10	0	y_0
F11	F10+0.5*(E10+E11)*$B$6	$y_{n+1} = y_n + \frac{1}{2}\left(v_n + v_{n-1}\right)\Delta t$
G10	0	y_0
G11	$E$10*D11–0.5*$B$4*D11^2	$v_0 t - \frac{1}{2} g t^2$

	A	B	C	D	E	F	G
4	g=	9.81	m/s^2				
5	vt=	41.7	m/s				
6	Δt=	0.1	s				
7							
8							
9				t	v	y	y no drag
10				0.0	41.70	0.00	0.00
11				0.1	39.74	4.07	4.12
12				0.2	37.87	7.95	8.14
40				3.0	3.01	60.13	81.00
41				3.1	2.03	60.39	82.18
42				3.2	1.05	60.54	83.26
43				3.3	0.07	60.60	84.25
44				3.4	−0.91	60.55	85.14
45				3.5	−1.89	60.41	85.93
46				3.6	−2.87	60.17	86.62
78				6.8	−28.34	6.26	56.98
79				6.9	−28.86	3.41	54.44
80				7.0	−29.37	0.49	51.80
81				7.1	−29.87	−2.47	49.06

From the table we can see that, after 3.5 s, the ball reaches a height of about 60.4 m. It reaches its peak a little earlier-at about 3.3 s, and its height at $t = 3.3$ s is 60.6 m. The ball hits the ground at about $t =$ 7 s —so it spends a little longer coming down than going up.

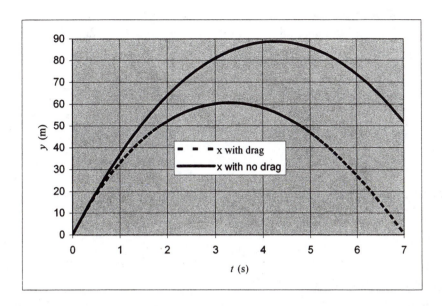

General Problems

*103   ••   An 800-N box rests on an incline making a 30° angle with the horizontal. A physics student finds that she can prevent the box from sliding if she pushes with a force of at least 200 N parallel to the incline. (*a*) What is the coefficient of static friction between the box and the incline? (*b*) What is the greatest force that can be applied to the box parallel to the incline before the box slides up the incline?

Picture the Problem The free-body diagram shows the forces acting on the box. If the student is pushing with a force of 200 N and the box is on the verge of moving, the static friction force must be at its maximum value. In part (*b*), the motion is impending up the incline; the direction of $f_{s,max}$ is down the incline.

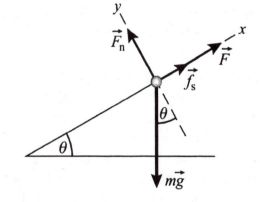

(*a*) Apply $\sum \vec{F} = m\vec{a}$ to the box:

$$\sum F_x = f_s + F - mg\sin\theta = 0$$
and
$$\sum F_y = F_n - mg\cos\theta = 0$$

Substitute $f_s = f_{s,max} = \mu_s F_n$, eliminate F_n between the two

$$\mu_s = \tan\theta - \frac{F}{mg\cos\theta}$$

equations, and solve for μ_s:

Substitute numerical values and evaluate μ_s:

$$\mu_s = \tan 30° - \frac{200\,N}{(800\,N)\cos 30°}$$

$$= \boxed{0.289}$$

(b) Find $f_{s,max}$ from the x-direction force equation:

$$f_{s,max} = mg\sin\theta - F$$

Substitute numerical values and evaluate $f_{s,max}$:

$$f_{s,max} = (800\,N)\sin 30° - 200\,N$$
$$= 200\,N$$

If the block is on the verge of sliding up the incline, $f_{s,max}$ must act down the incline. The x-direction equation becomes:

$$-f_{s,max} + F - mg\sin\theta = 0$$

Solve the x-direction force equation for F:

$$F = mg\sin\theta + f_{s,max}$$

Substitute numerical values and evaluate F:

$$F = (800\,N)\sin 30° + 200\,N = \boxed{600\,N}$$

*107 ••   A brick slides down an inclined plank at constant speed when the plank is inclined at an angle θ_0. If the angle is increased to θ_1, the brick accelerates down the plank with acceleration a. The coefficient of kinetic friction is the same in both cases. Given θ_0 and θ_1, calculate a.

Picture the Problem The free-body diagram shows the forces acting on the brick as it slides down the inclined plane. We'll apply Newton's 2nd law to the brick when it is sliding down the incline with constant speed to derive an expression for μ_k in terms of θ_0. We'll apply Newton's 2nd law a second time for $\theta = \theta_1$ and solve the equations simultaneously to obtain an expression for a as a function of θ_0 and θ_1.

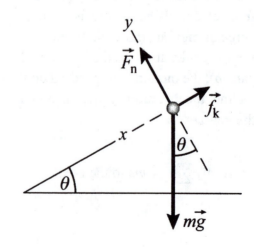

Apply $\sum\vec{F} = m\vec{a}$ to the brick when it is sliding with constant

$$\sum F_x = -f_k + mg\sin\theta_0 = 0$$

and

speed:

$$\sum F_y = F_n - mg\cos\theta_0 = 0$$

Solve the y equation for F_n and using $f_k = \mu_k F_n$, eliminate both F_n and f_k from the x equation and solve for μ_k:

$$\mu_k = \tan\theta_0$$

Apply $\sum \vec{F} = m\vec{a}$ to the brick when $\theta = \theta_1$:

$$\sum F_x = -f_k + mg\sin\theta_1 = ma$$

and

$$\sum F_y = F_n - mg\cos\theta_1 = 0$$

Solve the y equation for F_n, use $f_k = \mu_k F_n$ to eliminate both F_n and f_k from the x equation, and use the expression for μ_k obtained above to obtain:

$$\boxed{a = g(\sin\theta_1 - \tan\theta_0 \cos\theta_1)}$$

*110 •• A flat-topped toy cart moves on frictionless wheels, pulled by a rope under tension T. The mass of the cart is m_1. A load of mass m_2 rests on top of the cart with a coefficient of static friction μ_s. The cart is pulled up a ramp that is inclined at an angle θ above the horizontal. The rope is parallel to the ramp. What is the maximum tension T that can be applied without making the load slip?

Picture the Problem The pictorial representation to the right shows the cart and its load on the inclined plane. The load will not slip provided its maximum acceleration is not exceeded. We can find that maximum acceleration by applying Newton's 2nd law to the load. We can then apply Newton's 2nd law to the cart-plus-load system to determine the tension in the rope when the system is experiencing its maximum acceleration.

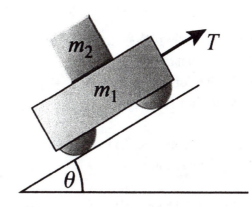

Draw the free-body diagram for
the cart and its load:

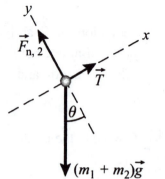

Apply $\sum F_x = ma_x$ to the cart
plus its load:

$$T - (m_1 + m_2)g \sin\theta = (m_1 + m_2)a_{max} \quad (1)$$

Draw the free-body diagram for
the load of mass m_2 on top of the
cart:

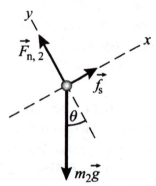

Apply $\sum \vec{F} = m\vec{a}$ to the load on
top of the cart:

$$\sum F_x = f_{s,max} - m_2 g \sin\theta = m_2 a_{max}$$
and
$$\sum F_y = F_{n,2} - m_2 g \cos\theta = 0$$

Using $f_{s,max} = \mu_s F_{n,2}$, eliminate
$F_{n,2}$ between the two equations
and solve for the maximum
acceleration of the load:

$$a_{max} = g(\mu_s \cos\theta - \sin\theta) \quad (2)$$

Substitute equation (2) in equation
(1) and solve for T :

$$T = \boxed{(m_1 + m_2)g\mu_s \cos\theta}$$

*114 •• The position of a particle of mass $m = 0.8$ kg as a function of time is
$$\vec{r} = x \,\hat{i} + y \,\hat{j} = R \sin\omega t \,\hat{i} + R \cos\omega t \,\hat{j},$$
where $R = 4.0$ m, and $\omega = 2\pi \mathrm{s}^{-1}$. (a) Show that this path of the particle is a circle
of radius R with its center at the origin. (b) Compute the velocity vector. Show
that $v_x/v_y = -y/x$. (c) Compute the acceleration vector and show that it is in the
radial direction and has the magnitude v^2/r. (d) Find the magnitude and direction
of the net force acting on the particle.

Picture the Problem The path of the particle is a circle if r is a constant. Once we have shown that it is, we can calculate its value from its components and determine the particle's velocity and acceleration by differentiation. The direction of the net force acting on the particle can be determined from the direction of its acceleration.

(a) Express the magnitude of $\vec{r}$ in terms of its components:

$$r = \sqrt{r_x^2 + r_y^2}$$

Evaluate r with $r_x = R\sin\omega t$ and $r_y = R\cos\omega t$:

$$r = \sqrt{[R\sin\omega t]^2 + [R\cos\omega t]^2}$$
$$= \sqrt{R^2(\sin^2\omega t + \cos^2\omega t)} = R = 4.0\,\mathrm{m}$$

$\therefore$ the path of the particle is a circle centered at the origin.

(b) Differentiate $\vec{r}$ with respect to time to obtain $\vec{v}$:

$$\vec{v} = d\vec{r}/dt = [R\omega\cos\omega t]\hat{i}$$
$$+ [-R\omega\sin\omega t]\hat{j}$$

$$= \begin{array}{l} [(8\pi\cos 2\pi t)\mathrm{m/s}]\hat{i} \\ -[(8\pi\sin 2\pi t)\mathrm{m/s}]\hat{j} \end{array}$$

Express the ratio $\dfrac{v_x}{v_y}$:

$$\frac{v_x}{v_y} = \frac{8\pi\cos\omega t}{-8\pi\sin\omega t} = -\cot\omega t$$

Express the ratio $-\dfrac{y}{x}$:

$$-\frac{y}{x} = -\frac{R\cos\omega t}{R\sin\omega t} = -\cot\omega t$$

$$\therefore \boxed{\frac{v_x}{v_y} = -\frac{y}{x}}$$

(c) Differentiate $\vec{v}$ with respect to time to obtain $\vec{a}$:

$$\vec{a} = d\vec{v}/dt$$

$$= \begin{array}{l} [(-16\pi^2\mathrm{m/s}^2)\sin\omega t]\hat{i} \\ +[(-16\pi^2\mathrm{m/s}^2)\cos\omega t]\hat{j} \end{array}$$

Factor $-4\pi^2/\mathrm{s}^2$ from $\vec{a}$ to obtain:

$$\vec{a} = (-4\pi^2/\mathrm{s}^2)(4\sin\omega t)\hat{i} + (4\cos\omega t)\hat{j}$$
$$= \boxed{(-4\pi^2/\mathrm{s}^2)\vec{r}}$$

Because $\vec{a}$ is in the opposite direction from $\vec{r}$, it is directed toward the center of the circle in which the particle is traveling.

Find the ratio $\dfrac{v^2}{r}$:

$$\dfrac{v^2}{r} = \dfrac{(8\pi \text{ m/s})^2}{4\text{ m}} = \boxed{16\pi^2 \text{ m/s}^2 = a}$$

(d) Apply $\sum \vec{F} = m\vec{a}$ to the particle:

$$F_{\text{net}} = ma = (0.8\,\text{kg})(16\pi^2 \text{ m/s}^2)$$
$$= \boxed{12.8\pi^2 \text{ N}}$$

Because the direction of $\vec{F}_{\text{net}}$ is the same as that of $\vec{a}$:

$\boxed{\vec{F}_{\text{net}} \text{ is toward the center of the circle.}}$

*117   ••• (a) Show that a point on the surface of the earth at latitude θ, shown in Figure 5-63, has an acceleration relative to a reference frame not rotating with the earth with a magnitude of $(3.37 \text{ cm/s}^2) \cos\theta$. What is the direction of this acceleration? (b) Discuss the effect of this acceleration on the apparent weight of an object near the surface of the earth. (c) The free-fall acceleration of an object at sea level measured *relative to the earth's surface* is 9.78 m/s^2 at the equator and 9.81 m/s^2 at latitude $\theta = 45°$. What are the values of the gravitational field g at these points?

Figure 5-63 Problem 117

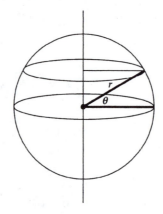

Picture the Problem The diagram shows a point on the surface of the earth at latitude θ. The distance R to the axis of rotation is given by $R = r\cos\theta$. We can use the definition of centripetal acceleration to express the centripetal acceleration of a point on the surface of the earth due to the rotation of the earth.

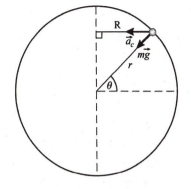

(*a*) Referring to the figure, express a_c for a point on the surface of the earth at latitude θ :

$$a_c = \frac{v^2}{R}\text{ where }R = r\cos\theta$$

Express the speed of the point due to the rotation of the earth:

$$v = \frac{2\pi R}{T}$$

where T is the time for one revolution.

Substitute for v in the expression for a_c and simplify to obtain:

$$a_c = \frac{4\pi^2 r\cos\theta}{T^2}$$

Substitute numerical values and evaluate a_c:

$$a_c = \frac{4\pi^2 (6370\,\text{km})\cos\theta}{[(24\,\text{h})(3600\,\text{s/h})]^2}$$

$$= \boxed{(3.37\,\text{cm/s}^2)\cos\theta,\text{ toward the earth's axis.}}$$

(*b*)

A stone dropped from a hand at a location on earth. The effective weight of the stone is equal to $m\vec{a}_{\text{st, surf}}$, where $\vec{a}_{\text{st, surf}}$ is the acceleration of the falling stone (neglecting air resistance) relative to the local surface of the earth. The gravitational force on the stone is equal to $m\vec{a}_{\text{st,iner}}$, where $\vec{a}_{\text{st, iner}}$ is the acceleration of the local surface of the earth relative to the inertial frame (the acceleration of the surface due to the rotation of the earth). Multiplying through this equation by m and rearranging gives $m\vec{a}_{\text{st, surf}} = m\vec{a}_{\text{st, iner}} - m\vec{a}_{\text{surf, iner}}$, which relates the apparent weight to the acceleration due to gravity and the acceleration due to the earth's rotation. A vector addition diagram can be used to show that the magnitude of $m\vec{a}_{\text{st, surf}}$ is slightly less than that of $m\vec{a}_{\text{st, iner}}$.

(*c*) At the equator, the gravitational acceleration and the radial acceleration are both directed toward the center of the earth. Therefore:

$$g = g_{eff} + a_c$$
$$= 978\,\text{cm/s}^2 + (3.37\,\text{cm/s}^2)\cos 0°$$
$$= \boxed{981.4\,\text{cm/s}^2}$$

At latitude θ the gravitational acceleration points toward the center of the earth whereas the centripetal acceleration points toward the axis of rotation. Use the law of cosines to relate g_{eff}, g, and a_c:

$$g_{eff}^2 = g^2 + a_c^2 - 2ga_c\cos\theta$$

Substitute for θ, g_{eff}, and a_c and simplify to obtain the quadratic equation:

$$g^2 - (4.75\,\text{cm/s}^2)g - 962350\,\text{cm}^2/\text{s}^4 = 0$$

Solve for the physically meaningful (i.e., positive) root:

$$g = \boxed{983\,\text{cm/s}^2}$$

*118 ••• A small block of mass 0.01 kg is at rest atop a smooth (frictionless) sphere of radius 0.8 m. The block is given a tiny nudge and starts to slide down the sphere. The block loses contact with the sphere when the angle between the vertical and the line from the center of the sphere to the position of the block is θ. Find the angle θ.

Picture the Problem The diagram shows the block in its initial position, an intermediate position and as it is separating from the sphere. Because the sphere is frictionless, the only forces acting on the block are the normal and gravitational forces. We'll apply Newton's 2nd law and set F_n equal to zero to determine the angle θ_c at which the block leaves the surface.

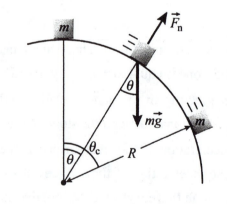

Taking the inward direction to be positive, apply $\sum F_r = ma_r$ to the block:

$$mg\cos\theta - F_n = m\frac{v^2}{R}$$

Apply the separation condition to obtain:	$$mg\cos\theta_c = m\frac{v^2}{R}$$
Solve for $\cos\theta_c$:	$$\cos\theta_c = \frac{v^2}{gR} \qquad (1)$$
Apply $\sum F_t = ma_t$ to the block:	$$mg\sin\theta = ma_t$$ or $$a_t = \frac{dv}{dt} = g\sin\theta$$ Note that a is not constant and, hence, we cannot use constant acceleration equations.
Multiply the left hand side of the equation by one in the form of $d\theta/d\theta$ and rearrange to obtain:	$$\frac{dv}{dt}\frac{d\theta}{d\theta} = g\sin\theta$$ and $$\frac{d\theta}{dt}\frac{dv}{d\theta} = g\sin\theta$$
Relate the arc distance s the block travels to the angle θ and the radius R of the sphere:	$$\theta = \frac{s}{R} \text{ and } \frac{d\theta}{dt} = \frac{1}{R}\frac{ds}{dt} = \frac{v}{R}$$ where v is the block's instantaneous speed.
Substitute to obtain:	$$\frac{v}{R}\frac{dv}{d\theta} = g\sin\theta$$
Separate the variables and integrate from $v' = 0$ to v and $\theta = 0$ to θ_c:	$$\int_0^v v'dv' = gR\int_0^{\theta_c}\sin\theta d\theta$$ or $$v^2 = 2gR(1-\cos\theta_c)$$
Substitute in equation (1) to obtain:	$$\cos\theta_c = \frac{2gR(1-\cos\theta_c)}{gR}$$ $$= 2(1-\cos\theta_c)$$
Solve for and evaluate θ_c:	$$\theta_c = \cos^{-1}\left(\frac{2}{3}\right) = \boxed{48.2°}$$

Chapter 6
Work and Energy

Conceptual Problems

*1 • True or false:

(*a*) Only the net force acting on an object can do work.
(*b*) No work is done on a particle that remains at rest.
(*c*) A force that is always perpendicular to the velocity of a particle never does work on the particle.

Determine the Concept A force does work on an object when its point of application moves through some distance and there is a component of the force along the line of motion.

(*a*) False. The *net* force acting on an object is the vector sum of all the forces acting on the object and is responsible for displacing the object. Any or all of the forces contributing to the net force may do work.

(*b*) True. The object could be at rest in one reference frame and moving in another. If we consider only the frame in which the object is at rest, then, because it must undergo a displacement in order for work to be done on it, we would conclude that the statement is true.

(*c*) True. A force that is always perpendicular to the velocity of a particle changes neither it's kinetic nor potential energy and, hence, does no work on the particle.

*4 • By what factor does the kinetic energy of a car change when its speed is doubled?

Determine the Concept The kinetic energy of any object is proportional to the square of its speed. Because $K \equiv \frac{1}{2}mv^2$, replacing v by $2v$ yields $K' = \frac{1}{2}m(2v)^2 = 4\left(\frac{1}{2}mv^2\right) = 4K$. Thus doubling the speed of a car quadruples its kinetic energy.

*7 • How does the work required to stretch a spring 2 cm from its natural length compare with that required to stretch it 1 cm from its natural length?

Determine the Concept The work required to stretch or compress a spring a distance x is given by $W = \frac{1}{2}kx^2$ where k is the spring's stiffness constant. Because $W \propto x^2$, doubling the distance the spring is stretched will require four times as much work.

*12 ••   Figure 6-30 shows a plot of a potential-energy function U versus x.
(*a*) At each point indicated, state whether the force F_x is positive, negative, or
zero. (*b*) At which point does the force have the greatest magnitude? (*c*) Identify
any equilibrium points, and state whether the equilibrium is stable, unstable, or
neutral.

Figure 6-30 Problem 12

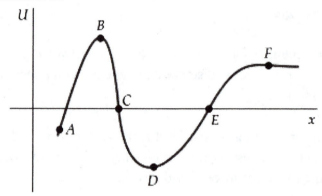

Picture the Problem F_x is defined to be the negative of the derivative of the
potential function with respect to x; i.e., $F_x = -dU/dx$.

(*a*) Examine the slopes of the
curve at each of the lettered
points, remembering that F_x is
the negative of the slope of the
potential energy graph, to
complete the table:

Point	dU/dx	F_x
A	+	−
B	0	0
C	−	+
D	0	0
E	+	−
F	0	0

(*b*) Find the point where the
slope is steepest:

At point C, $|F_x|$ is greatest.

(*c*) If $d^2U/dx^2 < 0$, then the
curve is concave downward and
the equilibrium is *unstable*.

At point B, the equilibrium is unstable.

If $d^2U/dx^2 > 0$, then the curve is
concave upward and the
equilibrium is *stable*.

At point D, the equilibrium is stable.

Remarks: At point F, $d^2U/dx^2 = 0$ and the equilibrium is neither *stable* nor *unstable*; it is said to be *neutral*.

Estimation and Approximation

*15 ••   A tightrope walker whose mass is 50 kg walks across a tightrope held between two supports 10 m apart; the tension in the rope is 5000 N. The height of the rope is 10 m above the circus floor. Estimate (*a*) the sag in the tightrope when she stands in the exact center and (*b*) the change in her gravitational potential energy just before stepping onto the tightrope to when she stands at its dead center.

Picture the Problem The diagram depicts the situation when the tightrope walker is at the center of rope. M represents her mass and the vertical components of tensions $\vec{T}_1$ and $\vec{T}_2$, equal in magnitude, support her weight. We can apply a condition for static equilibrium in the vertical direction to relate the tension in the rope to the angle θ and use trigonometry to find s as a function of θ.

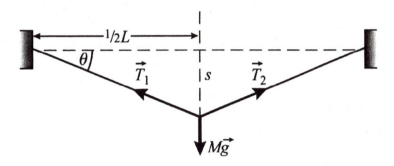

(*a*) Use trigonometry to relate the sag s in the rope to its length L and θ:	$\tan\theta = \dfrac{s}{\frac{1}{2}L}$ and $s = \dfrac{L}{2}\tan\theta$
Apply $\sum F_y = 0$ to the tightrope walker when she is at the center of the rope to obtain:	$2T\sin\theta - Mg = 0$ where T is the magnitude of $\vec{T}_1$ and $\vec{T}_2$.
Solve for θ to obtain:	$\theta = \sin^{-1}\dfrac{Mg}{2T}$
Substitute numerical values and evaluate θ:	$\theta = \sin^{-1}\left[\dfrac{(50\,\text{kg})(9.81\,\text{m/s}^2)}{2(5000\,\text{N})}\right] = 2.81°$
Substitute to obtain:	$s = \dfrac{10\,\text{m}}{2}\tan 2.81° = \boxed{0.245\,\text{m}}$

(b) Express the change in the tightrope walker's gravitational potential energy as the rope sags:

$$\Delta U = U_{\text{at center}} - U_{\text{end}} = Mg\Delta y$$

Substitute numerical values and evaluate ΔU:

$$\Delta U = (50\,\text{kg})(9.81\,\text{m/s}^2)(-0.245\,\text{m})$$
$$= \boxed{-120\,\text{J}}$$

*18 •• The mass of the Space Shuttle orbiter is about 8×10^4 kg and the period of its orbit is 90 min. Estimate (a) the kinetic energy of the orbiter, and (b) the change in its potential energy between resting on the surface of the earth and in its orbit, 200 mi above the surface of the earth. (c) Why is the change in potential energy much smaller than the shuttle's kinetic energy? Shouldn't they be equal?

Picture the Problem We can find the orbital speed of the Shuttle from the radius of its orbit and its period and its kinetic energy from $K = \frac{1}{2}mv^2$. We'll ignore the variation in the acceleration due to gravity to estimate the change in the potential energy of the orbiter between its value at the surface of the earth and its orbital value.

(a) Express the kinetic energy of the orbiter:

$$K = \tfrac{1}{2}mv^2$$

Relate the orbital speed of the orbiter to its radius r and period T:

$$v = \frac{2\pi r}{T}$$

Substitute to obtain:

$$K = \tfrac{1}{2}m\left(\frac{2\pi r}{T}\right)^2 = \frac{2\pi^2 mr^2}{T^2}$$

Substitute numerical values and evaluate K:

$$K = 2\pi^2\left(8 \times 10^4\,\text{kg}\right)$$
$$\times \frac{\left[(200\,\text{mi} + 3960\,\text{mi})(1.609\,\text{km/mi})\right]^2}{\left[(90\,\text{min})(60\,\text{s/min})\right]^2}$$
$$= \boxed{2.43\,\text{TJ}}$$

(b) Assuming the acceleration due to gravity to be constant over the 200 miles and equal to its value at the surface of the earth (actually it is closer to 9 m/s² at an elevation of 200 mi), express

$$\Delta U = mgh$$

the change in gravitational potential energy of the orbiter, relative to the surface of the earth, as the Shuttle goes into orbit:

Substitute numerical values and evaluate ΔU:

$$\Delta U = (8\times10^4 \, \text{kg})(9.81 \, \text{m/s}^2)$$
$$\times(200 \, \text{mi})(1.609 \, \text{km/mi})$$
$$= \boxed{0.253 \, \text{TJ}}$$

(c) No, they shouldn't be equal because there is more than just the force of gravity to consider here. When the shuttle is resting on the surface of the Earth, it is supported against the force of gravity by the normal force the Earth exerts upward on it. We would need to take into consideration the change in potential energy of the surface of Earth in its deformation under the weight of the shuttle to find the actual change in potential energy.

Work and Kinetic Energy

*20 •   A 15-g bullet has a speed of 1.2 km/s. (a) What is its kinetic energy in joules? (b) What is its kinetic energy if its speed is halved? (c) What is its kinetic energy if its speed is doubled?

Picture the Problem We can use $\frac{1}{2}mv^2$ to find the kinetic energy of the bullet.

(a) Use the definition of K:

$$K = \tfrac{1}{2}mv^2$$
$$= \tfrac{1}{2}(0.015 \, \text{kg})(1.2\times10^3 \, \text{m/s})^2$$
$$= \boxed{10.8 \, \text{kJ}}$$

(b) Because $K \propto v^2$:

$$K' = \tfrac{1}{4}K = \boxed{2.70 \, \text{kJ}}$$

(c) Because $K \propto v^2$:

$$K' = 4K = \boxed{43.2 \, \text{kJ}}$$

*24 ••   You run a race with a friend. At first you each have the same kinetic energy, but you find that she is beating you. When you increase your speed by 25 percent, you are running at the same speed she is. If your mass is 85 kg, what is her mass?

Picture the Problem We can use the definition of kinetic energy to find the mass of your friend.

Using the definition of kinetic energy and letting "1" denote your mass and speed and "2" your girlfriend's, express the equality of your kinetic energies and solve for your girlfriend's mass as a function of both your masses and speeds:

$$\tfrac{1}{2}m_1v_1^2 = \tfrac{1}{2}m_2v_2^2$$

and

$$m_2 = m_1\left(\frac{v_1}{v_2}\right)^2 \qquad (1)$$

Express the condition on your speed that enables you to run at the same speed as your girlfriend:

$$v_2 = 1.25v_1 \qquad (2)$$

Substitute (2) in (1) to obtain:

$$m_2 = m_1\left(\frac{v_1}{v_2}\right)^2 = (85\,\text{kg})\left(\frac{1}{1.25}\right)^2$$

$$= \boxed{54.4\,\text{kg}}$$

Work Done by a Variable Force

*26 ••   A force F_x acts on a particle. The force is related to the position of the particle by the formula $F_x = Cx^3$, where C is a constant. Find the work done by this force on the particle when the particle moves from $x = 1.5$ m to $x = 3$ m.

Picture the Problem The work done by this force as it displaces the particle is the area under the curve of F as a function of x. Note that the constant C has units of N/m^3.

Because F varies with position non-linearly, express the work it does as an integral and evaluate the integral between the limits $x = 1.5$ m and $x = 3$ m:

$$W = (C\,\text{N/m}^3)\int_{1.5\,\text{m}}^{3\,\text{m}} x'^3\, dx'$$

$$= (C\,\text{N/m}^3)\left[\tfrac{1}{4}x'^4\right]_{1.5\,\text{m}}^{3\,\text{m}}$$

$$= \frac{(C\,\text{N/m}^3)}{4}\left[(3\,\text{m})^4 - (1.5\,\text{m})^4\right]$$

$$= \boxed{19C\,\text{J}}$$

*29 ••   Near Margaret's cabin is a 20-m water tower that attracts many birds during the summer months. During the hot spell last year, the tower went dry, and Margaret had to have her water hauled in. She was lonesome without the birds visiting, so she decided to carry some water up the tower to attract them back. Her bucket has a mass of 10 kg and holds 30 kg of water when it is full. However, the bucket has a hole, and as Margaret climbed at a constant speed, water leaked out at a constant rate. Several birds took advantage of the shower below, but when she got to the top, only 10 kg of water remained for the birdbath. (*a*) Write an expression for the mass of the bucket plus water as a function of the height (*y*) climbed. (*b*) Find the work done by Margaret on the bucket.

Picture the Problem We can express the mass of the water in Margaret's bucket as the difference between its initial mass and the product of the rate at which it loses water and her position during her climb. Because Margaret must do work against gravity in lifting and carrying the bucket, the work she does is the integral of the product of the gravitational field and the mass of the bucket as a function of its position.

(*a*) Express the mass of the bucket and the water in it as a function of its initial mass, the rate at which it is losing water, and Margaret's position, *y*, during her climb:

$$m(y) = 40\,\text{kg} - ry$$

Find the rate, $r = \dfrac{\Delta m}{\Delta y}$, at which Margaret's bucket loses water:

$$r = \frac{\Delta m}{\Delta y} = \frac{20\,\text{kg}}{20\,\text{m}} = 1\,\text{kg/m}$$

Substitute to obtain:

$$m(y) = 40\,\text{kg} - ry = \boxed{40\,\text{kg} - \frac{1\,\text{kg}}{\text{m}}\,y}$$

(*b*) Integrate the force Margaret exerts on the bucket, *m*(*y*)*g*, between the limits of *y* = 0 and *y* = 20 m:

$$W = g \int_{0}^{20\,\text{m}} \left(40\,\text{kg} - \frac{1\,\text{kg}}{\text{m}}\,y' \right) dy'$$

$$= \left(9.81\,\text{m/s}^2 \right)\left[(40\,\text{kg})y' - \tfrac{1}{2}(1\,\text{kg/m})y'^2 \right]_{0}^{20\,\text{m}}$$

$$= \boxed{5.89\,\text{kJ}}$$

Remarks: We could also find the work Margaret did on the bucket, at least approximately, by plotting a graph of $m(y)g$ and finding the area under this curve between $y = 0$ and $y = 20$ m.

Work, Energy, and Simple Machines

*32 •   *Simple machines* are used for reducing the amount of force that must be supplied to perform a task such as lifting a heavy weight. Such machines include the screw, block-and-tackle systems, and levers, but the simplest of the simple machines is the inclined plane. In Figure 6-34 you are raising the heavy box to the height of the truck bed by pushing it up an inclined plane (a ramp). (*a*) We define the *mechanical advantage M* of the inclined plane as the ratio of the force it would take to lift the block into the truck directly from the ground (at constant speed) to the force it takes to push it up the ramp (at constant speed). If the plane is frictionless, show that $M = 1/\sin\theta = L/H$, where H is the height of the truck bed and L is the length of the ramp. (*b*) Show that the work you do by moving the block into the truck is the same whether you lift it directly into the truck or push it up the frictionless ramp.

Figure 6-34 Problem 32

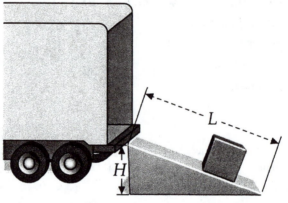

Picture the Problem The free-body diagram, with $\vec{F}$ representing the force required to move the block at constant speed, shows the forces acting on the block. We can apply Newton's 2nd law to the block to relate F to its weight w and then use the definition of the mechanical advantage of an inclined plane. In the second part of the problem we'll use the definition of work.

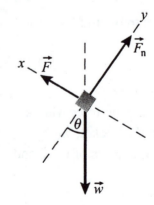

(*a*) Express the mechanical advantage M of the inclined plane:

$$M = \frac{w}{F}$$

Apply $\sum F_x = ma_x$ to the block:

$F - w\sin\theta = 0$ because $a_x = 0$.

Solve for F and substitute to obtain:

$$M = \frac{w}{w\sin\theta} = \frac{1}{\sin\theta}$$

Refer to the figure to obtain:

$$\sin\theta = \frac{H}{L}$$

Substitute to obtain:

$$M = \boxed{\frac{1}{\sin\theta} = \frac{L}{H}}$$

(b) Express the work done pushing the block up the ramp:

$$W_{ramp} = FL = mgL\sin\theta$$

Express the work done lifting the block into the truck:

$$W_{lifting} = mgH = mgL\sin\theta$$

and

$$\boxed{W_{ramp} = W_{lifting}}$$

Dot Products

*35 • What is the angle between the vectors $\vec{A}$ and $\vec{B}$ if $\vec{A}\cdot\vec{B} = -AB$?

Picture the Problem Because $\vec{A}\cdot\vec{B} \equiv AB\cos\theta$ we can solve for $\cos\theta$ and use the fact that $\vec{A}\cdot\vec{B} = -AB$ to find θ.

Solve for θ:

$$\theta = \cos^{-1}\frac{\vec{A}\cdot\vec{B}}{AB}$$

Substitute for $\vec{A}\cdot\vec{B}$ and evaluate θ:

$$\theta = \cos^{-1}(-1) = \boxed{180°}$$

*41 •• Given two vectors $\vec{A}$ and $\vec{B}$, show that if $\left|\vec{A}+\vec{B}\right| = \left|\vec{A}-\vec{B}\right|$, then $\vec{A}\perp\vec{B}$.

Picture the Problem We can use the definitions of the magnitude of a vector and the dot product to show that if $\left|\vec{A}+\vec{B}\right| = \left|\vec{A}-\vec{B}\right|$, then $\vec{A}\perp\vec{B}$.

Express $\left|\vec{A}+\vec{B}\right|^2$:

$$\left|\vec{A}+\vec{B}\right|^2 = \left(\vec{A}+\vec{B}\right)^2$$

Express $\left|\vec{A} - \vec{B}\right|$:

$$\left|\vec{A} - \vec{B}\right|^2 = \left(\vec{A} - \vec{B}\right)^2$$

Equate these expressions to obtain:

$$\left(\vec{A} + \vec{B}\right)^2 = \left(\vec{A} - \vec{B}\right)^2$$

Expand both sides of the equation to obtain:

$$A^2 + 2\vec{A} \cdot \vec{B} + B^2 = A^2 - 2\vec{A} \cdot \vec{B} + B^2$$

Simplify to obtain:

$$4\vec{A} \cdot \vec{B} = 0$$
or
$$\vec{A} \cdot \vec{B} = 0$$

From the definition of the dot product we have:

$$\vec{A} \cdot \vec{B} \equiv AB\cos\theta$$
where θ is the angle between $\vec{A}$ and $\vec{B}$.

Because neither $\vec{A}$ nor $\vec{B}$ is the zero vector:

$$\cos\theta = 0 \Rightarrow \theta = 90° \text{ and } \vec{A} \perp \vec{B}.$$

*45 •• When a particle moves in a circle centered at the origin with constant speed, the magnitudes of its position vector and velocity vectors are constant. (a) Differentiate $\vec{r} \cdot \vec{r} = r^2$ = constant with respect to time to show that $\vec{v} \cdot \vec{r} = 0$ and therefore $\vec{v} \perp \vec{r}$. (b) Differentiate $\vec{v} \cdot \vec{v} = v^2$ = constant with respect to time and show that $\vec{a} \cdot \vec{v} = 0$ and therefore $\vec{a} \perp \vec{v}$. What do the results of (a) and (b) imply about the direction of $\vec{a}$? (c) Differentiate $\vec{v} \cdot \vec{r} = 0$ with respect to time and show that $\vec{a} \cdot \vec{r} + v^2 = 0$ and therefore $a_r = -v^2/r$.

Picture the Problem The rules for the differentiation of vectors are the same as those for the differentiation of scalars and scalar multiplication is commutative.

(a) Differentiate $\vec{r} \cdot \vec{r} = r^2$ = constant:

$$\frac{d}{dt}\left(\vec{r} \cdot \vec{r}\right) = \vec{r} \cdot \frac{d\vec{r}}{dt} + \frac{d\vec{r}}{dt} \cdot \vec{r} = 2\vec{v} \cdot \vec{r}$$

$$= \frac{d}{dt}\left(\text{constant}\right) = 0$$

Because $\vec{v} \cdot \vec{r} = 0$:

$$\boxed{\vec{v} \perp \vec{r}}$$

(b) Differentiate $\vec{v} \cdot \vec{v} = v^2$ = constant with respect to time:

$$\frac{d}{dt}\left(\vec{v} \cdot \vec{v}\right) = \vec{v} \cdot \frac{d\vec{v}}{dt} + \frac{d\vec{v}}{dt} \cdot \vec{v} = 2\vec{a} \cdot \vec{v}$$

$$= \frac{d}{dt}\left(\text{constant}\right) = 0$$

Because $\vec{a} \cdot \vec{v} = 0$:

$$\boxed{\vec{a} \perp \vec{v}}$$

> The results of (a) and (b) tell us that $\vec{a}$ is perpendicular to $\vec{r}$ and and parallel (or antiparallel) to $\vec{r}$.

(c) Differentiate $\vec{v} \cdot \vec{r} = 0$ with respect to time:

$$\frac{d}{dt}(\vec{v} \cdot \vec{r}) = \vec{v} \cdot \frac{d\vec{r}}{dt} + \vec{r} \cdot \frac{d\vec{v}}{dt}$$

$$= v^2 + \vec{r} \cdot \vec{a} = \frac{d}{dt}(0) = 0$$

Because $v^2 + \vec{r} \cdot \vec{a} = 0$:

$$\boxed{\vec{r} \cdot \vec{a} = -v^2} \qquad (1)$$

Express a_r in terms of θ, where θ is the angle between $\vec{r}$ and $\vec{a}$:

$$a_r = a\cos\theta$$

Express $\vec{r} \cdot \vec{a}$:

$$\vec{r} \cdot \vec{a} = ra\cos\theta = ra_r$$

Substitute in equation (1) to obtain:

$$ra_r = -v^2$$

Solve for a_r:

$$a_r = \boxed{-\frac{v^2}{r}}$$

Power

*51 •   A small food service elevator (dumbwaiter) in a cafeteria is connected over a pulley system to a motor as shown in Figure 6-37; the motor raises and lowers the dumbwaiter. The mass of the dumbwaiter is 35 kg. In operation, it moves at a speed of 0.35 m/s upward, without accelerating (except for a brief initial period just after the motor is turned on.) If the output power from the motor is 27 percent of its input power, what is the input power to the motor? Assume that the pulleys are frictionless.

Figure 6-37 Problem 51

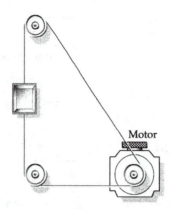

Picture the Problem Choose a coordinate system in which upward is the positive y direction. We can find P_{in} from the given information that $P_{out} = 0.27P_{in}$. We can express P_{out} as the product of the tension in the cable T and the constant speed v of the dumbwaiter. We can apply Newton's 2nd law to the dumbwaiter to express T in terms of its mass m and the gravitational field g.

Express the relationship between the motor's input and output power:	$P_{out} = 0.27P_{in}$ or $P_{in} = 3.7P_{out}$
Express the power required to move the dumbwaiter at a constant speed v:	$P_{out} = Tv$
Apply $\sum F_y = ma_y$ to the dumbwaiter:	$T - mg = ma_y$ or, because $a_y = 0$, $T = mg$
Substitute to obtain:	$P_{in} = 3.7Tv = 3.7mgv$
Substitute numerical values and evaluate P_{in}:	$P_{in} = 3.7(35\,\text{kg})(9.81\,\text{m/s}^2)(0.35\,\text{m/s})$ $= \boxed{445\,\text{W}}$

*53 •• A cannon placed at the top of a cliff of height H fires a cannonball into the air with an initial speed v_0, shooting directly upward. The cannonball rises, falls back down, missing the cannon by a little bit, and lands at the foot of the cliff. Neglecting air resistance, calculate the velocity $\vec{v}(t)$ for all times while the cannonball is in the air, and show explicitly that the integral of $\vec{F} \cdot \vec{v}$ over the time that the cannonball spends in the air is equal to the change in the kinetic energy of the cannonball.

Picture the Problem Because, in the absence of air resistance, the acceleration of the cannonball is constant, we can use a constant-acceleration equation to relate its velocity to the time it has been in flight. We can apply Newton's 2nd law to the cannonball to find the net force acting on it and then form the dot product of $\vec{F}$ and $\vec{v}$ to express the rate at which the gravitational field does work on the cannonball. Integrating this expression over the time-of-flight T of the ball will yield the desired result.

Express the velocity of the cannonball as a function of time while it is in the air:	$\vec{v}(t) = 0\hat{i} + (v_0 - gt)\hat{j}$
Apply $\sum \vec{F} = m\vec{a}$ to the cannonball to express the force acting on it while it is in the air:	$\vec{F} = -mg\,\hat{j}$
Evaluate $\vec{F} \cdot \vec{v}$:	$\vec{F} \cdot \vec{v} = -mg\,\hat{j} \cdot (v_0 - gt)\hat{j}$ $= -mgv_0 + mg^2 t$
Relate $\vec{F} \cdot \vec{v}$ to the rate at which work is being done on the cannonball:	$\dfrac{dW}{dt} = \vec{F} \cdot \vec{v} = -mgv_0 + mg^2 t$
Separate the variables and integrate over the time T that the cannonball is in the air:	$W = \displaystyle\int_0^T \left(-mgv_0 + mg^2 t\right) dt$ (1) $= \tfrac{1}{2} mg^2 T^2 - mgv_0 T$
Using a constant-acceleration equation, relate the speed v of the cannonball when it lands at the bottom of the cliff to its initial speed v_0 and the height of the cliff H:	$v^2 = v_0^2 + 2a\Delta y$ or, because $a = g$ and $\Delta y = H$, $v^2 = v_0^2 + 2gH$
Solve for v to obtain:	$\boxed{v = \sqrt{v_0^2 + 2gH}}$
Using a constant acceleration equation, relate the time-of-flight T to the initial and impact speeds of the cannonball:	$v = v_0 - gT$
Solve for T to obtain:	$T = \dfrac{v_0 - v}{g}$

Substitute for T in equation (1) and simplify to evaluate W:

$$W = \tfrac{1}{2}mg^2 \frac{v_0^2 - 2vv_0 + v^2}{g^2}$$

$$- mgv_0 \left(\frac{v_0 - v}{g} \right)$$

$$= \tfrac{1}{2}mv^2 - \tfrac{1}{2}mv_0^2 = \boxed{\Delta K}$$

Potential Energy

*60 ••   A simple Atwood's machine (Figure 6-38) uses two masses, m_1 and m_2. Starting from rest, the speed of the two masses is 4.0 m/s at the end of 3.0 s. At that time, the kinetic energy of the system is 80 J and each mass has moved a distance of 6.0 m. Determine the values of m_1 and m_2.

Figure 6-38 Problem 60

Picture the Problem In a simple Atwood's machine, the only effect of the pulley is to connect the motions of the two objects on either side of it; i.e., it could be replaced by a piece of polished pipe. We can relate the kinetic energy of the rising and falling objects to the mass of the system and to their common speed and relate their accelerations to the sum and difference of their masses ... leading to simultaneous equations in m_1 and m_2.

Use the definition of the kinetic energy of the system to determine the total mass being accelerated:

$$K = \tfrac{1}{2}(m_1 + m_2)v^2$$

and

$$m_1 + m_2 = \frac{2K}{v^2} = \frac{2(80\,\text{J})}{(4\,\text{m/s})^2} = 10.0\,\text{kg} \quad (1)$$

In Chapter 4, the acceleration of the masses was shown to be:

$$a = \frac{m_1 - m_2}{m_1 + m_2}g$$

Because $v(t) = at$, we can eliminate a in the previous equation to obtain:

$$v(t) = \frac{m_1 - m_2}{m_1 + m_2} gt$$

Solve for $m_1 - m_2$:

$$m_1 - m_2 = \frac{(m_1 + m_2)v(t)}{gt}$$

$$= \frac{(10\,\text{kg})(4\,\text{m/s})}{(9.81\,\text{m/s}^2)(3\,\text{s})} \quad (2)$$

$$= 1.36\,\text{kg}$$

Solve equations (1) and (2) simultaneously to obtain:

$$m_1 = \boxed{5.68\,\text{kg}} \text{ and } m_2 = \boxed{4.32\,\text{kg}}$$

Force, Potential Energy, and Equilibrium

*64 ••   On the potential-energy curve for U versus y shown in Figure 6-40, the segments AB and CD are straight lines. Sketch a plot of the force F_y versus y.

Figure 6-40 Problem 64

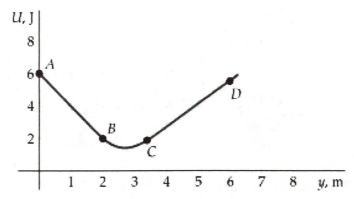

Picture the Problem F_y is defined to be the negative of the derivative of the potential function with respect to y, i.e. $F_y = -dU/dy$. Consequently, we can obtain F_y by examining the slopes of the graph of U as a function of y.

The table to the right summarizes the information we can obtain from Figure 6-40:

Interval	Slope (N)	F_y (N)
$A \rightarrow B$	-2	2
$B \rightarrow C$	transitional	$-2 \rightarrow 1.4$
$C \rightarrow D$	1.4	-1.4

The graph of F as a function of y is shown to the right:

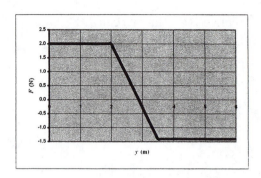

*71 •••   A novelty desk clock is shown in Figure 6-41: the clock (which has mass m) is supported by two light cables running over the two pulleys which are attached to counterweights that each have mass M. (*a*) Find the potential energy of the system as a function of the distance y. (*b*) Find the value of y for which the potential energy of the system is a minimum. (*c*) If the potential energy is a minimum then the system is in equilibrium. Apply Newton's second law to the clock and show that it is in equilibrium (the forces on it sum to zero) for the value of y obtained for part (*b*). Is this a point of stable or unstable equilibrium?

Figure 6-41 Problem 71

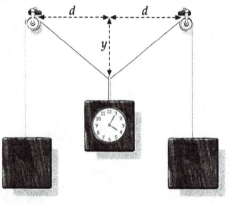

Picture the Problem Let L be the total length of one cable and the zero of gravitational potential energy be at the top of the pulleys. We can find the value of y for which the potential energy of the system is an extremum by differentiating $U(y)$ with respect to y and setting this derivative equal to zero. We can establish that this value corresponds to a minimum by evaluating the second derivative of $U(y)$ at the point identified by the first derivative. We can apply Newton's 2nd law to the clock to confirm the result we obtain by examining the derivatives of $U(y)$.

(*a*) Express the potential energy of the system as the sum of the potential energies of the clock and counterweights:

$$U(y) = U_{\text{clock}}(y) + U_{\text{weights}}(y)$$

Substitute to obtain:

$$U(y) = \boxed{- mgy - 2Mg\left(L - \sqrt{y^2 + d^2}\right)}$$

(b) Differentiate $U(y)$ with respect to y:

$$\frac{dU(y)}{dy} = -\frac{d}{dy}\left[mgy + 2Mg\left(L - \sqrt{y^2 + d^2}\right)\right]$$

$$= -\left[mg - 2Mg\frac{y}{\sqrt{y^2 + d^2}}\right]$$

or

$$mg - 2Mg\frac{y'}{\sqrt{y'^2 + d^2}} = 0 \text{ for extrema}$$

Solve for y' to obtain:

$$y' = d\sqrt{\frac{m^2}{4M^2 - m^2}}$$

Find $\dfrac{d^2U(y)}{dy^2}$:

$$\frac{d^2U(y)}{dy^2} = -\frac{d}{dy}\left[mg - 2Mg\frac{y}{\sqrt{y^2 + d^2}}\right]$$

$$= \frac{2Mgd^2}{\left(y^2 + d^2\right)^{3/2}}$$

Evaluate $\dfrac{d^2U(y)}{dy^2}$ at $y = y'$:

$$\left.\frac{d^2U(y)}{dy^2}\right|_{y'} = \left.\frac{2Mgd^2}{\left(y^2 + d^2\right)^{3/2}}\right|_{y'}$$

$$= \frac{2Mgd}{\left(\dfrac{m^2}{4M^2 - m^2} + 1\right)^{3/2}}$$

$$> 0$$

and the potential energy is a minimum at

$$\boxed{y = d\sqrt{\frac{m^2}{4M^2 - m^2}}}$$

(c) The FBD for the clock is shown to the right:

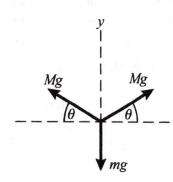

Apply $\sum F_y = 0$ to the clock:

$$2Mg\sin\theta - mg = 0 \Rightarrow \sin\theta = \frac{m}{2M}$$

Express $\sin\theta$ in terms of y and d:

$$\sin\theta = \frac{y}{\sqrt{y^2 + d^2}}$$

Substitute to obtain:

$$\boxed{\frac{m}{2M} = \frac{y}{\sqrt{y^2 + d^2}}}$$

which is equivalent to the first equation in part (b).

> This is a point of stable equilibrium. If the clock is displaced downward, θ increases, leading to a larger upward force on the clock. Similarly, if the clock is displaced upward, the net force from the cables decreases. Because of this, the clock will be pulled back toward the equilibrium point if it is displaced away from it.

Remarks: Because we've shown that the potential energy of the system is a minimum at $y = y'$ (i.e., $U(y)$ is concave upward at that point), we can conclude that this point is one of stable equilibrium.

General Problems

*72 •   In February 2002, a total of 60.7 billion kW-h of electrical energy was generated by nuclear power plants in the United States. At that time, the population of the United States was about 287 million people. If the average American has a mass of 60 kg, and if the entire energy output of all nuclear power plants was diverted to supplying energy for a single giant elevator, estimate the height h to which the entire population of the country could be lifted by the elevator. In your calculations, assume that 25 percent of the energy goes into lifting the people; assume also that g is constant over the entire height h.

Picture the Problem 25 percent of the electrical energy generated is to be diverted to do the work required to change the potential energy of the American people. We can calculate the height to which they can be lifted by equating the change in potential energy to the available energy.

Express the change in potential energy of the population of the United States in this process:

$$\Delta U = Nmgh$$

Letting E represent the total energy generated in February 2002, relate the change in

$$Nmgh = 0.25E$$

potential to the energy available
to operate the elevator:

Solve for h:

$$h = \frac{0.25E}{Nmg}$$

Substitute numerical values and
evaluate h:

$$h = \frac{0.25\left(60.7\times10^9\,\text{kW}\cdot\text{h}\right)\left(\dfrac{3600\,\text{s}}{1\,\text{h}}\right)}{\left(287\times10^6\right)\left(60\,\text{kg}\right)\left(9.81\,\text{m/s}^2\right)}$$

$$= \boxed{323\,\text{km}}$$

*76 •   The movie crew arrives in the Badlands ready to shoot a scene. The
script calls for a car to crash into a vertical rock face at 100 km/h. Unfortunately,
the car won't start, and there is no mechanic in sight. The crew are about to skulk
back to the studio to face the producer's wrath when the cameraman gets an idea.
They use a crane to lift the car by its rear end and then drop it vertically, filming
at an angle that makes the car appear to be traveling horizontally. How high
should the 800-kg car be lifted so that it reaches a speed of 100 km/h in the fall?

Picture the Problem We can solve this problem by equating the expression for
the gravitational potential energy of the elevated car and its kinetic energy when
it hits the ground.

Express the gravitational potential
energy of the car when it is at a
distance h above the ground:

$$U = mgh$$

Express the kinetic energy of the
car when it is about to hit the
ground:

$$K = \tfrac{1}{2}mv^2$$

Equate these two expressions
(because at impact, all the
potential energy has been
converted to kinetic energy) and
solve for h:

$$h = \frac{v^2}{2g}$$

Substitute numerical values and
evaluate h:

$$h = \frac{\left[\left(100\,\text{km/h}\right)\left(1\,\text{h}/3600\,\text{s}\right)\right]^2}{2\left(9.81\,\text{m/s}^2\right)} = \boxed{39.3\,\text{m}}$$

*79 ••   A horizontal force acts on a cart of mass m such that the speed v of the cart increases with distance x as $v = Cx$, where C is a constant. (*a*) Find the force acting on the cart as a function of position. (*b*) What is the work done by the force in moving the cart from $x = 0$ to $x = x_1$?

Picture the Problem We can use the definition of work to obtain an expression for the position-dependent force acting on the cart. The work done on the cart can be calculated from its change in kinetic energy.

(*a*) Express the force acting on the cart in terms of the work done on it:

$$F(x) = \frac{dW}{dx}$$

Because U is constant:

$$F(x) = \frac{d}{dx}\left(\tfrac{1}{2}mv^2\right) = \frac{d}{dx}\left[\tfrac{1}{2}m(Cx)^2\right]$$

$$= \boxed{mC^2x}$$

(*b*) The work done by this force changes the kinetic energy of the cart:

$$W = \Delta K = \tfrac{1}{2}mv_1^2 - \tfrac{1}{2}mv_0^2$$

$$= \tfrac{1}{2}mv_1^2 - 0 = \tfrac{1}{2}m(Cx_1)^2$$

$$= \boxed{\tfrac{1}{2}mC^2x_1^2}$$

*83 ••   The initial kinetic energy imparted to a 0.020-kg bullet is 1200 J. Neglecting air resistance, find the range of this projectile when it is fired at an angle such that the range equals the maximum height attained.

Picture the Problem We'll assume that the firing height is negligible and that the bullet lands at the same elevation from which it was fired. We can use the equation $R = \left(v_0^2/g\right)\sin 2\theta$ to find the range of the bullet and constant acceleration equations to find its maximum height. The bullet's initial speed can be determined from its initial kinetic energy.

Express the range of the bullet as a function of its firing speed and angle of firing:

$$R = \frac{v_0^2}{g}\sin 2\theta$$

Rewrite the range equation using the trigonometric identity $\sin 2\theta = 2\sin\theta\cos\theta$:

$$R = \frac{v_0^2 \sin 2\theta}{g} = \frac{2v_0^2 \sin\theta\cos\theta}{g}$$

Express the position coordinates of the projectile along its flight

$$x = (v_0\cos\theta)t$$

path in terms of the parameter t:	and $$y = (v_0 \sin \theta)t - \tfrac{1}{2}gt^2$$
Eliminate the parameter t and make use of the fact that the maximum height occurs when the projectile is at half the range to obtain:	$$h = \frac{(v_0 \sin \theta)^2}{2g}$$
Equate R and h and solve the resulting equation for θ.	$$\tan \theta = 4 \Rightarrow \theta = \tan^{-1} 4 = 76.0°$$
Relate the bullet's kinetic energy to its mass and speed and solve for the square of its speed:	$$K = \tfrac{1}{2}mv_0^2 \text{ and } v_0^2 = \frac{2K}{m}$$
Substitute for v_0^2 and θ and evaluate R:	$$R = \frac{2(1200\,\text{J})}{(0.02\,\text{kg})(9.81\,\text{m/s}^2)}\sin 2(76°)$$ $$= \boxed{5.74\,\text{km}}$$

*88 ••• A force in the xy plane is given by:

$$\vec{F} = -\left(\frac{b}{r^3}\right)(x\hat{i} + y\hat{j})$$

where b is a positive constant and $r = \sqrt{x^2 + y^2}$.

(a) Show that the magnitude of the force varies as the inverse of the square of the distance to the origin, and that its direction is antiparallel (opposite) to the radius vector $\vec{r} = x\hat{i} + y\hat{j}$. (b) If $b = 3$ N·m^2, find the work done by this force on a particle moving along a straight-line path between an initial position $x = 2$ m, $y = 0$ m and a final position $x = 5$ m, $y = 0$ m. (c) Find the work done by this force on a particle moving once around a circle of radius $r = 7$ m centered around the origin. (d) If this force is the only force acting on the particle, what is the particle's speed as it moves along this circular path? Assume that the particle's mass is $m = 2$ kg.

Picture the Problem We can substitute for r and $x\hat{i} + y\hat{j}$ in $\vec{F}$ to show that the magnitude of the force varies as the inverse of the square of the distance to the origin, and that its direction is opposite to the radius vector. We can find the work done by this force by evaluating the integral of F with respect to x from an initial position $x = 2$ m, $y = 0$ m to a final position $x = 5$ m, $y = 0$ m. Finally, we can apply Newton's 2nd law to the particle to relate its speed to its radius, mass, and the constant b.

(*a*) Substitute for r and $x\hat{i} + y\hat{j}$ in $\vec{F}$ to obtain:

$$\vec{F} = -\left(\frac{b}{\left(x^2 + y^2\right)^{3/2}}\right)\sqrt{x^2 + y^2}\,\hat{r}$$

where $\hat{r}$ is a unit vector pointing from the origin toward the point of application of $\vec{F}$.

Simplify to obtain:

$$\vec{F} = -b\left(\frac{1}{x^2 + y^2}\right)\hat{r} = \boxed{-\frac{b}{r^2}\hat{r}}$$

i.e., the magnitude of the force varies as the inverse of the square of the distance to the origin, and its direction is antiparallel (opposite) to the radius vector $\vec{r} = x\hat{i} + y\hat{j}$.

(*b*) Find the work done by this force by evaluating the integral of F with respect to x from an initial position $x = 2$ m, $y = 0$ m to a final position $x = 5$ m, $y = 0$ m:

$$W = -\int_{2\,\text{m}}^{5\,\text{m}} \frac{b}{x'^2}\,dx' = b\left[\frac{1}{x'}\right]_{2\,\text{m}}^{5\,\text{m}}$$

$$= 3\,\text{N}\cdot\text{m}^2\left(\frac{1}{5\,\text{m}} - \frac{1}{2\,\text{m}}\right) = \boxed{-0.900\,\text{J}}$$

(*c*)

$$\boxed{\begin{array}{l}\text{No work is done as the force is}\\\text{perpendicular to the velocity.}\end{array}}$$

(*d*) Because the particle is moving in a circle, the force on the particle must be supplying the centripetal acceleration keeping it moving in the circle. Apply $\sum F_r = ma_c$ to the particle:

$$\frac{b}{r^2} = m\frac{v^2}{r}$$

Solve for v:

$$v = \sqrt{\frac{b}{mr}}$$

Substitute numerical values and evaluate v:

$$v = \sqrt{\frac{3\,\text{N}\cdot\text{m}^2}{(2\,\text{kg})(7\,\text{m})}} = \boxed{0.463\,\text{m/s}}$$

*90 ••• A theoretical formula for the potential energy associated with the nuclear force between two protons, two neutrons, or a neutron and a proton is the Yukawa potential $U(r) = -U_0\left(a/r\right)e^{-r/a}$, where U_0 and a are constants, and r is the separation between the two nucleons. (*a*) Using a spreadsheet program such as

Microsoft Excel™, make a graph of U vs r, using $U_0 = 4$ pJ (a picojoule, pJ, is 1×10^{-12} J) and $a = 2.5$ fm (a femtometer, fm, is 1×10^{-15} m). (*b*) Find the force $F(r)$ as a function of the separation of the two nucleons. (*c*) Compare the magnitude of the force at the separation $r = 2a$ to that at $r = a$. (*d*) Compare the magnitude of the force at the separation $r = 5a$ to that at $r = a$.

Picture the Problem A spreadsheet program to plot the Yukawa potential is shown below. The constants used in the potential function and the formula used to calculate the Yukawa potential are as follows:

Cell	Content/Formula	Algebraic Form
B1	4	U_0
B2	2.5	a
D8	−$B$1*($B$2/C9)*EXP(−C9/$B$2)	$-U_0\left(\dfrac{a}{r}\right)e^{-r/a}$
C10	C9+0.1	$r + \Delta r$

(*a*)

	A	B	C	D
1	U0=	4	pJ	
2	a=	2.5	fm	
3				
7				
8			r	U
9			0.5	−16.37
10			0.6	−13.11
11			0.7	−10.80
12			0.8	−9.08
13			0.9	−7.75
14			1	−6.70
64			6	−0.15
65			6.1	−0.14
66			6.2	−0.14
67			6.3	−0.13
68			6.4	−0.12
69			6.5	−0.11
70			6.6	−0.11

A graph of U as a function of r follows.

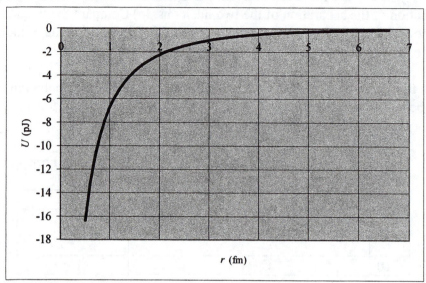

r (fm)

(b) Relate the force between the nucleons to the slope of the potential energy function:

$$F(r) = -\frac{dU(r)}{dr}$$

$$= -\frac{d}{dr}\left[-U_0\left(\frac{a}{r}\right)e^{-r/a}\right]$$

$$= \boxed{-U_0 e^{-r/a}\left(\frac{a}{r^2}+\frac{1}{r}\right)}$$

(c) Evaluate $F(2a)$:

$$F(2a) = -U_0 e^{-2a/a}\left(\frac{a}{(2a)^2}+\frac{1}{2a}\right)$$

$$= -U_0 e^{-2}\left(\frac{3}{4a}\right)$$

Evaluate $F(a)$:

$$F(a) = -U_0 e^{-a/a}\left(\frac{a}{(a)^2}+\frac{1}{a}\right)$$

$$= -U_0 e^{-1}\left(\frac{1}{a}+\frac{1}{a}\right) = -U_0 e^{-1}\left(\frac{2}{a}\right)$$

Express the ratio $F(2a)/F(a)$:

$$\frac{F(2a)}{F(a)} = \frac{-U_0 e^{-2}\left(\dfrac{3}{4a}\right)}{-U_0 e^{-1}\left(\dfrac{2}{a}\right)} = \frac{3}{8}e^{-1}$$

$$= \boxed{0.138}$$

(*d*) Evaluate $F(5a)$:

$$F(5a) = -U_0 e^{-5a/a}\left(\frac{a}{(5a)^2} + \frac{1}{5a}\right)$$

$$= -U_0 e^{-5}\left(\frac{6}{25a}\right)$$

Express the ratio $F(5a)/F(a)$:

$$\frac{F(5a)}{F(a)} = \frac{-U_0 e^{-5}\left(\dfrac{6}{25a}\right)}{-U_0 e^{-1}\left(\dfrac{2}{a}\right)} = \frac{3}{25}e^{-4}$$

$$= \boxed{2.20 \times 10^{-3}}$$

Chapter 7
Conservation of Energy

Conceptual Problems

*1 • Two objects of unequal mass are connected by a massless cord passing over a frictionless peg. After the objects are released from rest, which of the following statements are true? (U = gravitational potential energy, K = kinetic energy of the system.) (*a*) $\Delta U < 0$ and $\Delta K > 0$ (*b*) $\Delta U = 0$ and $\Delta K > 0$ (*c*) $\Delta U < 0$ and $\Delta K = 0$ (*d*) $\Delta U = 0$ and $\Delta K = 0$ (*e*) $\Delta U > 0$ and $\Delta K < 0$

Determine the Concept Because the peg is frictionless, mechanical energy is conserved as this system evolves from one state to another. The system moves and so we know that $\Delta K > 0$. Because $\Delta K + \Delta U = constant$, $\Delta U < 0$. $\boxed{(a) \text{ is correct.}}$

*5 • In *Surely you're joking, Mr. Feynman,*[†] Richard Feynman described his annoyance at how the concept of energy was portrayed in a children's textbook in the following way: "There was a book which started out with four pictures: first, there was a wind-up toy; then there was an automobile; then there was a boy riding a bicycle; then there was something else. And underneath each picture it said 'What makes it go?' ... I turned the page. The answer was ... for everything, 'Energy makes it go'... It's also not even true that 'energy makes it go' because if it stops, you could say 'energy makes it stop' just as well. ... Energy is neither increased nor decreased in these examples; it's just changed from one form to another." Describe how energy changes from one form to another when a little girl pedals her bike up a hill, then freewheels down the hill, and brakes to a stop.

Determine the Concept As she starts pedaling, chemical energy inside her body is converted into kinetic energy as the bike picks up speed. As she rides it up the hill, chemical energy is converted into gravitational potential and thermal energy. While freewheeling down the hill, potential energy is converted to kinetic energy, and while braking to a stop, kinetic energy is converted into thermal energy (a more random form of kinetic energy) by the frictional forces acting on the bike.

*6 • A body falling through the atmosphere (air resistance is present) gains 20 J of kinetic energy. The amount of gravitational potential energy that is lost is (*a*) 20 J, (*b*) more than 20 J, (*c*) less than 20 J, (*d*) impossible to tell without knowing the mass of the body, (*e*) impossible to tell without knowing how far the body falls.

Determine the Concept If we define the system to include the falling body and the earth, then no work is done by an external agent and $\Delta K + \Delta U_g + \Delta E_{therm} = 0$. Solving for the change in the gravitational potential energy we find $\Delta U_g = -(\Delta K +$

friction energy). $\boxed{(b) \text{ is correct.}}$

Estimation and Approximation

*10 ••   The *metabolic rate* is the rate at which the body uses chemical energy to sustain its life functions. Experimentally, the average metabolic rate is proportional to the total skin surface area of the body. The surface area for a 5-ft, 10-in male weighing 175 lb is just about 2.0 m^2, and for a 5-ft, 4-in female weighing 110 lb it is approximately 1.5 m^2. There is about a 1 percent change in surface area for every 3 lb above or below the weights quoted here and a 1 percent change for every inch above or below the heights quoted. (*a*) Estimate your average metabolic rate over the course of a day using the following guide for physical activity: sleeping, metabolic rate = 40 W/m^2; sitting, 60 W/m^2; walking, 160 W/m^2; moderate physical activity, 175 W/m^2; and moderate aerobic exercise, 300 W/m^2. How does it compare to the power of a 100-W light bulb? (*b*) Express your average metabolic rate in terms of kcal/day (1 kcal = 4190 J). (A kcal is the "food calorie" used by nutritionists.) (*c*) An estimate used by nutritionists is that the "average person" must eat roughly 12-15 kcal/lb of body weight a day to maintain his or her weight. From the calculations in part (*b*), are these estimates reasonable?

Picture the Problem We'll use the data for the "typical male" described above and assume that he spends 8 hours per day sleeping, 2 hours walking, 8 hours sitting, 1 hour in aerobic exercise, and 5 hours doing moderate physical activity. We can approximate his energy utilization using $E_{activity} = A P_{activity} \Delta t_{activity}$, where A is the surface area of his body, $P_{activity}$ is the rate of energy consumption in a given activity, and $\Delta t_{activity}$ is the time spent in the given activity. His total energy consumption will be the sum of the five terms corresponding to his daily activities.

(*a*) Express the energy consumption of the hypothetical male:	$E = E_{sleeping} + E_{walking} + E_{sitting}$ $\quad + E_{mod.\,act.} + E_{aerobic\,act.}$
Evaluate $E_{sleeping}$:	$E_{sleeping} = A P_{sleeping} \Delta t_{sleeping}$ $= (2\,m^2)(40\,W/m^2)(8\,h)(3600\,s/h)$ $= 2.30 \times 10^6\,J$
Evaluate $E_{walking}$:	$E_{walking} = A P_{walking} \Delta t_{walking}$ $= (2\,m^2)(160\,W/m^2)(2\,h)(3600\,s/h)$ $= 2.30 \times 10^6\,J$

Evaluate $E_{sitting}$:

$$E_{sitting} = AP_{sitting}\,\Delta t_{sitting}$$
$$= (2\,m^2)(60\,W/m^2)(8\,h)(3600\,s/h)$$
$$= 3.46\times10^6\,J$$

Evaluate $E_{mod.\,act.}$:

$$E_{mod.\,act.} = AP_{mod.\,act.}\,\Delta t_{mod.\,act.}$$
$$= (2\,m^2)(175\,W/m^2)(5\,h)(3600\,s/h)$$
$$= 6.30\times10^6\,J$$

Evaluate $E_{aerobic\,act.}$:

$$E_{aerobic\,act.} = AP_{aerobic\,act.}\,\Delta t_{aerobic\,act.}$$
$$= (2\,m^2)(300\,W/m^2)(1\,h)(3600\,s/h)$$
$$= 2.16\times10^6\,J$$

Substitute to obtain:

$$E = 2.30\times10^6\,J + 2.30\times10^6\,J + 3.46\times10^6\,J$$
$$+\,6.30\times10^6\,J + 2.16\times10^6\,J$$
$$= \boxed{16.5\times10^6\,J}$$

Find the average metabolic rate represented by this energy consumption:

$$P_{av} = \frac{E}{\Delta t} = \frac{16.5\times10^6\,J}{(24\,h)(3600\,s/h)} = \boxed{191\,W}$$

or about twice that of a 100 W light bulb.

(b) Express his average energy consumption in terms of kcal/day:

$$E = \frac{16.5\times10^6\,J/day}{4190\,J/kcal} = \boxed{3940\,kcal/day}$$

(c)

$$\frac{3940\,kcal}{175\,lb} = 22.5\,kcal/lb \text{ is higher than the}$$

estimate given in the statement of the problem. However, by adjusting the day's activities, the metabolic rate can vary by more than a factor of 2.

*13 •   The chemical energy released by burning a gallon of gasoline is approximately 2.6×10^5 kJ. Estimate the total energy used by all of the cars in the United States during the course of one year. What fraction does this represent of the total energy use by the United States in one year (about 5×10^{20} J)?

Picture the Problem There are about 3×10^8 people in the United States. On the assumption that the average family has 4 people in it and that they own two cars, we have a total of 1.5×10^8 automobiles on the road (excluding those used for industry). We'll assume that each car uses about 15 gal of fuel per week.

Calculate, based on the assumptions identified above, the total annual consumption of energy derived from gasoline:

$$\left(1.5\times10^8 \text{ auto}\right)\left(15\frac{\text{gal}}{\text{auto}\cdot\text{week}}\right)\left(52\frac{\text{weeks}}{y}\right)\left(2.6\times10^8\frac{\text{J}}{\text{gal}}\right) = \boxed{3.04\times10^{19} \text{ J/y}}$$

Express this rate of energy use as a fraction of the total annual energy use by the US:

$$\frac{3.04\times10^{19} \text{ J/y}}{5\times10^{20} \text{ J/y}} \approx \boxed{6\%}$$

Remarks: This is an average power expenditure of roughly 9×10^{11} watt, and a total cost (assuming \$1.15 per gallon) of about 140 billion dollars per year.

The Conservation of Mechanical Energy

*18 •   A pendulum of length L with a bob of mass m is pulled aside until the bob is a height $L/4$ above its equilibrium position. The bob is then released. Find the speed of the bob as it passes the equilibrium position.

Picture the Problem The diagram shows the pendulum bob in its initial position. Let the zero of gravitational potential energy be at the low point of the pendulum's swing, the equilibrium position.   We can find the speed of the bob at it passes through the equilibrium position by equating its initial potential energy to its kinetic energy as it passes through its lowest point.

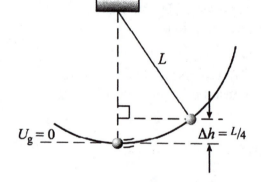

Equate the initial gravitational potential energy and the kinetic energy of the bob as it passes through its lowest point and solve for v:

$$mg\Delta h = \tfrac{1}{2}mv^2$$
and
$$v = \sqrt{2g\Delta h}$$

Express Δh in terms of the length L of the pendulum:

$$\Delta h = \frac{L}{4}$$

Substitute and simplify:

$$v = \sqrt{\dfrac{gL}{2}}$$

*24 •• The system shown in Figure 7-19 is initially at rest when the lower string is cut. Find the speed of the objects when they are at the same height. The frictionless pulley has negligible mass.

Figure 7-19 Problem 24

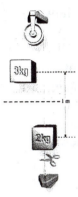

Picture the Problem Let the system include the two objects and the earth. Then $W_{ext} = 0$. Choose $U_g = 0$ at the elevation at which the two objects meet. With this choice, the initial potential energy of the 3-kg object is positive and that of the 2-kg object is negative. Their sum, however, is positive. Given our choice for $U_g = 0$, this initial potential energy is transformed entirely into kinetic energy.

Apply conservation of energy:

$$W_{ext} = \Delta K + \Delta U_g = 0$$
or, because $W_{ext} = 0$,
$$\Delta K = -\Delta U_g$$

Substitute for ΔK and solve for v_f, noting that m represents the sum of the masses of the objects as they are both moving in the final state:

$$\tfrac{1}{2}mv_f^2 - \tfrac{1}{2}mv_i^2 = -\Delta U_g$$
or, because $v_i = 0$,
$$v_f = \sqrt{\dfrac{-2\Delta U_g}{m}}$$

Express and evaluate ΔU_g:

$$\Delta U_g = U_{g,f} - U_{g,i} = 0 - (3\,\text{kg} - 2\,\text{kg})(0.5\,\text{m})(9.81\,\text{m/s}^2) = -4.91\,\text{J}$$

Substitute and evaluate v_f:

$$v_f = \sqrt{\dfrac{-2(-4.91\,\text{J})}{5\,\text{kg}}} = \boxed{1.40\,\text{m/s}}$$

*27 ••   A ball at the end of a string moves in a vertical circle with constant mechanical energy E. What is the difference between the tension at the bottom of the circle and the tension at the top?

Picture the Problem The diagram represents the ball traveling in a circular path with constant energy. U_g has been chosen to be zero at the lowest point on the circle and the superimposed free-body diagrams show the forces acting on the ball at the top and bottom of the circular path. We'll apply Newton's 2nd law to the ball at the top and bottom of its path to obtain a relationship between T_T and T_B and the conservation of mechanical energy to relate the speeds of the ball at these two locations.

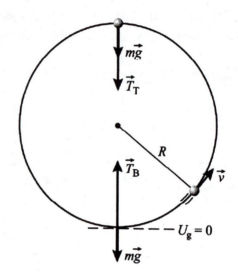

Apply $\sum F_{radial} = ma_{radial}$ to the ball at the bottom of the circle and solve for T_B:

$$T_B - mg = m\frac{v_B^2}{R}$$

and

$$T_B = mg + m\frac{v_B^2}{R} \qquad (1)$$

Apply $\sum F_{radial} = ma_{radial}$ to the ball at the top of the circle and solve for T_T:

$$T_T + mg = m\frac{v_T^2}{R}$$

and

$$T_T = -mg + m\frac{v_T^2}{R} \qquad (2)$$

Subtract equation (2) from equation (1) to obtain:

$$T_B - T_T = mg + m\frac{v_B^2}{R}$$

$$-\left(-mg + m\frac{v_T^2}{R}\right)$$

$$= m\frac{v_B^2}{R} - m\frac{v_T^2}{R} + 2mg \qquad (3)$$

Using conservation of energy, relate the mechanical energy of the ball at the bottom of its path

$$\tfrac{1}{2}mv_B^2 = \tfrac{1}{2}mv_T^2 + mg(2R)$$

$$m\frac{v_B^2}{R} - m\frac{v_T^2}{R} = 4mg$$

to its mechanical energy at the top of the circle and solve for $m\dfrac{v_B^2}{R} - m\dfrac{v_T^2}{R}$:

Substitute in equation (3) to obtain:

$$T_B - T_T = \boxed{6mg}$$

*32 ••   A stone is thrown upward at an angle of 53° above the horizontal. Its maximum height during the trajectory is 24 m. What was the stone's initial speed?

Picture the Problem Let the system consist of the stone and the earth and ignore the influence of air resistance. Then $W_{ext} = 0$. Choose $U_g = 0$ as shown in the figure. Apply the law of the conservation of mechanical energy to describe the energy transformations as the stone rises to the highest point of its trajectory.

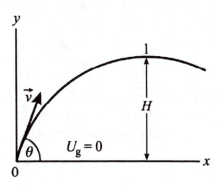

Apply conservation of energy:

$$W_{ext} = \Delta K + \Delta U = 0$$
and
$$K_1 - K_0 + U_1 - U_0 = 0$$

Because $U_0 = 0$:

$$K_1 - K_0 + U_1 = 0$$

Substitute to obtain:

$$\tfrac{1}{2}mv_x^2 - \tfrac{1}{2}mv^2 + mgH = 0$$

In the absence of air resistance, the horizontal component of $\vec{v}$ is constant and equal to $v_x = v\cos\theta$. Hence:

$$\tfrac{1}{2}m(v\cos\theta)^2 - \tfrac{1}{2}mv^2 + mgH = 0$$

Solve for v:

$$v = \sqrt{\dfrac{2gH}{1-\cos^2\theta}}$$

Substitute numerical values and evaluate v:

$$v = \sqrt{\dfrac{2(9.81\,\text{m/s}^2)(24\,\text{m})}{1-\cos^2 53°}} = \boxed{27.2\,\text{m/s}}$$

*35 ••   The Royal Gorge bridge over the Arkansas River is $L = 310$ m above the river. A bungee jumper of mass 60 kg has an elastic cord of length $d = 50$ m attached to her feet. Assume the cord acts like a spring of force constant k. The jumper leaps, barely touches the water, and after numerous ups and downs comes to rest at a height h above the water. (*a*) Find h. (*b*) Find the maximum speed of the jumper.

Picture the Problem Choose $U_g = 0$ at the bridge, and let the system be the earth, the jumper and the bungee cord. Then $W_{ext} = 0$. Use the conservation of mechanical energy to relate to relate her initial and final gravitational potential energies to the energy stored in the stretched bungee, U_s cord. In part (*b*), we'll use a similar strategy but include a kinetic energy term because we are interested in finding her maximum speed.

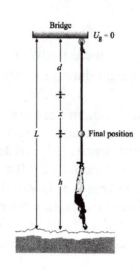

(*a*) Express her final height h above the water in terms of L, d and the distance x the bungee cord has stretched:

$$h = L - d - x \qquad (1)$$

Use the conservation of mechanical energy to relate her gravitational potential energy as she just touches the water to the energy stored in the stretched bungee cord:

$W_{ext} = \Delta K + \Delta U = 0$

Because $\Delta K = 0$ and $\Delta U = \Delta U_g + \Delta U_s$,
$-mgL + \tfrac{1}{2} kx^2 = 0,$

where x is the maximum distance the bungee cord has stretched.

Solve for k:

$$k = \frac{2mgL}{x^2}$$

Find the maximum distance the bungee cord stretches:

$x = 310 \text{ m} - 50 \text{ m} = 260 \text{ m}.$

Evaluate k:

$$k = \frac{2(60\,\text{kg})(9.81\,\text{m/s}^2)(310\,\text{m})}{(260\,\text{m})^2}$$

$$= 5.40\,\text{N/m}$$

Express the relationship between the forces acting on her when she has finally come to rest and solve for x:

$F_{net} = kx - mg = 0$

and

$x = \dfrac{mg}{k}$

Evaluate x:

$x = \dfrac{(60\,\text{kg})(9.81\,\text{m/s}^2)}{5.40\,\text{N/m}} = 109\,\text{m}$

Substitute in equation (1) and evaluate h:

$h = 310\,\text{m} - 50\,\text{m} - 109\,\text{m} = \boxed{151\,\text{m}}$

(b) Using conservation of energy, express her total energy E:

$E = K + U_g + U_s = E_i = 0$

Because v is a maximum when K is a maximum, solve for K and set its derivative with respect to x equal to zero:

$K = -U_g - U_s$

$\quad = mg(d + x) - \tfrac{1}{2}kx^2$ $\qquad$ (1)

$\dfrac{dK}{dx} = mg - kx = 0$ for extreme values

Solve for and evaluate x:

$x = \dfrac{mg}{k} = \dfrac{(60\,\text{kg})(9.81\,\text{m/s}^2)}{5.40\,\text{N/m}} = 109\,\text{m}$

From equation (1) we have:

$\tfrac{1}{2}mv^2 = mg(d + x) - \tfrac{1}{2}kx^2$

Solve for v to obtain:

$v = \sqrt{2g(d + x) - \dfrac{kx^2}{m}}$

Substitute numerical values and evaluate v for $x = 109$ m:

$v = \sqrt{2(9.81\,\text{m/s}^2)(50\,\text{m} + 109\,\text{m}) - \dfrac{(5.4\,\text{N/m})(109\,\text{m})^2}{60\,\text{kg}}} = \boxed{45.3\,\text{m/s}}$

Because $\dfrac{d^2K}{dx^2} = -k < 0$:

$x = 109$ m corresponds to K_{max} and so v is a maximum.

*39 •• Walking by a pond, you find a rope attached to a stout tree limb 5.2 m off the ground. You decide to use the rope to swing out over the pond. The rope is a bit frayed but supports your weight. You estimate that the rope might break if the tension is 80 N greater than your weight. You grab the rope at a point 4.6 m from the limb and move back to swing out over the pond. (*a*) What is the maximum safe initial angle between the rope and the vertical so that it will not break during the swing? (*b*) If you begin at this maximum angle, and the surface of the pond is 1.2 m below the level of the ground, with what speed will you enter the water if you let go of the rope when the rope is vertical?

Picture the Problem Let the system consist of you and the earth. Then there are no external forces to do work on the system and $W_{ext} = 0$. In the figure, your initial position is designated with the numeral 1, the point at which you release the rope and begin to fall with a 2, and your point of impact with the water is identified with a 3. Choose $U_g = 0$ at the water level. We can apply Newton's 2nd law to the forces acting on you at point 2 and apply conservation of energy between points 1 and 2 to determine the maximum angle at which you can begin your swing and then between points 1 and 3 to determine the speed with which you will hit the water.

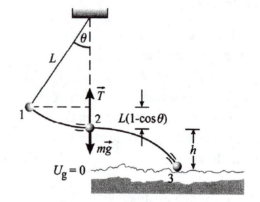

(*a*) Use conservation of energy to relate your speed at point 2 to your potential energy there and at point 1:

$W_{ext} = \Delta K + \Delta U = 0$

$K_2 - K_1 + U_2 - U_1 = 0$

Because $K_1 = 0$,

$\tfrac{1}{2}mv_2^2 + mgh$
$\quad - [mgL(1 - \cos\theta) + mgh] = 0$

Solve this equation for θ:

$$\theta = \cos^{-1}\left[1 - \frac{v_2^2}{2gL}\right] \qquad (1)$$

Apply $\sum F_{radial} = ma_{radial}$ yourself at point 2 and solve for T:

$T - mg = m\dfrac{v_2^2}{L}$

and

$T = mg + m\dfrac{v_2^2}{L}$

Because you've estimated that the rope might break if the tension in it exceeds your weight by 80 N, it must be that:

$$m\frac{v_2^2}{L} = 80\,\text{N}$$

or

$$v_2^2 = \frac{(80\,\text{N})L}{m}$$

Let's assume your weight is 650 N. Then your mass is 66.3 kg and:

$$v_2^2 = \frac{(80\,\text{N})(4.6\,\text{m})}{66.3\,\text{kg}} = 5.55\,\text{m}^2/\text{s}^2$$

Substitute in equation (1) to obtain:

$$\theta = \cos^{-1}\left[1 - \frac{5.55\,\text{m}^2/\text{s}^2}{2(9.81\,\text{m/s}^2)(4.6\,\text{m})}\right]$$

$$= \boxed{20.2°}$$

(b) Apply conservation of energy to the energy transformations between points 1 and 3:

$$W_{\text{ext}} = \Delta K + \Delta U = 0$$
$$K_3 - K_1 + U_3 - U_1 = 0 \text{ where } U_3 \text{ and } K_1\text{are zero}$$

Substitute for K_3 and U_1 to obtain:

$$\tfrac{1}{2}mv_3^2 - mg[h + L(1 - \cos\theta)] = 0$$

Solve for v_3:

$$v_3 = \sqrt{2g[h + L(1 - \cos\theta)]}$$

Substitute numerical values and evaluate v_3:

$$v_3 = \sqrt{2(9.81\,\text{m/s}^2)[1.8\,\text{m} + (4.6\,\text{m})(1 - \cos 20.2°)]} = \boxed{6.39\,\text{m/s}}$$

Kinetic Friction

*46 •   Returning to Problem 19 and Figure 7-17, suppose that the surfaces described are not frictionless and that the coefficient of kinetic friction between the block and the surfaces is 0.30. Find (a) the speed of the block when it reaches the ramp, and (b) the distance that the block slides up the ramp before coming momentarily to rest. (Neglect the energy dissipated along the transition curve.)

Picture the Problem Choose $U_g = 0$ at the foot of the ramp and let the system consist of the block, ramp, and the earth. Then the kinetic energy of the block at the foot of the ramp is equal to its initial kinetic energy less the energy dissipated by friction. The block's kinetic energy at the foot of the incline is partially converted to gravitational potential energy and partially dissipated by friction as the block slides up the incline. The free-body diagram shows the forces acting on the block as it slides up the incline. Applying Newton's 2nd law to the block will allow us to determine f_k and express the energy dissipated by friction.

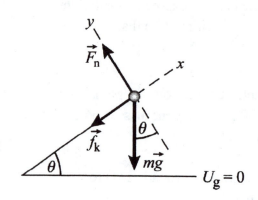

(*a*) Apply conservation of energy to the system while the block is moving horizontally:

$$W_{ext} = \Delta K + \Delta U$$
or, because $\Delta U = 0$,
$$W_{ext} = \Delta K = K_f - K_i$$

Solve for K_f:

$$K_f = K_i + W_{ext}$$

Express the work done by the friction force:

$$W_{ext} = \vec{F} \cdot \vec{s} = -f_k \Delta x = -\mu_k mg \Delta x$$

Substitute for W_{ext} to obtain:

$$\tfrac{1}{2}mv_f^2 = \tfrac{1}{2}mv_i^2 - \mu_k mg \Delta x$$

Solve for v_f:

$$v_f = \sqrt{v_i^2 - 2\mu_k g \Delta x}$$

Substitute numerical values and evaluate v_f:

$$v_f = \sqrt{(7\,\text{m/s})^2 - 2(0.3)(9.81\,\text{m/s}^2)(2\,\text{m})}$$
$$= \boxed{6.10\,\text{m/s}}$$

(*b*) Relate the initial kinetic energy of the block to its final potential energy and the energy dissipated by friction:

$$K_i = U_f + W_{ext}$$

Apply $\sum F_y = ma_y$ to the block:

$$F_n - mg\cos\theta = 0 \Rightarrow F_n = mg\cos\theta$$

Express W_{ext}:

$$W_{ext} = f_k L = \mu_k F_n L = \mu_k mgL \cos\theta$$

Express the final potential energy of the block:

$$U_f = mgL \sin\theta$$

Substitute for U_f and W_{ext} to obtain:

$$K_f = mgL \sin\theta + \mu_k mgL \cos\theta$$

Solve for L:

$$L = \frac{K_f}{mg(\sin\theta + \mu_k \cos\theta)}$$

Substitute numerical values and evaluate L:

$$L = \frac{\frac{1}{2}(6.10\,\text{m/s})^2}{(9.81\,\text{m/s}^2)(\sin 40° + (0.3)\cos 40°)}$$

$$= \boxed{2.17\,\text{m}}$$

*50 ••   A compact object of mass m moves in a horizontal circle of radius r on a rough table. It is attached to a horizontal string fixed at the center of the circle. The speed of the object is initially v_0. After completing one full trip around the circle, the speed of the object is $\frac{1}{2}v_0$. (a) Find the energy dissipated by friction during that one revolution in terms of m, v_0, and r. (b) What is the coefficient of kinetic friction? (c) How many more revolutions will the object make before coming to rest?

Picture the Problem Let the system consist of the particle and the earth. Then the friction force is external to the system and does work to change the energy of the system. The energy dissipated by friction during one revolution is the work done by the friction force.

(a) Relate the work done by friction to the change in energy of the system:

$$W_{ext} = W_f = \Delta K + \Delta U$$
$$= K_f - K_i, \text{ since } \Delta U = 0$$

Substitute for K_f and K_i and simplify to obtain:

$$W_f = \tfrac{1}{2}mv_f^2 - \tfrac{1}{2}mv_i^2$$
$$= \tfrac{1}{2}m(\tfrac{1}{2}v_0)^2 - \tfrac{1}{2}m(v_0)^2$$
$$= \boxed{\tfrac{3}{8}mv_0^2}$$

(b) Relate the work done by friction to the distance traveled and the coefficient of kinetic friction and solve for the latter:

$$W_f = \mu_k mg\Delta s$$
$$= \mu_k mg(2\pi r)$$

and

$$\mu_k = \frac{W_f}{2\pi mgr} = \frac{\frac{3}{8}mv_0^2}{2\pi mgr} = \boxed{\frac{3v_0^2}{16\pi gr}}$$

(c) | Since in one revolution it lost $\frac{3}{4}K_i$, it will only require another 1/3 revolution to lose the remaining $\frac{1}{4}K_i$.

Mass and Energy

*56 • If a black hole and a "normal" star orbit each other, gases from the star falling into the black hole can be heated millions of degrees by frictional heating in the black hole's accretion disk. When the gases are heated that much, they begin to radiate light in the X-ray region of the spectrum. Cygnus X-1, the second brightest source in the X-ray sky, is thought to be one such binary system; it radiates an estimated power of 4×10^{31} W. If we assume that 1 percent of the infalling mass-energy escapes as X rays, at what rate is the black hole gaining mass?

Picture the Problem We can differentiate the mass-energy equation to obtain an expression for the rate at which the black hole gains energy.

Using the mass-energy relationship, express the energy radiated by the black hole:

$$E = 0.01mc^2$$

Differentiate this expression to obtain an expression for the rate at which the black hole is radiating energy:

$$\frac{dE}{dt} = \frac{d}{dt}\left[0.01mc^2\right] = 0.01c^2\frac{dm}{dt}$$

Solve for dm/dt:

$$\frac{dm}{dt} = \frac{dE/dt}{0.01c^2}$$

Substitute numerical values and evaluate dm/dt:

$$\frac{dm}{dt} = \frac{4\times10^{31}\text{ watt}}{0.01(2.998\times10^8\text{ m/s})^2}$$

$$= \boxed{4.45\times10^{16}\text{ kg/s}}$$

General Problems

*62 ••   A block of mass m, starting from rest, is pulled up a frictionless inclined plane that makes an angle θ with the horizontal by a string parallel to the plane. The tension in the string is T. After traveling a distance L, the speed of the block is v. The work done by the tension T is (a) $mgL \sin \theta$, (b) $mgL \cos \theta + \frac{1}{2}mv^2$, (c) $mgL \sin \theta + \frac{1}{2}mv^2$, (d) $mgL \cos \theta$, (e) $TL \cos \theta$.

Picture the Problem Let the system consist of the block, the earth, and the incline. Then the tension in the string is an external force that will do work to change the energy of the system. Because the incline is frictionless; the work done by the tension in the string as it displaces the block on the incline is equal to the sum of the changes in the kinetic and gravitational potential energies.

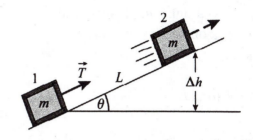

Relate the work done by the tension force to the changes in the kinetic and gravitational potential energies of the block:

$$W_{\text{tension force}} = W_{\text{ext}} = \Delta U + \Delta K$$

Referring to the figure, express the change in the potential energy of the block as it moves from position 1 to position 2:

$$\Delta U = mg\Delta h = mgL \sin \theta$$

Because the block starts from rest:

$$\Delta K = K_2 = \tfrac{1}{2}mv^2$$

Substitute to obtain:

$$W_{\text{tension force}} = mgL \sin \theta + \tfrac{1}{2}mv^2$$

and $\boxed{(c) \text{ is correct.}}$

*65 •   The average energy per unit time per unit area that reaches the upper atmosphere of the earth from the sun, called the solar constant, is 1.35 kW/m². Because of absorption and reflection by the atmosphere, about 1 kW/m² reaches the surface of the earth on a clear day. How much energy is collected during 8 h of daylight by a window 1 m by 2 m? The window is on a mount that rotates, keeping the window facing the sun so the sun's rays remain perpendicular to the window.

Picture the Problem The solar constant is the average energy per unit area and per unit time reaching the upper atmosphere. This physical quantity can be thought of as the power per unit area and is known as *intensity*.

Letting $I_{surface}$ represent the intensity of the solar radiation at the surface of the earth, express $I_{surface}$ as a function of power and the area on which this energy is incident:

$$I_{surface} = \frac{P}{A} = \frac{\Delta E / \Delta t}{A}$$

Solve for ΔE:

$$\Delta E = I_{surface} A \Delta t$$

Substitute numerical values and evaluate ΔE:

$$\Delta E = (1\,kW/m^2)(2\,m^2)(8\,h)(3600\,s/h)$$
$$= \boxed{57.6\,MJ}$$

*70 • A 1200-kg elevator can safely carry a maximum load of 800 kg. What is the power provided by the electric motor powering the elevator when the elevator ascends with a full load at a speed of 2.3 m/s?

Picture the Problem The power provided by a motor that is delivering sufficient energy to exert a force F on a load which it is moving at a speed v is Fv.

The power provided by the motor is given by:

$$P = Fv$$

Because the elevator is ascending with constant speed, the tension in the support cable(s) is:

$$F = (m_{elev} + m_{load})g$$

Substitute for F to obtain:

$$P = (m_{elev} + m_{load})gv$$

Substitute numerical values and evaluate P:

$$P = (2000\,kg)(9.81\,m/s^2)(2.3\,m/s)$$
$$= \boxed{45.1\,kW}$$

*73 •• In a volcanic eruption, a 2-kg piece of porous volcanic rock is thrown vertically upward with an initial speed of 40 m/s. It travels upward a distance of

50 m before it begins to fall back to the earth. (*a*) What is the initial kinetic energy of the rock? (*b*) What is the increase in thermal energy due to air friction during ascent? (*c*) If the increase in thermal energy due to air friction on the way down is 70% of that on the way up, what is the speed of the rock when it returns to its initial position?

Picture the Problem Let the system consist of the earth, rock, and air. Given this choice, there are no external forces to do work on the system and $W_{ext} = 0$. Choose $U_g = 0$ to be where the rock begins its upward motion. The initial kinetic energy of the rock is partially transformed into potential energy and partially dissipated by air resistance as the rock ascends. During its descent, its potential energy is partially transformed into kinetic energy and partially dissipated by air resistance.

(*a*) Using the definition of kinetic energy, calculate the initial kinetic energy of the rock:	$K_i = \frac{1}{2}mv_i^2 = \frac{1}{2}(2\,\text{kg})(40\,\text{m/s})^2$ $= \boxed{1.60\,\text{kJ}}$
(*b*) Apply the work-energy theorem with friction to relate the energies of the system as the rock ascends:	$\Delta K + \Delta U + W_f = 0$
Because $K_f = 0$:	$-K_i + \Delta U + W_f = 0$ and $W_f = K_i - \Delta U$
Substitute numerical values and evaluate W_f:	$W_f = 1600\,\text{J} - (2\,\text{kg})(9.81\,\text{m/s}^2)(50\,\text{m})$ $= \boxed{619\,\text{J}}$
(*c*) Apply the work-energy theorem with friction to relate the energies of the system as the rock descends:	$\Delta K + \Delta U + W_f = 0$
Because $K_i = U_f = 0$:	$K_f - U_i + W_f' = 0$ where $W_f' = 0.7W_f$.
Substitute for the energies to obtain:	$\frac{1}{2}mv_f^2 - mgh + 0.7W_f = 0$

Solve for v_f:

$$v_f = \sqrt{2gh - \frac{1.4W_f}{m}}$$

Substitute numerical values and evaluate v_f:

$$v_f = \sqrt{2(9.81\,\text{m/s}^2)(50\,\text{m}) - \frac{1.4(619\,\text{J})}{2\,\text{kg}}}$$

$$= \boxed{23.4\,\text{m/s}}$$

*75 •   A 1.5×10^4-kg stone slab rests on a steel girder. On a very hot day, you find that the girder has expanded, lifting the slab by 0.1 cm. (*a*) What work does the girder do on the slab? (*b*)Where does the energy come from to lift the slab? Give a microscopic picture of what is happening in the steel.

Picture the Problem We can find the work done by the girder on the slab by calculating the change in the potential energy of the slab.

(*a*) Relate the work the girder does on the slab to the change in potential energy of the slab:

$$W = \Delta U = mg\Delta h$$

Substitute numerical values and evaluate W:

$$W = (1.5\times10^4\,\text{kg})(9.81\,\text{m/s}^2)(0.001\,\text{m})$$

$$= \boxed{147\,\text{J}}$$

(*b*)

> The energy is transferred to the girder from its surroundings, which are warmer than the girder. As the temperature of the girder rises, the atoms in the girder vibrate with a greater average kinetic energy, leading to a larger average separation, which causes the girder's expansion.

*77 ••   A particle of mass m is suspended from the ceiling by a spring and is free to move vertically in the y direction as indicated (Figure 7-29). We are given that the potential energy as a function of position is $U = \frac{1}{2}ky^2 - mgy$. (*a*) Using a spreadsheet program or graphing calculator, make a graph of U as a function of y. For what value of y is the spring *unstretched*? (*b*) From the expression given for U, find the net force acting on m at any position y. (*c*) The particle is released from rest at $y = 0$; if there is no friction, what is the maximum value of y, y_{max} that will be reached by the mass? Indicate y_{max} on your graph. (*d*) Now consider the effect of friction. The mass ultimately settles into an equilibrium position y_{eq}. Find this point on your sketch. (*e*) Find the amount of thermal energy produced by friction from the start of the operation to the final equilibrium.

Figure 7-29 Problem 77

Picture the Problem Given the potential energy function as a function of y, we can find the net force acting on a given system from $F = -dU / dy$. The maximum extension of the spring, i.e., the lowest position of the mass on its end, can be found by applying the work-energy theorem. The equilibrium position of the system can be found by applying the work-energy theorem with friction … as can the amount of thermal energy produced as the system oscillates to its equilibrium position.

(*a*) The graph of U as a function of y is shown to the right. Because k and m are not specified, k has been set equal to 2 and mg to 1. The spring is unstretched when $y = y_0 = 0$. Note that the minimum value of U (a position of stable equilibrium) occurs near $y = 5$ m.

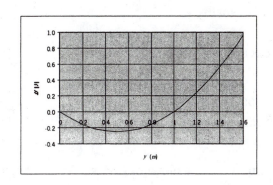

(*b*) Evaluate the negative of the derivative of U with respect to y:

$$F = -\frac{dU}{dy} = -\frac{d}{dy}\left(\tfrac{1}{2}ky^2 - mgy\right)$$

$$= \boxed{-ky + mg}$$

(*c*) Apply conservation of energy to the movement of the mass from $y = 0$ to $y = y_{max}$:

$$\Delta K + \Delta U + W_f = 0$$

Because $\Delta K = 0$ (the object starts from rest and is momentarily at rest at $y = y_{max}$) and $W_f = 0$ (no friction), it follows that:

$$\Delta U = U(y_{max}) - U(0) = 0$$

Because $U(0) = 0$, it follows that:

$$U(y_{max}) = 0 \Rightarrow \tfrac{1}{2}ky_{max}^2 - mgy_{max} = 0$$

Solve for y_{max}:

$$y_{max} = \boxed{\dfrac{2mg}{k}}$$

(d) Express the condition of F at equilibrium and solve for y_{eq}:

$$F_{eq} = 0 \Rightarrow -ky_{eq} + mg = 0$$

and

$$y_{eq} = \boxed{\dfrac{mg}{k}}$$

(e) Apply the conservation of energy to the movement of the mass from $y = 0$ to $y = y_{eq}$ and solve for W_f:

$$\Delta K + \Delta U + W_f = 0$$

or, because $\Delta K = 0$.

$$W_f = -\Delta U = U_i - U_f$$

Because $U_i = U(0) = 0$:

$$W_f = -U_f = -\left(\tfrac{1}{2}ky_{eq}^2 - mgy_{eq}\right)$$

Substitute for y_{eq} and simplify to obtain:

$$W_f = \boxed{\dfrac{m^2 g^2}{2k}}$$

*80 •   An elevator (mass $M = 2000$ kg) is moving downward at $v_0 = 1.5$ m/s. A braking system prevents the downward speed from increasing. (a) At what rate (in J/s) is the braking system converting mechanical energy to thermal energy? (b) While the elevator is moving downward at $v_0 = 1.5$ m/s, the braking system fails and the elevator is in free fall for a distance $d = 5$ m before hitting the top of a large safety spring with force constant $k = 1.5 \times 10^4$ N/m. After the elevator hits the top of the spring, we want to know the distance Δy that the spring is compressed before the elevator is brought to rest. Write an algebraic expression for the value of Δy in terms of the known quantities M, v_0, g, k, and d, and substitute the given values to find Δy.

Picture the Problem The rate of conversion of mechanical energy can be determined from $P = \vec{F} \cdot \vec{v}$. The pictorial representation shows the elevator moving downward just as it goes into freefall as state 1. In state 2 the elevator is moving faster and is about to strike the relaxed spring. The momentarily at rest elevator on the compressed spring is shown as state 3. Let $U_g = 0$ where the spring has its maximum compression and the system consist of the earth, the elevator, and the spring. Then $W_{ext} = 0$ and we can apply the conservation of mechanical energy to the analysis of the falling elevator and compressing spring.

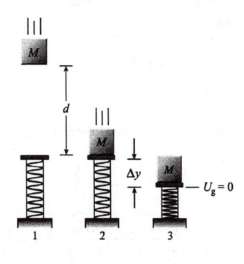

(a) Express the rate of conversion of mechanical energy to thermal energy as a function of the speed of the elevator and braking force acting on it:

$$P = F_{braking}\, v_0$$

Because the elevator is moving with constant speed, the net force acting on it is zero and:

$$F_{braking} = Mg$$

Substitute for $F_{braking}$ and evaluate P:

$$P = Mgv_0$$
$$= (2000\,\text{kg})(9.81\,\text{m/s}^2)(1.5\,\text{m/s})$$
$$= \boxed{29.4\,\text{kW}}$$

(b) Apply the conservation of energy to the falling elevator and compressing spring:

$$\Delta K + \Delta U_g + \Delta U_s = 0$$

or

$$K_3 - K_1 + U_{g,3} - U_{g,1} + U_{s,3} - U_{s,1} = 0$$

Because $K_3 = U_{g,3} = U_{s,1} = 0$:

$$-\tfrac{1}{2}Mv_0^2 - Mg(d + \Delta y) + \tfrac{1}{2}k(\Delta y)^2 = 0$$

Rewrite this equation as a quadratic equation in Δy, the maximum compression of the spring:

$$(\Delta y)^2 - \left(\frac{2Mg}{k}\right)\Delta y - \frac{M}{k}\left(2gd + v_0^2\right) = 0$$

Solve for Δy to obtain:

$$\Delta y = \frac{Mg}{k} \pm \sqrt{\frac{M^2 g^2}{k^2} + \frac{M}{k}\left(2gd + v_0^2\right)}$$

Substitute numerical values and evaluate Δy:

$$\Delta y = \frac{(2000\,\text{kg})(9.81\,\text{m/s}^2)}{1.5\times10^4\,\text{N/m}}$$

$$+ \sqrt{\frac{(2000\,\text{kg})^2(9.81\,\text{m/s}^2)^2}{(1.5\times10^4\,\text{N/m})^2} + \frac{2000\,\text{kg}}{1.5\times10^4\,\text{N/m}}\left[2(9.81\,\text{m/s}^2)(5\,\text{m}) + (1.5\,\text{m/s})^2\right]}$$

$$= \boxed{5.19\,\text{m}}$$

*84 ••   While driving, one expects to expend more power when accelerating than when driving at a constant speed. (*a*) Neglecting friction, calculate the energy required to give a 1200-kg car a speed of 50 km/h. (*b*) If friction (rolling friction and air drag) results in a retarding force of 300 N at a speed of 50 km/h, what is the energy needed to move the car a distance of 300 m at a constant speed of 50 km/h? (*c*) Assuming the energy losses due to friction in Part (*a*) are 75 percent of those found in Part (*b*), estimate the ratio of the energy consumption for the two cases considered.

Picture the Problem While on a horizontal surface, the work done by an automobile engine changes the kinetic energy of the car and does work against friction. These energy transformations are described by the work-energy theorem with friction. Let the system include the earth, the roadway, and the car *but not the car's engine*.

(*a*) The required energy equals the change in the kinetic energy of the car:

$$\Delta K = \tfrac{1}{2}mv^2$$

$$= \tfrac{1}{2}(1200\,\text{kg})\left(50\frac{\text{km}}{\text{h}} \times \frac{1\text{h}}{3600\,\text{s}}\right)^2$$

$$= \boxed{116\,\text{kJ}}$$

(*b*) The required energy equals the work done against friction:

$$W_f = F_f\Delta s$$

Substitute numerical values and evaluate W_f:

$$W_f = (300\,\text{N})(300\,\text{m}) = \boxed{90.0\,\text{kJ}}$$

(c) Apply the work-energy theorem with friction to express the required energy:

$$E' = W_{\text{ext}} = \Delta K + W_f$$
$$= \Delta K + 0.75E$$

Divide both sides of the equation by E to express the ratio of the two energies:

$$\frac{E'}{E} = \frac{\Delta K}{E} + 0.75$$

Substitute numerical values and evaluate E'/E:

$$\frac{E'}{E} = \frac{116\,\text{kJ}}{90\,\text{kJ}} + 0.75 = \boxed{2.04}$$

*85 •••   A pendulum consisting of a string of length L with a small bob on the end (mass M) is pulled horizontally, then released (Figure 7-32). At the lowest point of the swing, the string catches on a thin peg a distance R above the lowest point. Show that R must be smaller than $2L/5$ if the bob is to swing around the peg in a full circle.

Figure 7-32 Problem 85

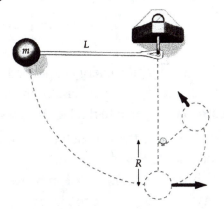

Picture the Problem Assume that the bob is moving with speed v as it passes the top vertical point when looping around the peg. There are two forces acting on the bob: the tension in the string (if any) and the force of gravity, Mg; both point downward when the ball is in the topmost position. The minimum possible speed for the bob to pass the vertical occurs when the tension is 0; from this, gravity must supply the centripetal force required to keep the ball moving in a circle. We

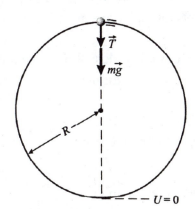

can use conservation of energy to relate v to L and R.

Express the condition that the bob swings around the peg in a full circle:	$M\dfrac{v^2}{R} > Mg$
Simplify to obtain:	$\dfrac{v^2}{R} > g$
Use conservation of energy to relate the kinetic energy of the bob at the bottom of the loop to its potential energy at the top of its swing:	$\tfrac{1}{2}Mv^2 = Mg(L - 2R)$
Solve for v^2:	$v^2 = 2g(L - 2R)$
Substitute to obtain:	$\dfrac{2g(L - 2R)}{R} > g$
Solve for R:	$\boxed{R < \dfrac{2}{5}L}$

*88 ••   In one model of jogging, the energy expended is assumed to go into accelerating and decelerating the legs. If the mass of the leg is m and the running speed is v, the energy needed to accelerate the leg from rest to v is $\tfrac{1}{2}mv^2$, and the same energy is needed to decelerate the leg back to rest for the next stride. Thus, the energy required for each stride is mv^2. Assume that the mass of a man's leg is 10 kg and that he jogs at a speed of 3 m/s with 1 m between one footfall and the next. Therefore, the energy he must provide to his legs in each second is $3 \times mv^2$. Calculate the rate of the man's energy expenditure using this model and assuming that his muscles have an efficiency of 20 percent.

Picture the Problem We're given $P = dW / dt$ and are asked to evaluate it under the assumed conditions.

Express the rate of energy expenditure by the man:	$P = 3mv^2 = 3(10\,\text{kg})(3\,\text{m/s})^2$ $= 270\,\text{W}$
Express the rate of energy expenditure P' assuming that his	$P = \tfrac{1}{5}P'$

muscles have an efficiency of
20%:

Solve for and evaluate P': $P' = 5P = 5(270\,\text{W}) = \boxed{1.35\,\text{kW}}$

*94 •••   A block of wood (mass M) is connected to two massless springs as
shown in Figure 7-36. Each spring has unstretched length L and spring constant
k. (*a*) If the block is displaced a distance x, as shown, what is the change in the
potential energy stored in the springs? (*b*) What is the magnitude of the force
pulling the block back toward the equilibrium position? (*c*) Using a spreadsheet
program or graphing calculator, make a graph of the potential energy U as a
function of x, for $0 \le x \le 0.2$ m. Assume $k = 1$ N/m, $L = 0.1$ m, and $M = 1$ kg.
(*d*) If the block is displaced a distance $x = 0.1$ m and released, what is its speed as
it passes the equilibrium point? Assume that the block is resting on a frictionless
surface.

Figure 7-36 Problem 94

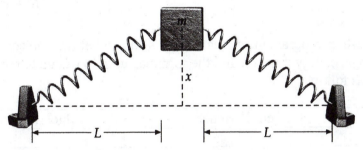

Picture the Problem The diagram to
the right shows the forces each of the
springs exerts on the block. The change
in the potential energy stored in the
springs is due to the elongation of both
springs when the block is displaced a
distance x from its equilibrium position
and we can find ΔU using $\frac{1}{2}k(\Delta L)^2$. We
can find the magnitude of the force
pulling the block back toward its
equilibrium position by finding the sum
of the magnitudes of the y components
of the forces exerted by the springs. In
Part (*d*) we can use conservation of
energy to find the speed of the block as
it passes through its equilibrium
position.

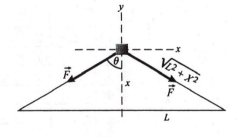

(*a*) Express the change in the
potential energy stored in the $\Delta U = 2\left[\frac{1}{2}k(\Delta L)^2\right] = k(\Delta L)^2$

springs when the block is displaced a distance x:

where ΔL is the change in length of a spring.

Referring to the force diagram, express ΔL:

$$\Delta L = \sqrt{L^2 + x^2} - L$$

Substitute to obtain:

$$\Delta U = \boxed{k\left(\sqrt{L^2 + x^2} - L\right)^2}$$

(*b*) Sum the forces acting on the block to express $F_{restoring}$:

$$F_{restoring} = 2F\cos\theta = 2k\Delta L\cos\theta$$

$$= 2k\Delta L\frac{x}{\sqrt{L^2 + x^2}}$$

Substitute for ΔL to obtain:

$$F_{restoring} = 2k\left(\sqrt{L^2 + x^2} - L\right)\frac{x}{\sqrt{L^2 + x^2}}$$

$$= \boxed{2kx\left(1 - \frac{L}{\sqrt{L^2 + x^2}}\right)}$$

(*c*) A spreadsheet program to calculate $U(x)$ is shown below. The constants used in the potential energy function and the formulas used to calculate the potential energy are as follows:

Cell	Content/Formula	Algebraic Form
B1	1	L
B2	1	k
B3	1	M
C8	C7+0.01	x

	A	B	C	D
1	L =	0.1	m	
2	k =	1	N/m	
3	M =	1	kg	
4				
5				
6			x	$U(x)$
7			0	0
8			0.01	2.49E–07
9			0.02	3.92E–06
10			0.03	1.94E–05
11			0.04	5.93E–05
12			0.05	1.39E–04
23			0.16	7.86E–03

24			0.17	9.45E–03
25			0.18	1.12E–02
26			0.19	1.32E–02
27			0.20	1.53E–02

The graph shown below was plotted using the data from columns C (x) and D ($U(x)$).

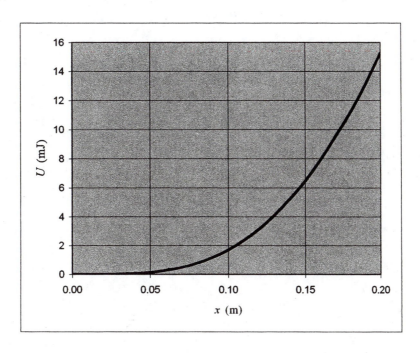

(*d*) Use conservation of energy to relate the kinetic energy of the block as it passes through the equilibrium position to the change in its potential energy as it returns to its equilibrium position:

$$K_{\text{equilibrium}} = \Delta U$$

or

$$\tfrac{1}{2}Mv^2 = \Delta U$$

Solve for v to obtain:

$$v = \sqrt{\frac{2\Delta U}{M}} = \sqrt{\frac{2k\left(\sqrt{L^2 + x^2} - L\right)^2}{M}}$$

$$= \left(\sqrt{L^2 + x^2} - L\right)\sqrt{\frac{2k}{M}}$$

Substitute numerical values and evaluate v:

$$v = \left(\sqrt{(0.1\,\text{m})^2 + (0.1\,\text{m})^2} - 0.1\,\text{m}\right)\sqrt{\frac{2(1\,\text{N/m})}{1\,\text{kg}}} = \boxed{5.86\,\text{cm/s}}$$

Chapter 8
Systems of Particles and Conservation
of Linear Momentum

Conceptual Problems

*2 • A cannonball is dropped off a high tower while, simultaneously, an identical cannonball is launched directly upward into the air. The center of mass of the two cannonballs (*a*) stays in the same place, (*b*) initially rises, then falls, but begins to fall *before* the cannonball launched into the air starts falling, (*c*) initially rises, then falls, but begins to fall *at the same time* as the cannonball launched upward begins to fall, (*d*) initially rises, then falls, but begins to fall *after* the cannonball launched upward begins to fall.

Determine the Concept The center of mass is midway between the two balls, and is in free-fall along with them (all forces can be thought to be concentrated at the center of mass.) The center of mass will initially rise, then fall.

Because the initial velocity of the center of mass is half of the initial velocity of the ball thrown upwards, the mass thrown upwards will rise for twice the time that the center of mass rises. Also, the center of mass will rise until the velocities of the two balls are equal but opposite. $\boxed{(b) \text{ is correct.}}$

*5 • If two particles have equal kinetic energies, are the magnitudes of their momenta necessarily equal? Explain and give an example.

Determine the Concept No. Consider a 1-kg block with a speed of 1 m/s and a 2-kg block with a speed of 0.707 m/s. The blocks have equal kinetic energies but momenta of magnitude 1 kg·m /s and 1.414 kg·m/s, respectively.

*8 • A child jumps from a small boat to a dock. Why does she have to jump with more energy than she would need if she were jumping the same distance from one dock to another?

Determine the Concept When she jumps from a boat to a dock, she must, in order for momentum to be conserved, give the boat a recoil momentum, i.e., her forward momentum must be the same as the boat's backward momentum. The energy she imparts to the boat is $E_{\text{boat}} = p_{\text{boat}}^2 / 2m_{\text{boat}}$.

$\boxed{\text{When she jumps from one dock to another, the mass of the dock plus the earth is so large that the energy she imparts to them is essentially zero.}}$

*9 •• Much early research in rocket motion was done by Robert Goddard, physics professor at Clark College in Worcester, Massachusetts. A quotation from a 1920 editorial in the *New York Times* illustrates the public opinion of his work: "That Professor Goddard with his 'chair' at Clark College and the countenance of the Smithsonian Institution does not know the relation between action and reaction, and the need to have something better than a vacuum against which to react—to say that would be absurd. Of course, he only seems to lack the knowledge ladled out daily in high schools."[†] The belief that a rocket needs something to push against was a prevalent misconception before rockets in space were commonplace. Explain why that belief is wrong.

Determine the Concept Conservation of momentum requires only that the net external force acting on the system be zero. It does not require the presence of a medium such as air.

*14 •• If only external forces can cause the center of mass of a system of particles to accelerate, how can a car move? We normally think of the car's engine as supplying the force needed to accelerate the car, but is this true? Where does the external force that accelerates the car come from?

Determine the Concept There is only one force which can cause the car to move forward–the friction of the road! The car's engine causes the tires to rotate, but if the road were frictionless (as is closely approximated by icy conditions) the wheels would simply spin without the car moving anywhere. Because of friction, the car's tire pushes backwards against the road–from Newton's third law, the frictional force acting on the tire must then push it forward. This may seem odd, as we tend to think of friction as being a retarding force only, but true.

*20 •• Under what conditions can all the initial kinetic energy of an isolated system consisting of two colliding bodies be lost in a collision?

Determine the Concept All the initial kinetic energy of the isolated system is lost in a perfectly inelastic collision in which the velocity of the center of mass is zero.

*22 •• A double-barreled pea shooter is shown in Figure 8-48, the diagram below. Air is blown from the left end of the straw, and identical peas A and B are positioned inside the straw as shown. If the shooter is held horizontally while the peas are shot off, which pea, A or B, will travel farther after leaving the straw? Why? *Hint: The answer has to do with the impulse-momentum theorem.*

Figure 8-48 Problem 22

Determine the Concept A will travel farther. Both peas are acted on by the same force, but pea A is acted on by that force for a longer time. By the impulse-momentum theorem, its momentum (and, hence, speed) will be higher than pea B's speed on leaving the shooter.

*28 •• A railroad car is passing by a grain elevator, which is dumping grain into it at a constant rate. (*a*) Does momentum conservation imply that the railroad car should be slowing down as it passes the grain elevator? Assume that the track is frictionless and perfectly level. (*b*) If the car is slowing down, this implies that there is some external force acting on the car to slow it down. Where does this force come from? (*c*) After passing the elevator, the railroad car springs a leak, and grain starts leaking out of a vertical hole in its floor at a constant rate. Should the car speed up as it loses mass?

Determine the Concept We can apply conservation of momentum and Newton's laws of motion to the analysis of these questions.

(*a*) Yes, the car should slow down. An easy way of seeing this is to imagine a "packet" of grain being dumped into the car all at once: This is a completely inelastic collision, with the packet having an initial horizontal velocity of 0. After the collision, it is moving with the same horizontal velocity that the car does, so the car must slow down.

(*b*) When the packet of grain lands in the car, it initially has a horizontal velocity of 0, so it must be accelerated to come to the same speed as the car of the train. Therefore, the train must exert a force on it to accelerate it. By Newton's 3rd law, the grain exerts an equal but opposite force on the car, slowing it down. In general, this is a frictional force which causes the grain to come to the same speed as the car.

(*c*) No it doesn't speed up. Imagine a packet of grain being "dumped" out of the railroad car. This can be treated as a collision, too. It has the same horizontal speed as the railroad car when it leaks out, so the train car doesn't have to speed up or slow down to conserve momentum.

*29 •• To show that even really bright people can make mistakes, consider the following problem which was given to the freshman class at Caltech on an exam (paraphrased): *A sailboat is sitting in the water on a windless day. In order to make the boat move, a misguided sailor sets up a fan in the back of the boat to blow into the sails to make the boat move forward. Explain why the boat won't move.* The idea was that the net force of the wind pushing the sail forward would be counteracted by the force pushing the fan back (Newton's third law). However, as one of the students pointed out to his professor, the sailboat *could* in fact move forward. Why is that?

Determine the Concept Think of the stream of air molecules hitting the sail. Imagine that they bounce off the sail elastically–their net change in momentum is then roughly twice the change in momentum that they experienced going through the fan. Another way of looking at it: Initially, the air is at rest, but after passing through the fan and bouncing off the sail, it is moving backward–therefore, the boat must exert a net force on the air pushing it backward, and there must be a force on the boat pushing it forward.

Estimation and Approximation

*32 ••   A counterintuitive physics demonstration can be performed by firing a rifle bullet into a melon. (Don't try this at home!) When hit, nine times out of ten the melon will jump backward, toward the rifle, *opposite* to the direction in which the bullet was moving. (The tenth time, the melon simply explodes.) Doesn't this violate the laws of conservation of momentum? It doesn't, because we're not dealing simply with a two-body collision. Instead, a significant fraction of the energy of the bullet can be dumped into a jet of melon that is violently ejected out of the front of the melon. This jet can have a momentum greater than the momentum of the bullet, so that the rest of the melon must jump backward to conserve momentum. Let's make the following assumptions:

1. The mass of the melon is 2.50 kg;
2. The mass of the rifle bullet is 10.4 g and its velocity is 1800 ft/s;
3. 10 percent of the energy of the bullet is deposited as kinetic energy into a jet flying out of the front of the melon;
4. The mass of the matter in the jet is 0.14 kg;
5. All collisions occur in a straight line.

What would be the speed of the melon's recoil? Compare this to a typical measured recoil speed of about 1.6 ft/s.

Picture the Problem The diagram depicts the bullet just before its collision with the melon and the motion of the melon-and-bullet-less-jet and the jet just after the collision. We'll assume that the bullet stays in the watermelon after the collision and use conservation of momentum to relate the mass of the bullet and its initial velocity to the momenta of the melon jet and the melon less the plug after the collision.

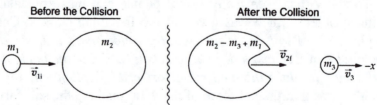

Apply conservation of momentum to the collision to obtain:

$$m_1 v_{1i} = (m_2 - m_3 + m_1)v_{2f} + \sqrt{2m_3 K_3}$$

Solve for v_{2f}:

$$v_{2f} = \frac{m_1 v_{1i} - \sqrt{2m_3 K_3}}{m_2 - m_3 + m_1}$$

Express the kinetic energy of the jet of melon in terms of the initial kinetic energy of the bullet:

$$K_3 = \tfrac{1}{10} K_1 = \tfrac{1}{10}\left(\tfrac{1}{2} m_1 v_{1i}^2\right) = \tfrac{1}{20} m_1 v_{1i}^2$$

Substitute and simplify to obtain:

$$v_{2f} = \frac{m_1 v_{1i} - \sqrt{2m_3\left(\tfrac{1}{20} m_1 v_{1i}^2\right)}}{m_2 - m_3 + m_1}$$

$$= \frac{v_{1i}\left(m_1 - \sqrt{0.1 m_1 m_3}\right)}{m_2 - m_3 + m_1}$$

Substitute numerical values and evaluate v_{2f}:

$$v_{2f} = \left(1800\,\frac{\text{ft}}{\text{s}} \times \frac{1\text{m}}{3.281\,\text{ft}}\right)\frac{\left(0.0104\,\text{kg} - \sqrt{0.1(0.0104\,\text{kg})(0.14\,\text{kg})}\right)}{2.50\,\text{kg} - 0.14\,\text{kg} + 0.0104\,\text{kg}} = -0.386\,\text{m/s}$$

$$= \boxed{-1.27\,\text{ft/s}}$$

Note that this result is in reasonably good agreement with experimental results.

Finding the Center of Mass

*34 • Alley Oop's club-ax consists of a symmetrical 8-kg stone attached to the end of a uniform 2.5-kg stick that is 80 cm long. The dimensions of the club-ax are shown in Figure 8-50. How far is the center of mass from the handle end of the club-ax?

Figure 8-50 Problem 34

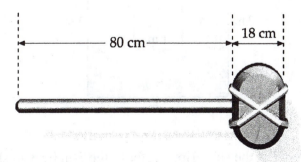

Picture the Problem Let the left end of the handle be the origin of our coordinate system. We can disassemble the club-ax, find the center of mass of each piece, and then use these coordinates and the masses of the handle and stone to find the center of mass of the club-ax.

Express the center of mass of the
handle plus stone system:

$$x_{cm} = \frac{m_{stick} x_{cm,stick} + m_{stone} x_{cm,stone}}{m_{stick} + m_{stone}}$$

Assume that the stone is drilled
and the stick passes through it.
Use symmetry considerations to
locate the center of mass of the
stick:

$$x_{cm,stick} = 45.0\,cm$$

Use symmetry considerations to
locate the center of mass of the
stone:

$$x_{cm,stone} = 89.0\,cm$$

Substitute numerical values and
evaluate x_{cm}:

$$x_{cm} = \frac{(2.5\,kg)(45\,cm) + (8\,kg)(89\,cm)}{2.5\,kg + 8\,kg}$$

$$= \boxed{78.5\,cm}$$

*37 ••   Find the center of mass of the uniform sheet of plywood in Figure
8-52. We shall consider this as two sheets, a square sheet of 3 m edge length and
mass m_1 and a rectangular sheet 1 m × 2 m with a mass of $-m_2$. Let the coordinate
origin be at the lower left-hand corner of the sheet.

Figure 8-52 Problem 37

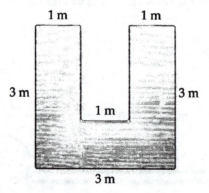

Picture the Problem Let the subscript 1 refer to the 3-m by 3-m sheet of
plywood before the 2-m by 1-m piece has been cut from it. Let the subscript 2
refer to 2-m by 1-m piece that has been removed and let σ be the area density of
the sheet. We can find the center-of-mass of these two regions; treating the
missing region as though it had negative mass, and then finding the center-of-
mass of the U-shaped region by applying its definition.

Express the coordinates of the center of mass of the sheet of plywood:

$$x_{cm} = \frac{m_1 x_{cm,1} - m_2 x_{cm,2}}{m_1 - m_2}$$

$$y_{cm} = \frac{m_1 y_{cm,1} - m_2 y_{cm,2}}{m_1 - m_2}$$

Use symmetry to find $x_{cm,1}, y_{cm,1}, x_{cm,2},$ and $y_{cm,2}$:

$x_{cm,1} = 1.5\,\text{m}, \; y_{cm,1} = 1.5\,\text{m}$

and

$x_{cm,2} = 1.5\,\text{m}, \; y_{cm,2} = 2.0\,\text{m}$

Determine m_1 and m_2:

$m_1 = \sigma A_1 = 9\sigma\,\text{kg}$

and

$m_2 = \sigma A_2 = 2\sigma\,\text{kg}$

Substitute and evaluate x_{cm}:

$$x_{cm} = \frac{(9\sigma\,\text{kg})(1.5\,\text{m}) - (2\sigma\,\text{kg})(1.5\,\text{kg})}{9\sigma\,\text{kg} - 2\sigma\,\text{kg}}$$

$$= 1.50\,\text{m}$$

Substitute and evaluate y_{cm}:

$$y_{cm} = \frac{(9\sigma\,\text{kg})(1.5\,\text{m}) - (2\sigma\,\text{kg})(2\,\text{m})}{9\sigma\,\text{kg} - 2\sigma\,\text{kg}}$$

$$= 1.36\,\text{m}$$

The center of mass of the U-shaped sheet of plywood is at $\boxed{(1.50\,\text{m}, 1.36\,\text{m})}$.

Finding the Center of Mass by Integration

*39 •• Show that the center of mass of a uniform semicircular disk of radius R is at a point $4R/(3\pi)$ from the center of the circle.

Picture the Problem A semicircular disk and a surface element of area dA is shown in the diagram. Because the disk is a continuous object, we'll use $M\vec{r}_{cm} = \int \vec{r}\,dm$ and symmetry to find its center of mass.

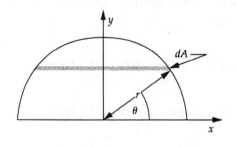

Express the coordinates of the center of mass of the semicircular disk:

$x_{cm} = 0$ by symmetry.

$$y_{cm} = \frac{\int y\sigma\,dA}{M}$$

Express y as a function of r and θ:	$y = r\sin\theta$

Express dA in terms of r and θ:	$dA = r\,d\theta\,dr$

Express M as a function of r and θ:	$M = \sigma A_{\text{half disk}} = \tfrac{1}{2}\sigma\pi R^2$

Substitute and evaluate y_{cm}:

$$y_{cm} = \frac{\sigma\int\limits_{0}^{R}\int\limits_{0}^{\pi} r^2\sin\theta\,d\theta\,dr}{M} = \frac{2\sigma}{M}\int\limits_{0}^{R} r^2\,dr$$

$$= \frac{2\sigma}{3M}R^3 = \boxed{\frac{4}{3\pi}R}$$

Motion of the Center of Mass

*44 • A 1500-kg car is moving westward with a speed of 20 m/s, and a 3000-kg truck is traveling east with a speed of 16 m/s. Find the velocity of the center of mass of the system.

Picture the Problem Choose a coordinate system in which east is the positive x direction and use the relationship $\vec{P} = \sum\limits_{i} m_i\vec{v}_i = M\vec{v}_{cm}$ to determine the velocity of the center of mass of the system.

Use the expression for the total momentum of a system to relate the velocity of the center of mass of the two-vehicle system to the momenta of the individual vehicles:	$\vec{v}_{cm} = \dfrac{\sum\limits_{i} m_i\vec{v}_i}{M}$ $= \dfrac{m_t\vec{v}_t + m_c\vec{v}_c}{m_t + m_c}$

Express the velocity of the truck:	$\vec{v}_t = (16\,\text{m/s})\hat{i}$

Express the velocity of the car:	$\vec{v}_c = (-20\,\text{m/s})\hat{i}$

Substitute numerical values and evaluate $\vec{v}_{cm}$:

$$\vec{v}_{cm} = \frac{(3000\,\text{kg})(16\,\text{m/s})\hat{i} + (1500\,\text{kg})(-20\,\text{m/s})\hat{i}}{3000\,\text{kg} + 1500\,\text{kg}} = \boxed{(4.00\,\text{m/s})\hat{i}}$$

*47 ••   A massless, vertical spring of force constant k is attached at the bottom to a platform of mass m_p, and at the top to a massless cup, as in Figure 8-54. The platform rests on a scale. A ball of mass m_b is dropped into the cup from a negligible height. What is the reading on the scale (a) when the spring is compressed an amount $d = m_b g/k$; (b) when the ball comes to rest momentarily with the spring compressed; (c) when the ball again comes to rest in its original position?

Figure 8-54 Problem 47

Picture the Problem The free-body diagram shows the forces acting on the platform when the spring is partially compressed. The scale reading is the force the scale exerts on the platform and is represented on the FBD by F_n. We can use Newton's 2nd law to determine the scale reading in part (a) and the work-energy theorem in conjunction with Newton's 2nd law in parts (b) and (c).

(a) Apply $\sum F_y = ma_y$ to the spring when it is compressed a distance d:

$$\sum F_y = F_n - m_p g - F_{\text{ball on spring}} = 0$$

Solve for F_n:

$$F_n = m_p g + F_{\text{ball on spring}}$$

$$= m_p g + kd = m_p g + k\left(\frac{m_b g}{k}\right)$$

$$= \boxed{m_p g + m_b g = \left(m_p + m_b\right)g}$$

(b) Use conservation of mechanical energy, with $U_g = 0$ at the position at which the spring is fully compressed, to relate the gravitational potential energy of the system to the energy stored in the fully compressed spring:

$$\Delta K + \Delta U_g + \Delta U_s = 0$$

Because $\Delta K = U_{g,f} = U_{s,i} = 0$,

$$U_{g,i} - U_{s,f} = 0$$

or

$$m_b gd - \tfrac{1}{2}kd^2 = 0$$

Solve for d:

$$d = \frac{2m_b g}{k}$$

Evaluate our force equation in (a) with $d = \dfrac{2m_b g}{k}$:

$$F_n = m_p g + F_{\text{ball on spring}}$$

$$= m_p g + kd = m_p g + k\left(\frac{2m_b g}{k}\right)$$

$$= \boxed{m_p g + 2m_b g = (m_p + 2m_b)g}$$

(c) When the ball is in its original position, the spring is relaxed and exerts no force on the ball. Therefore:

$$F_n = \text{scale reading}$$

$$= \boxed{m_p g}$$

*48 ••   In the Atwood's machine in Figure 8-55, the string passes over a fixed cylinder of mass m_c. The cylinder does not rotate. Instead, the string slides on its frictionless surface. (a) Find the acceleration of the center of mass of the two-block-and-cylinder system. (b) Use Newton's second law for systems to find the force F exerted by the support. (c) Find the tension T in the string connecting the blocks and show that $F = m_c g + 2T$.

Figure 8-55 Problem 48

Picture the Problem Assume that the object whose mass is m_1 is moving downward and take that direction to be the positive direction. We'll use Newton's 2nd law for a system of particles to relate the acceleration of the center of mass to the acceleration of the individual particles.

(a) Relate the acceleration of the center of mass to m_1, m_2, m_c and their accelerations:

$$M\vec{a}_{cm} = m_1\vec{a}_1 + m_2\vec{a}_2 + m_c\vec{a}_c$$

Because m_1 and m_2 have a common acceleration a and $a_c = 0$:

$$a_{cm} = a\frac{m_1 - m_2}{m_1 + m_2 + m_c}$$

From Problem 4-81 we have:

$$a = g\frac{m_1 - m_2}{m_1 + m_2}$$

Substitute to obtain:

$$a_{cm} = \left(\frac{m_1 - m_2}{m_1 + m_2}g\right)\left(\frac{m_1 - m_2}{m_1 + m_2 + m_c}\right)$$

$$= \boxed{\frac{(m_1 - m_2)^2}{(m_1 + m_2)(m_1 + m_2 + m_c)}g}$$

(b) Use Newton's 2nd law for a system of particles to obtain:

$$F - Mg = -Ma_{cm}$$

where $M = m_1 + m_2 + m_c$ and F is positive upwards.

Solve for F and substitute for a_{cm} from part (a):

$$F = Mg - Ma_{cm}$$

$$= Mg - \frac{(m_1 - m_2)^2}{m_1 + m_2}g$$

$$= \boxed{\left[\frac{4m_1m_2}{m_1 + m_2} + m_c\right]g}$$

(c) From Problem 4-81:

$$T = \frac{2m_1m_2}{m_1 + m_2}g$$

Substitute in our result from part (b) to obtain:

$$F = \left[2\frac{2m_1m_2}{m_1 + m_2} + m_c\right]g$$

$$= \left[2\frac{T}{g} + m_c\right]g = \boxed{2T + m_cg}$$

The Conservation of Momentum

*52 • Figure 8-56 shows the behavior of a projectile just after it has broken up into three pieces. What was the speed of the projectile the instant before it broke up? (a) v_3 (b) $v_3/3$ (c) $v_3/4$ (d) $4v_3$ (e) $(v_1 + v_2 + v_3)/4$

Figure 8-56 Problem 52

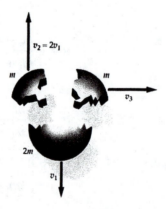

Picture the Problem This is an explosion-like event in which linear momentum is conserved. Thus we can equate the initial and final momenta in the x direction and the initial and final momenta in the y direction. Choose a coordinate system in the positive x direction is to the right and the positive y direction is upward.

Equate the momenta in the y direction before and after the explosion:

$$\sum p_{y,i} = \sum p_{y,f} = mv_2 - 2mv_1$$
$$= m(2v_1) - 2mv_1 = 0$$

We can conclude that the momentum was entirely in the x direction before the particle exploded.

Equate the momenta in the x direction before and after the explosion:	$\sum p_{x,i} = \sum p_{x,f}$ $\therefore 4mv_i = mv_3$
Solve for v_3:	$v_i = \frac{1}{4}v_3$ and $\boxed{(c) \text{ is correct.}}$

*54 •• A block and a gun are firmly fixed to opposite ends of a long glider mounted on a frictionless air track (Figure 8-57). The bullet leaves the muzzle with a velocity v_b and impacts the block, becoming imbedded in it. The mass of the bullet is m_b and the mass of the gun-platform-glider-block system is m_p. (a) What is the velocity of the platform immediately after the bullet leaves the muzzle? (b) What is the velocity of the platform immediately after the bullet comes to rest in the block? (c) How far does the platform move while the bullet is in transit between the gun at rest and the block at rest?

Figure 8-57 Problem 54

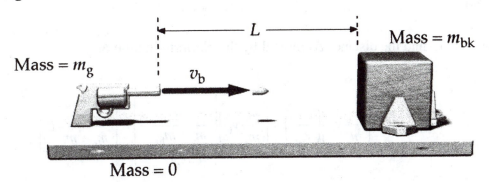

Picture the Problem Let the system include the earth and the platform, gun and block. Then $\vec{F}_{net,ext} = 0$ and momentum is conserved within the system.

(a) Apply conservation of momentum to the system just before and just after the bullet leaves the gun:	$\vec{P}_{before} = \vec{P}_{after}$ or $0 = \vec{P}_{bullet} + \vec{P}_{platform}$
Substitute for $\vec{P}_{bullet}$ and $\vec{P}_{platform}$ and solve for $\vec{v}_{platform}$:	$0 = m_b v_b \hat{i} + m_p \vec{v}_{platform}$ and

$$\vec{v}_{platform} = \boxed{-\frac{m_b}{m_p}v_b\hat{i}}$$

(b) Apply conservation of momentum to the system just before the bullet leaves the gun and just after it comes to rest in the block:

$$\vec{P}_{before} = \vec{P}_{after}$$

or

$$0 = \vec{P}_{platform} \Rightarrow \vec{v}_{platform} = 0$$

(c) Express the distance Δs traveled by the platform:

$$\Delta s = v_{platform}\,\Delta t$$

Express the velocity of the bullet relative to the platform:

$$v_{rel} = v_b - v_{platform} = v_b + \frac{m_b}{m_p}v_b$$

$$= \left(1 + \frac{m_b}{m_p}\right)v_b = \frac{m_p + m_b}{m_p}v_b$$

Relate the time of flight Δt to L and v_{rel}:

$$\Delta t = \frac{L}{v_{rel}}$$

Substitute to find the distance Δs moved by the platform in time Δt:

$$\Delta s = v_{platform}\,\Delta t = \left(\frac{m_b}{m_p}v_b\right)\left(\frac{L}{v_{rel}}\right) = \left(\frac{m_b}{m_p}v_b\right)\left(\frac{L}{\dfrac{m_p + m_b}{m_p}v_b}\right) = \boxed{\frac{m_b}{m_p + m_b}L}$$

*56 •• Each glider on the frictionless air track (Figure 8-58) supports a strong magnet. These magnets attract each other. The mass of glider 1 is 0.10 kg and the mass of glider 2 is 0.20 kg. (Each mass includes the mass of the glider plus the supported magnet.) The origin is at the left end of the track, the center of glider 1 is at $x_1 = 0.10$ m, and the center of glider 2 is at $x_2 = 1.60$ m. Glider 1 is 10-cm long, while glider 2 is 20-cm long. Each glider has its center of mass at its center. (a) When the two gliders are released from rest, they move toward each other and stick together. What is the position of the center of each glider when they first touch? (b) Will the two gliders continue to move after they stick together? Explain.

Figure 8-58 Problem 56

Picture the Problem Because no external forces act on either cart, the center of mass of the two-cart system can't move. We can use the data concerning the masses and separation of the gliders initially to calculate its location and then apply the definition of the center of mass a second time to relate the positions X_1 and X_2 of the centers of the carts when they first touch. We can also use the separation of the centers of the gliders when they touch to obtain a second equation in X_1 and X_2 that we can solve simultaneously with the equation obtained from the location of the center of mass.

(a) Apply its definition to find the center of mass of the 2-glider system:

$$x_{cm} = \frac{m_1 x_1 + m_2 x_2}{m_1 + m_2}$$

$$= \frac{(0.1\,\text{kg})(0.1\,\text{m}) + (0.2\,\text{kg})(1.6\,\text{m})}{0.1\,\text{kg} + 0.2\,\text{kg}}$$

$$= 1.10\,\text{m}$$

from the left end of the air track.

Use the definition of the center of mass to relate the coordinates of the centers of the two gliders when they first touch to the location of the center of mass:

$$1.10\,\text{m} = \frac{m_1 X_1 + m_2 X_2}{m_1 + m_2}$$

$$= \frac{(0.1\,\text{kg})X_1 + (0.2\,\text{kg})X_2}{0.1\,\text{kg} + 0.2\,\text{kg}}$$

$$= \tfrac{1}{3}X_1 + \tfrac{2}{3}X_2$$

Also, when they first touch, their centers are separated by half their combined lengths:

$$X_2 - X_1 = \tfrac{1}{2}(10\,\text{cm} + 20\,\text{cm}) = 0.15\,\text{m}$$

Thus we have:

$$0.333 X_1 + 0.667 X_2 = 1.10\,\text{m}$$
and
$$X_2 - X_1 = 0.15\,\text{m}$$

Solve these equations simultaneously to obtain:

$$X_1 = \boxed{1.00\,\text{m}}$$
and
$$X_2 = \boxed{1.15\,\text{m}}$$

(b) | No. The initial momentum of the system is zero, so it must be zero after the collision.

Kinetic Energy of a System of Particles

*57 •   A 3-kg block is traveling to the right at 5 m/s, and a second 3-kg block is traveling to the left at 2 m/s. (*a*) Find the total kinetic energy of the two blocks. (*b*) Find the velocity of the center of mass of the two-block system. (*c*) Find the velocities of each block relative to the center of mass. (*d*) Find the kinetic energy of the motion of the blocks relative to the center of mass. (*e*) Show that your answer for part (*a*) is greater than your answer for part (*d*) by an amount equal to the kinetic energy associated with the motion of the center of mass.

Picture the Problem Choose a coordinate system in which the positive *x* direction is to the right. Use the expression for the total momentum of a system to find the velocity of the center of mass and the definition of relative velocity to express the sum of the kinetic energies relative to the center of mass.

(*a*) Find the sum of the kinetic energies:

$$K = K_1 + K_2$$
$$= \tfrac{1}{2} m_1 v_1^2 + \tfrac{1}{2} m_2 v_2^2$$
$$= \tfrac{1}{2}(3\,\text{kg})(5\,\text{m/s})^2 + \tfrac{1}{2}(3\,\text{kg})(2\,\text{m/s})^2$$
$$= \boxed{43.5\,\text{J}}$$

(*b*) Relate the velocity of the center of mass of the system to its total momentum:

$$M\vec{v}_{\text{cm}} = m_1\vec{v}_1 + m_2\vec{v}_2$$

Solve for $\vec{v}_{\text{cm}}$:

$$\vec{v}_{\text{cm}} = \frac{m_1\vec{v}_1 + m_2\vec{v}_2}{m_1 + m_2}$$

Substitute numerical values and evaluate $\vec{v}_{\text{cm}}$:

$$\vec{v}_{\text{cm}} = \frac{(3\,\text{kg})(5\,\text{m/s})\hat{i} - (3\,\text{kg})(2\,\text{m/s})\hat{i}}{3\,\text{kg} + 3\,\text{kg}}$$
$$= \boxed{(1.50\,\text{m/s})\hat{i}}$$

(*c*) The velocity of an object relative to the center of mass is given by:

$$\vec{v}_{\text{rel}} = \vec{v} - \vec{v}_{\text{cm}}$$

Substitute numerical values and evaluate $\vec{v}_{1,\text{rel}}$ and $\vec{v}_{2,\text{rel}}$:

$$\vec{v}_{1,\text{rel}} = (5\,\text{m/s})\hat{i} - (1.5\,\text{m/s})\hat{i}$$
$$= \boxed{(3.50\,\text{m/s})\hat{i}}$$
$$\vec{v}_{2,\text{rel}} = (-2\,\text{m/s})\hat{i} - (1.5\,\text{m/s})\hat{i}$$
$$= \boxed{(-3.50\,\text{m/s})\hat{i}}$$

(d) Express the sum of the kinetic energies relative to the center of mass:

$$K_{\text{rel}} = K_{1,\text{rel}} + K_{2,\text{rel}} = \tfrac{1}{2}m_1 v_{1,\text{rel}}^2 + \tfrac{1}{2}m_2 v_{2,\text{rel}}^2$$

Substitute numerical values and evaluate K_{rel}:

$$K_{\text{rel}} = \tfrac{1}{2}(3\,\text{kg})(3.5\,\text{m/s})^2$$
$$+ \tfrac{1}{2}(3\,\text{kg})(-3.5\,\text{m/s})^2$$
$$= \boxed{36.75\,\text{J}}$$

(e) Find K_{cm}:

$$K_{\text{cm}} = \tfrac{1}{2}m_{\text{tot}} v_{\text{cm}}^2 = \tfrac{1}{2}(6\,\text{kg})(1.5\,\text{m/s})^2$$
$$= 6.75\,\text{J}$$
$$= 43.5\,\text{J} - 36.75\,\text{J}$$
$$= \boxed{K - K_{\text{rel}}}$$

Impulse and Average Force

*61 • A meteorite of mass 30.8 tonne (1 tonne = 1000 kg) is exhibited in the American Museum of Natural History in New York City. Suppose that the kinetic energy of the meteorite as it hit the ground was 617 MJ. Find the impulse I experienced by the meteorite up to the time its kinetic energy was halved (which took about $t = 3.0$ s). Find also the average force F exerted on the meteorite during this time interval.

Picture the Problem The impulse exerted by the ground on the meteorite equals the *change* in momentum of the meteorite and is also the product of the average force exerted by the ground on the meteorite and the time during which the average force acts.

Express the impulse exerted by the ground on the meteorite:

$$I = \Delta p_{\text{meteorite}} = p_{\text{f}} - p_{\text{i}}$$

Relate the kinetic energy of the meteorite to its initial momentum

$$K_{\text{i}} = \frac{p_{\text{i}}^2}{2m} \Rightarrow p_{\text{i}} = \sqrt{2mK_{\text{i}}}$$

and solve for its initial
momentum:

Express the ratio of the initial
and final kinetic energies of the
meteorite:

$$\frac{K_i}{K_f} = \frac{\frac{p_i^2}{2m}}{\frac{p_f^2}{2m}} = \frac{p_i^2}{p_f^2} = 2$$

Solve for p_f:

$$p_f = \frac{p_i}{\sqrt{2}}$$

Substitute in our expression for I
and simplify:

$$I = \frac{p_i}{\sqrt{2}} - p_i = p_i\left(\frac{1}{\sqrt{2}} - 1\right)$$

$$= \sqrt{2mK_i}\left(\frac{1}{\sqrt{2}} - 1\right)$$

Because our interest is in its magnitude, evaluate $|I|$:

$$|I| = \left|\sqrt{2(30.8\times10^3\,\text{kg})(617\times10^6\,\text{J})}\left(\frac{1}{\sqrt{2}} - 1\right)\right| = \boxed{1.81\,\text{MN}\cdot\text{s}}$$

Express the impulse delivered to
the meteorite as a function of the
average force acting on it and
solve for and evaluate F_{av}:

$$I = F_{av}\Delta t$$

and

$$F_{av} = \frac{I}{\Delta t} = \frac{1.81\,\text{MN}\cdot\text{s}}{3\,\text{s}} = \boxed{0.602\,\text{MN}}$$

*63 •• A 60-g handball moving with a speed of 5.0 m/s strikes the wall at an
angle of 40° and then bounces off with the same speed at the same angle. It is in
contact with the wall for 2 ms. What is the average force exerted by the ball on
the wall?

Picture the Problem The figure shows the handball just before and immediately after its collision with the wall. Choose a coordinate system in which the positive x direction is to the right. The wall changes the momentum of the ball by exerting a force on it during the ball's collision with it. The reaction to this force is the force the ball exerts on the wall. Because these action and reaction forces are equal in magnitude, we can find the average force exerted on the ball by finding the change in momentum of the ball.

Using Newton's 3rd law, relate the average force exerted by the ball on the wall to the average force exerted by the wall on the ball:

$$\vec{F}_{\text{av on wall}} = -\vec{F}_{\text{av on ball}}$$

and

$$F_{\text{av on wall}} = F_{\text{av on ball}} \qquad (1)$$

Relate the average force exerted by the wall on the ball to its change in momentum:

$$\vec{F}_{\text{av on ball}} = \frac{\Delta \vec{p}}{\Delta t} = \frac{m\Delta \vec{v}}{\Delta t}$$

Express $\Delta \vec{v}_x$ for the ball:

$$\Delta \vec{v}_x = v_{f,x}\hat{i} - v_{i,x}\hat{i}$$

or, because $v_{i,x} = v\cos\theta$ and $v_{f,x} = -v\cos\theta$,

$$\Delta \vec{v}_x = -v\cos\theta\,\hat{i} - v\cos\theta\,\hat{i} = -2v\cos\theta\,\hat{i}$$

Substitute in our expression for $\vec{F}_{\text{av on ball}}$:

$$\vec{F}_{\text{av on ball}} = \frac{m\Delta \vec{v}}{\Delta t} = -\frac{2mv\cos\theta}{\Delta t}\hat{i}$$

Evaluate the magnitude of $\vec{F}_{\text{av on ball}}$:

$$F_{\text{av on ball}} = \frac{2mv\cos\theta}{\Delta t}$$

$$= \frac{2(0.06\,\text{kg})(5\,\text{m/s})\cos 40°}{2\,\text{ms}}$$

$$= 230\,\text{N}$$

Substitute in equation (1) to obtain:

$$F_{\text{av on wall}} = \boxed{230\,\text{N}}$$

Collisions in One Dimension

*67 •   A 2000-kg car traveling to the right at 30 m/s is chasing a second car of the same mass that is traveling to the right at 10 m/s. (*a*) If the two cars collide and stick together, what is their speed just after the collision? (*b*) What fraction of the initial kinetic energy of the cars is lost during this collision? Where does it go?

Picture the Problem We can apply conservation of momentum to this perfectly inelastic collision to find the after-collision speed of the two cars. The ratio of the transformed kinetic energy to kinetic energy before the collision is the fraction of kinetic energy lost in the collision.

(*a*) Letting *V* be the velocity of the two cars after their collision, apply conservation of momentum to their perfectly inelastic collision:

$$p_{initial} = p_{final}$$
or
$$mv_1 + mv_2 = (m+m)V$$

Solve for and evaluate *V*:

$$V = \frac{v_1 + v_2}{2} = \frac{30\,\text{m/s} + 10\,\text{m/s}}{2}$$
$$= \boxed{20.0\,\text{m/s}}$$

(*b*) Express the ratio of the kinetic energy that is lost to the kinetic energy of the two cars before the collision:

$$\frac{\Delta K}{K_{initial}} = \frac{K_{final} - K_{initial}}{K_{initial}}$$

$$= \frac{K_{final}}{K_{initial}} - 1$$

$$= \frac{\frac{1}{2}(2m)V^2}{\frac{1}{2}mv_1^2 + \frac{1}{2}mv_2^2} - 1$$

$$= \frac{2V^2}{v_1^2 + v_2^2} - 1$$

Substitute numerical values to obtain:

$$\frac{\Delta K}{K_{initial}} = \frac{2(20\,\text{m/s})^2}{(30\,\text{m/s})^2 + (10\,\text{m/s})^2} - 1$$
$$= -0.200$$

> 20% of the initial kinetic energy is transformed into heat, sound, and the deformation of metal.

*71 ••   A proton of mass *m* undergoes a head-on elastic collision with a stationary carbon nucleus of mass 12*m*. The speed of the proton is 300 m/s.

(*a*) Find the velocity of the center of mass of the system. (*b*) Find the velocity of the proton after the collision.

Picture the Problem Let the direction the proton is moving before the collision be the positive *x* direction. We can use both conservation of momentum and conservation of mechanical energy to obtain an expression for velocity of the proton after the collision.

(*a*) Use the expression for the total momentum of a system to find v_{cm}:

$$\vec{P} = \sum_i m_i \vec{v}_i = M\vec{v}_{cm}$$

and

$$\vec{v}_{cm} = \frac{m\vec{v}_{p,i}}{m+12m} = \tfrac{1}{13}(300\,\text{m/s})\hat{i}$$

$$= \boxed{(23.1\,\text{m/s})\hat{i}}$$

(*b*) Use conservation of momentum to obtain one relation for the final velocities:

$$m_p v_{p,i} = m_p v_{p,f} + m_{nuc} v_{nuc,f} \qquad (1)$$

Use conservation of mechanical energy to set the velocity of recession equal to the negative of the velocity of approach:

$$v_{nuc,f} - v_{p,f} = -\left(v_{nuc,i} - v_{p,i}\right) = v_{p,i} \quad (2)$$

To eliminate $v_{nuc,f}$, solve equation (2) for $v_{nuc,f}$, and substitute the result in equation (1):

$$v_{nuc,f} = v_{p,i} + v_{p,f}$$
$$m_p v_{p,i} = m_p v_{p,f} + m_{nuc}\left(v_{p,i} + v_{p,f}\right)$$

Solve for and evaluate $v_{p,f}$:

$$v_{p,f} = \frac{m_p - m_{nuc}}{m_p + m_{nuc}} v_{p,i}$$

$$= \frac{m-12m}{13m}(300\,\text{m/s}) = \boxed{-254\,\text{m/s}}$$

*74 •• A bullet of mass *m* is fired vertically from below into a thin sheet of plywood of mass *M* that is initially at rest, supported by a thin sheet of paper. The bullet blasts through the plywood, which rises to a height *H* above the paper before falling back down. The bullet continues rising to a height *h* above the paper. (*a*) Express the upward velocity of the bullet and the plywood immediately after the bullet exits the plywood in terms of *h* and *H*. (*b*) Use conservation of momentum to express the speed of the bullet before it enters the sheet of plywood in terms of *m, h, M,* and *H*. (*c*) Obtain expressions for the mechanical energy of the system before and after the inelastic collision. (*d*) Express the energy dissipated in terms of *m*, *h*, *M*, and *H*.

Picture the Problem Let the system include the earth, the bullet, and the sheet of plywood. Then $W_{\text{ext}} = 0$. Choose the zero of gravitational potential energy to be where the bullet enters the plywood. We can apply both conservation of energy and conservation of momentum to obtain the various physical quantities called for in this problem.

(*a*) Use conservation of mechanical energy after the bullet exits the sheet of plywood to relate its exit speed to the height to which it rises:

$$\Delta K + \Delta U = 0$$
or, because $K_f = U_i = 0$,
$$-\tfrac{1}{2}mv_m^2 + mgh = 0$$

Solve for v_m:

$$v_m = \boxed{\sqrt{2gh}}$$

Proceed similarly to relate the initial velocity of the plywood to the height to which it rises:

$$v_M = \boxed{\sqrt{2gH}}$$

(*b*) Apply conservation of momentum to the collision of the bullet and the sheet of plywood:

$$\vec{p}_i = \vec{p}_f$$
or
$$mv_{mi} = mv_m + Mv_M$$

Substitute for v_m and v_M and solve for v_{mi}:

$$v_{mi} = \boxed{\sqrt{2gh} + \frac{M}{m}\sqrt{2gH}}$$

(*c*) Express the initial mechanical energy of the system (i.e., just before the collision):

$$E_i = \tfrac{1}{2}mv_{mi}^2$$

$$= \boxed{mg\left[h + \frac{2M}{m}\sqrt{hH} + \left(\frac{M}{m}\right)^2 H\right]}$$

Express the final mechanical energy of the system (i.e., when the bullet and block have reached their maximum heights):

$$E_f = mgh + MgH = \boxed{g(mh + MH)}$$

(*d*) Use the work-energy theorem with $W_{\text{ext}} = 0$ to find the energy dissipated by friction in the

$$E_f - E_i + W_{\text{friction}} = 0$$
and

inelastic collision:

$$W_{\text{friction}} = E_i - E_f$$

$$= \boxed{gMH\left[2\sqrt{\dfrac{h}{H}} + \dfrac{M}{m} - 1\right]}$$

*78 •• Show that in a one-dimensional elastic collision, if the mass and velocity of object 1 are m_1 and v_{1i}, and if the mass and velocity of object 2 are m_2 and v_{2i}, then the final velocities v_{1f} and v_{2f} are given by:

$$v_{1f} = \frac{m_1 - m_2}{m_1 + m_2}v_{1i} + \frac{2m_2}{m_1 + m_2}v_{2i}$$

$$v_{2f} = \frac{2m_1}{m_1 + m_2}v_{1i} + \frac{m_2 - m_1}{m_1 + m_2}v_{2i}$$

Picture the Problem We can apply conservation of momentum and the definition of an elastic collision to obtain equations relating the initial and final velocities of the colliding objects that we can solve for v_{1f} and v_{2f}.

Apply conservation of momentum to the elastic collision of the particles to obtain:

$$m_1 v_{1f} + m_2 v_{2f} = m_1 v_{1i} + m_2 v_{2i} \qquad (1)$$

Relate the initial and final kinetic energies of the particles in an elastic collision:

$$\tfrac{1}{2}m_1 v_{1f}^2 + \tfrac{1}{2}m_2 v_{2f}^2 = \tfrac{1}{2}m_1 v_{1i}^2 + \tfrac{1}{2}m_2 v_{2i}^2$$

Rearrange this equation and factor to obtain:

$$m_2\left(v_{2f}^2 - v_{2i}^2\right) = m_1\left(v_{1i}^2 - v_{1f}^2\right)$$

or

$$m_2\left(v_{2f} - v_{2i}\right)\left(v_{2f} + v_{2i}\right)$$
$$= m_1\left(v_{1i} - v_{1f}\right)\left(v_{1i} + v_{1f}\right) \qquad (2)$$

Rearrange equation (1) to obtain:

$$m_2\left(v_{2f} - v_{2i}\right) = m_1\left(v_{1i} - v_{1f}\right) \qquad (3)$$

Divide equation (2) by equation (3) to obtain:

$$v_{2f} + v_{2i} = v_{1i} + v_{1f}$$

Rearrange this equation to obtain equation (4):

$$v_{1f} - v_{2f} = v_{2i} - v_{1i} \qquad (4)$$

Multiply equation (4) by m_2 and add it to equation (1) to obtain:

$$(m_1 + m_2)v_{1f} = (m_1 - m_2)v_{1i} + 2m_2 v_{2i}$$

Solve for v_{1f} to obtain:

$$\boxed{v_{1f} = \frac{m_1 - m_2}{m_1 + m_2}v_{1i} + \frac{2m_2}{m_1 + m_2}v_{2i}}$$

Multiply equation (4) by m_1 and subtract it from equation (1) to obtain:

$$(m_1 + m_2)v_{2f} = (m_2 - m_1)v_{2i} + 2m_1v_{1i}$$

Solve for v_{2f} to obtain:

$$v_{2f} = \boxed{\frac{2m_1}{m_1 + m_2}v_{1i} + \frac{m_2 - m_1}{m_1 + m_2}v_{2i}}$$

Remarks: Note that the velocities satisfy the condition that $v_{2f} - v_{1f} = -(v_{2i} - v_{1i})$. This verifies that the speed of recession equals the speed of approach.

Perfectly Inelastic Collisions and the Ballistic Pendulum

*81 •• A bullet of mass m_1 is fired with a speed v into the bob of a ballistic pendulum of mass m_2. Find the maximum height h attained by the bob if the bullet passes through the bob and emerges with a speed $v/2$.

Picture the Problem Choose $U_g = 0$ at the equilibrium position of the ballistic pendulum. Momentum is conserved in the collision of the bullet with the bob and kinetic energy is transformed into gravitational potential energy as the bob swings up to its maximum height.

Letting V represent the initial speed of the bob as it begins its upward swing, use conservation of momentum to relate this speed to the speeds of the bullet just before and after its collision with the bob:

$$m_1v = m_1\left(\tfrac{1}{2}v\right) + m_2V$$

Solve for the speed of the bob:

$$V = \frac{m_1}{2m_2}v \qquad (1)$$

Use conservation of energy to relate the initial kinetic energy of the bob to its potential energy at its maximum height:

$\Delta K + \Delta U = 0$
or, because $K_f = U_i = 0$,
$-K_i + U_f = 0$

Substitute for K_i and U_f:

$$-\tfrac{1}{2}m_2V^2 + m_2gh = 0$$

Solve for h:

$$h = \frac{V^2}{2g} \qquad (2)$$

Substitute V from equation (1) in equation (2) and simplify to obtain:

$$h = \frac{\left(\dfrac{m_1}{2m_2}v\right)^2}{2g} = \boxed{\frac{v^2}{8g}\left(\frac{m_1}{m_2}\right)^2}$$

*84 ••   Tarzan is in the path of a pack of stampeding elephants when Jane swings in to the rescue on a rope vine, hauling him off to safety. The length of the vine is 25 m, and Jane starts her swing with the rope horizontal. If Jane's mass is 54 kg, and Tarzan's is 82 kg, to what height above the ground will the pair swing after she grabs him?

Picture the Problem Jane's collision with Tarzan is a perfectly inelastic collision. We can find her speed v_1 just before she grabs Tarzan from conservation of energy and their speed V just after she grabs him from conservation of momentum. Their kinetic energy just after their collision will be transformed into gravitational potential energy when they have reached their greatest height h.

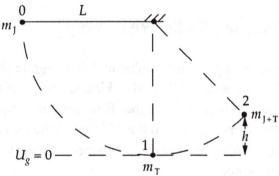

Use conservation of energy to relate the potential energy of Jane and Tarzan at their highest point (2) to their kinetic energy immediately after Jane grabbed Tarzan:

$$U_2 = K_1$$

or

$$m_{J+T}\,gh = \tfrac{1}{2}m_{J+T}V^2$$

Solve for h to obtain:

$$h = \frac{V^2}{2g} \qquad (1)$$

Use conservation of momentum to relate Jane's velocity just before she collides with Tarzan to their velocity just after their perfectly inelastic collision:

$$m_J v_1 = m_{J+T}V$$

Solve for V:

$$V = \frac{m_J}{m_{J+T}} v_1 \qquad (2)$$

Apply conservation of energy to relate Jane's kinetic energy at 1 to her potential energy at 0:

$$K_1 = U_0$$
or
$$\tfrac{1}{2} m_J v_1^2 = m_J g L$$

Solve for v_1:

$$v_1 = \sqrt{2gL}$$

Substitute in equation (2) to obtain:

$$V = \frac{m_J}{m_{J+T}} \sqrt{2gL}$$

Substitute in equation (1) and simplify:

$$h = \frac{1}{2g}\left(\frac{m_J}{m_{J+T}}\right)^2 2gL = \left(\frac{m_J}{m_{J+T}}\right)^2 L$$

Substitute numerical values and evaluate h:

$$h = \left(\frac{54\,\text{kg}}{54\,\text{kg} + 82\,\text{kg}}\right)^2 (25\,\text{m}) = \boxed{3.94\,\text{m}}$$

Exploding Objects and Radioactive Decay

*88 •••   The boron isotope ^{9}B is unstable and disintegrates into a proton and two α particles. The total energy released as kinetic energy of the decay products is 4.4×10^{-14} J. In one such event, with the ^{9}B nucleus at rest prior to decay, the velocity of the proton is measured as 6.0×10^6 m/s. If the two α particles have equal energies, find the magnitude and the direction of their velocities with respect to that of the proton.

Picture the Problem This nuclear reaction is ^{9}B $\rightarrow 2\alpha + p + 4.4 \times 10^{-14}$ J. Assume that the proton moves in the $-x$ direction as shown in the figure. The sum of the kinetic energies of the decay products equals the energy released in the decay. We'll use conservation of momentum to find the angle between the velocities of the proton and the alpha particles. Note that $v_\alpha = v'_\alpha$.

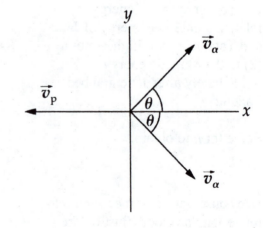

Express the energy released to

$$K_p + 2K_\alpha = E_{rel}$$

the kinetic energies of the decay products:

or

$$\tfrac{1}{2}m_p v_p^2 + 2\left(\tfrac{1}{2}m_\alpha v_\alpha^2\right) = E_{rel}$$

Solve for v_α:

$$v_\alpha = \sqrt{\frac{E_{rel} - \tfrac{1}{2}m_p v_p^2}{m_\alpha}}$$

Substitute numerical values and evaluate v_α:

$$v_\alpha = \sqrt{\frac{4.4\times10^{-14}\ J}{6.68\times10^{-27}\ kg} - \frac{\tfrac{1}{2}\left(1.67\times10^{-27}\ kg\right)\left(6\times10^6\ m/s\right)^2}{6.68\times10^{-27}\ kg}} = \boxed{1.44\times10^6\ m/s}$$

Given that the boron isotope was at rest prior to the decay, use conservation of momentum to relate the momenta of the decay products:

$$\vec{p}_f = \vec{p}_i = 0 \;\Rightarrow\; p_{xf} = 0$$
$$\therefore\; 2\left(m_\alpha v_\alpha \cos\theta\right) - m_p v_p = 0$$

or

$$2\left(4m_p v_\alpha \cos\theta\right) - m_p v_p = 0$$

Solve for θ:

$$\theta = \cos^{-1}\left[\frac{v_p}{8v_\alpha}\right]$$

$$= \cos^{-1}\left[\frac{6\times10^6\ m/s}{8\left(1.44\times10^6\ m/s\right)}\right] = \pm 58.7°$$

Let θ' equal the angle the velocities of the alpha particles make with that of the proton:

$$\theta' = \pm\left(180° - 58.7°\right)$$
$$= \boxed{\pm 121°}$$

Coefficient of Restitution

*90 • According to the official rules of racquetball, a ball acceptable for tournament play must bounce to a height of between 173 and 183 cm when dropped from a height of 254 cm at room temperature. What is the acceptable range of values for the coefficient of restitution for the racquetball-floor system?

Picture the Problem The coefficient of restitution is defined as the ratio of the velocity of recession to the velocity of approach. These velocities can be determined from the heights from which an object was dropped and the height to which it rebounded by using conservation of mechanical energy.

Use its definition to relate the coefficient of restitution to the velocities of approach and recession:

$$e = \frac{v_{rec}}{v_{app}} \qquad (1)$$

Letting $U_g = 0$ at the surface of the steel plate, apply conservation of energy to express the velocity of approach:

$\Delta K + \Delta U = 0$

Because $K_i = U_f = 0$,

$K_f - U_i = 0$

or

$\frac{1}{2} m v_{app}^2 - m g h_{app} = 0$

Solve for v_{app}:

$$v_{app} = \sqrt{2 g h_{app}}$$

In like manner, show that:

$$v_{rec} = \sqrt{2 g h_{rec}}$$

Substitute in equation (1) to obtain:

$$e = \frac{\sqrt{2 g h_{rec}}}{\sqrt{2 g h_{app}}} = \sqrt{\frac{h_{rec}}{h_{app}}}$$

Find e_{min}:

$$e_{min} = \sqrt{\frac{173\,cm}{254\,cm}} = 0.825$$

Find e_{max}:

$$e_{max} = \sqrt{\frac{183\,cm}{254\,cm}} = 0.849$$

and $\boxed{0.825 \le e \le 0.849}$

Collisions in Three Dimensions

*94 ••   In Section 8-6 it was proven by geometrical means that when a particle elastically collides with another particle of equal mass that is initially at rest, the two separate at right angles. In this problem we will examine another way of proving this, one that illustrates the power of abstract vector notation. (*a*) Given $\vec{A} = \vec{B} + \vec{C}$, show that $A^2 = B^2 + C^2 + 2\vec{B} \cdot \vec{C}$. (*b*) Let the momentum of the initially moving particle be $\vec{P}$ and the momenta of the particles after the collision be $\vec{p}_1$ and $\vec{p}_2$. By writing the vector equation for the conservation of momentum and obtaining the dot product of each side with itself, and then comparing it to the equation for the conservation of energy, show that $\vec{p}_1 \cdot \vec{p}_2 = 0$.

Picture the Problem We can use the definition of the magnitude of a vector and the definition of the dot product to establish the result called for in (*a*). In part (*b*) we can use the result of part (a), the conservation of momentum, and the definition of an elastic collision (kinetic energy is conserved) to show that the particles separate at right angles.

(a) Find the dot product of $\vec{B} + \vec{C}$ with itself:

$$\left(\vec{B} + \vec{C}\right) \cdot \left(\vec{B} + \vec{C}\right)$$
$$= B^2 + C^2 + 2\vec{B} \cdot \vec{C}$$

Because $\vec{A} = \vec{B} + \vec{C}$:

$$A^2 = \left|\vec{B} + \vec{C}\right|^2 = \left(\vec{B} + \vec{C}\right) \cdot \left(\vec{B} + \vec{C}\right)$$

Substitute to obtain:

$$\boxed{A^2 = B^2 + C^2 + 2\vec{B} \cdot \vec{C}}$$

(b) Apply conservation of momentum to the collision of the particles:

$$\vec{p}_1 + \vec{p}_2 = \vec{P}$$

Form the dot product of each side of this equation with itself to obtain:

$$\left(\vec{p}_1 + \vec{p}_2\right) \cdot \left(\vec{p}_1 + \vec{p}_2\right) = \vec{P} \cdot \vec{P}$$

or

$$p_1^2 + p_2^2 + 2\vec{p}_1 \cdot \vec{p}_2 = P^2 \qquad (1)$$

Apply the definition of an elastic collision to obtain:

$$\frac{p_1^2}{2m} + \frac{p_2^2}{2m} = \frac{P^2}{2m}$$

or

$$p_1^2 + p_2^2 = P^2 \qquad (2)$$

Subtract equation (1) from equation (2) to obtain:

$$2\vec{p}_1 \cdot \vec{p}_2 = 0 \text{ or } \boxed{\vec{p}_1 \cdot \vec{p}_2 = 0}$$

i.e., the particles move apart along paths that are at right angles to each other.

*97 ••   A puck of mass 0.5 kg moving at 2 m/s approaches an identical puck that is stationary on frictionless ice. After the collision, the first puck leaves with a speed v_1 at 30° to the original line of motion; the second puck leaves with speed v_2 at 60°, as in Figure 8-61. (a) Calculate v_1 and v_2. (b) Was the collision elastic?

Figure 8-61 Problem 97

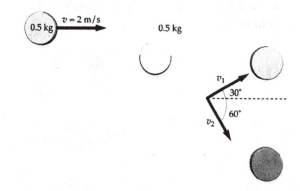

Picture the Problem Let the direction of motion of the puck that is moving before the collision be the positive x direction. Applying conservation of momentum to the collision in both the x and y directions will lead us to two equations in the unknowns v_1 and v_2 that we can solve simultaneously. We can decide whether the collision was elastic by either calculating the system's kinetic energy before and after the collision or by determining whether the angle between the final velocities is 90°.

(*a*) Use conservation of momentum in the x direction to relate the velocities of the collision participants before and after the collision:	$p_{xi} = p_{xf}$ or $mv = mv_1 \cos 30° + mv_2 \cos 60°$ or $v = v_1 \cos 30° + v_2 \cos 60°$
Use conservation of momentum in the y direction to obtain a second equation relating the velocities of the collision participants before and after the collision:	$p_{yi} = p_{yf}$ or $0 = mv_1 \sin 30° - mv_2 \sin 60°$ or $0 = v_1 \sin 30° - v_2 \sin 60°$
Solve these equations simultaneously to obtain:	$v_1 = \boxed{1.73 \,\text{m/s}}$ and $v_2 = \boxed{1.00 \,\text{m/s}}$

(*b*) $\boxed{\text{Because the angle between } \vec{v}_1 \text{ and } \vec{v}_2 \text{ is } 90°, \text{ the collision was } \textit{elastic.}}$

*99 ••   A ball moving at 10 m/s makes an off-center elastic collision with another ball of equal mass that is initially at rest. The incoming ball is deflected at an angle of 30° from its original direction of motion. Find the velocity of each ball after the collision.

Picture the Problem Let the direction of motion of the ball that is moving before the collision be the positive x direction. Let v represent the velocity of the ball that is moving before the collision, v_1 its velocity after the collision and v_2 the velocity of the initially-at-rest ball after the collision. We know that because the collision is elastic and the balls have the same mass, v_1 and v_2 are 90° apart. Applying conservation of momentum to the collision in both the x and y directions will lead us to two equations in the unknowns v_1 and v_2 that we can solve simultaneously.

Noting that the angle of deflection for the recoiling ball is 60°, use conservation of momentum in the x direction to relate the velocities of the collision participants before and after the collision:

$$P_{xi} = P_{xf}$$

or

$$mv = mv_1 \cos 30° + mv_2 \cos 60°$$

or

$$v = v_1 \cos 30° + v_2 \cos 60°$$

Use conservation of momentum in the y direction to obtain a second equation relating the velocities of the collision participants before and after the collision:

$$P_{yi} = P_{yf}$$

or

$$0 = mv_1 \sin 30° - mv_2 \sin 60°$$

or

$$0 = v_1 \sin 30° - v_2 \sin 60°$$

Solve these equations simultaneously to obtain:

$$v_1 = \boxed{8.66\,\text{m/s}} \text{ and } v_2 = \boxed{5.00\,\text{m/s}}$$

Center-of-Mass Frame

*102 •• A 3-kg block is traveling in the negative x direction at 5 m/s, and a 1-kg block is traveling in the positive x direction at 3 m/s. (*a*) Find the velocity v_{cm} of the center of mass. (*b*) Subtract v_{cm} from the velocity of each block to find the velocity of each block in the center-of-mass reference frame. (*c*) After they make a head-on elastic collision, the velocity of each block is reversed (in this frame). Find the velocity of each block in the center-of-mass frame after the collision. (*d*) Transform back into the original frame by adding v_{cm} to the velocity of each block. (*e*) Check your result by finding the initial and final kinetic energies of the blocks in the original frame and comparing them.

Picture the Problem Let the numerals 3 and 1 denote the blocks whose masses are 3 kg and 1 kg respectively. We can use $\sum_i m_i \vec{v}_i = M\vec{v}_{cm}$ to find the velocity of the center-of-mass of the system and simply follow the directions in the problem step by step.

(a) Express the total momentum of this two-particle system in terms of the velocity of its center of mass:

$$\vec{P} = \sum_i m_i \vec{v}_i = m_1 \vec{v}_1 + m_3 \vec{v}_3$$
$$= M \vec{v}_{cm} = (m_1 + m_3) \vec{v}_{cm}$$

Solve for $\vec{v}_{cm}$:

$$\vec{v}_{cm} = \frac{m_3 \vec{v}_3 + m_1 \vec{v}_1}{m_3 + m_1}$$

Substitute numerical values and evaluate $\vec{v}_{cm}$:

$$\vec{v}_{cm} = \frac{(3\,\text{kg})(-5\,\text{m/s})\hat{i} + (1\,\text{kg})(3\,\text{m/s})\hat{i}}{3\,\text{kg} + 1\,\text{kg}}$$
$$= \boxed{(-3.00\,\text{m/s})\hat{i}}$$

(b) Find the velocity of the 3-kg block in the center of mass reference frame:

$$\vec{u}_3 = \vec{v}_3 - \vec{v}_{cm} = (-5\,\text{m/s})\hat{i} - (-3\,\text{m/s})\hat{i}$$
$$= \boxed{(-2.00\,\text{m/s})\hat{i}}$$

Find the velocity of the 1-kg block in the center of mass reference frame:

$$\vec{u}_1 = \vec{v}_1 - \vec{v}_{cm} = (3\,\text{m/s})\hat{i} - (-3\,\text{m/s})\hat{i}$$
$$= \boxed{(6.00\,\text{m/s})\hat{i}}$$

(c) Express the after-collision velocities of both blocks in the center of mass reference frame:

$$\vec{u}_3' = \boxed{(2.00\,\text{m/s})\hat{i}}$$
and
$$\vec{u}_1' = \boxed{(-6.00\,\text{m/s})\hat{i}}$$

(d) Transform the after-collision velocity of the 3-kg block from the center of mass reference frame to the original reference frame:

$$\vec{v}_3' = \vec{u}_3' + \vec{v}_{cm} = (2\,\text{m/s})\hat{i} + (-3\,\text{m/s})\hat{i}$$
$$= \boxed{(-1.00\,\text{m/s})\hat{i}}$$

Transform the after-collision velocity of the 1-kg block from the center of mass reference frame to the original reference frame:

$$\vec{v}_1' = \vec{u}_1' + \vec{v}_{cm} = (-6\,\text{m/s})\hat{i} + (-3\,\text{m/s})\hat{i}$$
$$= \boxed{(-9.00\,\text{m/s})\hat{i}}$$

(e) Express K_i in the original frame of reference:

$$K_i = \tfrac{1}{2}m_3 v_3^2 + \tfrac{1}{2}m_1 v_1^2$$

Substitute numerical values and evaluate K_i:	$K_i = \frac{1}{2}\left[(3\,\text{kg})(5\,\text{m/s})^2 + (1\,\text{kg})(3\,\text{m/s})^2\right]$ $= \boxed{42.0\,\text{J}}$
Express K_f in the original frame of reference:	$K_f = \frac{1}{2}m_3 v_3'^2 + \frac{1}{2}m_1 v_1'^2$
Substitute numerical values and evaluate K_f:	$K_f = \frac{1}{2}\left[(3\,\text{kg})(1\,\text{m/s})^2 + (1\,\text{kg})(9\,\text{m/s})^2\right]$ $= \boxed{42.0\,\text{J}}$

Systems With Continuously Varying Mass: Rocket Propulsion

*106 •• The *specific impulse* of a rocket propellant is defined as $I_{sp} = F_{th}/Rg$, where F_{th} is the thrust of the propellant, g the magnitude of free-fall acceleration at the surface of the earth, and R the rate at which the propellant is burned. The rate depends predominantly on the type and exact mixture of the propellant. (a) Show that specific impulse has the dimension of time. (b) Show that $u_{ex} = gI_{sp}$, where u_{ex} is the exhaust velocity of the propellant. (c) What is the specific impulse (in seconds) of the propellant used in the Saturn V rocket of Example 8-21?

Picture the Problem We can use the dimensions of thrust, burn rate, and acceleration to show that the dimension of specific impulse is time. Combining the definitions of rocket thrust and specific impulse will lead us to $u_{ex} = gI_{sp}$.

(a) Express the dimension of specific impulse in terms of the dimensions of F_{th}, R, and g:	$[I_{sp}] = \dfrac{[F_{th}]}{[R][g]} = \dfrac{\dfrac{M \cdot L}{T^2}}{\dfrac{M}{T} \cdot \dfrac{L}{T^2}} = \boxed{T}$
(b) From the definition of rocket thrust we have:	$F_{th} = Ru_{ex}$
Solve for u_{ex}:	$u_{ex} = \dfrac{F_{th}}{R}$
Substitute for F_{th} to obtain:	$u_{ex} = \dfrac{RgI_{sp}}{R} = \boxed{gI_{sp}}$
(c) From Example 8-21 we have:	$R = 1.384 \times 10^4$ kg/s and $F_{th} = 3.4 \times 10^6$ N

Substitute numerical values and evaluate I_{sp}:

$$I_{sp} = \frac{3.4\times10^6 \text{ N}}{(1.384\times10^4 \text{ kg/s})(9.81\text{m/s}^2)}$$

$$= \boxed{25.0\text{s}}$$

*107 ••• The initial *thrust to weight ratio* τ_0 of a rocket is $\tau_0 = F_{th}/(m_0 g)$, where F_{th} is the rocket's thrust and m_0 the initial mass of the rocket. (*a*) For a rocket launched straight up from the earth's surface, show that $\tau_0 = 1 + (a_0/g)$, where a_0 is the initial acceleration of the rocket. For manned rocket flight, τ_0 cannot be made much larger than 4 for the comfort and safety of the astronauts. (The astronauts will feel that their weight as the rocket lifts off is equal to τ_0 times their normal weight.) (*b*) Show that the final velocity of a rocket launched from the earth's surface, in terms of τ_0 and I_{sp} (see Problem 106) can be written as

$$v_f = gI_{sp}\left(\ln\left(\frac{m_0}{m_f}\right) - \frac{1}{\tau_0}\left(1 - \frac{m_f}{m_0}\right) \right)$$

where m_0 is the initial mass of payload and propellant and m_f is the mass of the payload. (*c*) Using a spreadsheet program or graphing calculator, graph v_f as a function of the mass ratio m_0/m_f for $I_{sp} = 250$ s and $\tau_0 = 2$ for values of the mass ratio from 2 to 10. (Note that the mass ratio cannot be less than 1.) (*d*) To lift a rocket into orbit, a final velocity after burnout of $v_f = 7$ km/s is needed. Calculate the mass ratio required of a single stage rocket to do this, using the values of specific impulse and thrust ratio given in Part (*b*). For engineering reasons, it is difficult to make a rocket with a mass ratio much greater than 10. Can you see why multistage rockets are usually used to put payloads into orbit around the earth?

Picture the Problem We can use the rocket equation and the definition of rocket thrust to show that $\tau_0 = 1 + a_0/g$. In part (*b*) we can express the burn time t_b in terms of the initial and final masses of the rocket and the rate at which the fuel burns, and then use this equation to express the rocket's final velocity in terms of I_{sp}, τ_0, and the mass ratio m_0/m_f. In part (*d*) we'll need to use trial-and-error methods or a graphing calculator to solve the transcendental equation giving v_f as a function of m_0/m_f.

(*a*) Express the rocket equation:

$$-mg + Ru_{ex} = ma$$

From the definition of rocket thrust we have:

$$F_{th} = Ru_{ex}$$

Substitute to obtain:

$$-mg + F_{th} = ma$$

Solve for F_{th} at takeoff:

$$F_{th} = m_0 g + m_0 a_0$$

Divide both sides of this equation by $m_0 g$ to obtain:

$$\frac{F_{th}}{m_0 g} = 1 + \frac{a_0}{g}$$

Because $\tau_0 = F_{th}/(m_0 g)$:

$$\boxed{\tau_0 = 1 + \frac{a_0}{g}}$$

(b) Use equation 8-42 to express the final speed of a rocket that starts from rest with mass m_0:

$$v_f = u_{ex} \ln\frac{m_0}{m_f} - g t_b, \qquad (1)$$

where t_b is the burn time.

Express the burn time in terms of the burn rate R (assumed constant):

$$t_b = \frac{m_0 - m_f}{R} = \frac{m_0}{R}\left(1 - \frac{m_f}{m_0}\right)$$

Multiply t_b by one in the form gT/gT and simplify to obtain:

$$t_b = \frac{gF_{th}}{gF_{th}}\frac{m_0}{R}\left(1 - \frac{m_f}{m_0}\right)$$

$$= \frac{gm_0}{F_{th}}\frac{F_{th}}{gR}\left(1 - \frac{m_f}{m_0}\right)$$

$$= \frac{I_{sp}}{\tau_0}\left(1 - \frac{m_f}{m_0}\right)$$

Substitute in equation (1):

$$v_f = u_{ex} \ln\frac{m_0}{m_f} - \frac{gI_{sp}}{\tau_0}\left(1 - \frac{m_f}{m_0}\right)$$

From Problem 32 we have:

$$u_{ex} = gI_{sp},$$

where u_{ex} is the exhaust velocity of the propellant.

Substitute and factor to obtain:

$$v_f = gI_{sp}\ln\frac{m_0}{m_f} - \frac{gI_{sp}}{\tau_0}\left(1 - \frac{m_f}{m_0}\right)$$

$$\boxed{= gI_{sp}\left[\ln\left(\frac{m_0}{m_f}\right) - \frac{1}{\tau_0}\left(1 - \frac{m_f}{m_0}\right)\right]}$$

(c) A spreadsheet program to calculate the final velocity of the rocket as a function of the mass ratio m_0/m_f is shown below. The constants used in the velocity function and the formulas used to calculate the final velocity are as follows:

Cell	Content/Formula	Algebraic Form
B1	250	I_{sp}
B2	9.81	g
B3	2	τ
D9	D8 + 0.25	m_0/m_f

	A	B	C	D	E
1	Isp =	250	s		
2	g =	9.81	m/s^2		
3	tau =	2			
4					
5					
6					
7				mass ratio	vf
8				2.00	1.252E+02
9				2.25	3.187E+02
10				2.50	4.854E+02
11				2.75	6.316E+02
12				3.00	7.614E+02
36				9.00	2.204E+03
37				9.25	2.237E+03
38				9.50	2.269E+03
39				9.75	2.300E+03
40				10.00	2.330E+03

A graph of final velocity as a function of mass ratio is shown below.

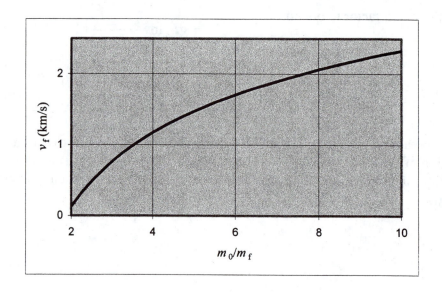

(d) Substitute the data given in part (c) in the equation derived in part (b) to obtain:

$$7\,\text{km/s} = \left(9.81\,\text{m/s}^2\right)\left(250\,\text{s}\right)\left(\ln\frac{m_0}{m_f} - \frac{1}{2}\left(1 - \frac{m_f}{m_0}\right)\right)$$

or

$$2.854 = \ln x - 0.5 + \frac{0.5}{x} \quad \text{where } x = m_0/m_f.$$

Use trial-and-error methods or a graphing calculator to solve this transcendental equation for the root greater than 1:

$$x = \boxed{28.1},$$

a value considerably larger than the practical limit of 10 for single-stage rockets.

General Problems

*111 • A 4-kg fish is swimming at 1.5 m/s to the right. He swallows a 1.2-kg fish swimming toward him at 3 m/s. Neglecting water resistance, what is the velocity of the larger fish immediately after his lunch?

Picture the Problem Let the direction the 4-kg fish is swimming be the positive x direction and the system include the fish, the water, and the earth. The velocity of the larger fish immediately after its lunch is the velocity of the center of mass in this perfectly inelastic collision.

Relate the velocity of the center of mass to the total momentum of the system:

$$\vec{P} = \sum_i m_i \vec{v}_i = m\vec{v}_{cm}$$

Solve for v_{cm}:

$$v_{cm} = \frac{m_4 v_4 - m_{1.2} v_{1.2}}{m_4 + m_{1.2}}$$

Substitute numerical values and evaluate v_{cm}:

$$v_{cm} = \frac{\left(4\,\text{kg}\right)\left(1.5\,\text{m/s}\right) - \left(1.2\,\text{kg}\right)\left(3\,\text{m/s}\right)}{4\,\text{kg} + 1.2\,\text{kg}}$$

$$= \boxed{0.462\,\text{m/s}}$$

*114 •• A 60-kg woman stands on the back of a 6-m-long, 120-kg raft that is floating at rest in still water with no friction. The raft is 0.5 m from a fixed pier, as shown in Figure 8-63. (a) The woman walks to the front of the raft and stops. How far is the raft from the pier now? (b) While the woman walks, she maintains a constant speed of 3 m/s relative to the raft. Find the total kinetic energy of the system (woman plus raft), and compare with the kinetic energy if the woman walked at 3 m/s on a raft tied to the pier. (*c*) Where does this energy come from, and where does it go when the woman stops at the front of the raft? (*d*) On land, the woman puts a lead shot 6 m. She stands at the back of the raft, aims forward, and puts the shot so that just after it leaves her hand, it has the same velocity relative to her as it did when she threw it from the ground. Where does the shot land?

Figure 8-63 Problem 114

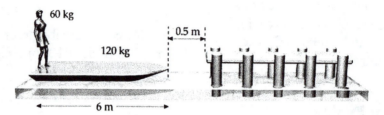

Picture the Problem Take the origin to be at the initial position of the right-hand end of raft and let the positive *x* direction be to the left. Let "*w*" denote the woman and "*r*" the raft, *d* be the distance of the end of the raft from the pier after the woman has walked to its front. The raft moves to the left as the woman moves to the right; with the center of mass of the woman-raft system remaining fixed (because $F_{\text{ext,net}} = 0$). The diagram shows the initial ($x_{\text{w,i}}$) and final ($x_{\text{w,f}}$) positions of the woman as well as the initial ($x_{\text{r\_cm,i}}$) and final ($x_{\text{r\_cm,f}}$) positions of the center of mass of the raft both before and after the woman has walked to the front of the raft.

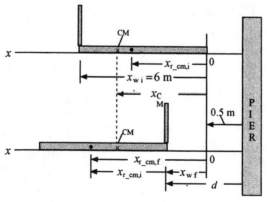

(*a*) Express the distance of the raft from the pier after the woman has walked to the front of the raft:

$$d = 0.5\,\text{m} + x_{\text{f,w}} \qquad (1)$$

Express x_{cm} before the woman has walked to the front of the raft:

$$x_{cm} = \frac{m_w x_{w,i} + m_r x_{r\_cm,i}}{m_w + m_r}$$

Express x_{cm} after the woman has walked to the front of the raft:

$$x_{cm} = \frac{m_w x_{w,f} + m_r x_{r\_cm,f}}{m_w + m_r}$$

Because $F_{ext,net} = 0$, the center of mass remains fixed and we can equate these two expressions for x_{cm} to obtain:

$$m_w x_{w,i} + m_r x_{r\_cm,i} = m_w x_{w,f} + m_r x_{r\_cm,f}$$

Solve for $x_{w,f}$:

$$x_{w,f} = x_{w,i} - \frac{m_r}{m_w}\left(x_{r\_cm,f} - x_{r\_cm,i}\right)$$

From the figure it can be seen that $x_{r\_cm,f} - x_{r\_cm,i} = x_{w,f}$. Substitute $x_{w,f}$ for $x_{r\_cm,f} - x_{r\_cm,i}$ and to obtain:

$$x_{w,f} = \frac{m_w x_{w,i}}{m_w + m_r}$$

Substitute numerical values and evaluate $x_{w,f}$:

$$x_{w,f} = \frac{(60\,\text{kg})(6\,\text{m})}{60\,\text{kg} + 120\,\text{kg}} = 2.00\,\text{m}$$

Substitute in equation (1) to obtain:

$$d = 2.00\,\text{m} + 0.5\,\text{m} = \boxed{2.50\,\text{m}}$$

(b) Express the total kinetic energy of the system:

$$K_{tot} = \tfrac{1}{2} m_w v_w^2 + \tfrac{1}{2} m_r v_r^2$$

Noting that the elapsed time is 2 s, find v_w and v_r:

$$v_w = \frac{x_{w,f} - x_{w,i}}{\Delta t} = \frac{2\,\text{m} - 6\,\text{m}}{2\,\text{s}} = -2\,\text{m/s}$$

relative to the dock, and

$$v_r = \frac{x_{r,f} - x_{r,i}}{\Delta t} = \frac{2.50\,\text{m} - 0.5\,\text{m}}{2\,\text{s}} = 1\,\text{m/s},$$

also relative to the dock.

Substitute numerical values and evaluate K_{tot}:

$$K_{tot} = \tfrac{1}{2}(60\,\text{kg})(-2\,\text{m/s})^2$$
$$+ \tfrac{1}{2}(120\,\text{kg})(1\,\text{m/s})^2$$
$$= \boxed{180\,\text{J}}$$

Evaluate K with the raft tied to the pier:

$$K_{tot} = \tfrac{1}{2}m_w v_w^2 = \tfrac{1}{2}(60\,kg)(3\,m/s)^2$$
$$= \boxed{270\,J}$$

(c)	All the kinetic energy derives from the chemical energy of the woman and, assuming she stops via static friction, the kinetic energy is transformed into her internal energy.

(d)	After the shot leaves the woman's hand, the raft - woman system constitutes an inertial reference frame. In that frame the shot has the same initial velocity as did the shot that had a range of 6 m in the reference frame of the land. Thus, in the raft - woman frame, the shot also has a range of 6 m and lands at the front of the raft.

*116   •• Figure 8-65 shows a World War I cannon mounted on a railcar so that it will project a shell at an angle of 30°. With the car initially at rest, the cannon fires a 200-kg projectile at 125 m/s. (All values are for the frame of reference of the track.) Now consider a system composed of a cannon, shell, and railcar, all on a frictionless track. (*a*) Will the total vector momentum of that system be the same (that is, "conserved") before and after the shell is fired? Explain your answer. (*b*) If the mass of the railcar plus cannon is 5000 kg, what will be the recoil velocity of the car along the track after the firing? (*c*) The shell is observed to rise to a maximum height of 180 m as it moves through its trajectory. At this point, its speed is 80 m/s. On the basis of this information, calculate the amount of thermal energy produced by air friction on the shell on its way from firing to this maximum height.

Figure 8-60 Problem 116

Picture the Problem We can use conservation of momentum in the horizontal direction to find the recoil velocity of the car along the track after the firing. Because the shell will neither rise as high nor be moving as fast at the top of its trajectory as it would be in the absence of air friction, we can apply the work-energy theorem to find the amount of thermal energy produced by the air friction.

(a) | No. The vertical reaction force of the rails is an external force and so the momentum of the system will not be conserved.

(b) Use conservation of momentum in the horizontal (x) direction to obtain:

$$\Delta p_x = 0$$
or
$$mv\cos 30° - Mv_{recoil} = 0$$

Solve for and evaluate v_{recoil}:

$$v_{recoil} = \frac{mv\cos 30°}{M}$$

Substitute numerical values and evaluate v_{recoil}:

$$v_{recoil} = \frac{(200\,kg)(125\,m/s)\cos 30°}{5000\,kg}$$

$$= \boxed{4.33\,m/s}$$

(c) Using the work-energy theorem, relate the thermal energy produced by air friction to the change in the energy of the system:

$$W_{ext} = W_f = \Delta E_{sys} = \Delta U + \Delta K$$

Substitute for ΔU and ΔK to obtain:

$$W_{ext} = mgy_f - mgy_i + \tfrac{1}{2}mv_f^2 - \tfrac{1}{2}mv_i^2$$
$$= mg(y_f - y_i) + \tfrac{1}{2}m(v_f^2 - v_i^2)$$

Substitute numerical values and evaluate W_{ext}:

$$W_{ext} = (200\,kg)(9.81\,m/s^2)(180\,m) + \tfrac{1}{2}(200\,kg)[(80\,m/s)^2 - (125\,m/s)^2] = \boxed{-569\,kJ}$$

*121 •• In the "slingshot effect," the transfer of energy in an elastic collision is used to boost the energy of a space probe so that it can escape from the solar system. All speeds are relative to an inertial frame in which the center of the sun remains at rest. Figure 8-68 shows a space probe moving at 10.4 km/s toward Saturn, which is moving at 9.6 km/s toward the probe. Because of the gravitational attraction between Saturn and the probe, the probe swings around Saturn and heads back in the opposite direction with speed v_f. (a) Assuming this collision to be a one-dimensional elastic collision with the mass of Saturn much greater than that of the probe, find v_f. (b) By what factor is the kinetic energy of the probe increased? Where does this energy come from?

Figure 8-68 Problem 121

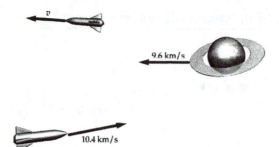

Picture the Problem Let the direction the probe is moving after its elastic collision with Saturn be the positive direction. The probe gains kinetic energy at the expense of the kinetic energy of Saturn. We'll relate the velocity of approach relative to the center of mass to u_{rec} and then to v.

(*a*) Relate the velocity of recession to the velocity of recession relative to the center of mass:	$v = u_{rec} + v_{cm}$
Find the velocity of approach:	$u_{app} = -9.6\,\text{km/s} - 10.4\,\text{km/s}$ $= -20.0\,\text{km/s}$
Relate the relative velocity of approach to the relative velocity of recession for an elastic collision:	$u_{rec} = -u_{app} = 20.0\,\text{km/s}$
Because Saturn is so much more massive than the space probe:	$v_{cm} = v_{Saturn} = 9.6\,\text{km/s}$
Substitute and evaluate v:	$v = u_{rec} + v_{cm} = 20\,\text{km/s} + 9.6\,\text{km/s}$ $= \boxed{29.6\,\text{km/s}}$
(*b*) Express the ratio of the final kinetic energy to the initial kinetic energy:	$\dfrac{K_f}{K_i} = \dfrac{\frac{1}{2}Mv_{rec}^2}{\frac{1}{2}Mv_i^2} = \left(\dfrac{v_{rec}}{v_i}\right)^2$ $= \left(\dfrac{29.6\,\text{km/s}}{10.4\,\text{km/s}}\right)^2 = \boxed{8.10}$

The energy comes from an immeasurably small slowing of Saturn.

*122 •• Imagine that a flashlight is floating in intergalactic space, far from any planet or star. It has batteries in it that are charged with a total energy $E = 1.5$ kJ. When the flashlight is turned on, it loses energy in the form of light; because of this, it also loses mass slightly, from $E = mc^2$. Because the lost "mass" is traveling away at the speed of light, the flashlight should start moving in the opposite direction. (*a*) If the flashlight loses energy ΔE, argue that the change in momentum of the flashlight should be $P = \Delta E/c$. (*b*) If the mass of the flashlight is 1.5 kg, and it is turned on and left on until the batteries are discharged, what is its final velocity? Assume that the flashlight is 100 percent efficient in converting battery power into radiant power, and that it is initially at rest.

Picture the Problem We can use the relationships $P = c\Delta m$ and $\Delta E = \Delta mc^2$ to show that $P = \Delta E/c$. We can then equate this expression with the change in momentum of the flashlight to find the latter's final velocity.

(*a*) Express the momentum of the mass lost (i.e., carried away by the light) by the flashlight:

$$P = c\Delta m$$

Relate the energy carried away by the light to the mass lost by the flashlight:

$$\Delta m = \frac{\Delta E}{c^2}$$

Substitute to obtain:

$$P = c\frac{\Delta E}{c^2} = \boxed{\frac{\Delta E}{c}}$$

(*b*) Relate the final momentum of the flashlight to ΔE:

$$\frac{\Delta E}{c} = \Delta p = mv$$

because the flashlight is initially at rest.

Solve for v:

$$v = \frac{\Delta E}{mc}$$

Substitute numerical values and evaluate v:

$$v = \frac{1.5 \times 10^3 \, \text{J}}{(1.5\,\text{kg})(2.998 \times 10^8 \, \text{m/s})}$$

$$= 3.33 \times 10^{-6} \, \text{m/s}$$

$$= \boxed{3.33 \, \mu\text{m/s}}$$

*127 •• A careless driver rear-ends a car that is halted at a stop sign. Just before impact, the driver slams on his brakes, locking the wheels. The driver of the struck car also has his foot solidly on the brake pedal, locking his brakes. The mass of the struck car is 900 kg, and that of the initially moving vehicle is 1200 kg. On collision, the bumpers of the two cars mesh. Police determine from the skid marks that after the collision the two cars moved 0.76 m together. Tests revealed that the coefficient of sliding friction between the tires and pavement was 0.92. The driver of the moving car claims that he was traveling at less than 15 km/h as he approached the intersection. Is he telling the truth?

Picture the Problem Let the direction the moving car was traveling before the collision be the positive x direction. Let the numeral 1 denote this car and the numeral 2 the car that is stopped at the stop sign and the system include both cars and the earth. We can use conservation of momentum to relate the speed of the initially-moving car to the speed of the meshed cars immediately after their perfectly inelastic collision and conservation of energy to find the initial speed of the meshed cars.

Using conservation of momentum, relate the before-collision velocity to the after-collision velocity of the meshed cars:

$$p_i = p_f$$

or

$$m_1 v_1 = (m_1 + m_2)V$$

Solve for v_1:

$$v_1 = \frac{m_1 + m_2}{m_1}V = \left(1 + \frac{m_2}{m_1}\right)V$$

Using conservation of energy, relate the initial kinetic energy of the meshed cars to the work done by friction in bringing them to a stop:

$$\Delta K + \Delta E_{thermal} = 0$$

or, because $K_f = 0$ and $\Delta E_{thermal} = f\Delta s$,

$$-K_i + f_k \Delta s = 0$$

Substitute for Ki and, using $f_k = \mu_k F_n = \mu_k Mg$, eliminate f_k to obtain:

$$-\tfrac{1}{2}MV^2 + \mu_k Mg\Delta x = 0$$

Solve for V:

$$V = \sqrt{2\mu_k g\Delta x}$$

Substitute to obtain:

$$v_1 = \left(1 + \frac{m_2}{m_1}\right)\sqrt{2\mu_k g\Delta x}$$

Substitute numerical values and evaluate v_1:

$$v_1 = \left(1 + \frac{900\,\text{kg}}{1200\,\text{kg}}\right)\sqrt{2(0.92)(9.81\,\text{m/s}^2)(0.76\,\text{m})} = 6.48\,\text{m/s} = 23.3\,\text{km/h}$$

The driver was not telling the truth. He was traveling at 23.3 km/h.

*130 •• A circular plate of radius r has a circular hole of radius $r/2$ cut out of it (Figure 8-72). Find the center of mass of the plate. *Hint: The hole can be represented by two disks superimposed, one of mass m and the other of mass –m.*

Figure 8-72 Problem 130

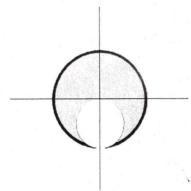

Picture the Problem By symmetry, $x_{\text{cm}} = 0$. Let σ be the mass per unit area of the disk. The mass of the modified disk is the difference between the mass of the whole disk and the mass that has been removed.

Start with the definition of y_{cm}:

$$y_{\text{cm}} = \frac{\sum_i m_i y_i}{M - m_{\text{hole}}}$$

$$= \frac{m_{\text{disk}} y_{\text{disk}} - m_{\text{hole}} y_{\text{hole}}}{M - m_{\text{hole}}}$$

Express the mass of the complete disk:

$$M = \sigma A = \sigma \pi r^2$$

Express the mass of the material removed:

$$m_{\text{hole}} = \sigma \pi \left(\tfrac{1}{2}r\right)^2 = \tfrac{1}{4}\sigma \pi r^2 = \tfrac{1}{4}M$$

Substitute and simplify to obtain:

$$y_{\text{cm}} = \frac{M(0) - \left(\tfrac{1}{4}M\right)\left(-\tfrac{1}{2}r\right)}{M - \tfrac{1}{4}M} = \boxed{\tfrac{1}{6}r}$$

*132 •• A neutron of mass m makes an elastic head-on collision with a stationary nucleus of mass M. (*a*) Show that the energy of the nucleus after the collision is $K_{nucleus} = [4mM/(m + M)^2]K_n$, where K_n is the initial energy of the neutron. (*b*) Show that the fraction of energy lost by the neutron in this collision is

$$\frac{-\Delta K_n}{K_n} = \frac{4mM}{(m+M)^2} = \frac{4(m/M)}{(1+m/M)^2}$$

Picture the Problem In this elastic head-on collision, the kinetic energy of recoiling nucleus is the difference between the initial and final kinetic energies of the neutron. We can derive the indicated results by using both conservation of energy and conservation of momentum and writing the kinetic energies in terms of the momenta of the particles before and after the collision.

(*a*) Use conservation of energy to relate the kinetic energies of the particles before and after the collision:

$$\frac{p_{ni}^2}{2m} = \frac{p_{nf}^2}{2m} + \frac{p_{nucleus}^2}{2M} \qquad (1)$$

Apply conservation of momentum to obtain a second relationship between the initial and final momenta:

$$p_{ni} = p_{nf} + p_{nucleus} \qquad (2)$$

Eliminate p_{nf} in equation (1) using equation (2):

$$\frac{p_{nucleus}}{2M} + \frac{p_{nucleus}}{2m} - \frac{p_{ni}}{m} = 0 \qquad (3)$$

Use equation (3) to write $p_{ni}^2/2m$ in terms of $p_{nucleus}$:

$$\frac{p_{ni}^2}{2m} = K_n = \frac{p_{nucleus}^2(M+m)^2}{8M^2m} \qquad (4)$$

Use equation (4) to express $K_{nucleus} = p_{nucleus}^2/2M$ in terms of K_n:

$$K_{nucleus} = \boxed{K_n\left[\frac{4Mm}{(M+m)^2}\right]} \qquad (5)$$

(*b*) Relate the *change* in the kinetic energy of the neutron to the after-collision kinetic energy of the nucleus:

$$\Delta K_n = -K_{nucleus}$$

Using equation (5), express the fraction of the energy lost in the collision:

$$\frac{-\Delta K_n}{K_n} = \boxed{\frac{4Mm}{(M+m)^2} = \frac{4\dfrac{m}{M}}{\left(1+\dfrac{m}{M}\right)^2}}$$

*138 •• The ratio of the mass of the earth to the mass of the moon is $M_e/m_m = 81.3$. The radius of the earth is about 6370 km, and the distance from the earth to the moon is about 384,000 km. (*a*) Locate the center of mass of the earth–moon system. Is it above or below the surface of the earth? (*b*) What external forces act on the earth-moon system? (*c*) In what direction is the acceleration of the center of mass of this system? (*d*) Assume that the center of mass of this system moves in a circular orbit around the sun. How far must the center of the earth move in the radial direction (toward or away from the sun) during the 14 days between the time the moon is farthest from the sun (full moon) and the time it is closest to the sun (new moon)?

Picture the Problem We can use the definition of the center of mass of a system containing multiple objects to locate the center of mass of the earth–moon system. Any object external to the system will exert accelerating forces on the system.

(*a*) Express the center of mass of the earth–moon system relative to the center of the earth:

$$M\vec{r}_{cm} = \sum_i m_i \vec{r}_i$$

or

$$r_{cm} = \frac{M_e(0) + m_m r_{em}}{M_e + m_m} = \frac{m_m r_{em}}{M_e + m_m}$$

$$= \frac{r_{em}}{\dfrac{M_e}{m_m} + 1}$$

Substitute numerical values and evaluate r_{cm}:

$$r_{cm} = \frac{3.84 \times 10^5 \text{ km}}{81.3 + 1} = \boxed{4670 \text{ km}}$$

Because this distance is less than the radius of the earth, the position of the center of mass of the earth – moon system is below the surface of the earth.

(*b*) Any object not in the earth – moon system exerts forces on the system; e.g., the sun and other planets.

(*c*) Because the sun exerts the dominant external force on the earth – moon system, the acceleration of the system is toward the sun.

(*d*) Because the center of mass is at a fixed distance from the sun, the distance *d* moved by the earth in this time interval is:

$$d = 2r_{em} = 2(4670 \text{ km}) = \boxed{9340 \text{ km}}$$

*140 ••• A stream of glass beads, each with a mass of 0.5 g, comes out of a
horizontal tube at a rate of 100 per second (Figure 8-74). The beads fall a distance
of 0.5 m to a balance pan and bounce back to their original height. How much
mass must be placed in the other pan of the balance to keep the pointer at zero?

Figure 8-74 Problem 140

Picture the Problem Take the zero of gravitational potential energy to be at the
elevation of the pan and let the system include the balance, the beads, and the
earth. We can use conservation of energy to find the vertical component of the
velocity of the beads as they hit the pan and then calculate the net downward
force on the pan from Newton's 2nd law.

Use conservation of energy to relate the y component of the bead's velocity as it hits the pan to its height of fall:	$\Delta K + \Delta U = 0$ or, because $K_i = U_f = 0$, $\frac{1}{2}mv_y^2 - mgh = 0$
Solve for v_y:	$v_y = \sqrt{2gh}$
Substitute numerical values and evaluate v_y:	$v_y = \sqrt{2(9.81\,\text{m/s}^2)(0.5\,\text{m})} = 3.13\,\text{m/s}$
Express the change in momentum in the y direction per bead:	$\Delta p_y = p_{yf} - p_{yi} = mv_y - (-mv_y) = 2mv_y$
Use Newton's second law to express the net force in the y direction exerted on the pan by the beads:	$F_{net,y} = N\dfrac{\Delta p_y}{\Delta t}$
Letting M represent the mass to be placed on the other pan, equate its	$Mg = N\dfrac{\Delta p_y}{\Delta t}$

weight to the net force exerted by the beads, substitute for Δp_y, and solve for M:

and

$$M = \frac{N}{\Delta t} \frac{2mv_y}{g}$$

Substitute numerical values and evaluate M:

$$M = (100/s)\frac{[2(0.0005\,\text{kg})(3.13\,\text{m/s})]}{9.81\,\text{m/s}^2}$$

$$= \boxed{31.9\,\text{g}}$$

Chapter 9
Rotation

Conceptual Problems

*1 •   Two points are on a disk turning at constant angular velocity, one point on the rim and the other halfway between the rim and the axis. Which point moves the greater distance in a given time? Which turns through the greater angle? Which has the greater speed? The greater angular velocity? The greater tangential acceleration? The greater angular acceleration? The greater centripetal acceleration?

Determine the Concept Because r is greater for the point on the rim, it moves the greater distance. Both turn through the same angle. Because r is greater for the point on the rim, it has the greater speed. Both have the same angular velocity. Both have zero tangential acceleration. Both have zero angular acceleration. Because r is greater for the point on the rim, it has the greater centripetal acceleration.

The distance traveled along the rim is related to the angle through which the disk turns (the same for both points) and the radial distance from the center through $s = r\theta$.	Because r is greater for the point on the rim, it moves the greater distance.
	Both turn through the same angle.
The speed of both points is related to their common angular speed according to $v = r\omega$.	Because r is greater for the point on the rim, it has the greater speed.
	Both have the same angular velocity.
Because the angular velocity is constant, the tangential acceleration is zero.	Both have zero tangential acceleration.
Because the angular velocity is constant, the angular acceleration is zero.	Both have zero angular acceleration.
The centripetal acceleration is directly proportional to r ($a_c = r\omega^2$).	Because r is greater for the point on the rim, it has the greater centripetal

259

acceleration.

*4 •   The dimension of torque is the same as that of (*a*) impulse, (*b*) energy, (*c*) momentum, (*d*) none of these.

Determine the Concept Torque has the dimension $\left[\dfrac{ML^2}{T^2}\right]$.

(*a*) Impulse has the dimension $\left[\dfrac{ML}{T}\right]$.

(*b*) Energy has the dimension $\left[\dfrac{ML^2}{T^2}\right]$. | (b) is correct. |

(*c*) Momentum has the dimension $\left[\dfrac{ML}{T}\right]$.

*6 •   Can an object continue to rotate in the absence of torque?

Determine the Concept Yes. A net torque is required to *change* the rotational state of an object. In the absence of a net torque an object continues in whatever state of rotational motion it was at the instant the net torque became zero.

*10 •   The moment of inertia of an object about an axis that does not pass through its center of mass is _____ the moment of inertia about a parallel axis through its center of mass. (*a*) always less than, (*b*) sometimes less than, (*c*) sometimes equal to, (*d*) always greater than.

Determine the Concept From the parallel-axis theorem we know that $I = I_{\text{cm}} + Mh^2$ where I_{cm} is the moment of inertia of the object with respect to an axis through its center of mass, M is the mass of the object, and h is the distance between the parallel axes.

Therefore, I is always greater than I_{cm} by Mh^2. | (d) is correct. |

*14 •   A wheel of radius R is rolling without slipping on a flat stationary surface. The velocity of the point on the rim that is in contact with the surface is (*a*) equal to $R\omega$ in the direction of motion of the center of mass, (*b*) equal to $R\omega$ opposite to the direction of motion of the center of mass, (*c*) zero, (*d*) equal to the velocity of the center of mass and in the same direction, (*e*) equal to the velocity of the center of mass but in the opposite direction.

Determine the Concept If the wheel is rolling without slipping, a point at the top of the wheel moves with a speed twice that of the center of mass of the wheel, but the bottom of the wheel is momentarily at rest. $\boxed{(c) \text{ is correct.}}$

*16 • Two identical-looking 1-m-long pipes enclose slugs of lead whose total mass is 10 kg (much larger than the mass of the pipe). In the first pipe the lead is concentrated at the center of the pipe, while in the second the lead is divided into two equal masses placed at opposite ends of the pipe. Without opening either pipe, how could you determine which is which?

Determine the Concept You could spin the pipes about their center. The one which is easier to spin has its mass concentrated closer to the center of mass and, hence, has a smaller moment of inertia.

Estimation and Approximation

*25 •• Consider the moment of inertia of an average adult man about an axis running vertically through the center of his body when he is standing straight up with arms flat at his sides and again when he is standing straight up holding his arms straight out. Estimate the ratio of his moment of inertia with his arms straight out to the moment of inertia with his arms flat against his sides.

Picture the Problem Assume that the mass of an average adult male is about 80 kg, and that we can model his body when he is standing straight up with his arms at his sides as a cylinder. From experience in men's clothing stores, a man's average waist circumference seems to be about 34 inches, and the average chest circumference about 42 inches. We'll also assume that about 20% of the body's mass is in the two arms, and each has a length $L = 1$ m, so that each arm has a mass of about $m = 8$ kg.

Letting I_{out} represent his moment of inertia with his arms straight out and I_{in} his moment of inertia with his arms at his side, the ratio of these two moments of inertia is:

$$\frac{I_{out}}{I_{in}} = \frac{I_{body} + I_{arms}}{I_{in}} \qquad (1)$$

Express the moment of inertia of the "man as a cylinder":

$$I_{in} = \tfrac{1}{2}MR^2$$

Express the moment of inertia of his arms:

$$I_{arms} = 2\left(\tfrac{1}{3}\right)mL^2$$

Express the moment of inertia of his body-less-arms:

$$I_{body} = \tfrac{1}{2}(M - m)R^2$$

Substitute in equation (1) to obtain:

$$\frac{I_{out}}{I_{in}} = \frac{\frac{1}{2}(M-m)R^2 + 2\left(\frac{1}{3}\right)mL^2}{\frac{1}{2}MR^2}$$

Assume the circumference of the cylinder to be the average of the average waist circumference and the average chest circumference:

$$c_{av} = \frac{34\,\text{in} + 42\,\text{in}}{2} = 38\,\text{in}$$

Find the radius of a circle whose circumference is 38 in:

$$R = \frac{c_{av}}{2\pi} = \frac{38\,\text{in} \times \dfrac{2.54\,\text{cm}}{\text{in}} \times \dfrac{1\,\text{m}}{100\,\text{cm}}}{2\pi}$$

$$= 0.154\,\text{m}$$

Substitute numerical values and evaluate I_{out}/I_{in}:

$$\frac{I_{out}}{I_{in}} = \frac{\frac{1}{2}(80\,\text{kg} - 16\,\text{kg})(0.154\,\text{m})^2 + \frac{2}{3}(8\,\text{kg})(1\,\text{m})^2}{\frac{1}{2}(80\,\text{kg})(0.154\,\text{m})^2} = \boxed{6.42}$$

Angular Velocity and Angular Acceleration

*28 •   When a turntable rotating at $33\frac{1}{3}$ rev/min is shut off, it comes to rest in 26 s. Assuming constant angular acceleration, find (*a*) the angular acceleration, (*b*) the average angular velocity of the turntable, and (*c*) the number of revolutions it makes before stopping.

Picture the Problem Because we're assuming constant angular acceleration; we can find the various physical quantities called for in this problem by using constant acceleration equations.

(*a*) Using its definition, find the angular acceleration of the turntable:

$$\alpha = \frac{\Delta\omega}{\Delta t} = \frac{\omega - \omega_0}{\Delta t}$$

$$= \frac{0 - 33\frac{1}{3}\dfrac{\text{rev}}{\text{min}} \times \dfrac{2\pi\,\text{rad}}{\text{rev}} \times \dfrac{1\,\text{min}}{60\,\text{s}}}{26\,\text{s}}$$

$$= \boxed{0.134\,\text{rad/s}^2}$$

(b) Because the angular acceleration is constant, the average angular velocity is given by:

$$\omega_{av} = \frac{\omega_0 + \omega}{2}$$

$$= \frac{33\frac{1}{3}\frac{rev}{min} \times \frac{2\pi\,rad}{rev} \times \frac{1\,min}{60\,s}}{2}$$

$$= \boxed{1.75\,rad/s}$$

(c) Using the definition of ω_{av}, find the number or revolutions the turntable makes before stopping:

$$\Delta\theta = \omega_{av}\Delta t = (1.75\,rad/s)(26\,s)$$

$$= 45.4\,rad \times \frac{1\,rev}{2\pi\,rad}$$

$$= \boxed{7.22\,rev}$$

*35 •• The tape in a standard VHS videotape cassette has a length $L = 246$ m; the tape plays for 2.0 h (Figure 9-41). As the tape starts, the full reel has an outer radius of about $R = 45$ mm and an inner radius of about $r = 12$ mm. At some point during the play, both reels have the same angular speed. Calculate this angular speed in radians per second and revolutions per minute.

Figure 9-41 Problem 35

45 mm
12 mm

Picture the Problem The two tapes will have the same tangential and angular velocities when the two reels are the same size, i.e., have the same area. We can calculate the tangential speed of the tape from its length and running time and relate the angular velocity to the constant tangential speed and the radius of the reels when they are turning with the same angular velocity.

Relate the angular velocity of the tape to its tangential speed:

$$\omega = \frac{v}{r} \qquad (1)$$

Letting R_f represent the outer radius of the reel when the reels have the same area, express the condition that they have the same speed:

$$\pi R_f^2 - \pi r^2 = \frac{1}{2}\left(\pi R^2 - \pi r^2\right)$$

Solve for R_f:

$$R_f = \sqrt{\frac{R^2 + r^2}{2}} = \sqrt{\frac{(45\,\text{mm})^2 + (12\,\text{mm})^2}{2}}$$

$$= 32.9\,\text{mm}$$

Find the tangential speed of the tape from its length and running time:

$$v = \frac{L}{\Delta t} = \frac{246\,\text{m}}{2\,\text{h} \times \dfrac{3600\,\text{s}}{\text{h}}} = 3.42\,\text{cm/s}$$

Substitute in equation (1) and evaluate ω:

$$\omega = \frac{v}{R_f} = \frac{3.42\,\text{cm/s}}{32.9\,\text{mm}} = \boxed{1.04\,\text{rad/s}}$$

Convert 1.04 rad/s to rev/min:

$$1.04\,\text{rad/s} = 1.04\,\frac{\text{rad}}{\text{s}} \times \frac{1\,\text{rev}}{2\pi\,\text{rad}} \times \frac{60\,\text{s}}{\text{min}}$$

$$= \boxed{9.92\,\text{rev/min}}$$

Torque, Moment of Inertia, and Newton's Second Law for Rotation

*37 •   A 2.5-kg cylinder of radius 11 cm, is initially at rest, is free to rotate about the axis of the cylinder. A rope of negligible mass is wrapped around it and pulled with a force of 17 N. Find (*a*) the torque exerted by the rope, (*b*) the angular acceleration of the cylinder, and (*c*) the angular velocity of the cylinder at $t = 5$ s.

Picture the Problem We can find the torque exerted by the 17-N force from the definition of torque. The angular acceleration resulting from this torque is related to the torque through Newton's 2nd law in rotational form. Once we know the angular acceleration, we can find the angular velocity of the cylinder as a function of time.

(*a*) Calculate the torque from its definition:

$$\tau = F\ell = (17\,\text{N})(0.11\,\text{m}) = \boxed{1.87\,\text{N}\cdot\text{m}}$$

(*b*) Use Newton's 2nd law in rotational form to relate the acceleration resulting from this torque to the torque:

$$\alpha = \frac{\tau}{I}$$

Express the moment of inertia of the cylinder with respect to its

$$I = \tfrac{1}{2}MR^2$$

axis of rotation:

Substitute to obtain:

$$\alpha = \frac{2\tau}{MR^2}$$

Substitute numerical values and evaluate α:

$$\alpha = \frac{2(1.87\,\text{N}\cdot\text{m})}{(2.5\,\text{kg})(0.11\,\text{m})^2} = \boxed{124\,\text{rad/s}^2}$$

(c) Using a constant acceleration equation, express the angular velocity of the cylinder as a function of time:

$$\omega = \omega_0 + \alpha t$$
or, because $\omega_0 = 0$,
$$\omega = \alpha t$$

Evaluate ω (5 s):

$$\omega(5\,\text{s}) = (124\,\text{rad/s}^2)(5\,\text{s}) = \boxed{620\,\text{rad/s}}$$

*40 ••• A uniform rod of mass M and length L is pivoted at one end and hangs as in Figure 9-42 so that it is free to rotate without friction about its pivot. It is struck by a horizontal force F_0 for a short time Δt at a distance x below the pivot as shown. (a) Show that the speed of the center of mass of the rod just after being struck is given by $v_0 = 3F_0 x \Delta t/2ML$. (b) Find the horizontal component of the force delivered by the pivot, and show that this force component is zero if $x = 2L/3$. (Note: The point $x = 2L/3$ is called the center of percussion of the rod.)

Figure 9-42 Problem 40

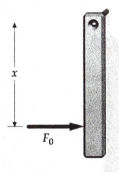

Picture the Problem We can express the velocity of the center of mass of the rod in terms of its distance from the pivot point and the angular velocity of the rod. We can find the angular velocity of the rod by using Newton's 2nd law to find its angular acceleration and then a constant acceleration equation that relates ω to α. We'll use the impulse-momentum relationship to derive the expression for the force delivered to the rod by the pivot. Finally, the location of the *center of percussion* of the rod will be verified by setting the force exerted by the pivot to zero.

(*a*) Relate the velocity of the center of mass to its distance from the pivot point:

$$v_{cm} = \frac{L}{2}\omega \qquad (1)$$

Express the torque due to F_0:

$$\tau = F_0 x = I\alpha$$

Express the moment of inertia of the rod with respect to an axis through its pivot point:

$$I_{pivot} = \tfrac{1}{3}ML^2$$

Substitute and solve for α:

$$\alpha = \frac{3F_0 x}{ML^2}$$

Express the angular velocity of the rod in terms of its angular acceleration:

$$\omega = \alpha\,\Delta t = \frac{3F_0 x\Delta t}{ML^2}$$

Substitute in equation (1) to obtain:

$$\boxed{v_{cm} = \frac{3F_0 x\Delta t}{2ML}}$$

(*b*) Let I_P be the impulse exerted by the pivot on the rod. Then the total impulse (equal to the change in momentum of the rod) exerted on the rod is:

$$I_P + F_0\Delta t = Mv_{cm}$$
and
$$I_P = Mv_{cm} - F_0\Delta t$$

Substitute our result from (a) to obtain:

$$I_P = \frac{3F_0 x\Delta t}{2L} - F_0\Delta t = F_0\Delta t\left(\frac{3x}{2L} - 1\right)$$

Because $I_P = F_P\Delta t$:

$$F_P = \boxed{F_0\left(\frac{3x}{2L} - 1\right)}$$

In order for F_P to be zero:

$$\frac{3x}{2L} - 1 = 0 \text{ or } x = \boxed{\frac{2L}{3}}$$

Calculating the Moment of Inertia

*43 •   Four particles at the corners of a square with side length $L = 2$ m are connected by massless rods (Figure   9-43). The masses of the particles are $m_1 = m_3 = 3$ kg and $m_2 = m_4 = 4$ kg. Find the moment of inertia of the system about the z axis.

Figure 9-43 Problem 43-45

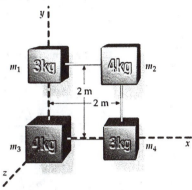

Picture the Problem The moment of inertia of a system of particles with respect to a given axis is the sum of the products of the mass of each particle and the square of its distance from the given axis.

Use the definition of the moment of inertia of a system of particles:

$$I = \sum_i m_i r_i^2$$

$$= m_1 r_1^2 + m_2 r_2^2 + m_3 r_3^2 + m_4 r_4^2$$

Substitute numerical values and evaluate I:

$$I = (3\,\text{kg})(2\,\text{m})^2 + (4\,\text{kg})(2\sqrt{2}\,\text{m})^2$$
$$+ (4\,\text{kg})(0)^2 + (3\,\text{kg})(2\,\text{m})^2$$
$$= \boxed{56.0\,\text{kg} \cdot \text{m}^2}$$

*48 ••   Two point masses m_1 and m_2 are separated by a massless rod of length L. (*a*) Write an expression for the moment of inertia about an axis perpendicular to the rod and passing through it at a distance x from mass m_1. (*b*) Calculate dI/dx and show that I is at a minimum when the axis passes through the center of mass of the system.

Picture the Problem The moment of inertia of a system of particles depends on the axis with respect to which it is calculated. Once this choice is made, the moment of inertia is the sum of the products of the mass of each particle and the square of its distance from the chosen axis.

(*a*) Apply the definition of the moment of inertia of a system of particles:

$$I = \sum_i m_i r_i^2 = \boxed{m_1 x^2 + m_2 (L - x)^2}$$

(*b*) Set the derivative of *I* with respect to *x* equal to zero in order to identify values for *x* that correspond to either maxima or minima:

$$\frac{dI}{dx} = 2m_1 x + 2m_2 (L - x)(-1)$$
$$= 2(m_1 x + m_2 x - m_2 L)$$
$$= 0 \text{ for extrema}$$

If $\dfrac{dI}{dx} = 0$, then:

$$m_1 x + m_2 x - m_2 L = 0$$

Solve for *x*:

$$x = \frac{m_2 L}{m_1 + m_2}$$

Convince yourself that you've found a minimum by showing that $\dfrac{d^2 I}{dx^2}$ is positive at this point.

$\boxed{x = \dfrac{m_2 L}{m_1 + m_2} \text{ is, by definition, the}}$
$\boxed{\text{distance of the center of mass from } m.}$

*50 •• Tracey and Corey are doing intensive research on baton-twirling. Each is using "The Beast" as a model baton: two uniform spheres, each of mass 500 g and radius 5 cm, mounted at the ends of a 30-cm uniform rod of mass 60 g (Figure 9-45). They want to calculate the moment of inertia of The Beast about an axis perpendicular to the rod and passing through its center. Corey uses the approximation that the two spheres can be treated as point particles that are 20 cm from the axis of rotation, and that the mass of the rod is negligible. Tracey, however, makes her calculations without approximations. (*a*) Compare the two results. (*b*) If the spheres retained the same mass but were hollow, would the rotational inertia increase or decrease? Justify your choice with a sentence or two. It is not necessary to calculate the new value of *I*.

Figure 9-45 Problem 50

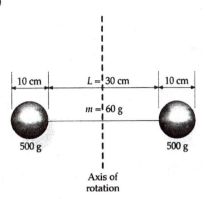

Picture the Problem Corey will use the point particle relationship
$I_{app} = \sum_i m_i r_i^2 = m_1 r_1^2 + m_2 r_2^2$ for his calculation whereas Tracey's calculation will
take into account not only the rod but also the fact that the spheres are not point
particles.

(a) Using the point-mass approximation and the definition of the moment of inertia of a system of particles, express I_{app}:

$$I_{app} = \sum_i m_i r_i^2 = m_1 r_1^2 + m_2 r_2^2$$

Substitute numerical values and evaluate I_{app}:

$$I_{app} = (0.5\,\text{kg})(0.2\,\text{m})^2 + (0.5\,\text{kg})(0.2\,\text{m})^2$$
$$= \boxed{0.0400\,\text{kg}\cdot\text{m}^2}$$

Express the moment of inertia of the two spheres and connecting rod system:

$$I = I_{spheres} + I_{rod}$$

Use Table 9-1 to find the moments of inertia of a sphere (with respect to its center of mass) and a rod (with respect to an axis through its center of mass):

$$I_{sphere} = \tfrac{2}{5} M_{sphere} R^2$$
and
$$I_{rod} = \tfrac{1}{12} M_{rod} L^2$$

Because the spheres are not on the axis of rotation, use the parallel axis theorem to express their moment of inertia with respect to the axis of rotation:

$$I_{sphere} = \tfrac{2}{5} M_{sphere} R^2 + M_{sphere} h^2$$
where h is the distance from the center of mass of a sphere to the axis of rotation.

Substitute to obtain:

$$I = 2\{\tfrac{2}{5}M_{sphere}R^2 + M_{sphere}h^2\} + \tfrac{1}{12}M_{rod}L^2$$

Substitute numerical values and evaluate I:

$$I = 2\{\tfrac{2}{5}(0.5\,kg)(0.05\,m)^2 + (0.5\,kg)(0.2\,m)^2\} + \tfrac{1}{12}(0.06\,kg)(0.3\,m)^2$$
$$= \boxed{0.0415\,kg \cdot m^2}$$

Compare I and I_{app} by taking their ratio: $\dfrac{I_{app}}{I} = \dfrac{0.0400\,kg \cdot m^2}{0.0415\,kg \cdot m^2} = \boxed{0.964}$

(b)
> The rotational inertia would increase because I_{cm} of a hollow sphere is greater than I_{cm} of a solid sphere.

*54 •••   The density of the earth is not quite uniform. It varies with the distance r from the center of the earth as $\rho = C(1.22 - r/R)$, where R is the radius of the earth and C is a constant. (a) Find C in terms of the total mass M and the radius R. (b) Find the moment of inertia of the earth. (See Problem 53.)

Picture the Problem We can find C in terms of M and R by integrating a spherical shell of mass dm with the given density function to find the mass of the earth as a function of M and then solving for C. In part (b), we'll start with the moment of inertia of the same spherical shell, substitute the earth's density function, and integrate from 0 to R.

(a) Express the mass of the earth using the given density function:

$$M = \int dm = \int_0^R 4\pi \rho r^2 dr$$

$$= 4\pi C \int_0^R 1.22 r^2 dr - \frac{4\pi C}{R}\int_0^R r^3 dr$$

$$= \frac{4\pi}{3}1.22CR^3 - \pi CR^3$$

Solve for C as a function of M and R to obtain:

$$C = \boxed{0.508\frac{M}{R^3}}$$

(b) From Problem 9-40 we have: $dI = \tfrac{8}{3}\pi \rho R^4 dR$

Integrate to obtain:

$$I = \tfrac{8}{3}\pi \int_0^R \rho\, R^4\, dR$$

$$= \frac{8\pi(0.508)M}{3R^3}\left[\int_0^R 1.22r^4\, dr - \frac{1}{R}\int_0^R r^5\, dr\right]$$

$$= \frac{4.26M}{R^3}\left[\frac{1.22}{5}R^5 - \frac{1}{6}R^5\right]$$

$$= \boxed{0.329MR^2}$$

Rotational Kinetic Energy

*59 • A solid ball of mass 1.4 kg and diameter 15 cm is rotating about its diameter at 70 rev/min. (*a*) What is its kinetic energy? (*b*) If an additional 2 J of energy are supplied to the rotational energy, what is the new angular speed of the ball?

Picture the Problem We can find the kinetic energy of this rotating ball from its angular speed and its moment of inertia. We can use the same relationship to find the new angular speed of the ball when it is supplied with additional energy.

(*a*) Express the kinetic energy of the ball:

$$K = \tfrac{1}{2}I\omega^2$$

Express the moment of inertia of ball with respect to its diameter:

$$I = \tfrac{2}{5}MR^2$$

Substitute and eliminate *I*:

$$K = \tfrac{1}{5}MR^2\omega^2$$

$$= \tfrac{1}{5}(1.4\,\text{kg})(0.075\,\text{m})^2$$

$$\times\left(70\,\frac{\text{rev}}{\text{min}}\times\frac{2\pi\,\text{rad}}{\text{rev}}\times\frac{1\,\text{min}}{60\,\text{s}}\right)^2$$

$$= \boxed{84.6\,\text{mJ}}$$

(*b*) Express the new kinetic energy with $K' = 2.0846$ J:

$$K' = \tfrac{1}{2}I\omega'^2$$

Express the ratio of *K* to *K'*:

$$\frac{K'}{K} = \frac{\tfrac{1}{2}I\omega'^2}{\tfrac{1}{2}I\omega'^2} = \left(\frac{\omega'}{\omega}\right)^2$$

Solve for ω':

$$\omega' = \omega\sqrt{\frac{K'}{K}}$$

Substitute numerical values and evaluate ω':

$$\omega' = (70\,\text{rev/min})\sqrt{\frac{2.0846\,\text{J}}{0.0846\,\text{J}}}$$

$$= \boxed{347\,\text{rev/min}}$$

*63 ••   A 2000-kg block is lifted at a constant speed of 8 cm/s by a steel cable that passes over a massless pulley to a motor-driven winch (Figure 9-48). The radius of the winch drum is 30 cm. (*a*) What force must be exerted by the cable? (*b*) What torque does the cable exert on the winch drum? (*c*) What is the angular velocity of the winch drum? (*d*) What power must be developed by the motor to drive the winch drum?

Figure 9-44 Problem 63

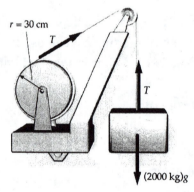

$r = 30$ cm
T
T
$(2000\,\text{kg})g$

Picture the Problem Because the load is not being accelerated, the tension in the cable equals the weight of the load. The role of the massless pulley is to change the direction the force (tension) in the cable acts.

(*a*) Because the block is lifted at constant speed:

$$T = mg = (2000\,\text{kg})(9.81\,\text{m/s}^2)$$

$$= \boxed{19.6\,\text{kN}}$$

(*b*) Apply the definition of torque at the winch drum:

$$\tau = Tr = (19.6\,\text{kN})(0.30\,\text{m})$$

$$= \boxed{5.89\,\text{kN}\cdot\text{m}}$$

(*c*) Relate the angular speed of the winch drum to the rate at which the load is being lifted (the tangential speed of the cable on the drum):

$$\omega = \frac{v}{r} = \frac{0.08\,\text{m/s}}{0.30\,\text{m}} = \boxed{0.267\,\text{rad/s}}$$

(d) Express the power developed by the power in terms of the tension in the cable and the speed with which the load is being lifted:

$$P = Tv = (19.6\,\text{kN})(0.08\,\text{m/s})$$
$$= \boxed{1.57\,\text{kW}}$$

Pulleys, Yo-Yos, and Hanging Things

*68 •• A 4-kg block resting on a frictionless horizontal ledge is attached to a string that passes over a pulley and is attached to a hanging 2-kg block (Figure 9-50). The pulley is a uniform disk of radius 8 cm and mass 0.6 kg. (a) Find the speed of the 2-kg block after it falls from rest a distance of 2.5 m. (b) What is the angular velocity of the pulley at this time?

Figure 9-50 Problems 68-70

Picture the Problem We'll solve this problem for the general case in which the mass of the block on the ledge is M, the mass of the hanging block is m, and the mass of the pulley is M_p, and R is the radius of the pulley. Let the zero of gravitational potential energy be 2.5 m below the initial position of the 2-kg block and R represent the radius of the pulley. Let the system include both blocks, the shelf and pulley, and the earth. The initial potential energy of the 2-kg block will be transformed into the translational kinetic energy of both blocks plus rotational kinetic of the pulley.

(a) Use energy conservation to relate the speed of the 2 kg block when it has fallen a distance Δh to its initial potential energy and the kinetic energy of the system:

$\Delta K + \Delta U = 0$
or, because $K_i = U_f = 0$,
$\frac{1}{2}(m + M)v^2 + \frac{1}{2}I_{\text{pulley}}\omega^2 - mgh = 0$

Substitute for I_{pulley} and ω to obtain:

$$\tfrac{1}{2}(m+M)v^2 + \tfrac{1}{2}\left(\tfrac{1}{2}MR^2\right)\frac{v^2}{R^2} - mgh = 0$$

Solve for v:

$$v = \sqrt{\frac{2mgh}{M + m + \tfrac{1}{2}M_p}}$$

Substitute numerical values and evaluate v:

$$v = \sqrt{\frac{2(2\,\text{kg})(9.81\,\text{m/s}^2)(2.5\,\text{m})}{4\,\text{kg} + 2\,\text{kg} + \tfrac{1}{2}(0.6\,\text{kg})}}$$

$$= \boxed{3.95\,\text{m/s}}$$

(b) Find the angular velocity of the pulley from its tangential speed:

$$\omega = \frac{v}{R} = \frac{3.95\,\text{m/s}}{0.08\,\text{m}} = \boxed{49.3\,\text{rad/s}}$$

*72 ••   The system in Figure 9-52 is released from rest. The 30-kg block is 2 m above the ledge. The pulley is a uniform disk with a radius of 10 cm and mass of 5 kg. Find (a) the speed of the 30-kg block just before it hits the ledge, (b) the angular speed of the pulley at that time, (c) the tensions in the strings, and (d) the time it takes for the 30-kg block to reach the ledge. Assume that the string does not slip on the pulley.

Figure 9-52 Problem 72

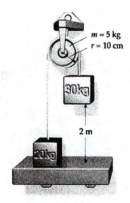

m = 5 kg
r = 10 cm
30 kg
2 m
20 kg

Picture the Problem Let the system include the blocks, the pulley and the earth. Choose the zero of gravitational potential energy to be at the ledge and apply energy conservation to relate the impact speed of the 30-kg block to the initial potential energy of the system. We can use a constant acceleration equations and Newton's 2nd law to find the tensions in the strings and the descent time.

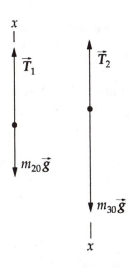

(a) Use energy conservation to relate the impact speed of the 30-kg block to the initial potential energy of the system:

$\Delta K + \Delta U = 0$

or, because $K_i = U_f = 0$,

$\frac{1}{2}m_{30}v^2 + \frac{1}{2}m_{20}v^2 + \frac{1}{2}I_p\omega_p^2$

$\quad + m_{20}g\Delta h - m_{30}g\Delta h = 0$

Substitute for ω_p and I_p to obtain:

$\frac{1}{2}m_{30}v^2 + \frac{1}{2}m_{20}v^2 + \frac{1}{2}\left(\frac{1}{2}M_p r^2\right)\left(\frac{v^2}{r^2}\right)$

$\quad + m_{20}g\Delta h - m_{30}g\Delta h = 0$

Solve for v:

$v = \sqrt{\dfrac{2g\Delta h(m_{30} - m_{20})}{m_{20} + m_{30} + \frac{1}{2}M_p}}$

Substitute numerical values and evaluate v:

$v = \sqrt{\dfrac{2(9.81\,\text{m/s}^2)(2\,\text{m})(30\,\text{kg} - 20\,\text{kg})}{20\,\text{kg} + 30\,\text{kg} + \frac{1}{2}(5\,\text{kg})}}$

$= \boxed{2.73\,\text{m/s}}$

(b) Relate the angular speed at impact to the speed at impact and the radius of the pulley:

$\omega = \dfrac{v}{r} = \dfrac{2.73\,\text{m/s}}{0.1\,\text{m}} = \boxed{27.3\,\text{rad/s}}$

(c) Apply Newton's 2nd law to the blocks:

$\sum F_x = T_1 - m_{20}g = m_{20}a \qquad (1)$

$\sum F_x = m_{30}g - T_2 = m_{30}a \qquad (2)$

Using a constant acceleration equation, relate the speed at impact to the fall distance and the

$v^2 = v_0^2 + 2a\Delta h$

or, because $v_0 = 0$,

acceleration and solve for a:

$$a = \frac{v^2}{2\Delta h}$$

Substitute numerical values and evaluate a:

$$a = \frac{(2.73 \, \text{m/s})^2}{2(2 \, \text{m})} = 1.87 \, \text{m/s}^2$$

Substitute in equation (1) to find T_1:

$$T_1 = m_{20}(g + a)$$
$$= (20 \, \text{kg})(9.81 \, \text{m/s}^2 + 1.87 \, \text{m/s}^2)$$
$$= \boxed{234 \, \text{N}}$$

Substitute in equation (2) to find T_2:

$$T_2 = m_{30}(g - a)$$
$$= (30 \, \text{kg})(9.81 \, \text{m/s}^2 - 1.87 \, \text{m/s}^2)$$
$$= \boxed{238 \, \text{N}}$$

(d) Noting that the initial speed of the 30-kg block is zero, express the time-of-fall in terms of the fall distance and the block's average speed:

$$\Delta t = \frac{\Delta h}{v_{av}} = \frac{\Delta h}{\frac{1}{2} v} = \frac{2\Delta h}{v}$$

Substitute numerical values and evaluate Δt:

$$\Delta t = \frac{2(2 \, \text{m})}{2.73 \, \text{m/s}} = \boxed{1.47 \, \text{s}}$$

*75 •• Two objects are attached to ropes that are attached to wheels on a common axle as shown in Figure 9-55. The two wheels are glued together so that they form a single object. The total moment of inertia of the object is 40 kg·m². The radii of the wheels are $R_1 = 1.2$ m and $R_2 = 0.4$ m. (a) If $m_1 = 24$ kg, find m_2 such that there is no angular acceleration of the wheels. (b) If 12 kg is gently added to the top of m_1, find the angular acceleration of the wheels and the tensions in the ropes.

Figure 9-55 Problem 75

Picture the Problem The diagram shows the forces acting on both objects and the pulley. By applying Newton's 2nd law of motion, we can obtain a system of three equations in the unknowns T_1, T_2, and α that we can solve simultaneously.

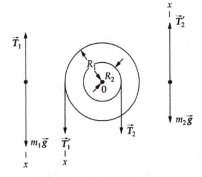

(a) Express the condition that the system does not accelerate:

$$\tau_{net} = m_1 g R_1 - m_2 g R_2 = 0$$

Solve for m_2:

$$m_2 = m_1 \frac{R_1}{R_2}$$

Substitute numerical values and evaluate m_2:

$$m_2 = (24\,\text{kg})\frac{1.2\,\text{m}}{0.4\,\text{m}} = \boxed{72.0\,\text{kg}}$$

(b) Apply Newton's 2nd law to the objects and the pulley:

$$\sum F_x = m_1 g - T_1 = m_1 a, \quad (1)$$
$$\sum \tau_0 = T_1 R_1 - T_2 R_2 = I_0 \alpha, \quad (2)$$
and
$$\sum F_x = T_2 - m_2 g = m_2 a \quad (3)$$

Eliminate a in favor of α in equations (1) and (3) and solve for T_1 and T_2:

$$T_1 = m_1 (g - R_1 \alpha) \quad (4)$$
and
$$T_2 = m_2 (g + R_2 \alpha) \quad (5)$$

Substitute for T_1 and T_2 in equation (2) and solve for α to obtain:

$$\alpha = \frac{\left(m_1 R_1 - m_2 R_2\right)g}{m_1 R_1^2 + m_2 R_2^2 + I_0}$$

Substitute numerical values and evaluate α:

$$\alpha = \frac{\left[(36\,\text{kg})(1.2\,\text{m}) - (72\,\text{kg})(0.4\,\text{m})\right]\left(9.81\,\text{m/s}^2\right)}{(36\,\text{kg})(1.2\,\text{m})^2 + (72\,\text{kg})(0.4\,\text{m})^2 + 40\,\text{kg}\cdot\text{m}^2} = \boxed{1.37\,\text{rad/s}^2}$$

Substitute in equation (4) to find T_1:

$$T_1 = (36\,\text{kg})\left[9.81\,\text{m/s}^2 - (1.2\,\text{m})\left(1.37\,\text{rad/s}^2\right)\right] = \boxed{294\,\text{N}}$$

Substitute in equation (5) to find T_2:

$$T_2 = (72\,\text{kg})\left[9.81\,\text{m/s}^2 + (0.4\,\text{m})\left(1.37\,\text{rad/s}^2\right)\right] = \boxed{746\,\text{N}}$$

*78 ••   A device for measuring the moment of inertia of an object is shown in Figure 9-58. A circular platform has a concentric drum of radius 10 cm about which a string is wound. The string passes over a frictionless pulley to a weight of mass M. The weight is released from rest, and the time required for it to drop a distance D is measured. The system is then rewound, the object placed on the platform, and the system is again released from rest. The time required for the weight to drop the same distance D then provides the data needed to calculate I. With $M = 2.5$ kg and $D = 1.8$ m, the time is 4.2 s. (*a*) Find the combined moment of inertia of the platform, drum, shaft, and pulley. (*b*) With the object placed on the platform, the time is 6.8 s for $D = 1.8$ m. Find I of that object about the axis of the platform.

Figure 9-58 Problem 78

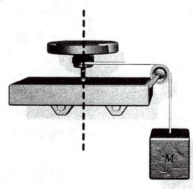

Picture the Problem Let r be the radius of the concentric drum (10 cm) and let I_0 be the moment of inertia of the drum plus platform. We can use Newton's 2nd law in both translational and rotational forms to express I_0 in terms of a and a constant acceleration equation to express a and then find I_0. We can use the same equation to find the total moment of inertia when the object is placed on the platform and then subtract to find its moment of inertia.

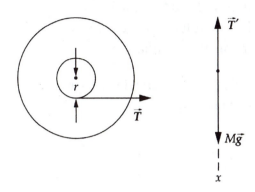

(a) Apply Newton's 2nd law to the platform and the weight:

$$\sum \tau_0 = Tr = I_0 \alpha \qquad (1)$$
$$\sum F_x = Mg - T = Ma \qquad (2)$$

Substitute a/r for α in equation (1) and solve for T:

$$T = \frac{I_0}{r^2} a$$

Substitute for T in equation (2) and solve for a to obtain:

$$I_0 = \frac{Mr^2(g-a)}{a} \qquad (3)$$

Using a constant-acceleration equation, relate the distance of fall to the acceleration of the weight and the time of fall and solve for the acceleration:

$$\Delta x = v_0 \Delta t + \tfrac{1}{2} a(\Delta t)^2$$
or, because $v_0 = 0$ and $\Delta x = D$,
$$a = \frac{2D}{(\Delta t)^2}$$

Substitute for a in equation (3) to obtain:

$$I_0 = Mr^2\left(\frac{g}{a} - 1\right) = Mr^2\left(\frac{g(\Delta t)^2}{2D} - 1\right)$$

Substitute numerical values and evaluate I_0:

$$I_0 = (2.5\,\text{kg})(0.1\,\text{m})^2$$
$$\times \left[\frac{(9.81\,\text{m/s}^2)(4.2\,\text{s})^2}{2(1.8\,\text{m})} - 1\right]$$
$$= \boxed{1.177\,\text{kg} \cdot \text{m}^2}$$

(b) Relate the moments of inertia of the platform, drum, shaft, and pulley (I_0) to the moment of inertia of the object and the total moment of inertia:

$$I_{tot} = I_0 + I = Mr^2 \left(\frac{g}{a} - 1 \right)$$

$$= Mr^2 \left(\frac{g(\Delta t)^2}{2D} - 1 \right)$$

Substitute numerical values and evaluate I_{tot}:

$$I_{tot} = (2.5\,\text{kg})(0.1\,\text{m})^2$$

$$\times \left[\frac{(9.81\,\text{m/s}^2)(6.8\,\text{s})^2}{2(1.8\,\text{m})} - 1 \right]$$

$$= \boxed{3.125\,\text{kg} \cdot \text{m}^2}$$

Solve for and evaluate I:

$$I = I_{tot} - I_0 = 3.125\,\text{kg} \cdot \text{m}^2$$

$$-1.177\,\text{kg} \cdot \text{m}^2$$

$$= \boxed{1.948\,\text{kg} \cdot \text{m}^2}$$

Objects Rolling Without Slipping

*79 ••   In 1993, a giant yo-yo of mass 400 kg measuring about 1.5 m in radius was dropped from a crane 57 m high. One end of the string was tied to the top of the crane, so the yo-yo unwound as it descended. Assuming that the axle of the yo-yo had a radius of $r = 0.1$ m, find the velocity of descent at the end of the fall.

Picture the Problem The forces acting on the yo-yo are shown in the figure. We can use a constant acceleration equation to relate the velocity of descent at the end of the fall to the yo-yo's acceleration and Newton's 2nd law in both translational and rotational form to find the yo-yo's acceleration.

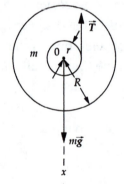

Using a constant acceleration equation, relate the yo-yo's final speed to its acceleration and fall distance:

$$v^2 = v_0^2 + 2a\Delta h$$

or, because $v_0 = 0$,

$$v = \sqrt{2a\Delta h} \qquad (1)$$

Use Newton's 2nd law to relate the forces that act on the yo-yo to its acceleration:

$$\sum F_x = mg - T = ma \qquad (2)$$

and

$$\sum \tau_0 = Tr = I_0 \alpha \qquad (3)$$

Use $a = r\alpha$ to eliminate α in equation (3)	$$Tr = I_0 \frac{a}{r} \qquad (4)$$

Eliminate T between equations (2) and (4) to obtain:	$$mg - \frac{I_0}{r^2} a = ma \qquad (5)$$

Substitute $\frac{1}{2} mR^2$ for I_0 in equation (5):	$$mg - \frac{\frac{1}{2} mR^2}{r^2} a = ma$$

Solve for a:	$$a = \frac{g}{1 + \dfrac{R^2}{2r^2}}$$

Substitute numerical values and evaluate a:	$$a = \frac{9.81 \, \text{m/s}^2}{1 + \dfrac{(1.5 \, \text{m})^2}{2(0.1 \, \text{m})^2}} = 0.0864 \, \text{m/s}^2$$

Substitute in equation (1) and evaluate v:	$$v = \sqrt{2(0.0864 \, \text{m/s}^2)(57 \, \text{m})}$$ $$= \boxed{3.14 \, \text{m/s}}$$

*82 • A homogeneous solid cylinder rolls without slipping on a horizontal surface. The total kinetic energy is K. The kinetic energy due to rotation about its center of mass is (a) $\frac{1}{2} K$, (b) $\frac{1}{3} K$, (c) $\frac{4}{7} K$, (d) none of these.

Picture the Problem We can determine the kinetic energy of the cylinder that is due to its rotation about its center of mass by examining the ratio K_{rot}/K.

Express the rotational kinetic energy of the homogeneous solid cylinder:	$$K_{rot} = \tfrac{1}{2} I_{cyl} \omega^2 = \tfrac{1}{2}\left(\tfrac{1}{2} mr^2\right)\frac{v^2}{r^2} = \tfrac{1}{4} mv^2$$

Express the total kinetic energy of the homogeneous solid cylinder:	$$K = K_{rot} + K_{trans} = \tfrac{1}{4} mv^2 + \tfrac{1}{2} mv^2 = \tfrac{3}{4} mv^2$$

Express the ratio $\dfrac{K_{rot}}{K}$:	$$\frac{K_{rot}}{K} = \frac{\tfrac{1}{4} mv^2}{\tfrac{3}{4} mv^2} = \tfrac{1}{3} \text{ and } \boxed{(b) \text{ is correct.}}$$

*86 ••   A uniform sphere rolls without slipping down an incline. What must be the angle of the incline if the linear acceleration of the center of mass of the sphere is $0.2g$?

Picture the Problem From Newton's 2nd law, the acceleration of the center of mass equals the net force divided by the mass. The forces acting on the sphere are its weight $m\vec{g}$ downward, the normal force $\vec{F}_n$ that balances the normal component of the weight, and the force of friction $\vec{f}$ acting up the incline. As the sphere accelerates down the incline, the angular velocity of rotation must increase to maintain the nonslip condition. We can apply Newton's 2nd law for rotation about a horizontal axis through the center of mass of the sphere to find α, which is related to the acceleration by the nonslip condition. The only torque about the center of mass is due to $\vec{f}$ because both $m\vec{g}$ and $\vec{F}_n$ act through the center of mass. Choose the positive direction to be down the incline.

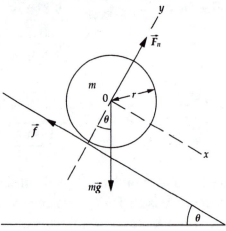

Apply $\sum \vec{F} = m\vec{a}$ to the sphere:

$$mg\sin\theta - f = ma_{cm} \qquad (1)$$

Apply $\sum \tau = I_{cm}\alpha$ to the sphere:

$$fr = I_{cm}\alpha$$

Use the nonslip condition to eliminate α and solve for f:

$$fr = I_{cm}\frac{a_{cm}}{r}$$

and

$$f = \frac{I_{cm}}{r^2}a_{cm}$$

Substitute this result for f in equation (1) to obtain:

$$mg\sin\theta - \frac{I_{cm}}{r^2}a_{cm} = ma_{cm}$$

From Table 9-1 we have, for a solid sphere:

$$I_{cm} = \tfrac{2}{5}mr^2$$

Substitute in equation (1) and simplify to obtain:

$$mg\sin\theta - \tfrac{2}{5}a_{cm} = ma_{cm}$$

Solve for and evaluate θ :

$$\theta = \sin^{-1}\left(\frac{7a_{cm}}{5g}\right)$$

$$= \sin^{-1}\left[\frac{7(0.2g)}{5g}\right] = \boxed{16.3°}$$

*90 ••   A hollow sphere and uniform sphere of the same mass m and radius R roll down an inclined plane from the same height H without slipping (Figure 9-59). Each is moving horizontally as it leaves the ramp. When the spheres hit the ground, the range of the hollow sphere is L. Find the range L' of the uniform sphere.

Figure 9-59 Problem 90

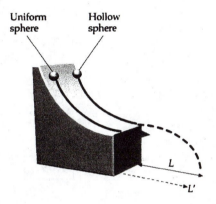

Picture the Problem Let the zero of gravitational potential energy be at the elevation where the spheres leave the ramp. The distances the spheres will travel are directly proportional to their speeds when they leave the ramp.

Express the ratio of the distances traveled by the two spheres in terms of their speeds when they leave the ramp:	$\dfrac{L'}{L} = \dfrac{v'\Delta t}{v\Delta t} = \dfrac{v'}{v}$	(1)
Use conservation of mechanical energy to find the speed of the spheres when they leave the ramp:	$\Delta K + \Delta U = 0$ or, because $K_i = U_f = 0$, $K_f - U_i = 0$	(2)

Express K_f for the spheres:

$$K_f = K_{trans} + K_{rot}$$
$$= \tfrac{1}{2}mv^2 + \tfrac{1}{2}I_{cm}\omega^2$$
$$= \tfrac{1}{2}mv^2 + \tfrac{1}{2}\left(kmR^2\right)\frac{v^2}{R^2}$$
$$= \tfrac{1}{2}mv^2 + \tfrac{1}{2}kmv^2$$
$$= \left(1+k\right)\tfrac{1}{2}mv^2$$

where k is 2/3 for the spherical shell and 2/5 for the uniform sphere.

Substitute in equation (2) to obtain:

$$\left(1+k\right)\tfrac{1}{2}mv^2 = mgH$$

Solve for v:

$$v = \sqrt{\frac{2gH}{1+k}}$$

Substitute in equation (1) to obtain:

$$\frac{L'}{L} = \sqrt{\frac{1+k}{1+k'}} = \sqrt{\frac{1+\frac{2}{3}}{1+\frac{2}{5}}} = 1.09$$

or

$$L' = \boxed{1.09L}$$

*96 ••• A uniform cylinder of mass M and radius R is at rest on a block of mass m, which in turn rests on a horizontal, frictionless table (Figure 9-61). If a horizontal force $\vec{F}$ is applied to the block, it accelerates and the cylinder rolls without slipping. Find the acceleration of the block.

Figure 9-61 Problems 96-98

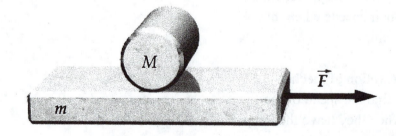

Picture the Problem Let the letter B identify the block and the letter C the cylinder. We can find the accelerations of the block and cylinder by applying Newton's 2nd law and solving the resulting equations simultaneously.

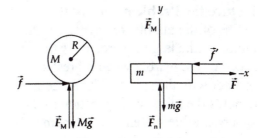

Apply $\sum F_x = ma_x$ to the block: $F - f = ma_B$ (1)

Apply $\sum F_x = ma_x$ to the cylinder: $f = Ma_C$,   (2)

Apply $\sum \tau_{CM} = I_{CM}\alpha$ to the cylinder: $fR = I_{CM}\alpha$ (3)

Substitute for I_{CM} in equation (5) and solve for f to obtain: $f = \tfrac{1}{2}MR\alpha$ (4)

Relate the acceleration of the block to the acceleration of the cylinder:

$a_C = a_B + a_{CB}$
or, because $a_{CB} = -R\alpha$,
$a_C = a_B - R\alpha$
and
$R\alpha = a_B - a_C$ (5)

Equate equations (2) and (4) and substitute from (5) to obtain: $a_B = 3a_C$

Substitute equation (4) in equation (1) and substitute for a_C to obtain: $F - \tfrac{1}{3}Ma_B = ma_B$

Solve for a_B: $$a_B = \boxed{\frac{3F}{M + 3m}}$$

*100 •• A marble of radius 1 cm rolls from rest from the top of a large sphere of radius 80 cm, which is held fixed. (*a*) Assuming that the marble rolls without slipping while it is in contact with the sphere (which is unrealistic), find the angle from the top of the sphere to the point where the marble breaks contact with the sphere. (*b*) Why is it unrealistic to assume that the marble rolls without slipping all the way down to the point where it breaks contact?

Picture the Problem Let r be the radius of the marble, m its mass, R the radius of the large sphere, and v the speed of the marble when it breaks contact with the sphere. The numeral 1 denotes the initial configuration of the sphere-marble system and the numeral 2 is configuration as the marble separates from the sphere. We can use conservation of energy to relate the initial potential energy of the marble to the sum of its translational and rotational kinetic energies as it leaves the sphere. Our choice of the zero of potential energy is shown on the diagram.

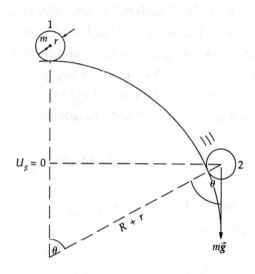

(a) Apply conservation of energy:

$$\Delta U + \Delta K = 0$$

or

$$U_2 - U_1 + K_2 - K_1 = 0$$

Since $U_2 = K_1 = 0$:

$$-mg[R + r - (R + r)\cos\theta]$$
$$+ \tfrac{1}{2}mv^2 + \tfrac{1}{2}I\omega^2 = 0$$

or

$$-mg[(R + r)(1 - \cos\theta)]$$
$$+ \tfrac{1}{2}mv^2 + \tfrac{1}{2}I\omega^2 = 0$$

Use the rolling-without-slipping condition to eliminate ω:

$$-mg[(R + r)(1 - \cos\theta)]$$
$$+ \tfrac{1}{2}mv^2 + \tfrac{1}{2}I\frac{v^2}{r^2} = 0$$

From Table 9-1 we have:

$$I = \tfrac{2}{5}mr^2$$

Substitute to obtain:

$$-mg[(R + r)(1 - \cos\theta)]$$
$$+ \tfrac{1}{2}mv^2 + \tfrac{1}{2}\left(\tfrac{2}{5}mr^2\right)\frac{v^2}{r^2} = 0$$

or

$$-mg[(R + r)(1 - \cos\theta)]$$
$$+ \tfrac{1}{2}mv^2 + \tfrac{1}{5}mv^2 = 0$$

Solve for v^2 to obtain:

$$v^2 = \frac{10}{7}g(R + r)(1 - \cos\theta)$$

Apply $\sum F_r = ma_r$ to the marble as it separates from the sphere:

$$mg\cos\theta = m\frac{v^2}{R+r}$$

or

$$\cos\theta = \frac{v^2}{g(R+r)}$$

Substitute for v^2:

$$\cos\theta = \frac{1}{g(R+r)}\left[\frac{10}{7}g(R+r)(1-\cos\theta)\right]$$

$$= \left[\frac{10}{7}(1-\cos\theta)\right]$$

Solve for and evaluate θ:

$$\theta = \cos^{-1}\left(\frac{10}{17}\right) = \boxed{54.0°}$$

(b) The force of friction is always less than μ_s multiplied by the normal force on the marble. However, the normal force decreases to 0 at the point where the ball leaves the sphere, meaning that the force of friction must be less than the force needed to keep the ball rolling without slipping before it leaves the sphere.

Rolling With Slipping

*102 •• A cue ball of radius r is initially at rest on a horizontal pool table (Figure 9-63). It is struck by a horizontal cue stick that delivers a force of magnitude P_0 for a very short time Δt. The stick strikes the ball at a point h above the ball's point of contact with the table. Show that the ball's initial angular velocity ω_0 is related to the initial linear velocity of its center of mass v_0 by $\omega_0 = 5v_0(h - r)/2r^2$.

Figure 9-63 Problem 102

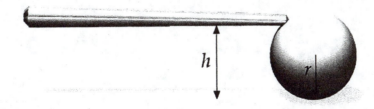

Picture the Problem The cue stick's blow delivers a rotational impulse as well as a translational impulse to the cue ball. The rotational impulse changes the angular momentum of the ball and the translational impulse changes its linear momentum.

Express the rotational impulse P_{rot} as the product of the average torque and the time during which the rotational impulse acts:	$P_{rot} = \tau_{av}\Delta t$
Express the average torque it produces about an axis through the center of the ball:	$\tau_{av} = P_0(h-r)\sin\theta° = P_0(h-r)$ where θ ($= 90°$) is the angle between F and the lever arm $h - r$.
Substitute in the expression for P_{rot} to obtain:	$P_{rot} = P_0(h-r)\Delta t = (P_0\Delta t)(h-r)$ $= P_{trams}(h-r) = \Delta L = I\omega_0$
The translational impulse is also given by:	$P_{trans} = P_0\Delta t = \Delta p = mv_0$
Substitute to obtain:	$mv_0(h-r) = \frac{2}{5}mr^2\omega_0$
Solve for ω_0:	$$\boxed{\omega_0 = \frac{5v_0(h-r)}{2r^2}}$$

*108   •• A solid cylinder of mass M resting on its side on a horizontal surface is given a sharp blow by a cue stick. The applied force is horizontal and passes through the center of the cylinder so that the cylinder begins translating with initial velocity v_0. The coefficient of sliding friction between the cylinder and surface is μ_k. (a) What is the translational velocity of the cylinder when it is rolling without slipping? (b) How far does the cylinder travel before it rolls without slipping? (c) What fraction of its initial mechanical energy is dissipated in friction?

Picture the Problem The figure shows the forces acting on the cylinder during the sliding phase of its motion. The friction force will cause the cylinder's translational speed to decrease and eventually satisfy the condition for rolling without slipping. We'll use Newton's 2nd law to find the linear and rotational velocities and accelerations of the ball and constant acceleration equations to relate these quantities to each other and to the distance traveled and the elapsed time until the satisfaction of the condition for rolling without slipping.

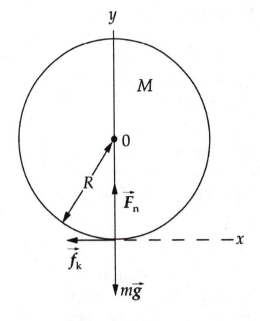

(*a*) Apply Newton's 2nd law to the cylinder:

$$\sum F_x = -f_k = Ma, \qquad (1)$$

$$\sum F_y = F_n - Mg = 0, \qquad (2)$$

and

$$\sum \tau_0 = f_k R = I_0 \alpha \qquad (3)$$

Eliminate f_k between equations (1) and (2) and solve for a:

$$a = -\mu_k g$$

Using a constant acceleration equation, relate the speed of the cylinder to its acceleration and the elapsed time:

$$v = v_0 + a\Delta t = v_0 - \mu_k g \Delta t$$

Eliminate f_k between equations (2) and (3) and solve for α:

$$\alpha = \frac{2\mu_k g}{R}$$

Using a constant acceleration equation, relate the angular speed of the cylinder to its acceleration and the elapsed time:

$$\omega = \omega_0 + \alpha \Delta t = \frac{2\mu_k g}{R} \Delta t$$

Apply the condition for rolling without slipping:

$$v = v_0 - \mu_k g \Delta t = R\omega = R\left(\frac{2\mu_k g}{R}\Delta t\right)$$

$$= 2\mu_k g \Delta t$$

Solve for Δt:

$$\Delta t = \frac{v_0}{3\mu_k g}$$

Substitute for Δt in the expression for v:

$$v = v_0 - \mu_k g \frac{v_0}{3\mu_k g} = \boxed{\frac{2}{3}v_0}$$

(b) Relate the distance the cylinder travels to its average speed and the elapsed time:

$$\Delta x = v_{av}\Delta t = \frac{1}{2}\left(v_0 + \frac{2}{3}v_0\right)\left(\frac{v_0}{3\mu_k g}\right)$$

$$= \boxed{\frac{5}{18}\frac{v_0^2}{\mu_k g}}$$

(c) Express the ratio of the energy dissipated in friction to the cylinder's initial mechanical energy:

$$\frac{W_{fr}}{K_i} = \frac{K_i - K_f}{K_i}$$

Express the kinetic energy of the cylinder as it begins to roll without slipping:

$$K_f = \frac{1}{2}Mv^2 + \frac{1}{2}I_{cm}\omega^2$$

$$= \frac{1}{2}Mv^2 + \frac{1}{2}\left(\frac{1}{3}MR^2\right)\frac{v^2}{R^2}$$

$$= \frac{3}{4}Mv^2 = \frac{3}{4}M\left(\frac{2}{3}v_0\right)^2 = \frac{1}{3}Mv_0^2$$

Substitute for K_i and K_f and simplify to obtain:

$$\frac{W_{fr}}{K_i} = \frac{\frac{1}{2}Mv_0^2 - \frac{1}{3}Mv_0^2}{\frac{1}{2}Mv^2} = \boxed{\frac{1}{3}}$$

General Problems

*110 •   The moon rotates as it revolves around the earth so that we always see the same side. Use this fact to find the angular velocity of the moon about its axis. (The period of revolution of the moon about the earth is 27.3 days.)

Picture the Problem The angular velocity of an object is the ratio of the number of revolutions it makes in a given period of time to the elapsed time.

The moon's angular velocity is:

$$\omega = \frac{1\,rev}{27.3\,days}$$

$$= \frac{1\,rev}{27.3\,days} \times \frac{2\pi\,rad}{rev} \times \frac{1\,day}{24\,h} \times \frac{1\,h}{3600\,s}$$

$$= \boxed{2.66 \times 10^{-6}\ rad/s}$$

*117 •• A uniform disk with a mass of 120 kg and a radius of 1.4 m rotates initially with an angular speed of 1100 rev/min. (a) A constant tangential force is applied at a radial distance of 0.6 m. How much work must this force do to stop the wheel? (b) If the wheel is brought to rest in 2.5 min, what torque does the force produce? What is the magnitude of the force? (c) How many revolutions does the wheel make in these 2.5 min?

Picture the Problem To stop the wheel, the tangential force will have to do an amount of work equal to the initial rotational kinetic energy of the wheel. We can find the stopping torque and the force from the average power delivered by the force during the slowing of the wheel. The number of revolutions made by the wheel as it stops can be found from a constant acceleration equation.

(a) Relate the work that must be done to stop the wheel to its kinetic energy:

$$W = \tfrac{1}{2}I\omega^2 = \tfrac{1}{2}\left(\tfrac{1}{2}mr^2\right)\omega^2 = \tfrac{1}{4}mr^2\omega^2$$

Substitute numerical values and evaluate W:

$$W = \tfrac{1}{4}(120\,kg)(1.4\,m)^2$$

$$\times \left[1100\frac{rev}{min} \times \frac{2\pi\,rad}{rev} \times \frac{1\,min}{60\,s}\right]^2$$

$$= \boxed{780\,kJ}$$

(b) Express the stopping torque is terms of the average power required:

$$P_{av} = \tau\omega_{av}$$

Solve for τ:

$$\tau = \frac{P_{av}}{\omega_{av}}$$

Substitute numerical values and evaluate τ:

$$\tau = \dfrac{\dfrac{780\,\text{kJ}}{(2.5\,\text{min})(60\,\text{s/min})}}{\dfrac{(1100\,\text{rev/min})(2\pi\,\text{rad/rev})(1\,\text{min/60\,s})}{2}}$$

$$= \boxed{90.3\,\text{N}\cdot\text{m}}$$

Relate the stopping torque to the magnitude of the required force and solve for F:

$$F = \dfrac{\tau}{R} = \dfrac{90.3\,\text{N}\cdot\text{m}}{0.6\,\text{m}} = \boxed{151\,\text{N}}$$

(c) Using a constant-acceleration equation, relate the angular displacement of the wheel to its average angular velocity and the stopping time:

$$\Delta\theta = \omega_{\text{av}}\Delta t$$

Substitute numerical values and evaluate $\Delta\theta$:

$$\Delta\theta = \left(\dfrac{1100\,\text{rev/min}}{2}\right)(2.5\,\text{min})$$

$$= \boxed{1380\,\text{rev}}$$

*121 •• Consider two uniform blocks of wood, identical in shape and composition, where one is larger than the other by a factor S in all dimensions. (a) What is the ratio of the surface areas of the two blocks? (b) What is the ratio of the masses of the two blocks? (c) What is the ratio of the moments of inertia about some axis running through the block (in the same relative position and orientation in each)? These are examples of *scaling laws*: How do surface area, mass, and moment of inertia vary with the size of an object?

Picture the Problem Let the smaller block have the dimensions shown in the diagram. Then the length, height, and width of the larger block are $S\ell, Sh,$ and Sw, respectively . Let the numeral 1 denote the smaller block and the numeral 2 the larger block and express the ratios of the surface areas, masses, and moments of inertia of the two blocks.

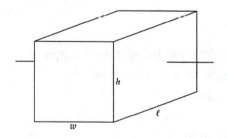

(a) Express the ratio of the surface areas of the two blocks:

$$\frac{A_2}{A_1} = \frac{2(Sw)(S\ell) + 2(S\ell)(Sh) + (Sw)(Sh)}{2w\ell + 2\ell h + 2wh}$$

$$= \frac{S^2(2w\ell + 2\ell h + 2wh)}{2w\ell + 2\ell h + 2wh}$$

$$= \boxed{S^2}$$

(b) Express the ratio of the masses of the two blocks:

$$\frac{M_2}{M_1} = \frac{\rho V_2}{\rho V_1} = \frac{V_2}{V_1} = \frac{(Sw)(S\ell)(Sh)}{w\ell h}$$

$$= \frac{S^3(w\ell h)}{w\ell h} = \boxed{S^3}$$

(c) Express the ratio of the moments of inertia, about the axis shown in the diagram, of the two blocks:

$$\frac{I_2}{I_1} = \frac{\frac{1}{12}M_2\left[(S\ell)^2 + (Sh)^2\right]}{\frac{1}{12}M_1\left[\ell^2 + h^2\right]}$$

$$= \frac{M_2}{M_1}\frac{S^2\left[\ell^2 + h^2\right]}{\left[\ell^2 + h^2\right]} = \left(\frac{M_2}{M_1}\right)(S^2)$$

In part (b) we showed that:

$$\frac{M_2}{M_1} = S^3$$

Substitute to obtain:

$$\frac{I_2}{I_1} = (S^3)(S^2) = \boxed{S^5}$$

*127 •• A popular classroom demonstration involves taking a meterstick and holding it horizontally at one end with a number of pennies spaced evenly along the stick. If the hand is relaxed so that the ruler pivots about the hand under the influence of gravity, an interesting thing is seen: Pennies near the pivot point stay on the ruler, while those farther away than a certain distance from the pivot are left behind by the falling meterstick. (This is often called the "faster than gravity" demonstration.) (a) What is the acceleration of the far end of the meterstick? (b) How far should a penny be from the end of the stick for it to be "left behind"?

Picture the Problem The diagram shows the force the hand supporting the meterstick exerts at the pivot point and the force the earth exerts on the meterstick acting at the center of mass. We can relate the angular acceleration to the acceleration of the end of the meterstick using $a = L\alpha$ and use Newton's 2nd law in rotational form to relate α to the moment of inertia of the meterstick.

(a) Relate the acceleration of the far end of the meterstick to the angular acceleration of the meterstick:

$$a = L\alpha \qquad (1)$$

Apply $\sum \tau_P = I_P \alpha$ to the meterstick:

$$Mg\left(\frac{L}{2}\right) = I_P \alpha$$

Solve for α:

$$\alpha = \frac{MgL}{2I_P}$$

From Table 9-1, for a rod pivoted at one end, we have:

$$I_P = \frac{1}{3}ML^2$$

Substitute to obtain:

$$\alpha = \frac{3MgL}{2ML^2} = \frac{3g}{2L}$$

Substitute in equation (1) to obtain:

$$a = \frac{3g}{2} = \frac{3(9.81\,\text{m/s}^2)}{2} = \boxed{14.7\,\text{m/s}^2}$$

(b) Express the acceleration of a point on the meterstick a distance x from the pivot point:

$$a = \alpha x = \frac{3g}{2L}x$$

Express the condition that the meterstick leaves the penny behind:

$$a > g$$

Substitute to obtain:

$$\frac{3g}{2L}x > g$$

Solve for and evaluate x:

$$x > \frac{2L}{3} = \frac{2(1\,m)}{3} = \boxed{66.7\,\text{cm}}$$

*133   •• In problems dealing with a pulley with a non-zero moment of inertia, the magnitude of the tensions in the ropes hanging on either side of the pulley are not equal. The difference in the tension is due to the static frictional force between the rope and the pulley; however, the static frictional force cannot be made arbitrarily large. If you consider a massless rope wrapped partly around a cylinder through an angle $\Delta\theta$ (measured in radians), then you can show that if the tension on one side of the pulley is T, while the tension on the other side is T' ($T' < T$), the maximum value of T' in relation to T that can be maintained without the rope slipping is $T'_{max} = Te^{\mu_s\Delta\theta}$, where μ_s is the coefficient of static friction. Consider the Atwood's machine in Figure 9-73: the pulley has a radius $r = 0.15$ m, moment of inertia $I = 0.35$ kg·m^2 and coefficient of static friction $\mu_s = 0.30$. (*a*) If the tension on one side of the pulley is 10 N, what is the maximum tension on the other side that will prevent the rope from slipping on the pulley? (*b*) If the mass of one of the hanging blocks is 1 kg, what is the maximum mass of the other block if, after the blocks are released, the pulley is to rotate without slipping? (*c*) What is the acceleration of the blocks in this case?

Figure 9-73 Problem 133

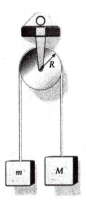

Picture the Problem Free-body diagrams for the pulley and the two blocks are shown to the right. Choose a coordinate system in which the direction of motion of the block whose mass is M (downward) is the positive y direction. We can use the given relationship $T'_{max} = Te^{\mu_s\Delta\theta}$ to relate the tensions in the rope on either side of the pulley and apply Newton's 2nd law in both rotational form (to the pulley) and translational form (to the blocks) to obtain a system of equations that we can solve simultaneously for a, T_1, T_2, and M.

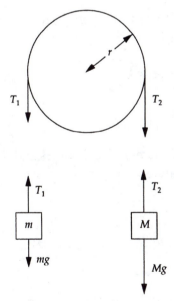

(*a*) Use $T'_{max} = Te^{\mu_s\Delta\theta}$ to evaluate the maximum tension required to

$T'_{max} = (10\,\text{N})e^{(0.3)\pi} = \boxed{25.7\,\text{N}}$

prevent the rope from slipping on the pulley:

(c) Given that the angle of wrap is π radians, express T_2 in terms of T_1:

$$T_2 = T_1 e^{0.3\pi} = 2.57 T_1 \qquad (1)$$

Because the rope doesn't slip, we can relate the angular acceleration, α, of the pulley to the acceleration, a, of the hanging masses by:

$$\alpha = \frac{a}{r}$$

Apply $\sum F_y = ma_y$ to the two blocks to obtain:

$$T_1 - mg = ma \qquad (2)$$
and
$$Mg - T_2 = Ma \qquad (3)$$

Apply $\sum \tau = I\alpha$ to the pulley to obtain:

$$(T_2 - T_1)r = I\frac{a}{r} \qquad (4)$$

Substitute for T_2 from equation (1) in equation (4) to obtain:

$$(2.57T_1 - T_1)r = I\frac{a}{r}$$

Solve for T_1 and substitute numerical values to obtain:

$$T_1 = \frac{I}{1.57r^2}a = \frac{0.35\,\text{kg}\cdot\text{m}^2}{1.57(0.15\,\text{m})^2}a \quad (5)$$
$$= (9.91\,\text{kg})a$$

Substitute in equation (2) to obtain:

$$(9.91\,\text{kg})a - mg = ma$$

Solve for a:

$$a = \frac{mg}{9.91\,\text{kg} - m} = \frac{g}{\dfrac{9.91\,\text{kg}}{m} - 1}$$

Substitute numerical values and evaluate a:

$$a = \frac{9.81\,\text{m/s}^2}{\dfrac{9.91\,\text{kg}}{1\,\text{kg}} - 1} = \boxed{1.10\,\text{m/s}^2}$$

(b) Solve equation (3) for M:

$$M = \frac{T_2}{g - a}$$

Substitute in equation (5) to find T_1:

$$T_1 = (9.91\,\text{kg})(1.10\,\text{m/s}^2) = 10.9\,\text{N}$$

Substitute in equation (1) to find T_2:

$$T_2 = 2.57(10.9\,\text{N}) = 28.0\,\text{N}$$

Evaluate M:

$$M = \frac{28.0\,\text{N}}{9.81\,\text{m/s}^2 - 1.10\,\text{m/s}^2} = \boxed{3.21\,\text{kg}}$$

Chapter 10
Conservation of Angular Momentum

Conceptual Problems

*1 • True or false: (*a*) If two vectors are parallel, their cross product must be zero. (*b*) When a disk rotates about its symmetry axis, $\vec{\omega}$ is along the axis. (*c*) The torque exerted by a force is always perpendicular to the force.

(*a*) True. The cross product of the vectors $\vec{A}$ and $\vec{B}$ is defined to be $\vec{A} \times \vec{B} = AB\sin\phi\,\hat{n}$. If $\vec{A}$ and $\vec{B}$ are parallel, $\sin\phi = 0$.

(*b*) True. By definition, $\vec{\omega}$ is along the axis.

(*c*) True. The direction of a torque exerted by a force is determined by the definition of the cross product.

*5 •• A particle travels in a circular path and point *P* is at the center of the circle. (*a*) If its linear momentum $\vec{p}$ is doubled, how is its angular momentum about *P* affected? (*b*) If the radius of the circle is doubled but the speed is unchanged, how is the angular momentum of the particle about *P* affected?

Determine the Concept $\vec{L}$ and $\vec{p}$ are related according to $\vec{L} = \vec{r} \times \vec{p}$.

(*a*) Because $\vec{L}$ is directly proportional to $\vec{p}$:

> Doubling $\vec{p}$ doubles $\vec{L}$.

(*b*) Because $\vec{L}$ is directly proportional to $\vec{r}$:

> Doubling $\vec{r}$ doubles $\vec{L}$.

*8 •• Standing on a turntable that is initially not rotating, can you rotate yourself through 180°? Assume that no external torques act on the you-turntable system. *Hint: While you cannot change your mass easily, there are ways to change your moment of inertia.*

Determine the Concept Yes, you can. Imagine rotating the top half of your body with arms flat at sides through a (roughly) 90° angle. Because the net angular momentum of the system is 0, the bottom half of your body rotates in the opposite direction. Now extend your arms out and rotate the top half of your body back. Because the moment of inertia of the top half of your body is larger than it was previously, the angle which the bottom half of your body rotates through will be smaller, leading to a net rotation. You can repeat this process as necessary to rotate through any arbitrary angle.

*12 ••   A block sliding on a frictionless table is attached to a string that passes through a hole in the table. Initially, the block is sliding with speed v_0 in a circle of radius r_0. A student under the table pulls slowly on the string. What happens as the block spirals inward? Give supporting arguments for your choice. (*a*) Its energy and angular momentum are conserved. (*b*) Its angular momentum is conserved and its energy increases. (*c*) Its angular momentum is conserved and its energy decreases. (*d*) Its energy is conserved and its angular momentum increases. (*e*) Its energy is conserved and its angular momentum decreases.

Determine the Concept The pull that the student exerts on the block is at right angles to its motion and exerts no torque (recall that $\vec{\tau} = \vec{r} \times \vec{F}$ and $\tau = rF \sin \theta$). Therefore, we can conclude that the angular momentum of the block is conserved. The student does, however, do work in displacing the block in the direction of the radial force and so the block's energy increases. $\boxed{(b) \text{ is correct.}}$

*13 ••   How can you tell a hardboiled egg from an uncooked one without breaking it? One way is to lay the egg flat on a hard surface and try to spin it. A hardboiled egg will spin easily, while it takes a lot of effort to make an uncooked egg spin. However, once spinning, the uncooked egg will do something unusual: if you stop it with your finger, it may start spinning again. Explain the difference in the behavior of the two types of eggs.

Determine the Concept The hardboiled egg is solid inside, so everything rotates with a uniform velocity. By contrast, it is difficult to get the viscous fluid inside a raw egg to start rotating; however, once it is rotating, stopping the shell will not stop the motion of the interior fluid, and the egg may start rotating again after momentarily stopping for this reason.

*22 ••   In tetherball, a ball is attached to a string that is attached to a pole. When the ball is hit, the string wraps around the pole and the ball spirals inward. Neglecting air resistance, what happens as the ball swings around the pole? Give supporting arguments for your choice. (*a*) The mechanical energy and angular momentum of the ball are conserved. (*b*) The angular momentum of the ball is conserved, but the mechanical energy of the ball increases. (*c*) The angular momentum of the ball is conserved, and the mechanical energy of the ball decreases. (*d*) The mechanical energy of the ball is conserved and the angular momentum of the ball increases. (*e*) The mechanical energy of the ball is conserved and the angular momentum of the ball decreases.

Determine the Concept Consider the overhead view of a tether pole and ball shown in the adjoining figure. The ball rotates counterclockwise. The torque about the center of the pole is clockwise and of magnitude RT, where R is the pole's radius and T is the tension. So L must decrease and $\boxed{(e)}$ is correct.

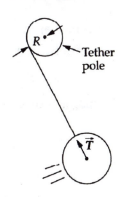

Estimation and Approximation

*25 •• An ice skater starts her pirouette with arms outstretched, rotating at 1.5 rev/s. Estimate her rotational speed (in revolutions per second) when she brings her arms flat against her body.

Picture the Problem Because we have no information regarding the mass of the skater, we'll assume that her body mass (not including her arms) is 50 kg and that each arm has a mass of 4 kg. Let's also assume that her arms are 1 m long and that her body is cylindrical with a radius of 20 cm. Because the net external torque acting on her is zero, her angular momentum will remain constant during her pirouette.

Express the conservation of her angular momentum during her pirouette:

$$L_i = L_f$$
or
$$I_{arms\,out}\omega_{arms\,out} = I_{arms\,in}\omega_{arms\,in} \qquad (1)$$

Express her total moment of inertia with her arms out:

$$I_{arms\,out} = I_{body} + I_{arms}$$

Treating her body as though it is cylindrical, calculate its moment of inertia of her body, minus her arms:

$$I_{body} = \tfrac{1}{2}mr^2 = \tfrac{1}{2}(50\,\text{kg})(0.2\,\text{m})^2$$
$$= 1.00\,\text{kg}\cdot\text{m}^2$$

Modeling her arms as though they are rods, calculate their moment of inertia when she has them out:

$$I_{arms} = 2\left[\tfrac{1}{3}(4\,\text{kg})(1\,\text{m})^2\right]$$
$$= 2.67\,\text{kg}\cdot\text{m}^2$$

Substitute to determine her total moment of inertia with her arms out:

$$I_{\text{arms out}} = 1.00\,\text{kg}\cdot\text{m}^2 + 2.67\,\text{kg}\cdot\text{m}^2$$
$$= 3.67\,\text{kg}\cdot\text{m}^2$$

Express her total moment of inertia with her arms in:

$$I_{\text{arms in}} = I_{\text{body}} + I_{\text{arms}}$$
$$= 1.00\,\text{kg}\cdot\text{m}^2 + 2\left[(4\,\text{kg})(0.2\,\text{m})^2\right]$$
$$= 1.32\,\text{kg}\cdot\text{m}^2$$

Solve equation (1) for $\omega_{\text{arms in}}$ and substitute to obtain:

$$\omega_{\text{arms in}} = \frac{I_{\text{arms out}}}{I_{\text{arms in}}}\,\omega_{\text{arms out}}$$
$$= \frac{3.67\,\text{kg}\cdot\text{m}^2}{1.32\,\text{kg}\cdot\text{m}^2}(1.5\,\text{rev/s})$$
$$= \boxed{4.17\,\text{rev/s}}$$

*28 •• One problem in astrophysics in the 1960s was explaining pulsars – extremely regular astronomical sources of radio pulses whose periods ranged from seconds to milliseconds. At one point, these radio sources were given the acronym LGM, standing for "Little Green Men", a reference to the idea that they might be signals of extraterrestrial civilizations. The explanation given today is no less interesting: The sun, which is a fairly typical star, has a mass of 1.99×10^{30} kg and a radius of 6.96×10^8 m. While it doesn't rotate uniformly, because it isn't a solid body, its average rate of rotation can be taken as about 1 revolution every 25 days. Stars somewhat larger than the sun can end their life in spectacular explosions – supernovae – leaving behind a collapsed remnant of the star called a neutron star. These neutron-star remnants have masses comparable to the original mass of the star, but radii of only a few kilometers! The high rotation rate is due to the conservation of angular momentum during the collapse. These stars emit beams of radio waves. Because of the rapid angular speed of the stars, the beam sweeps past the earth at regular intervals. To produce the observed radio wave pulses, the star has to rotate at rates from about one rev/s to one rev/ms. (a) Using data from the textbook, estimate the rotation rate of the sun if it were to collapse into a neutron star of radius 10 km. Because the sun is not a uniform sphere of gas, its moment of inertia is given by the formula $I = 0.059MR^2$. Assume that the neutron star is spherical and has a uniform mass distribution. (b) Is the rotational kinetic energy of the sun greater or smaller after the collapse? By what factor does it change, and where does the energy go to or come from?

Picture the Problem We can use conservation of angular momentum in part (a) to relate the before-and-after collapse rotation rates of the sun. In part (b), we can express the fractional change in the rotational kinetic energy of the sun as it collapses into a neutron star to decide whether its rotational kinetic energy is greater initially or after the collapse.

(a) Use conservation of angular momentum to relate the angular momenta of the sun before and after its collapse:

$$I_b\omega_b = I_a\omega_a \qquad (1)$$

Using the given formula, approximate the moment of inertia I_b of the sun before collapse:

$$I_b = 0.059MR_{sun}^2$$
$$= 0.059(1.99\times10^{30}\text{ kg})(6.96\times10^5\text{ km})^2$$
$$= 5.69\times10^{46}\text{ kg}\cdot\text{m}^2$$

Find the moment of inertia I_a of the sun when it has collapsed into a spherical neutron star of radius 10 km and uniform mass distribution:

$$I_a = \tfrac{2}{5}MR^2$$
$$= \tfrac{2}{5}(1.99\times10^{30}\text{ kg})(10\text{ km})^2$$
$$= 7.96\times10^{37}\text{ kg}\cdot\text{m}^2$$

Substitute in equation (1) and solve for ω_a to obtain:

$$\omega_a = \frac{I_b}{I_a}\omega_b = \frac{5.69\times10^{46}\text{ kg}\cdot\text{m}^2}{7.96\times10^{37}\text{ kg}\cdot\text{m}^2}\omega_b$$
$$= 7.15\times10^8\,\omega_b$$

Given that $\omega_b = 1$ rev/25 d, evaluate ω_a:

$$\omega_a = 7.15\times10^8\left(\frac{1\text{rev}}{25\text{d}}\right)$$
$$= \boxed{2.86\times10^7\text{ rev/d}}$$

> The additional rotational kinetic energy comes at the expense of gravitational potential energy, which decreases as the sun gets smaller.

Note that the rotational period decreases by the same factor of I_b/I_a and becomes:

$$T_a = \frac{2\pi}{\omega_a} = \frac{2\pi}{2.86\times10^7\,\dfrac{\text{rev}}{\text{d}}\times\dfrac{2\pi\text{ rad}}{\text{rev}}\times\dfrac{1\text{d}}{24\text{h}}\times\dfrac{1\text{h}}{3600\text{s}}} = 3.02\times10^{-3}\text{ s}$$

(b) Express the fractional change in the sun's rotational kinetic energy as a consequence of its collapse and simplify to obtain:

$$\frac{\Delta K}{K_b} = \frac{K_a - K_b}{K_b} = \frac{K_a}{K_b} - 1$$
$$= \frac{\tfrac{1}{2}I_a\omega_a^2}{\tfrac{1}{2}I_b\omega_b^2} - 1$$
$$= \frac{I_a\omega_a^2}{I_b\omega_b^2} - 1$$

Substitute numerical values and evaluate $\Delta K/K_b$:

$$\frac{\Delta K}{K_b} = \left(\frac{1}{7.15\times10^8}\right)\left(\frac{2.86\times10^7\,\text{rev/d}}{1\,\text{rev/25d}}\right)^2 - 1 = \boxed{7.15\times10^8}\ \text{(i.e., the rotational kinetic}$$

energy *increases* by a factor of approximately 7×10^8.)

*30 ••   Estimate the angular velocity and angular momentum of the diver in Figure 10-24 about his center of mass. Make any approximations that you think reasonable.

Picture the Problem Let's estimate that the diver with arms extended over head is about 2.5 m long and has a mass $M = 80$ kg. We'll also assume that it is reasonable to model the diver as a uniform stick rotating about its center of mass. From the photo, it appears that he sprang about 3 m in the air, and that the diving board was about 3 m high. We can use these assumptions and estimated quantities, together with their definitions, to estimate ω and L.

Express the diver's angular velocity ω and angular momentum L:	$\omega = \dfrac{\Delta\theta}{\Delta t}$ (1) and $L = I\omega$ (2)
Using a constant-acceleration equation, express his time in the air:	$\Delta t = \Delta t_{\text{rise 3 m}} + \Delta t_{\text{fall 6 m}}$ $= \sqrt{\dfrac{2\Delta y_{\text{up}}}{g}} + \sqrt{\dfrac{2\Delta y_{\text{down}}}{g}}$
Substitute numerical values and evaluate Δt:	$\Delta t = \sqrt{\dfrac{2(3\,\text{m})}{9.81\,\text{m/s}^2}} + \sqrt{\dfrac{2(6\,\text{m})}{9.81\,\text{m/s}^2}} = 1.89\,\text{s}$
Estimate the angle through which he rotated in 1.89 s:	$\Delta\theta \approx 0.5\,\text{rev} = \pi\,\text{rad}$
Substitute in equation (1) and evaluate ω:	$\omega = \dfrac{\pi\,\text{rad}}{1.89\,\text{s}} = \boxed{1.66\,\text{rad/s}}$
Use the "stick rotating about an axis through its center of mass" model to approximate the moment of inertia of the diver:	$I = \tfrac{1}{12}ML^2$
Substitute in equation (2) to obtain:	$L = \tfrac{1}{12}ML^2\omega$

Substitute numerical values and evaluate L:

$$L = \tfrac{1}{12}(80\,\text{kg})(2.5\,\text{m})^2(1.66\,\text{rad/s})$$

$$= 69.2\,\text{kg}\cdot\text{m}^2/\text{s} \approx \boxed{70\,\text{kg}\cdot\text{m}^2/\text{s}}$$

Remarks: We can check the reasonableness of this estimation in another way. Because the diver rose about 3 m in the air, the initial impulse acting on him must be about 600 kg·m/s (i.e., $I = \Delta p = Mv_i$). If we estimate that the lever arm of the force is roughly $\ell = 1.5$ m, and the angle between the force exerted by the board and a line running from his feet to the center of mass is about 5°, we obtain $L = I\ell\sin5° \approx 78$ kg·m²/s, which is not too bad considering the approximations made here.

*32 •• Estimate Timothy Goebel's initial takeoff speed, rotational velocity, and angular momentum when he performs a quadruple Lutz (Figure 10-38). Make any assumptions you think reasonable, but be prepared to justify them. Goebel's mass is about 60 kg and the height of the jump is about 0.6 m. Note that the angular velocity will change quite a bit during the jump, as he begins with arms outstretched and pulls them in. Your answer should be accurate to within a factor of 2 if you're careful.

Figure 10-38 Problem 32

Picture the Problem We'll assume that he launches himself at an angle of 45° with the horizontal with his arms spread wide, and then pulls them in to increase his rotational speed during the jump. We'll also assume that we can model him as a 2-m long cylinder with an average radius of 0.15 m and a mass of 60 kg. We can then find his take-off speed and "air time" using constant-acceleration equations, and use the latter, together with the definition of rotational velocity, to find his initial rotational velocity. Finally, we can apply conservation of angular momentum to find his initial angular momentum.

Using a constant-acceleration equation, relate his takeoff speed v_0 to his maximum elevation Δy:

$$v^2 = v_{0y}^2 + 2a_y\Delta y$$

or, because $v_{0y} = v_0\sin45°$, $v = 0$, and

$$a_y = -g,$$
$$0 = v_0^2 \sin^2 45° - 2g\Delta y$$

Solve for v_0 to obtain:

$$v_0 = \sqrt{\frac{2g\Delta y}{\sin^2 45°}} = \frac{\sqrt{2g\Delta y}}{\sin 45°}$$

Substitute numerical values and evaluate v_0:

$$v_0 = \frac{\sqrt{2(9.81\,\text{m/s}^2)(0.6\,\text{m})}}{\sin 45°} = \boxed{4.85\,\text{m/s}}$$

Use its definition to express Goebel's angular velocity:

$$\omega = \frac{\Delta\theta}{\Delta t}$$

Use a constant-acceleration equation to express Goebel's "air time" Δt:

$$\Delta t = 2\Delta t_{\text{rise 0.6 m}} = 2\sqrt{\frac{2\Delta y}{g}}$$

Substitute numerical values and evaluate Δt:

$$\Delta t = 2\sqrt{\frac{2(0.6\,\text{m})}{9.81\,\text{m/s}^2}} = 0.699\,\text{s}$$

Substitute numerical values and evaluate ω:

$$\omega = \frac{4\,\text{rev}}{0.699\,\text{s}} \cdot \frac{2\pi\,\text{rad}}{\text{rev}} = \boxed{36.0\,\text{rad/s}}$$

Use conservation of angular momentum to relate his take-off angular velocity ω_0 to his average angular velocity ω as he performs a quadruple Lutz:

$$I_0\omega_0 = I\omega$$

Assuming that he can change his angular momentum by a factor of 2 by pulling his arms in, solve for and evaluate ω_0:

$$\omega_0 = \frac{I}{I_0}\omega = \frac{1}{2}(36\,\text{rad/s}) = \boxed{18.0\,\text{rad/s}}$$

Express his take-off angular momentum:

$$L_0 = I_0\omega_0$$

Assuming that we can model him as a solid cylinder of length ℓ with an average radius r and mass m, express his moment of inertia with arms drawn in (his take-off configuration):

$$I_0 = 2(\tfrac{1}{2}mr^2) = mr^2$$
where the factor of 2 represents our assumption that he can double his moment of inertia by extending his arms.

Substitute to obtain:

$$L_0 = mr^2\omega_0$$

Substitute numerical values and evaluate L_0:

$$L_0 = (60\,\text{kg})(0.15\,\text{m})^2(18\,\text{rad/s})$$
$$= \boxed{24.3\,\text{kg}\cdot\text{m}^2/\text{s}}$$

Vector Nature of Rotation

*36 •   Under what conditions is the magnitude of $\vec{A}\times\vec{B}$ equal to $\vec{A}\cdot\vec{B}$?

Picture the Problem The magnitude of $\vec{A}\times\vec{B}$ is given by $|AB\sin\theta|$.

Equate the magnitudes of $\vec{A}\times\vec{B}$ and $\vec{A}\cdot\vec{B}$:

$$|AB\sin\theta| = |AB\cos\theta|$$
$$\therefore |\sin\theta| = |\cos\theta|$$

or

$$\tan\theta = \pm1$$

Solve for θ to obtain:

$$\theta = \tan^{-1}\pm1 = \boxed{\pm45° \text{ or } \pm135°}$$

*42 ••   Using the cross product, prove the *law of sines* for the triangle shown in Figure 10-40: if A, B, and C are the lengths of each side of the triangle, show that $A/\sin a = B/\sin b = C/\sin c$.

Figure 10-40 Problem 42

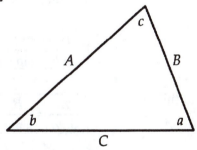

Picture the Problem Draw the triangle using the three vectors as shown below. Note that $\vec{A}+\vec{B}=\vec{C}$. We can find the magnitude of the cross product of $\vec{A}$ and $\vec{B}$ and of $\vec{A}$ and $\vec{C}$ and then use the cross product of $\vec{A}$ and $\vec{C}$, using $\vec{A}+\vec{B}=\vec{C}$, to show that $AC\sin b = AB\sin c$ or $B/\sin b = C/\sin c$. Proceeding similarly, we can extend the law of sines to the third side of the triangle and the angle opposite it.

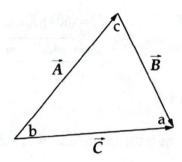

Express the magnitude of the cross product of $\vec{A}$ and $\vec{B}$:	$\left|\vec{A} \times \vec{B}\right| = AB\sin c$

Express the magnitude of the cross product of $\vec{A}$ and $\vec{C}$:	$\left|\vec{A} \times \vec{C}\right| = AC\sin b$

Form the cross product of $\vec{A}$ with $\vec{C}$ to obtain:	$\vec{A} \times \vec{C} = \vec{A} \times \left(\vec{A} + \vec{B}\right)$

$$= \vec{A} \times \vec{A} + \vec{A} \times \vec{B}$$
$$= \vec{A} \times \vec{B}$$

because $\vec{A} \times \vec{A} = 0$.

Because $\vec{A} \times \vec{C} = \vec{A} \times \vec{B}$:	$\left|\vec{A} \times \vec{C}\right| = \left|\vec{A} \times \vec{B}\right|$

and
$$AC\sin b = AB\sin c$$

Simplify and rewrite this expression to obtain:	$\dfrac{B}{\sin b} = \dfrac{C}{\sin c}$

Proceed similarly to extend this result to the law of sines:	$\dfrac{A}{\sin a} = \dfrac{B}{\sin b} = \dfrac{C}{\sin c}$

Angular Momentum

*46 ••   A particle is traveling with a constant velocity $\vec{v}$ along a line that is a distance b from the origin O (Figure 10-41). Let dA be the area swept out by the position vector from O to the particle in time dt. Show that dA/dt is constant and is equal to $\frac{1}{2}L/m$, where L is the angular momentum of the particle about the origin.

Figure 10-41 Problem 46

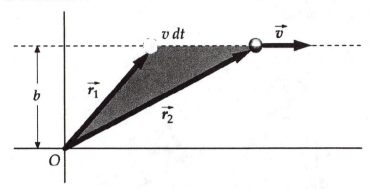

Picture the Problem We can use the formula for the area of a triangle to find the area swept out at $t = t_1$, add this area to the area swept out in time dt, and then differentiate this expression with respect to time to obtain the given expression for dA/dt.

Express the area swept out at $t = t_1$:

$A_1 = \frac{1}{2}br_1 \cos\theta_1 = \frac{1}{2}bx_1$

where θ_1 is the angle between $\vec{r}_1$ and $\vec{v}$ and x_1 is the component of $\vec{r}_1$ in the direction of $\vec{v}$.

Express the area swept out at $t = t_1 + dt$:

$A = A_1 + dA = \frac{1}{2}b(x_1 + dx)$
$= \frac{1}{2}b(x_1 + vdt)$

Differentiate with respect to t:

$\dfrac{dA}{dt} = \frac{1}{2}b\dfrac{dx}{dt} = \frac{1}{2}bv = \text{constant}$

Because $r\sin\theta = b$:

$\frac{1}{2}bv = \frac{1}{2}(r\sin\theta)v = \dfrac{1}{2m}(rp\sin\theta)$

$= \boxed{\dfrac{L}{2m}}$

Torque and Angular Momentum

*51 •• In Figure 10-43, the incline is frictionless and the string passes through the center of mass of each block. The pulley has a moment of inertia I and a radius R. (*a*) Find the net torque acting on the system (the two masses, string, and pulley) about the center of the pulley. (*b*) Write an expression for the total angular momentum of the system about the center of the pulley when the masses are moving with a speed v. (*c*) Find the acceleration of the masses from your results for Parts (*a*) and (*b*) by setting the net torque equal to the rate of change of the angular momentum of the system.

Figure 10-43 Problem 51

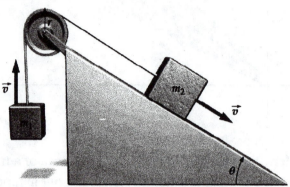

Picture the Problem Let the system include the pulley, string, and the blocks and assume that the mass of the string is negligible. The angular momentum of this system changes because a *net* torque acts on it.

(*a*) Express the net torque about an axis through the center of mass of the pulley:

$$\tau_{net} = Rm_2 g \sin\theta - Rm_1 g$$
$$= \boxed{Rg(m_2 \sin\theta - m_1)}$$

where we have taken clockwise to be positive to be consistent with a positive upward velocity of the block whose mass is m_1 as indicated in the figure.

(*b*) Express the total angular momentum of the system about an axis through the center of the pulley:

$$L = I\omega + m_1 vR + m_2 vR$$
$$= \boxed{vR\left(\frac{I}{R^2} + m_1 + m_2\right)}$$

(*c*) Express τ as the time derivative of the angular momentum:

$$\tau = \frac{dL}{dt} = \frac{d}{dt}\left[vR\left(\frac{I}{R^2} + m_1 + m_2\right)\right]$$
$$= aR\left(\frac{I}{R^2} + m_1 + m_2\right)$$

Equate this result to that of part (*a*) and solve for *a* to obtain:

$$a = \boxed{\dfrac{g(m_2 \sin\theta - m_1)}{\dfrac{I}{R^2} + m_1 + m_2}}$$

Conservation of Angular Momentum

*54 •   A planet moves in an elliptical orbit about the sun with the sun at one focus of the ellipse as in Figure 10-45. (*a*) What is the torque about the center of the sun due to the gravitational force of attraction of the sun for the planet? (*b*) At position *A*, the planet is a distance r_1 from the sun and is moving with a speed v_1 perpendicular to the line from the sun to the planet. At position *B*, it is at distance r_2 and is moving with speed v_2, again perpendicular to the line from the sun to the planet. What is the ratio of v_1 to v_2 in terms of r_1 and r_2?

Figure 10-45 Problem 54

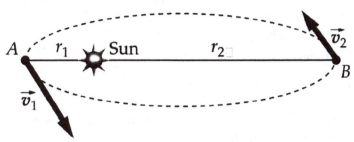

Picture the Problem Let *m* represent the mass of the planet and apply the definition of torque to find the torque produced by the gravitational force of attraction. We can use Newton's 2nd law of motion in the form $\vec{\tau} = d\vec{L}/dt$ to show that $\vec{L}$ is constant and apply conservation of angular momentum to the motion of the planet at points *A* and *B*.

(*a*) Express the torque produced by the gravitational force of attraction of the sun for the planet:

$\vec{\tau} = \vec{r} \times \vec{F} = \boxed{0}$ because $\vec{F}$ acts along the direction of $\vec{r}$.

(*b*) Because $\vec{\tau} = 0$:

$$\frac{d\vec{L}}{dt} = 0 \Rightarrow \vec{L} = \vec{r} \times m\vec{v} = \text{constant}$$

Noting that at points *A* and *B* $\left| \vec{r} \times \vec{v} \right| = rv$, express the relationship between the distances from the sun and the speeds of the planets:

$r_1 v_1 = r_2 v_2$

or

$$\frac{v_1}{v_2} = \boxed{\frac{r_2}{r_1}}$$

*56 ••   A small blob of putty of mass *m* falls from the ceiling and lands on the outer rim of a turntable of radius *R* and moment of inertia I_0 that is rotating freely with angular speed ω_i about its vertical fixed symmetry axis. (a) What is the post collision angular speed of the turntable plus putty? (b) After several turns, the blob flies off the edge of the turntable. What is the angular speed of the turntable after the blob flies off?

Picture the Problem Let the system consist of the blob of putty and the turntable. Because the net external torque acting on this system is zero, its angular momentum remains constant when the blob of putty falls onto the turntable.

(a) Using conservation of angular momentum, relate the initial and final angular speeds of the turntable to its initial and final moments of inertia and solve for ω_f:

$$I_0\omega_i = I_f\omega_f$$

and

$$\omega_f = \frac{I_0}{I_f}\omega_i$$

Express the final rotational inertia of the turntable-plus-blob:

$$I_f = I_0 + I_{blob} = I_0 + mR^2$$

Substitute and simplify to obtain:

$$\omega_f = \frac{I_0}{I_0 + mR^2}\omega_i = \boxed{\frac{1}{1 + \dfrac{mR^2}{I_0}}\omega_i}$$

(b) If the blob flies off tangentially to the turntable, its angular momentum doesn't change (with respect to an axis through the center of turntable). Because there is no external torque acting on the blob-turntable system, the total angular momentum of the system will remain constant and the angular momentum of the turntable will not change. Because the moment of inertia of the table hasn't changed either, the turntable will continue to spin at $\boxed{\omega' = \omega_f}$.

*58 ••   Two disks of identical mass but different radii (r and 2r) are spinning on frictionless bearings at the same angular speed ω_0 but in opposite directions (Figure 10-46). The two disks are brought slowly together. The resulting frictional force between the surfaces eventually brings them to a common angular velocity. What is the magnitude of that final angular velocity in terms of ω_0?

Figure 10-46 Problem 58

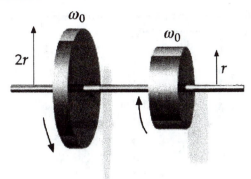

Picture the Problem The net external torque acting on this system is zero and so we know that angular momentum is conserved as these disks are brought together. Let the numeral 1 refer to the disk to the left and the numeral 2 to the disk to the right. Let the angular momentum of the disk with the larger radius be positive.

Using conservation of angular momentum, relate the initial angular speeds of the disks to their common final speed and to their moments of inertia:

$$I_i\omega_i = I_f\omega_f$$
or
$$I_1\omega_0 - I_2\omega_0 = (I_1 + I_2)\omega_f$$

Solve for ω_f:

$$\omega_f = \frac{I_1 - I_2}{I_1 + I_2}\omega_0$$

Express I_1 and I_2:

$$I_1 = \tfrac{1}{2}m(2r)^2 = 2mr^2$$
and
$$I_2 = \tfrac{1}{2}mr^2$$

Substitute and simplify to obtain:

$$\omega_f = \frac{2mr^2 - \tfrac{1}{2}mr^2}{2mr^2 + \tfrac{1}{2}mr^2}\omega_0 = \boxed{\tfrac{3}{5}\omega_0}$$

*60 •• A 0.2-kg point mass moving on a frictionless horizontal surface is attached to a rubber band whose other end is fixed at point P. The rubber band exerts a force $F = bx$ toward P, where x is the length of the rubber band and b is an unknown constant. The mass moves along the dotted line in Figure 10-47. When it passes point A, its velocity is 4 m/s, directed as shown. The distance AP is 0.6 m and BP is 1.0 m. (*a*) Find the velocity of the mass at points B and C. (*b*) Find b.

Figure 10-47 Problem 60

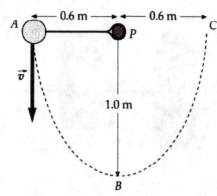

Picture the Problem Because the force exerted by the rubber band is parallel to the position vector of the point mass, the net external torque acting on it is zero and we can use the conservation of angular momentum to determine the speeds of the ball at points B and C. We'll use mechanical energy conservation to find b by relating the kinetic and elastic potential energies at A and B.

(a) Use conservation of momentum to relate the angular momenta at points A, B and C:

$$L_A = L_B = L_C$$
or
$$mv_A r_A = mv_B r_B = mv_C r_C$$

Solve for v_B in terms of v_A:

$$v_B = v_A \frac{r_A}{r_B}$$

Substitute numerical values and evaluate v_B:

$$v_B = (4\,\text{m/s})\frac{0.6\,\text{m}}{1\text{m}} = \boxed{2.40\,\text{m/s}}$$

Solve for v_C in terms of v_A:

$$v_C = v_A \frac{r_A}{r_C}$$

Substitute numerical values and evaluate v_C:

$$v_C = (4\,\text{m/s})\frac{0.6\,\text{m}}{0.6\,\text{m}} = \boxed{4.00\,\text{m/s}}$$

(b) Use conservation of mechanical energy between points A and B to relate the kinetic energy of the point mass and the energy stored in the stretched rubber band:

$$E_A = E_B$$
or
$$\tfrac{1}{2}mv_A^2 + \tfrac{1}{2}br_A^2 = \tfrac{1}{2}mv_B^2 + \tfrac{1}{2}br$$

Solve for b:

$$b = \frac{m\left(v_B^2 - v_A^2\right)}{r_A^2 - r_B^2}$$

Substitute numerical values and evaluate b:

$$b = \frac{(0.2\,\text{kg})\left[(2.4\,\text{m/s})^2 - (4\,\text{m/s})^2\right]}{(0.6\,\text{m})^2 - (1\,\text{m})^2}$$

$$= \boxed{3.20\,\text{N/m}}$$

Quantization of Angular Momentum

*61 •   The z component of the spin of an electron is $\frac{1}{2}\hbar$, but the magnitude of the spin vector is $\sqrt{0.75}\hbar$. What is the angle between the electron's spin angular momentum vector and the z axis?

Picture the Problem The electron's spin angular momentum vector is related to its z component as shown in the diagram.

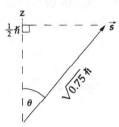

Using trigonometry, relate the magnitude of $\vec{s}$ to its z component:

$$\theta = \cos^{-1}\left(\frac{\frac{1}{2}\hbar}{\sqrt{0.75}\hbar}\right) = \boxed{54.7°}$$

*65 ••   How fast would a nitrogen molecule have to be moving for its translational kinetic energy to be equal to the rotational kinetic energy of its $\ell = 1$ quantum state?

Picture the Problem We can obtain an expression for the speed of the nitrogen molecule by equating its translational and rotational kinetic energies and solving for v. Because this expression includes the moment of inertia I of the nitrogen molecule, we can use the definition of the moment of inertia to express I for a dumbbell model of the nitrogen molecule. The rotational energies of a nitrogen molecule depend on the quantum number ℓ according to $E_\ell = L^2/2I = \ell(\ell+1)\hbar^2/2I$.

Equate the rotational kinetic energy of the nitrogen molecule in its $\ell = 1$ quantum state and its translational kinetic energy:

$$E_1 = \tfrac{1}{2}m_N v^2 \qquad\qquad (1)$$

Express the rotational energy levels of the nitrogen molecule:	$$E_\ell = \frac{L^2}{2I} = \frac{\ell(\ell+1)\hbar^2}{2I}$$
For $\ell = 1$:	$$E_1 = \frac{1(1+1)\hbar^2}{2I} = \frac{\hbar^2}{I}$$
Substitute in equation (1):	$$\frac{\hbar^2}{I} = \tfrac{1}{2}m_N v^2$$
Solve for v to obtain:	$$v = \sqrt{\frac{2\hbar^2}{m_N I}} \qquad\qquad (2)$$
Using a rigid dumbbell model, express the moment of inertia of the nitrogen molecule about its center of mass:	$$I = \sum_i m_i r_i^2 = m_N r^2 + m_N r^2 = 2m_N r^2$$ and $$m_N I = 2m_N^2 r^2$$
Substitute in equation (2):	$$v = \sqrt{\frac{2\hbar^2}{2m_N^2 r^2}} = \frac{\hbar}{m_N r}$$
Substitute numerical values and evaluate v:	$$v = \frac{1.055 \times 10^{-34}\ \text{J}\cdot\text{s}}{14(1.66 \times 10^{-27}\ \text{kg})(5.5 \times 10^{-11}\ \text{m})}$$ $$= \boxed{82.5\,\text{m/s}}$$

Collision Problems

*67 ••   Figure 10-49 shows a thin bar of length L and mass M and a small blob of putty of mass m. The system is supported on a frictionless horizontal surface. The putty moves to the right with velocity v, strikes the bar at a distance d from the center of the bar, and sticks to the bar at the point of contact. Obtain expressions for the velocity of the system's center of mass and for the angular velocity of the system about its center of mass.

Figure 10-49 Problems 67, 68

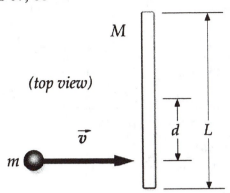

(top view)

M

$\vec{v}$

m

d L

Picture the Problem Because there are no external forces or torques acting on the system defined in the problem statement, both linear and angular momentum are conserved in the collision and the velocity of the center of mass after the collision is the same as before the collision. Let the direction the blob of putty is moving initially be the positive x direction and toward the top of the page in the figure be the positive y direction.

Using its definition, express the location of the center of mass relative to the center of the bar:	$y_{cm} = \dfrac{md}{M+m}$ below the center of the bar.
Using its definition, express the velocity of the center of mass:	$\boxed{v_{cm} = \dfrac{mv}{M+m}}$
Using the definition of L in terms of I and ω, express ω:	$\omega = \dfrac{L_{cm}}{I_{cm}}$ (1)
Express the angular momentum about the center of mass:	$L_{cm} = mv(d - y_{cm})$ $= mv\left(d - \dfrac{md}{M+m}\right) = \dfrac{mMvd}{M+m}$
Using the parallel axis theorem, express the moment of inertia of the system relative to its center of mass:	$I_{cm} = \tfrac{1}{12}ML^2 + My_{cm}^2 + m(d - y_{cm})^2$

Substitute for y_{cm} and simplify to obtain:

$$I_{cm} = \tfrac{1}{12}ML^2 + M\left(\frac{md}{M+m}\right)^2 + m\left(d - \frac{md}{M+m}\right)^2$$

$$= \tfrac{1}{12}ML^2 + \frac{Mm^2d^2}{(M+m)^2} + m\left(\frac{d(M+m)-md}{M+m}\right)^2$$

$$= \tfrac{1}{12}ML^2 + \frac{Mm^2d^2}{(M+m)^2} + \frac{mM^2d^2}{(M+m)^2} = \tfrac{1}{12}ML^2 + \frac{(M+m)mMd^2}{(M+m)^2}$$

$$= \tfrac{1}{12}ML^2 + \frac{mMd^2}{M+m}$$

Substitute for I_{cm} and L_{cm} in equation (1) and simplify to obtain:

$$\boxed{\omega = \frac{mMvd}{\tfrac{1}{12}ML^2(M+m) + Mmd^2}}$$

Remarks: You can verify the expression for I_{cm} by letting $m \to 0$ to obtain $I_{cm} = \tfrac{1}{12}ML^2$ and letting $M \to 0$ to obtain $I_{cm} = 0$.

*73 ••   A uniform rod of length L_1 and mass $M = 0.75$ kg is supported by a hinge at one end and is free to rotate in the vertical plane (Figure 10-52). The rod is released from rest in the position shown. A particle of mass $m = 0.5$ kg is supported by a thin string of length L_2 from the hinge. The particle sticks to the rod on contact. What should be the ratio L_2/L_1 so that $\theta_{max} = 60°$ after the collision?

Figure 10-52 Problems 73-76

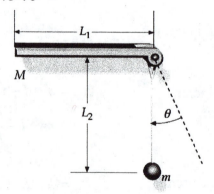

Picture the Problem Because the net external torque acting on the system is zero, angular momentum is conserved in this perfectly inelastic collision. The rod, on its downward swing, acquires rotational kinetic energy. Angular momentum is conserved in the perfectly inelastic collision with the particle and the rotational kinetic of the after-collision system is then transformed into gravitational potential energy as the rod-plus-particle swing upward. Let the zero of gravitational potential energy be at a distance L_1 below the pivot and use both angular momentum and mechanical energy conservation to relate the distances L_1 and L_2 and the masses M and m.

Use conservation of energy to relate the initial and final potential energy of the rod to its rotational kinetic energy just before it collides with the particle:	$K_f - K_i + U_f - U_i = 0$ or, because $K_i = 0$, $K_f + U_f - U_i = 0$
Substitute for K_f, U_f, and U_i to obtain:	$\frac{1}{2}\left(\frac{1}{3}ML_1^2\right)\omega^2 + Mg\frac{L_1}{2} - MgL_1 = 0$
Solve for ω.	$\omega = \sqrt{\frac{3g}{L_1}}$
Letting ω' represent the angular speed of the rod-and-particle system just after impact, use conservation of angular momentum to relate the angular momenta before and after the collision:	$L_i = L_f$ or $\left(\frac{1}{3}ML_1^2\right)\omega = \left(\frac{1}{3}ML_1^2 + mL_2^2\right)\omega'$
Solve for ω':	$\omega' = \dfrac{\frac{1}{3}ML_1^2}{\frac{1}{3}ML_1^2 + mL_2^2}\omega$
Use conservation of energy to relate the rotational kinetic energy of the rod-plus-particle just after their collision to their potential energy when they have swung through an angle θ_{max}:	$K_f - K_i + U_f - U_i = 0$ or, because $K_f = 0$, $-\frac{1}{2}I\omega'^2 + Mg\left(\frac{1}{2}L_1\right)\left(1 - \cos\theta_{max}\right)$ $\qquad + mgL_2\left(1 - \cos\theta_{max}\right) = 0$ (1)
Express the moment of inertia of the system with respect to the pivot:	$I = \frac{1}{3}ML_1^2 + mL_2^2$

Substitute for θ_{max}, I and ω' in equation (1):

$$\frac{3\frac{g}{L_1}\left(\frac{1}{3}ML_1^2\right)^2}{\frac{1}{3}ML_1^2 + mL_2^2} = Mg\left(\frac{1}{2}L_1\right) + mgL_2$$

Simplify to obtain:

$$L_1^3 = 2\frac{m}{M}L_1^2L_2 + 3L_2^2L_1 + 6\frac{m}{M}L_2^3 \quad (2)$$

Simplify equation (2) by letting $\alpha = m/M$ and $\beta = L_2/L_1$ to obtain:

$$6\alpha^2\beta^3 + 3\beta^2 + 2\alpha\beta - 1 = 0$$

Substitute for α and simplify to obtain the cubic equation in β:

$$12\beta^3 + 9\beta^2 + 4\beta - 3 = 0$$

Use the solver function* of your calculator to find the only real value of β:

$$\beta = \boxed{0.349}$$

*Remarks: Most graphing calculators have a "solver" feature. One can solve the cubic equation using either the "graph" and "trace" capabilities or the "solver" feature. The root given above was found using SOLVER on a TI-85.

*78 ••   A uniform disk of mass 2.5 kg and radius 6.4 cm is mounted in the center of a 10-cm-long axle and spun at 700 rev/min. The axle is then placed in a horizontal position with one end resting on a pivot. The other end is given an initial horizontal velocity such that the precession is smooth with no nutation. (*a*) What is the angular velocity of precession? (*b*) What is the speed of the center of mass during the precession? (*c*) What are the magnitude and direction of the acceleration of the center of mass? (*d*) What are the vertical and horizontal components of the force exerted by the pivot?

Picture the Problem The angular velocity of precession can be found from its definition. Both the speed and acceleration of the center of mass during precession are related to the angular velocity of precession. We can use Newton's 2nd law to find the vertical and horizontal components of the force exerted by the pivot.

(*a*) Using its definition, express the angular velocity of precession:

$$\omega_p = \frac{d\phi}{dt} = \frac{MgD}{I_s\omega_s} = \frac{MgD}{\frac{1}{2}MR^2\omega_s} = \frac{2gD}{R^2\omega_s}$$

Substitute numerical values and
evaluate ω_p:

$$\omega_p = \frac{2(9.81\,\text{m/s}^2)(0.05\,\text{m})}{(0.064\,\text{m})^2 \left(700\,\dfrac{\text{rev}}{\text{min}} \times \dfrac{2\pi\,\text{rad}}{\text{rev}} \times \dfrac{1\,\text{min}}{60\,\text{s}}\right)}$$

$$= \boxed{3.27\,\text{rad/s}}$$

(b) Express the speed of the
center of mass in terms of its
angular velocity of precession:

$$v_{cm} = D\omega_p = (0.05\,\text{m})(3.27\,\text{rad/s})$$

$$= \boxed{0.164\,\text{m/s}}$$

(c) Relate the acceleration of
the center of mass to its
angular velocity of precession:

$$a_{cm} = D\omega_p^2 = (0.05\,\text{m})(3.27\,\text{rad/s})^2$$

$$= \boxed{0.535\,\text{m/s}^2}$$

(d) Use Newton's 2nd law to
relate the vertical component
of the force exerted by the
pivot to the weight of the disk:

$$F_v = Mg = (2.5\,\text{kg})(9.81\,\text{m/s}^2)$$

$$= \boxed{24.5\,\text{N}}$$

Relate the horizontal
component of the force exerted
by the pivot to the acceleration
of the center of mass:

$$F_v = Ma_{cm} = (2.5\,\text{kg})(0.535\,\text{m/s}^2)$$

$$= \boxed{1.34\,\text{N}}$$

General Problems

*82 ••   A 2-kg ball attached to a string of length 1.5 m moves in a horizontal
circle as a conical pendulum (Figure 10-53). The string makes an angle $\theta = 30°$
with the vertical. (a) Show that the angular momentum of the ball about the point
of support P has a horizontal component toward the center of the circle as well as
a vertical component, and find these components. (b) Find the magnitude
of $d\vec{L}/dt$ and show that it equals the magnitude of the torque exerted by gravity
about the point of support.

Figure 10-53 Problem 82

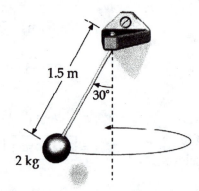

Picture the Problem Let the origin of the coordinate system be at the pivot (point P). The diagram shows the forces acting on the ball. We'll apply Newton's 2nd law to the ball to determine its speed. We'll then use the derivative of its position vector to express its velocity and the definition of angular momentum to show that $\vec{L}$ has both horizontal and vertical components. We can use the derivative of $\vec{L}$ with respect to time to show that the rate at which the angular momentum of the ball changes is equal to the torque, relative to the pivot point, acting on it.

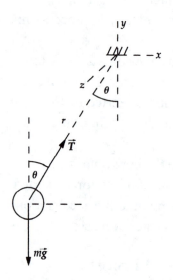

(*a*) Express the angular momentum of the ball about the point of support:

$$\vec{L} = \vec{r} \times \vec{p} = m\vec{r} \times \vec{v} \qquad (1)$$

Apply Newton's 2nd law to the ball:

$$\sum F_x = T \sin\theta = m\frac{v^2}{r\sin\theta}$$
and
$$\sum F_z = T\cos\theta - mg = 0$$

Eliminate T between these equations and solve for v:

$$v = \sqrt{rg\sin\theta\tan\theta}$$

Substitute numerical values and evaluate v:

$$v = \sqrt{(1.5\,\text{m})(9.81\,\text{m/s}^2)\sin 30^\circ \tan 30^\circ}$$
$$= 2.06\,\text{m/s}$$

Express the position vector of the ball:

$$\vec{r} = (1.5\,\text{m})\sin 30^\circ\left(\cos\omega t\,\hat{i} + \sin\omega t\hat{j}\right)$$
$$- (1.5\,\text{m})\cos 30^\circ\hat{k}$$

where $\vec{\omega} = \omega\hat{k}$.

Find the velocity of the ball:

$$\vec{v} = \frac{d\vec{r}}{dt}$$
$$= (0.75\omega\,\text{m/s})\left(-\sin\omega t\,\hat{i} + \cos\omega t\,\hat{j}\right)$$

Evaluate ω:

$$\omega = \frac{2.06\,\text{m/s}}{(1.5\,\text{m})\sin 30^\circ} = 2.75\,\text{rad/s}$$

Substitute for ω to obtain:

$$\vec{v} = (2.06\,\text{m/s})\left(-\sin\omega t\,\hat{i} + \cos\omega t\hat{j}\right)$$

Substitute in equation (1) and evaluate $\vec{L}$:

$$\vec{L} = (2\,\text{kg})\left[(1.5\,\text{m})\sin 30^\circ\left(\cos\omega t\,\hat{i} + \sin\omega t\,\hat{j}\right) - (1.5\,\text{m})\cos 30^\circ\hat{k}\right]$$
$$\times\left[(2.06\,\text{m/s})\left(-\sin\omega t\,\hat{i} + \cos\omega t\,\hat{j}\right)\right]$$
$$= \left[5.36\left(\cos\omega t\,\hat{i} + \sin\omega t\,\hat{j}\right) + 3.09\,\hat{k}\right]\text{J}\cdot\text{s}$$

The horizontal component of $\vec{L}$ is:

$$\boxed{5.36\left(\cos\omega t\,\hat{i} + \sin\omega t\,\hat{j}\right)\text{J}\cdot\text{s}}$$

The vertical component of $\vec{L}$ is:

$$\boxed{3.09\,\hat{k}\,\text{J}\cdot\text{s}}$$

(b) Evaluate $\dfrac{d\vec{L}}{dt}$:

$$\frac{d\vec{L}}{dt} = \left[5.36\omega\left(-\sin\omega t\,\hat{i} + \cos\omega t\hat{j}\right)\right]\text{J}$$

Evaluate the magnitude of $\dfrac{d\vec{L}}{dt}$:

$$\left|\frac{d\vec{L}}{dt}\right| = (5.36\,\text{N}\cdot\text{m}\cdot\text{s})(2.75\,\text{rad/s})$$
$$= \boxed{14.7\,\text{N}\cdot\text{m}}$$

Express the magnitude of the torque exerted by gravity about the point of support:

$$\tau = mgr\sin\theta$$

Substitute numerical values and evaluate τ:

$$\tau = (2\,\text{kg})(9.81\,\text{m/s}^2)(1.5\,\text{m})\sin 30°$$

$$= \boxed{14.7\,\text{N}\cdot\text{m}}$$

*89 ••   Kepler's second law states: *The radius vector from the sun to a planet sweeps out equal areas in equal times.* Show that this law follows directly from the law of conservation of angular momentum and the fact that the force of gravitational attraction between a planet and the sun acts along the line joining the two celestial objects.

Picture the Problem The drawing shows an elliptical orbit. The triangular element of the area is $dA = \frac{1}{2}r(rd\theta) = \frac{1}{2}r^2 d\theta.$

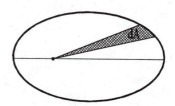

Differentiate dA with respect to t to obtain:

$$\frac{dA}{dt} = \frac{1}{2}r^2\frac{d\theta}{dt} = \frac{1}{2}r^2\omega$$

Because the gravitational force acts along the line joining the two objects, $\tau = 0$ and:

$$L = mr^2\omega = \text{constant}$$

Eliminate $r^2\omega$ between the two equations to obtain:

$$\frac{dA}{dt} = \boxed{\frac{L}{2m} = \text{constant}}$$

*93 ••   The Precession of the Equinoxes refers to the fact that the direction of earth's spin axis does not stay fixed in the sky, but moves in a circle of radius 23° with a period of about 26,000 y. This is why our pole star, Polaris, will not remain the pole star forever. The reason for this is that the earth is a giant gyroscope, with the torque on the earth is provided by the gravitational forces of the sun and moon. Calculate an approximate value for this torque, given that the period of rotation of the earth is 1 day, and its moment of inertia is 8.03×10^{37} kg·m^2.

Picture the Problem Let ω_P be the angular velocity of precession of the earth-as-gyroscope, ω_s its angular velocity about its spin axis, and I its moment of inertia with respect to an axis through its poles, and relate ω_P to ω_s and I using its definition.

Use its definition to express the precession rate of the earth as a giant gyroscope:

$$\omega_\text{P} = \frac{\tau}{L}$$

Substitute for I and solve for τ.
$$\tau = L\omega_p = I\omega\omega_p$$

Express the angular velocity ω_s of the earth about its spin axis:
$$\omega = \frac{2\pi}{T} \text{ where } T \text{ is the period of}$$
rotation of the earth.

Substitute to obtain:
$$\tau = \frac{2\pi I\omega_p}{T}$$

Substitute numerical values and evaluate τ:

$$\tau = \frac{2\pi\left(8.03\times10^{37} \text{ kg}\cdot\text{m}^2\right)\left(7.66\times10^{-12} \text{ s}^{-1}\right)}{1\text{d}\times\dfrac{24\text{h}}{\text{d}}\times\dfrac{3600\text{s}}{\text{h}}} = \boxed{4.47\times10^{22} \text{ N}\cdot\text{m}}$$

*97 ••• Figure 10-58 shows a pulley in the form of a uniform disk with a heavy rope hanging over it. The circumference of the pulley is 1.2 m and its mass is 2.2 kg. The rope is 8.0 m long and its mass is 4.8 kg. At the instant shown in the figure, the system is at rest and the difference in height of the two ends of the rope is 0.6 m. (*a*) What is the angular velocity of the pulley when the difference in height between the two ends of the rope is 7.2 m? (*b*) Obtain an expression for the angular momentum of the system as a function of time while neither end of the rope is above the center of the pulley. There is no slippage between rope and pulley wheel.

Figure 10-58 Problem 97

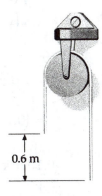

0.6 m

Picture the Problem Let the origin of the coordinate system be at the center of the pulley with the upward direction positive. Let λ be the linear density (mass per unit length) of the rope and L_1 and L_2 the lengths of the hanging parts of the rope. We can use conservation of mechanical energy to find the angular velocity

of the pulley when the difference in height between the two ends of the rope is
7.2 m.

(*a*) Apply conservation of energy to relate the final kinetic energy of the system to the change in potential energy:	$\Delta K + \Delta U = 0$ or, because $K_i = 0$, $K + \Delta U = 0$ $\qquad$ (1)

Express the change in potential
energy of the system:

$$\Delta U = U_f - U_i$$
$$= -\tfrac{1}{2}L_{1f}(L_{1f}\lambda)g - \tfrac{1}{2}L_{2f}(L_{2f}\lambda)g$$
$$\quad -\left[-\tfrac{1}{2}L_{1i}(L_{1i}\lambda)g - \tfrac{1}{2}L_{2i}(L_{2i}\lambda)g\right]$$
$$= -\tfrac{1}{2}\left(L_{1f}^2 + L_{2f}^2\right)\lambda g + \tfrac{1}{2}\left(L_{1i}^2 + L_{2i}^2\right)\lambda g$$
$$= -\tfrac{1}{2}\lambda g\left[\left(L_{1f}^2 + L_{2f}^2\right) - \left(L_{1i}^2 + L_{2i}^2\right)\right]$$

Because $L_1 + L_2 = 7.4$ m,
$L_{2i} - L_{1i} = 0.6$ m, and
$L_{2f} - L_{1f} = 7.2$ m, we obtain:

$L_{1i} = 3.4$ m, $L_{2i} = 4.0$ m,
$L_{1f} = 0.1$ m, and $L_{2f} = 7.3$ m.

Substitute numerical values and
evaluate ΔU:

$$\Delta U = -\tfrac{1}{2}(0.6\,\text{kg/m})(9.81\,\text{m/s}^2)$$
$$\quad \times \left[(0.1\,\text{m})^2 + (7.3\,\text{m})^2\right.$$
$$\quad \left. - (3.4\,\text{m})^2 - (4\,\text{m})^2\right]$$
$$= -75.75\,\text{J}$$

Express the kinetic energy of the
system when the difference in
height between the two ends of
the rope is 7.2 m:

$$K = \tfrac{1}{2}I_p\omega^2 + \tfrac{1}{2}Mv^2$$
$$= \tfrac{1}{2}\left(\tfrac{1}{2}M_pR^2\right)\omega^2 + \tfrac{1}{2}MR^2\omega^2$$
$$= \tfrac{1}{2}\left(\tfrac{1}{2}M_p + M\right)R^2\omega^2$$

Substitute numerical values and
simplify:

$$K = \tfrac{1}{2}\left[\tfrac{1}{2}(2.2\,\text{kg}) + 4.8\,\text{kg}\right]\left(\frac{1.2\,\text{m}}{2\pi}\right)^2\omega^2$$
$$= \left(0.1076\,\text{kg}\cdot\text{m}^2\right)\omega^2$$

Substitute in equation (1) and
solve for ω:

$$\left(0.1076\,\text{kg}\cdot\text{m}^2\right)\omega^2 - 75.75\,\text{J} = 0$$
and
$$\omega = \sqrt{\frac{75.75\,\text{J}}{0.1076\,\text{kg}\cdot\text{m}^2}} = \boxed{26.5\,\text{rad/s}}$$

(b) Noting that the moment arm of each portion of the rope is the same, express the total angular momentum of the system:

$$L = L_p + L_r = I_p \omega + M_r R^2 \omega$$
$$= \left(\tfrac{1}{2} M_p R^2 + M_r R^2\right)\omega \qquad (2)$$
$$= \left(\tfrac{1}{2} M_p + M_r\right)R^2 \omega$$

Letting θ be the angle through which the pulley has turned, express $U(\theta)$:

$$U(\theta) = -\tfrac{1}{2}\left[(L_{1i} - R\theta)^2 + (L_{2i} + R\theta)^2\right]\lambda g$$

Express ΔU and simplify to obtain:

$$\Delta U = U_f - U_i = U(\theta) - U(0)$$
$$= -\tfrac{1}{2}\left[(L_{1i} - R\theta)^2 + (L_{2i} + R\theta)^2\right]\lambda g$$
$$+ \tfrac{1}{2}\left(L_{1i}^2 + L_{2i}^2\right)\lambda g$$
$$= -R^2\theta^2\lambda g + (L_{1i} - L_{2i})R\theta\lambda g$$

Assume that, at $t = 0$, $L_{1i} \approx L_{2i}$. Then:

$$\Delta U \approx -R^2\theta^2\lambda g$$

Substitute for K and ΔU in equation (1) to obtain:

$$\left(0.1076\,\text{kg}\cdot\text{m}^2\right)\omega^2 - R^2\theta^2\lambda g = 0$$

Solve for ω.

$$\omega = \sqrt{\frac{R^2\theta^2\lambda g}{0.1076\,\text{kg}\cdot\text{m}^2}}$$

Substitute numerical values to obtain:

$$\omega = \sqrt{\frac{\left(\dfrac{1.2\,\text{m}}{2\pi}\right)^2 (0.6\,\text{kg/m})(9.81\,\text{m/s}^2)}{0.1076\,\text{kg}\cdot\text{m}^2}}\,\theta$$
$$= \left(1.41\,\text{s}^{-1}\right)\theta$$

Express ω as the rate of change of θ:

$$\frac{d\theta}{dt} = \left(1.41\,\text{s}^{-1}\right)\theta \Rightarrow \frac{d\theta}{\theta} = \left(1.41\,\text{s}^{-1}\right)dt$$

Integrate θ from 0 to θ to obtain:

$$\ln\theta = \left(1.41\,\text{s}^{-1}\right)t$$

Transform from logarithmic to exponential form to obtain:

$$\theta(t) = e^{\left(1.41\,\text{s}^{-1}\right)t}$$

Differentiate to express ω as a function of time:

$$\omega(t) = \frac{d\theta}{dt} = \left(1.41\,\text{s}^{-1}\right)e^{\left(1.41\,\text{s}^{-1}\right)t}$$

Substitute for ω in equation (2) to obtain:

$$L = \left(\tfrac{1}{2}M_p + M_r\right)R^2\left(1.41\,\mathrm{s}^{-1}\right)e^{\left(1.41\,\mathrm{s}^{-1}\right)t}$$

Substitute numerical values and evaluate L:

$$L = \left[\tfrac{1}{2}(2.2\,\mathrm{kg}) + (4.8\,\mathrm{kg})\right]\left(\frac{1.2\,\mathrm{m}}{2\pi}\right)^2\left[\left(1.41\,\mathrm{s}^{-1}\right)e^{\left(1.41\,\mathrm{s}^{-1}\right)t}\right] = \boxed{\left(0.303\,\mathrm{kg}\cdot\mathrm{m}^2/\mathrm{s}\right)e^{\left(1.41\,\mathrm{s}^{-1}\right)t}}$$

Chapter R
Relativity

Conceptual Problems

*2 • If event A occurs before event B in some frame, might it be possible for there to be a reference frame in which event B occurs before event A?

Determine the Concept Yes. If two events occur at the same time *and* place in one reference frame they occur at the same time *and* place in all reference frames. (Any pair of events that occur at the same time *and* at the same place in one reference frame are called a space-time coincidence.) Consider two clocks, C_1 and C_2, which are at rest relative to each other. In the rest frame of the two clocks the clocks are synchronized and are separated by distance L. Let event A be that clock C_1 reads 10 s and let event B be that Clock C_2 reads 10 s. Both events will occur at the same time in this rest frame. Now consider the clocks from the reference frame of a space ship traveling to the left at speed v relative to the clocks. In this frame the clocks are traveling to the right at speed v, with clock C_1 trailing clock C_2. The trailing clock is ahead of the leading clock by Lv/c^2, so in this frame event A occurs before event B. Next, consider the clocks from the reference frame of a space ship traveling to the right at speed v relative to the clocks. In this frame the clocks are traveling to the left at speed v, with clock C_2 trailing clock C_1. The trailing clock is ahead of the leading clock by Lv/c^2, so in this frame event B occurs before event A.

*6 • True or false:

(*a*) The speed of light is the same in all reference frames.
(*b*) Proper time is the shortest time interval between two events.
(*c*) Absolute motion can be determined by means of length contraction.
(*d*) The light-year is a unit of distance.
(*e*) For two events to form a spacetime coincidence they must occur at the same place.
(*f*) If two events are not simultaneous in one frame, they cannot be simultaneous in any other frame.

(*a*) True. This is Einstein's 2nd postulate.

(*b*) True. The time between events that happen at the *same place* in a reference frame is called the proper time and the time interval Δt measured in any other reference frame is always longer than the proper time.

(c) False. Absolute motion cannot be detected.

(d) True. A light-year is the distance light travels (in a vacuum) in one year.

(e) True. Two events that occur at the same time and at the same location are referred to as a spacetime coincidence.

(f) False. The fact that two events are not simultaneous in one frame tells us nothing about their simultaneity in any other frame.

Estimation and Approximation

*9 •• In 1975, an airplane carrying an atomic clock flew back and forth for 15 hours at an average speed of 140 m/s as part of a time-dilation experiment. The time on the clock was compared to the time on an atomic clock kept on the ground. How much time did the airborne clock "lose" with respect to the clock on the ground?

Picture the Problem We can use the time dilation equation to relate the elapsed time in frame of reference of airborne clock to the elapsed time in the frame of reference of the atomic clock kept on the ground.

Use the time dilation equation to relate the elapsed time Δt according to the clock on the ground to the elapsed time Δt_0 according to the airborne atomic clock:

$$\Delta t = \gamma \Delta t_0 \qquad (1)$$

Because $v \ll c$, we can use the approximation $\dfrac{1}{\sqrt{1-x}} \approx 1 + \dfrac{1}{2}x$ to obtain:

$$\gamma \approx 1 + \frac{1}{2}\left(\frac{v}{c}\right)^2$$

Substitute in equation (1):

$$\Delta t = \left[1 + \frac{1}{2}\left(\frac{v}{c}\right)^2\right]\Delta t_0$$

$$= \Delta t_0 + \frac{1}{2}\left(\frac{v}{c}\right)^2 \Delta t_0$$

$\qquad (2)$

where the second term represents the additional time measured by the clock on the ground.

Evaluate the proper elapsed time according to the clock on the airplane:

$$\Delta t_0 = (15\,h)\left(3600\,\frac{s}{h}\right) = 5.40\times10^4\,s$$

Substitute numerical values and evaluate the second term in equation (2):

$$\Delta t' = \frac{1}{2}\left(\frac{140\,m/s}{2.998\times10^8\,m/s}\right)^2(5.40\times10^4\,s)$$

$$= 5.89\times10^{-9}\,s \approx \boxed{6.00\,ns}$$

Length Contraction and Time Dilation

*12 •   The proper mean lifetime of a subnuclear particle called a muon is 2 μs. Muons in a beam are traveling at $0.999c$ relative to a laboratory. (a) What is their mean lifetime as measured in the laboratory? (b) How far do they travel, on average, before they decay?

Picture the Problem We can express the mean lifetimes of the muons in the laboratory in terms of their proper lifetimes using $\Delta t = \gamma \Delta t_p$. The average distance the muons will travel before they decay is related to their speed and mean lifetime in the laboratory frame according to $\Delta x = v\Delta t$.

(a) Use the time-dilation equation to relate the mean lifetime of a muon in the laboratory Δt to their proper mean lifetime Δt_0:

$$\Delta t = \gamma \Delta t_0 = \frac{\Delta t_0}{\sqrt{1-\left(\dfrac{v}{c}\right)^2}}$$

Substitute numerical values and evaluate Δt:

$$\Delta t = \frac{2\times10^{-6}\,s}{\sqrt{1-\left(\dfrac{0.999c}{c}\right)^2}} = \boxed{44.7\,\mu s}$$

(b) Express the average distance muons travel before they decay in terms of their speed and mean lifetime in the laboratory frame of reference:

$$\Delta x = v\Delta t$$

Substitute numerical values and evaluate Δx:

$$\Delta x = (0.999c)(44.7\,\mu s) = \boxed{13.4\,km}$$

*15 •   A spaceship travels from earth to a star 95 light-years away at a speed of 2.2×10^8 m/s. How long does it take to get there (*a*) as measured on the earth and (*b*) as measured by a passenger on the spaceship?

Picture the Problem We can use $\Delta x = v\Delta t$ to find the time for the trip as measured on earth and $\Delta t_0 = \Delta t / \gamma$ to find the time measured by a passenger on the spaceship.

(*a*) Express the elapsed time, as measured on earth, in terms of the distance traveled and the speed of the spaceship:

$$\Delta t = \frac{\Delta x}{v}$$

Substitute numerical values and evaluate Δt:

$$\Delta t = \frac{95\,c \cdot y}{2.2 \times 10^8 \text{ m/s}} \times \frac{9.461 \times 10^{15} \text{ m}}{c \cdot y}$$

$$= 4.09 \times 10^9 \text{ s} \times \frac{1\,y}{31.56\,\text{Ms}}$$

$$= \boxed{129\,y}$$

(*b*) A passenger on the spaceship will measure the proper time:

$$\Delta t_0 = \frac{\Delta t}{\gamma} = \Delta t \sqrt{1 - \left(\frac{v}{c}\right)^2}$$

Substitute numerical values and evaluate the proper time:

$$\Delta t_p = (129\,y)\sqrt{1 - \frac{(2.2 \times 10^8 \text{ m/s})^2}{(2.998 \times 10^8 \text{ m/s})^2}}$$

$$= \boxed{87.6\,y}$$

*20 ••   Two spaceships pass each other traveling in opposite directions. A passenger in ship A, who happens to know that her ship is 100 m long, notes that ship B is moving with a speed of $0.92c$ relative to A and that the length of B is 36 m. What are the lengths of the two spaceships as measured by a passenger in ship B?

Picture the Problem We can use the relationship between the measured length L of the spaceships and their proper lengths L_0 to find the lengths of the two spaceships as measured by a passenger in ship B.

Relate the measured length L_A of ship A to its proper length:

$$L_A = \frac{L_{0,A}}{\gamma} = L_{0,A}\sqrt{1 - \left(\frac{v}{c}\right)^2}$$

Substitute numerical values and evaluate L_A:

$$L_A = (100\,\text{m})\sqrt{1 - \frac{(0.92c)^2}{c^2}}$$

$$= \boxed{39.2\,\text{m}}$$

Relate the proper length $L_{0,B}$ of ship B to its measured length L_B:

$$L_{0,B} = \gamma L_B = \frac{L_B}{\sqrt{1 - \left(\dfrac{v}{c}\right)^2}}$$

Substitute numerical values and evaluate $L_{0,B}$:

$$L_{0,B} = \frac{36\,\text{m}}{\sqrt{1 - \dfrac{(0.92c)^2}{c^2}}} = \boxed{91.9\,\text{m}}$$

The Relativity of Simultaneity

Figure R-12

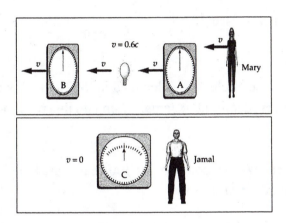

*22 • According to Jamal, (*a*) what is the distance between the flashbulb and clock A, (*b*) how far does the flash travel to reach clock A, and (*c*) how far does clock A travel while the flash is traveling from the flashbulb to it?

Picture the Problem The proper distance L_0 between the flashbulb and clock A is 50 c·min and is related to the distance L measured by Jamal according to $L = L_0/\gamma$. Because clock A and the flashbulb are at rest relative to each other, we can find the distance between them using the same relationship with $L_0 = 50$ c·min.

(*a*) Express L in terms of L_0:

$$L = \frac{L_0}{\gamma} = L_0\sqrt{1 - \left(\frac{v}{c}\right)^2}$$

Substitute numerical values and evaluate L:

$$L = (50c \cdot \text{min})\sqrt{1 - \left(\frac{0.6c}{c}\right)^2}$$

$$= \boxed{40.0c \cdot \text{min}}$$

(b) Jamal sees the flash traveling away from him at $1.0c$ and clock A approaching at $0.6c$. So:

$$\Delta t = \frac{40.0c \cdot \text{min}}{1.6c} = 25.0\,\text{min}$$

and so

$$d = \boxed{25.0c \cdot \text{min}}$$

(c) Express the distance d clock A travels while the flash is traveling to it in terms of the speed of clock A and the time Δt it takes the flash to reach it:

$$d = v\Delta t = v\frac{d}{c}$$

Substitute numerical values and evaluate L:

$$L = (0.6c)\left(\frac{25c \cdot \text{min}}{c}\right) = \boxed{15.0c \cdot \text{min}}$$

*26 ••   The advance of clock A calculated in Problem 24 is the amount that clock A leads clock B according to Jamal. Compare this result with vL_0/c^2, where $v = 0.6c$.

Picture the Problem We can compare the time calculated in Problem 24 with L_0v/c^2 by evaluating their ratio.

Express the ratio r of L_0v/c^2 to $\Delta t'$:

$$r = \frac{\dfrac{L_0 v}{c^2}}{\Delta t'} = \frac{\dfrac{L_0 v}{c^2}}{\dfrac{\Delta t}{\gamma}} = \frac{\gamma L_0 v}{c^2 \Delta t}$$

Substitute for γ to obtain:

$$r = \frac{L_0 v}{c^2 \Delta t \sqrt{1 - \left(\dfrac{v}{c}\right)^2}}$$

Substitute numerical values and evaluate r:

$$r = \frac{(100c \cdot \text{min})(0.6c)}{c^2(75\,\text{min})\sqrt{1 - \left(\dfrac{0.6c}{c}\right)^2}} = \boxed{1}$$

*30 ••   Al and Bert are twins. Al travels at $0.6c$ to Alpha Centauri (which is 4 $c \cdot y$ from the earth as measured in the reference frame of the earth) and returns immediately. Each twin sends the other a light signal every 0.01 y as measured in his own reference frame. (*a*) At what rate does Bert receive signals as Al is moving away from him? (*b*) How many signals does Bert receive at this rate? (*c*) How many total signals are received by Bert before Al has returned? (*d*) At what rate does Al receive signals as Bert is receding from him? (*e*) How many signals does Al receive at this rate? (*f*) How many total signals are received by Al? (*g*) Which twin is younger at the end of the trip and by how many years?

Picture the Problem We can use the relativistic Doppler shift equations to find the rates at which Bert and Al receive signals from each other. In parts (*b*) and (*c*) we can use $N = f_0 \Delta t_{Al}$, $\Delta t_{Al} = \Delta t_{Bert}/\gamma$, together with the definition of γ, to find the number of signals received by Bert. In parts (*d*) and (*e*) we can proceed similarly to find the number of signals received by Al and the total number he receives before he returns to Bert. Finally, we can use the number of signals each received to find the difference in their ages resulting from Al's trip.

(*a*) Apply the Doppler shift equation, for a receding source, to obtain:	$$f_{Bert} = f_0 \sqrt{\frac{1 - v/c}{1 + v/c}}$$
Substitute numerical values and evaluate f_{Bert}:	$$f_{Bert} = \left(100 \text{ y}^{-1}\right)\sqrt{\frac{1 - 0.6}{1 + 0.6}} = \boxed{50.0 \text{ y}^{-1}}$$
(*b*) Express the number of signals N received by Bert in terms of the number of signals sent by Al:	$$N = f_0 \Delta t_{Al}$$
Express the elapsed time in Al's frame of reference in terms of the elapsed time in Bert's frame (the proper elapsed time):	$$\Delta t_{Al} = \frac{\Delta t_{Bert}}{\gamma}$$
Express and evaluate γ:	$$\gamma = \frac{1}{\sqrt{1 - \dfrac{v^2}{c^2}}} = \frac{1}{\sqrt{1 - \dfrac{(0.6c)^2}{c^2}}} = 1.25$$
Substitute to obtain:	$$N = f_0 \frac{\Delta t_{Bert}}{\gamma}$$

Find Δt_{Bert}:

$$\Delta t_{\text{Bert}} = \frac{4c \cdot y}{0.6c} = 6.67\,y$$

Evaluate Δt_{Al}:

$$\Delta t_{\text{Al}} = \frac{6.67\,y}{1.25} = 5.34\,y$$

Substitute numerical values and evaluate N:

$$N = \left(100\,y^{-1}\right)\frac{6.67\,y}{1.25} = \boxed{534}$$

(c) Express the number of signals N received by Bert in terms of the number of signals sent by Al before he returns:

$$N = f_0 \Delta T_{\text{Al}}$$

Find the time in Al's frame for the round trip:

$$\Delta T_{\text{Al}} = 2\Delta t_{\text{Al}} = 2(5.34\,y) = 10.68\,y$$

Substitute numerical values and evaluate N:

$$N = \left(100\,y^{-1}\right)(10.68\,y) = \boxed{1068}$$

(d) Proceed as in (a) to obtain:

$$f_{\text{Al}} = \left(100\,y^{-1}\right)\sqrt{\frac{1-0.6}{1+0.6}} = \boxed{50.0\,y^{-1}}$$

(e) Express the number of signals N received by Al:

$$N = f_{\text{Al}} \Delta t_{\text{Al}}$$

Substitute numerical values and evaluate N:

$$N = \left(50\,y^{-1}\right)\frac{6.67y}{1.25} = \boxed{267}$$

(f) Express the total number of signals received by Al:

$$N_{\text{tot}} = N_{\text{outbound}} + N_{\text{return}}$$
$$= 267 + N_{\text{return}}$$

Using the relativistic Doppler effect equation for approach, express the rate at which signals are received by Al on the return trip:

$$f_{\text{Al, return}} = f_0 \sqrt{\frac{1+\dfrac{v}{c}}{1-\dfrac{v}{c}}}$$

Substitute numerical values and evaluate $f_{\text{Al, return}}$:

$$f_{\text{Al, return}} = \left(100\,y^{-1}\right)\sqrt{\frac{1+0.6}{1-0.6}} = 200\,y^{-1}$$

Find the number, N_{return}, of signals received by Al on his return:	$N_{return} = f_{Al,\,return}\,\Delta t_{Al}$ $= (200\,y^{-1})(5.34\,y) = 1068$
Substitute to obtain:	$N_{tot} = 267 + 1068 = \boxed{1335}$

(g) Express their age difference:	$\Delta t = \Delta t_{Bert} - \Delta t_{Al}$
Express the elapsed time in Al's frame in terms of the number of signals he sent:	$\Delta t_{Al} = \dfrac{1068}{100\,y^{-1}} = 10.68\,y$
Express the elapsed time in Bert's frame in terms of the number of signals he sent:	$\Delta t_{Bert} = \dfrac{1335}{100\,y^{-1}} = 13.35\,y$
Substitute to obtain:	$\Delta t = 13.35\,y - 10.68\,y = 2.67\,y$ and $\boxed{\text{Al is 2.67 y younger than Bert.}}$

Relativistic Energy and Momentum

*33 • How much energy would be required to accelerate a particle of mass m_0 from rest to (a) $0.5c$, (b) $0.9c$, and (c) $0.99c$? Express your answers as multiples of the rest energy.

Picture the Problem We can use $K = (\gamma - 1)m_0 c^2 = (\gamma - 1)E_0$ and the definition of γ to find the kinetic energy required to accelerate this particle from rest to the given speeds.

Express the relativistic kinetic energy of the particle:	$K = (\gamma - 1)m_0 c^2 = (\gamma - 1)E_0$
(a) Express and evaluate γ for $u = 0.5c$:	$\gamma = \dfrac{1}{\sqrt{1 - \dfrac{v^2}{c^2}}} = \dfrac{1}{\sqrt{1 - \dfrac{(0.5c)^2}{c^2}}} = 1.155$
Substitute numerical values and evaluate $K(0.5c)$:	$K(0.5c) = (1.155 - 1)E_0 = \boxed{0.155E_0}$

(b) Evaluate γ for $v = 0.9c$:

$$\gamma = \frac{1}{\sqrt{1 - \frac{(0.9c)^2}{c^2}}} = 2.294$$

Substitute numerical values and evaluate $K(0.9c)$:

$$K(0.9c) = (2.294 - 1)E_0 = \boxed{1.29E_0}$$

(c) Evaluate γ for $v = 0.99c$:

$$\gamma = \frac{1}{\sqrt{1 - \frac{(0.99c)^2}{c^2}}} = 7.089$$

Substitute numerical values and evaluate $K(0.99c)$:

$$K(0.99c) = (7.089 - 1)E_0 = \boxed{6.09E_0}$$

*36 ••   Using a spreadsheet program or graphing calculator, make a graph of the kinetic energy of a particle with mass $m = 100$ MeV/c^2 for speeds between 0 and c. On the same graph, plot $\frac{1}{2}mv^2$ by way of comparison. Using the graph, estimate at about what velocity is this formula no longer a good approximation to the kinetic energy. As a suggestion, plot the energy in units of MeV and the velocity in the dimensionless form v/c.

Picture the Problem We can create a spreadsheet program to plot both the classical and relativistic kinetic energy of the particle.

The relativistic kinetic energy of the particle is given by:

$$K_{\text{relativistic}} = m_0 c^2 \left(\frac{1}{\sqrt{1 - (v/c)^2}} - 1 \right)$$

A spreadsheet program to graph $K_{\text{relativistic}}$ and $K_{\text{classical}}$ is shown below. The formulas used to calculate the quantities in the columns are as follows:

Cell	Formula/Content	Algebraic Form
A8	A7+0.05	$v/c + 0.05$
B7	0.5*\$B\$3*A7^2	$\frac{1}{2}mv^2$
C7	\$B\$3*(1/((1−A7^2)^0.5)−1)	$m_0 c^2 \left(\dfrac{1}{\sqrt{1 - (v/c)^2}} - 1 \right)$

	A	B	C
1			
2			
3	m_0c^2=	100	MeV

	v/c	.5*mv^2	gamma*mc^2
4			
5	v/c	.5*mv^2	gamma*mc^2
6		Kclassical	Krelativistic
7	0.00	0.00	0.00
8	0.05	0.13	0.13
9	0.10	0.50	0.50
10	0.15	1.13	1.14
23	0.80	32.00	66.67
24	0.85	36.13	89.83
25	0.90	40.50	129.42
26	0.95	45.13	220.26

The solid curve is the graph of the relativistic kinetic energy.

> The relativistic formula, represented by the dashed curve, begins to deviate from the non - relativistic curve around $v/c \approx 0.4$

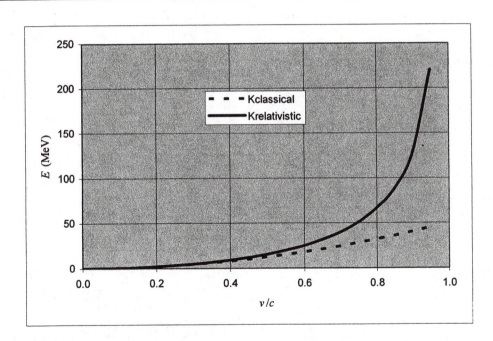

*40 •• The rest energy of a proton is about 938 MeV. If its kinetic energy is also 938 MeV, find (*a*) its momentum and (*b*) its speed.

Picture the Problem We can solve the relation for total energy, momentum, and rest energy of the proton for its momentum and evaluate this expression for $K = E_0$. In part (*b*) we can start with the expression for the relativistic energy of the proton, solve for u, and evaluate the resulting expression for $K = E_0$.

(*a*) Relate the total energy, momentum, and rest energy of a

$$E^2 = p^2c^2 + \left(m_0c^2\right)^2$$

proton:

Solve for p to obtain:

$$p = \frac{\sqrt{E^2 - E_0^2}}{c} = \frac{\sqrt{(E_0 + K)^2 - E_0^2}}{c}$$

$$= \frac{\sqrt{K^2 + 2E_0 K}}{c}$$

If $K = E_0$:

$$p_{K=E_0} = \frac{\sqrt{E_0^2 + 2E_0^2}}{c} = \frac{\sqrt{3}E_0}{c}$$

Substitute numerical values and evaluate p:

$$p_{K=E_0} = \frac{\sqrt{3}(938\,\text{MeV})}{c} = \boxed{1.62\,\text{GeV}/c}$$

(b) Express the relativistic energy of a particle:

$$E = \frac{m_0 c^2}{\sqrt{1 - \dfrac{v^2}{c^2}}}$$

Solve for $1 - v^2/c^2$:

$$1 - \frac{v^2}{c^2} = \frac{(m_0 c^2)^2}{E^2}$$

Solve for v to obtain:

$$v = c\sqrt{1 - \frac{E_0^2}{E^2}}$$

Because $E = E_0 + K$:

$$v = c\sqrt{1 - \frac{E_0^2}{(E_0 + K)^2}}$$

For $K = E_0$:

$$v_{K=E_0} = c\sqrt{1 - \frac{E_0^2}{(E_0 + E_0)^2}} = c\sqrt{1 - \frac{1}{4}}$$

$$= \sqrt{0.75}c = \boxed{0.866c}$$

General Problems

*45 ••   The rest mass of the neutrino (the "ghost particle" of physics) is known to have a small but as yet unmeasured value. Both high-energy and light neutrinos are produced by a supernova explosion, so it is possible to estimate the neutrino's mass by timing the relative arrival of light versus neutrinos from a supernova. If a supernova explodes 100,000 light-years from the earth, calculate the rest mass necessary for a neutrino with total energy of 100 MeV to trail the light by (a) 1 min, (b) 1 s, and (c) 0.01 s.

Picture the Problem The neutrino will be traveling at very nearly the speed of light. Let its speed be $c - \delta$. Then we can express the delay time τ between the neutrino arrival time and the photon arrival time in terms of the distance D to the supernova and the speeds of light and of the neutrinos.

Express the total energy of a neutrino:

$$E = \gamma mc^2$$

Solve for mc^2:

$$mc^2 = \frac{E}{\gamma} \qquad (1)$$

Express γ for a neutrino:

$$\gamma = \frac{1}{\sqrt{1 - ((c - \delta)/c)^2}}$$

Simplify to obtain:

$$\gamma = \frac{1}{\sqrt{1 - \left(\dfrac{c - \delta}{c}\right)^2}} = \frac{1}{\sqrt{1 - \left(1 - \dfrac{\delta}{c}\right)^2}}$$

$$= \frac{1}{\sqrt{1 - \left(1 - 2\dfrac{\delta}{c} + \dfrac{\delta^2}{c^2}\right)}} \approx \frac{1}{\sqrt{2\dfrac{\delta}{c}}} \qquad (2)$$

$$= \sqrt{\frac{c}{2\delta}}$$

Express the delay time between the arrival times of the neutrinos and the photons:

$$\tau = \frac{D}{c - \delta} - \frac{D}{c} = D\left(\frac{1}{c - \delta} - \frac{1}{c}\right)$$

$$= \frac{D}{c}\left(\frac{1}{1 - \delta/c} - 1\right)$$

Use the approximation $1/(1 - x) \approx 1 + x$ for $x \ll 1$ to obtain:

$$\tau \approx \frac{D}{c}\left(1 + \frac{\delta}{c} - 1\right) = D\frac{\delta}{c^2}$$

Solve for δ/c:

$$\frac{\delta}{c} = \frac{c\tau}{D}$$

Substitute in equation (2) to obtain:

$$\gamma \approx \sqrt{\frac{D}{2c\tau}}$$

Substitute in equation (1) to obtain:

$$mc^2 = E\sqrt{\frac{2c\tau}{D}}$$

Evaluate mc^2 for $\tau = 1$ min:

$$mc^2 = (100\,\text{MeV})\sqrt{\frac{2(1c\cdot\text{min})}{(10^5\,c\cdot\text{y})(365.24\,\text{d/y})(24\,\text{h/d})(60\,\text{min/h})}} = \boxed{617\,\text{eV}}$$

Evaluate mc^2 for $\tau = 1$ s:

$$mc^2 = (100\,\text{MeV})\sqrt{\frac{2(1c\cdot\text{s})}{(10^5\,c\cdot\text{y})(365.24\,\text{d/y})(24\,\text{h/d})(3600\,\text{s/h})}} = \boxed{79.6\,\text{eV}}$$

Evaluate mc^2 for $\tau = 0.01$ s:

$$mc^2 = (100\,\text{MeV})\sqrt{\frac{2(0.01c\cdot\text{s})}{(10^5\,c\cdot\text{y})(365.24\,\text{d/y})(24\,\text{h/d})(3600\,\text{s/h})}} = \boxed{7.96\,\text{eV}}$$

*47 ••• Keisha and Ernie are trying to fit a 15-foot-long ladder into a 10-foot-long shed with doors at each end. Recalling her physics lessons, Keisha suggests to Ernie that they open the front door to the shed and have Ernie run toward it with the ladder at a speed such that the length contraction of the ladder shortens it enough so that it fits in the shed. As soon as the back end of the ladder passes through the door, Keisha will slam it shut. (*a*) What is the minimum speed at which Ernie must run to fit the ladder into the shed? Express it as a fraction of the speed of light. (*b*) As Ernie runs toward the shed at a speed of 0.866c, he realizes that in the reference frame of the ladder and himself, it is the *shed* which is shorter, not the ladder. How long is the shed in the rest frame of the ladder? (*c*) In the reference frame of the ladder is there any instant that both ends of the ladder are simultaneously inside the shed? Examine this from the standpoint of the point of view of relativistic simultaneity.

Picture the Problem Let the letter "L" denote the ladder and the letter "S" the shed. We can apply the length contraction equation to the determination of the minimum speed at which Ernie must run to fit the ladder into the shed as well as the length of the shed in rest frame of the ladder.

(*a*) Express the length L_L of the shed in Ernie's frame of reference in terms of its proper length $L_{L,0}$:

$$L_L = L_{L,0}\sqrt{1-\left(\frac{v}{c}\right)^2}$$

Solve for v to obtain:

$$v = c\sqrt{1 - \frac{L_L^2}{L_{L,0}^2}}$$

Substitute numerical values and evaluate v:

$$v = c\sqrt{1 - \left(\frac{10\,\text{ft}}{15\,\text{ft}}\right)^2} = \boxed{0.745c}$$

(b) Express the length L_S of the shed in the rest frame of the ladder in terms of its proper length $L_{L,0}$:

$$L_S = L_{S,0}\sqrt{1 - \frac{v^2}{c^2}}$$

Substitute numerical values and evaluate L_S:

$$L_S = (10\,\text{ft})\sqrt{1 - \left(\frac{0.866c}{c}\right)^2} = \boxed{5.00\,\text{ft}}$$

(c)
> No. In Keisha's rest frame, the back end of the ladder will clear the door before the front end hits the wall of the shed, while in Ernie's rest frame, the front end will hit the wall of the shed while the back end has yet to clear the door.

Let the ladder be traveling from left to right. To "explain" the simultaneity issue we first describe the situation in the reference frame of the shed. In this frame the ladder has length $L_L = 7.5$ m and the shed has length $L_{S,0} = 10$ m. We (mentally) put a clock at each end of the shed. Let both clocks read zero at the instant the left end of the ladder enters the shed. At this instant the right end of the ladder is a distance $L_{S,0} - L_L = 2.5$ m from the right end of the shed. At the instant the right end of the ladder exits the shed both clocks read Δt, where $\Delta t = (L_{S,0} - L_L)/v = 9.62\,\text{ns}$. There are two spacetime coincidences to consider: the left end of the train enters the shed and the clock at the left end of the shed reads zero, and the right end of the ladder exits the shed and the clock on the right end of the shed reads $(L_{S,0} - L_L)/v = 9.62\,\text{ns}$.

In the reference frame of the ladder the two clocks are moving to the left at speed $v = 0.866c$. In this frame the clock on the right (the trailing clock) is ahead of the clock on the left by $vL_{S,0}/c^2 = 28.9$ ns, so when the clock on the right reads 9.62 ns the one on the left reads -19.2 ns. This means the left end of the ladder is yet to enter the shed when the right end of the ladder is exiting the shed. This is consistent with the assertion that in the rest frame of the ladder, the ladder is longer than the shed, so the entire ladder is never entirely inside the shed.

Chapter 11
Gravity

Conceptual Problems

*1 • True or false: (*a*) Kepler's law of equal areas implies that gravity varies inversely with the square of the distance. (*b*) The planet closest to the sun, on the average, has the shortest orbital period.

(*a*) False. Kepler's law of equal areas is a consequence of the fact that the gravitational force acts along the line joining two bodies but is independent of the manner in which the force varies with distance.

(*b*) True. The periods of the planets vary with the three-halves power of their distances from the sun. So the shorter the distance from the sun, the shorter the period of the planet's motion.

*8 • If K is the kinetic energy of Mercury in its orbit around the sun, and U is the potential energy of the Mercury-sun system, what is the relationship between K and U?

Determine the Concept Let m represent the mass of Mercury, M_S the mass of the sun, v the orbital speed of Mercury, and R the mean orbital radius of Mercury. We can use Newton's 2nd law of motion to relate the gravitational force acting on the Mercury to its orbital speed.

Use Newton's 2nd law to relate the gravitational force acting on Mercury to its orbital speed:

$$F_{net} = \frac{GM_S m}{R^2} = m\frac{v^2}{R}$$

Simplify to obtain:

$$\tfrac{1}{2}mv^2 = \tfrac{1}{2}\frac{GM_S m}{R} = -\tfrac{1}{2}\left(-\frac{GM_S m}{R}\right)$$

$$= -\tfrac{1}{2}U$$

or $\boxed{K = -\tfrac{1}{2}U}$

Estimation and Approximation

*12 ••• One of the great advances in astronomy made possible by the Hubble Space Telescope has been in the detection of planets outside the solar system. Between 1996, 76 planets have been detected orbiting stars other than our sun. While the planets themselves cannot be seen, the Hubble can detect the small periodic motion of the star as star and planet orbit around their common center of mass. (This is measured using the *Doppler effect*, which will be discussed later in the book.) Both the period of the motion and the velocity of the star over the course of time can be determined observationally, and the mass of the star can be found from its observed luminance and the theory of stellar structure. Iota Draconis is the eighth brightest star in the constellation Draco: Hubble observations show that there is a planet orbiting the star that has an orbital period of 1.50 years. The mass of Iota Draconis is 1.05 M_{sun}. (*a*) What is the size (in AU) of the semimajor axis of this planet's orbit? (*b*) The maximum measured radial velocity for the star is 296 m/s. Using the conservation of momentum, find the mass of the planet, assuming that the orbit is perfectly circular (which it is not). Express the mass as a multiple of the mass of Jupiter.

Picture the Problem We can use Kepler's 3rd law to find the size of the semi-major axis of the planet's orbit and the conservation of momentum to find its mass.

(*a*) Using Kepler's 3rd law, relate the period of this planet T to the length r of its semi-major axis:

$$T^2 = \frac{4\pi^2}{GM_{\text{Iota Draconis}}} r^3$$

$$= \frac{\dfrac{4\pi^2}{M_s}}{G\dfrac{M_{\text{Iota Draconis}}}{M_s}} r^3$$

$$= \frac{\dfrac{4\pi^2}{GM_s}}{\dfrac{M_{\text{Iota Draconis}}}{M_s}} r^3$$

If we measure time in years, distances in AU, and masses in terms of the mass of the sun:

$$\frac{4\pi^2}{MG_s} = 1 \text{ and } T^2 = \frac{1}{\dfrac{M_{\text{Iota Draconis}}}{M_s}} r^3$$

Solve for r to obtain:

$$r = \sqrt[3]{\frac{M_{\text{Iota Draconis}}}{M_s} T^2}$$

Substitute numerical values and evaluate r:

$$r = \sqrt[3]{\left(\frac{1.05 M_s}{M_s}\right)(1.5\,\text{y})^2} = \boxed{1.33\,\text{AU}}$$

(*b*) Apply conservation of momentum to the planet (mass m and speed v) and the star (mass $M_{\text{Iota Draconis}}$ and speed V) to obtain:

$$mv = M_{\text{Iota Draconis}} V$$

Solve for m to obtain:

$$m = M_{\text{Iota Draconis}} \frac{V}{v}$$

Use its definition to find the speed of the orbiting planet:

$$v = \frac{\Delta d}{\Delta t} = \frac{2\pi r}{T}$$

$$= \frac{2\pi \left(1.33\,\text{AU} \times \dfrac{1.5 \times 10^{11}\,\text{m}}{\text{AU}}\right)}{1.50\,\text{y} \cdot \dfrac{365.25\,\text{d}}{\text{y}} \times \dfrac{24\,\text{h}}{\text{d}} \times \dfrac{3600\,\text{s}}{\text{h}}}$$

$$= 2.65 \times 10^4\,\text{m/s}$$

Substitute numerical values and evaluate v:

$$m = M_{\text{Iota Draconis}} \left(\frac{296\,\text{m/s}}{2.65 \times 10^4\,\text{m/s}}\right)$$

$$= 0.0112 M_{\text{Iota Draconis}}$$
$$= 0.0112\left(1.05 M_{\text{sun}}\right)$$
$$= 0.0112(1.05)\left(1.99 \times 10^{30}\,\text{kg}\right)$$
$$= 2.34 \times 10^{28}\,\text{kg}$$

Express m in terms of the mass M_J of Jupiter:

$$\frac{m}{M_J} = \frac{2.34 \times 10^{28}\,\text{kg}}{1.90 \times 10^{27}\,\text{kg}} = 12.3$$

or

$$m = \boxed{12.3 M_J}$$

Remarks: A more sophisticated analysis, using the eccentricity of the orbit, leads to a lower bound of 8.7 Jovian masses. (Only a lower bound can be established, as the plane of the orbit is not known.)

Kepler's Laws

*19 •• Kepler determined distances in the solar system from his data. For example, he found the relative distance from the sun to Venus (as compared to the distance from the sun to earth) as follows. Because Venus's orbit is closer to the sun than is earth's, Venus is a morning or evening star – its position in the sky is never very far from the sun (see Figure 11-22). If we consider the orbit of Venus as a perfect circle, then consider the relative orientation of Venus, earth, and the sun at maximum extension – when Venus is farthest from the sun in the sky. (*a*) Under this condition, show that angle *b* in Figure 11-22 is 90°. (*b*) If the maximum elongation angle between Venus and the sun is 47°, what is the distance between Venus and the sun in AU?

Figure 11-22 Problem 19

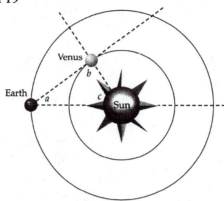

Picture the Problem We can use a property of lines tangent to a circle and radii drawn to the point of contact to show that $b = 90°$. Once we've established that b is a right angle we can use the definition of the sine function to relate the distance from the sun to Venus to the distance from the sun to the earth.

(a) The line from earth to Venus' orbit is tangent to the orbit of Venus at the point of maximum extension. Venus will appear closer to the sun in earth's sky when it passes the line drawn from earth and tangent to its orbit. Hence:	$b = \boxed{90°}$

(b) Using trigonometry, relate the distance from the sun to Venus d_{SV} to the angle a:	$\sin a = \dfrac{d_{SV}}{d_{SE}}$
Solve for d_{SV}:	$d_{SV} = d_{SE} \sin a$
Substitute numerical values and evaluate d_{SV}:	$d_{SV} = (1\,\text{AU})\sin 47° = \boxed{0.731\,\text{AU}}$

Remarks: The correct distance from the sun to Venus is closer to 0.723 AU.

Newton's Law of Gravity

*21 ••   Jupiter's satellite Europa orbits Jupiter with a period of 3.55 d, at an average distance of 6.71×10^8 m. (*a*) Assuming that the orbit is circular, determine the mass of Jupiter from the data given. (*b*) Another satellite of Jupiter, Callisto, orbits at an average distance of 18.8×10^8 m with an orbital period of 16.7 d. Determine the acceleration of Callisto and Europa from the data given, and show that this data is consistent with an inverse square law for gravity (NOTE: Do NOT use the value of G anywhere in part (b)).

Picture the Problem We can use Kepler's 3rd law to find the mass of Jupiter in part (a). In part (b) we can express the centripetal accelerations of Europa and Callisto and compare their ratio to the square of the ratio of their distances from the center of Jupiter to show that the given data is consistent with an inverse square law for gravity.

(a) Assuming a circular orbit, apply Kepler's 3rd law to the motion of Europa to obtain:

$$T_E^2 = \frac{4\pi^2}{GM_J} R_E^3$$

Solve for the mass of Jupiter:

$$M_J = \frac{4\pi^2}{GT_E^2} R_E^3$$

Substitute numerical values and evaluate M_J:

$$M_J = \frac{4\pi^2}{\left(6.673\times10^{-11}\,\text{N}\cdot\text{m}^2/\text{kg}^2\right)}$$

$$\times \frac{\left(6.71\times10^8\,\text{m}\right)^3}{\left(3.55\,\text{d}\times\dfrac{24\,\text{h}}{\text{d}}\times\dfrac{3600\,\text{s}}{\text{h}}\right)^2}$$

$$= \boxed{1.90\times10^{27}\,\text{kg}}\,,\text{a result in}$$

excellent agreement with the accepted value of 1.902×10^{27} kg.

(b) Express the centripetal acceleration of both of the moons to obtain:

$$\frac{v^2}{R} = \frac{\left(\dfrac{2\pi R}{T}\right)^2}{R} = \frac{4\pi^2 R}{T^2}$$

where R and T are the radii and periods of their motion.

Using this result, express the centripetal accelerations of Europa and Callisto:

$$a_E = \frac{4\pi^2 R_E}{T_E^2} \text{ and } a_C = \frac{4\pi^2 R_C}{T_C^2}$$

Substitute numerical values and evaluate a_E:

$$a_E = \frac{4\pi^2\left(6.71\times10^8\,\text{m}\right)}{\left[\left(3.55\,\text{d}\right)\left(24\,\text{h/d}\right)\left(3600\,\text{s/h}\right)\right]^2}$$

$$= \boxed{0.282\,\text{m/s}^2}$$

Substitute numerical values and evaluate a_C:

$$a_C = \frac{4\pi^2 \left(18.8 \times 10^8 \text{ m}\right)}{\left[(16.7\,\text{d})(24\,\text{h/d})(3600\,\text{s/h})\right]^2}$$

$$= \boxed{0.0356 \text{ m/s}^2}$$

Evaluate the ratio of these accelerations:

$$\frac{a_E}{a_C} = \frac{0.282 \text{ m/s}^2}{0.0356 \text{ m/s}^2} = 7.91$$

Evaluate the square of the ratio of the distance of Callisto divided by the distance of Europa to obtain:

$$\left(\frac{R_C}{R_E}\right)^2 = \left(\frac{18.8 \times 10^8 \text{ m}}{6.71 \times 10^8 \text{ m}}\right)^2 = 7.85$$

> The close agreement (within 1%) of our last two calculations strongly supports the conclusion that the gravitational force varies inversely with the square of the distance.

*22 • Some people think that shuttle astronauts are "weightless" because they are "beyond the pull of earth's gravity." In fact, this is completely untrue. (a) What is the magnitude of the acceleration of gravity for shuttle astronauts? A shuttle orbit is about 400 km above the ground. (b) Given the answer in part (a), why are shuttle astronauts "weightless"?

Determine the Concept The weight of anything, including astronauts, is the reading of a scale from which the object is suspended or on which it rests. If the scale reads zero, then we say the object is "weightless." The pull of the earth's gravity, on the other hand, depends on the local value of the acceleration of gravity and we can use Newton's law of gravity to find this acceleration at the elevation of the shuttle.

(a) Apply Newton's law of gravitation to an astronaut of mass m in a shuttle at a distance h above the surface of the earth:

$$mg_{\text{shuttle}} = \frac{GmM_E}{\left(h + R_E\right)^2}$$

Solve for g_{shuttle}:

$$g_{\text{shuttle}} = \frac{GM_E}{\left(h + R_E\right)^2}$$

Substitute numerical values and evaluate g_{shuttle}:

$$g_{\text{shuttle}} = \frac{\left(6.673 \times 10^{-11} \text{ N} \cdot \text{m}^2/\text{kg}^2\right)\left(5.98 \times 10^{24} \text{ kg}\right)}{\left(400 \text{ km} + 6370 \text{ km}\right)^2} = \boxed{8.71 \text{ m/s}^2}$$

Because they are in "free fall" everything on the shuttle is falling toward
(b) the center of the earth with exactly the same acceleration, so the astronauts will seem to be "weightless."

*26 • An object is dropped from a height of 6.37×10^6 m above the surface of the earth. What is its initial acceleration?

Picture the Problem We can relate the acceleration of an object at any elevation to its acceleration at the surface of the earth through the law of gravity and Newton's 2nd law of motion.

Letting a represent the acceleration due to gravity at this altitude (R_E) and m the mass of the object, apply Newton's 2nd law and the law of gravity to obtain:

$$\sum F_{radial} = \frac{GmM_E}{(2R_E)^2} = ma$$

and

$$a = \frac{GM_E}{(2R_E)^2} \qquad (1)$$

Apply Newton's 2nd law to the same object when it is at the surface of the earth:

$$\sum F_{radial} = \frac{GmM_E}{R_E^2} = mg$$

and

$$g = \frac{GM_E}{R_E^2} \qquad (2)$$

Divide equation (1) by equation (2) and solve for a:

$$\frac{a}{g} = \frac{R_E^2}{4R_E^2}$$

and

$$a = \tfrac{1}{4}g = \tfrac{1}{4}(9.81\,\text{m/s}^2) = \boxed{2.45\,\text{m/s}^2}$$

*31 •• The speed of an asteroid is 20 km/s at perihelion and 14 km/s at aphelion. Determine the ratio of the aphelion to perihelion distances.

Picture the Problem We can use conservation of angular momentum to relate the asteroid's aphelion and perihelion distances.

Using conservation of angular momentum, relate the angular momenta of the asteroid at aphelion and perihelion:

$$L_a = L_p$$

or

$$mv_p r_p = mv_a r_a$$

Solve for and evaluate the ratio of the asteroid's aphelion and perihelion distances:

$$\frac{r_a}{r_p} = \frac{v_p}{v_a} = \frac{20\,\text{km/s}}{14\,\text{km/s}} = \boxed{1.43}$$

*33 ••   A superconducting gravity meter can measure changes in gravity of the order $\Delta g/g = 10^{-11}$. (*a*) You are hiding behind a tree holding the meter, and your 80-kg friend approaches the tree from the other side. How close to you can your friend get before the meter detects a change in *g* due to his presence. (*b*) You are in hot air balloon and are using the meter to determine the rate of ascent (presumed constant). What is the smallest change in altitude that results in a detectable change in the gravitational field of the earth?

Picture the Problem We can determine the maximum range at which an object with a given mass can be detected by substituting the equation for the gravitational field in the expression for the resolution of the meter and solving for the distance. Differentiating $g(r)$ with respect to *r*, separating variables to obtain dg/g, and approximating Δr with dr will allow us to determine the vertical change in the position of the gravity meter in the earth's gravitational field is detectable.

(*a*) Express the gravitational field of the earth:

$$g_E = \frac{GM_E}{R_E^2}$$

Express the gravitational field due to the mass *m* (assumed to be a point mass) of your friend and relate it to the resolution of the meter:

$$g(r) = \frac{Gm}{r^2} = 10^{-11}\, g_E = 10^{-11} \frac{GM_E}{R_E^2}$$

Solve for *r*:

$$r = R_E \sqrt{\frac{10^{11}\, m}{M_E}}$$

Substitute numerical values and evaluate *r*:

$$r = \left(6.37 \times 10^6\,\text{m}\right)\sqrt{\frac{10^{11}\left(80\,\text{kg}\right)}{5.98 \times 10^{24}\,\text{kg}}}$$

$$= \boxed{7.37\,\text{m}}$$

(*b*) Differentiate $g(r)$ and simplify to obtain:

$$\frac{dg}{dr} = \frac{-2Gm}{r^3} = -\frac{2}{r}\left(\frac{Gm}{r^2}\right) = -\frac{2}{r}\,g$$

Separate variables to obtain:

$$\frac{dg}{g} = -2\frac{dr}{r} = 10^{-11}$$

Approximating dr with Δr, evaluate Δr with $r = R_E$:

$$\Delta r = \left| -\tfrac{1}{2}\left(10^{-11}\right)\left(6.37 \times 10^6 \text{ m}\right) \right|$$

$$= 3.19 \times 10^{-5} \text{ m}$$

$$= \boxed{0.0319 \,\text{mm}}$$

*35 •• The mass of the earth is 5.97×10^{24} kg and its radius is 6370 km. The radius of the moon is 1738 km. The acceleration of gravity at the surface of the moon is 1.62 m/s². What is the ratio of the average density of the moon to that of the earth?

Picture the Problem We can use the definitions of the gravitational fields at the surfaces of the earth and the moon to express the accelerations due to gravity at these locations in terms of the average densities of the earth and the moon. Expressing the ratio of these accelerations will lead us to the ratio of the densities.

Express the acceleration due to gravity at the surface of the earth in terms of the earth's average density:

$$g_E = \frac{GM_E}{R_E^2} = \frac{G\rho_E V_E}{R_E^2} = \frac{G\rho_E \tfrac{4}{3}\pi R_E^3}{R_E^2}$$

$$= \tfrac{4}{3} G\rho_E \pi R_E$$

Express the acceleration due to gravity at the surface of the moon in terms of the moon's average density:

$$g_M = \tfrac{4}{3} G\rho_M \pi R_M$$

Divide the second of these equations by the first to obtain:

$$\frac{g_M}{g_E} = \frac{\rho_M R_M}{\rho_E R_E}$$

Solve for $\dfrac{\rho_M}{\rho_E}$:

$$\frac{\rho_M}{\rho_E} = \frac{g_M R_E}{g_E R_M}$$

Substitute numerical values and evaluate $\dfrac{\rho_M}{\rho_E}$:

$$\frac{\rho_M}{\rho_E} = \frac{\left(1.62 \,\text{m/s}^2\right)\left(6.37 \times 10^6 \text{ m}\right)}{\left(9.81 \,\text{m/s}^2\right)\left(1.738 \times 10^6 \text{ m}\right)}$$

$$= \boxed{0.605}$$

Gravitational and Inertial Mass

*39 •   The principle of equivalence states that the free-fall acceleration of any object in a gravitational field is independent of the mass of the object. This can be seen from the form of the law of universal gravitation – but how well does it hold experimentally? The Roll-Krotkov-Dicke experiment performed in the 1960's indicates that the free-fall acceleration is independent of mass to at least 1 part in 10^{12}. Suppose two objects are simultaneously released from rest in a uniform gravitational field. Also, suppose one of the objects falls with a constant acceleration of exactly 9.8 m/s^2 while the other falls with a constant acceleration that is greater than 9.8 m/s^2 by 1 part in 10^{12}. How far will the first object have fallen when the second object has fallen 1 mm further than it has? Note that this estimate is an upper bound on the difference in the accelerations; most physicists believe that there is no difference whatsoever.

Picture the Problem Noting that $g_1 \sim g_2 \sim g$, let the acceleration of gravity on the first object be g_1, and on the second be g_2. We can use a constant-acceleration equation to express the difference in the distances fallen by each object and then relate the average distance fallen by the two objects to obtain an expression from which we can approximate the distance they would have to fall before we might measure a difference in their fall distances greater than 1 mm.

Express the difference Δd in the distances fallen by the two objects in time t:	$\Delta d = d_1 - d_2$
Express the distances fallen by each of the objects in time t:	$d_1 = \tfrac{1}{2} g_1 t^2$ and $d_2 = \tfrac{1}{2} g_2 t^2$
Substitute to obtain:	$\Delta d = \tfrac{1}{2} g_1 t^2 - \tfrac{1}{2} g_2 t^2 = \tfrac{1}{2}\left(g_1 - g_2\right) t^2$
Relate the average distance d fallen by the two objects to their time of fall:	$d = \tfrac{1}{2} g t^2$ or $t^2 = \dfrac{2d}{g}$
Substitute to obtain:	$\Delta d \approx \tfrac{1}{2} \Delta g \, \dfrac{2d}{g} = d \, \dfrac{\Delta g}{g}$
Solve for d to obtain:	$d = \Delta d \, \dfrac{g}{\Delta g}$

Substitute numerical values and evaluate d:

$$d = (10^{-3}\,\text{m})(10^{12}) = \boxed{10^9\,\text{m}}$$

Gravitational Potential Energy

*43 •• An object is dropped from rest from a height of 4×10^6 m above the surface of the earth. If there is no air resistance, what is its speed when it strikes the earth?

Picture the Problem Let the zero of gravitational potential energy be at infinity and let m represent the mass of the object. We'll use conservation of energy to relate the initial potential energy of the object-earth system to the final potential and kinetic energies.

Use conservation of energy to relate the initial potential energy of the system to its energy as the object is about to strike the earth:

$$K_f - K_i + U_f - U_i = 0$$
or, because $K_i = 0$,
$$K(R_E) + U(R_E) - U(R_E + h) = 0 \quad (1)$$
where h is the initial height above the earth's surface.

Express the potential energy of the object-earth system when the object is at a distance r from the surface of the earth:

$$U(r) = -\frac{GM_E m}{r}$$

Substitute in equation (1) to obtain:

$$\tfrac{1}{2}mv^2 - \frac{GM_E m}{R_E} + \frac{GM_E m}{R_E + h} = 0$$

Solve for v:

$$v = \sqrt{2\left(\frac{GM_E}{R_E} - \frac{GM_E}{R_E + h}\right)}$$

$$= \sqrt{2gR_E\left(\frac{h}{R_E + h}\right)}$$

Substitute numerical values and evaluate v:

$$v = \sqrt{\frac{2(9.81\,\text{m/s}^2)(6.37 \times 10^6\,\text{m})(4 \times 10^6\,\text{m})}{6.37 \times 10^6\,\text{m} + 4 \times 10^6\,\text{m}}} = \boxed{6.94\,\text{km/s}}$$

*48 •   The science fiction writer Robert Heinlein once said, "If you can get into orbit, then you're halfway to anywhere." Justify this statement by comparing the kinetic energy needed to place a satellite into low earth orbit ($h = 400$ km) to set it completely free from the bonds of earth's gravity.

Picture the Problem We'll consider a rocket of mass m which is initially on the surface of the earth (mass M and radius R) and compare the kinetic energy needed to get the rocket to its escape velocity with its kinetic energy in a low circular orbit around the earth. We can use conservation of energy to find the escape kinetic energy and Newton's law of gravity to derive an expression for the low earth-orbit kinetic energy.

Apply conservation of energy to relate the initial energy of the rocket to its escape kinetic energy:	$$K_f - K_i + U_f - U_i = 0$$
Letting the zero of gravitational potential energy be at infinity we have $U_f = K_f = 0$ and:	$$-K_i - U_i = 0$$ or $$K_e = -U_i = \frac{GMm}{R}$$
Apply Newton's law of gravity to the rocket in orbit at the surface of the earth to obtain:	$$\frac{GMm}{R^2} = m\frac{v^2}{R}$$
Rewrite this equation to express the low-orbit kinetic energy E_o of the rocket:	$$K_o = \tfrac{1}{2}mv^2 = \frac{GMm}{2R}$$
Express the ratio of K_o to K_e:	$$\frac{K_o}{K_e} = \frac{\dfrac{GMm}{2R}}{\dfrac{GMm}{R}} = \frac{1}{2} \Rightarrow K_e = \boxed{2K_o}\,,\text{ as}$$

asserted by Heinlein.

Orbits

*54 •   We normally say that the moon orbits the earth, but this is not quite true: instead, both the moon and the earth orbit their common center of mass, which is not at the center of the earth. (*a*) The mass of the earth is 5.98×10^{24} kg and the mass of the moon is 7.36×10^{22} kg. The mean distance from the center of the earth to the center of the moon is 3.82×10^8 m. How far above the surface of the earth is the center of mass of the earth-moon system? (*b*) Estimate the mean "orbital speed" of the earth as it orbits the center of mass of the earth-moon system, using the moon's mass, period, and average distance from the earth. Ignore any external forces acting on the system. The orbital period of the moon is 27.3 d.

Picture the Problem Let the origin of our coordinate system be at the center of the earth and let the positive x direction be toward the moon. We can apply the definition of center of mass to find the center of mass of the earth-moon system and find the "orbital" speed of the earth using x_{cm} as the radius of its motion and the period of the moon as the period of this motion of the earth.

(a) Using its definition, express the x coordinate of the center of mass of the earth-moon system:

$$x_{cm} = \frac{M_E x_E + m_{moon} x_{moon}}{M_E + m_{moon}}$$

Substitute numerical values and evaluate x_{cm}:

$$x_{cm} = \frac{M_E(0) + (7.36 \times 10^{22} \text{ kg})(3.82 \times 10^8 \text{ m})}{5.98 \times 10^{24} \text{ kg} + 7.36 \times 10^{22} \text{ kg}} = \boxed{4.64 \times 10^6 \text{ m}}$$

Note that, because the radius of the earth is 6.37×10^6 m, the center of mass is actually located about 1700 km *below* the surface of the earth.

(b) Express the "orbital" speed of the earth in terms of the radius of its circular orbit and its period of rotation:

$$v = \frac{2\pi x_{cm}}{T}$$

Substitute numerical values and evaluate v:

$$v = \frac{2\pi(4.64 \times 10^6 \text{ m})}{27.3 \text{d} \times \dfrac{24\text{h}}{\text{d}} \times \dfrac{3600\text{s}}{\text{h}}} = \boxed{12.4 \text{ m/s}}$$

The Gravitational Field

*58 • The gravitational field at some point is given by $\vec{g} = 2.5 \times 10^{-6}\,\text{N/kg}\,\hat{j}$. What is the gravitational force on a mass of 4 g at that point?

Picture the Problem The gravitational field at any point is defined by $\vec{g} = \vec{F}/m$.

Using its definition, express the gravitational field at a point in space:

$$\vec{g} = \frac{\vec{F}}{m}$$

Solve for $\vec{F}$ and substitute to obtain:

$$\vec{F} = m\vec{g}$$
$$= (0.004\,\text{kg})(2.5\times10^{-6}\,\text{N/kg})\hat{j}$$
$$= \boxed{(10^{-8}\,N)\hat{j}}$$

$\vec{g}$ due to Spherical Objects

*67 • Two solid spheres, S_1 and S_2, have equal radii R and equal masses M. The density of sphere S_1 is constant, whereas that of sphere S_2 depends on the radial distance according to $p(r) = C/r$. If the acceleration of gravity at the surface of sphere S_1 is g_1, what is the acceleration of gravity at the surface of sphere S_2?

Picture the Problem The gravitational field and acceleration of gravity at the surface of a sphere given by $g = GM/R^2$, where R is the radius of the sphere and M is its mass.

Express the acceleration of gravity on the surface of S_1:

$$g_1 = \frac{GM}{R^2}$$

Express the acceleration of gravity on the surface of S_2:

$$g_2 = \frac{GM}{R^2}$$

Divide the second of these equations by the first to obtain:

$$\frac{g_2}{g_1} = \frac{\dfrac{GM}{R^2}}{\dfrac{GM}{R^2}} = 1 \text{ or } \boxed{g_1 = g_2}$$

$\vec{g}$ Inside Solid Sphere

*71 •• Is your "weight" (as measured on a spring scale) at the bottom of a deep mine shaft greater than or less than your weight at the surface of the earth? Model the earth as a homogeneous sphere. Consider the effects in (a) and (b). (a) Show that the force of gravity on you from a uniform-density, perfectly spherical planet is proportional to your distance from the center of the planet. (b) Show that your effective "weight" increases linearly due to the effects of rotation as you approach the center. (Consider a mine shaft located on the equator.) (c) Which effect is more important for the earth? Use $M = 5.98\text{x}10^{24}$ kg, $R = 6370$ km, and $T = 24$ h.

Picture the Problem The "weight" as measured by a spring scale will be the normal force which the spring scale presses up against you. There are two forces acting on you as you stand at a distance r from the center of the planet: the normal force (F_N) and the force of gravity (mg). Because you are in equilibrium under the influence of these forces, your weight (the scale reading or normal force) will be equal to the gravitational force acting on you. We can use Newton's law of gravity to express this force.

(*a*) Express the force of gravity acting on you when you are a distance r from the center of the earth:

$$F_g = \frac{GM(r)m}{r^2} \qquad (1)$$

Using the definition of density, express the density of the earth between you and the center of the earth and the density of the earth as a whole:

$$\rho = \frac{M(r)}{V(r)} = \frac{M(r)}{\frac{4}{3}\pi r^3}$$

and

$$\rho = \frac{M_E}{V_E} = \frac{M_E}{\frac{4}{3}\pi R^3}$$

Because we're assuming the earth to of uniform-density and perfectly spherical:

$$\frac{M(r)}{\frac{4}{3}\pi r^3} = \frac{M_E}{\frac{4}{3}\pi R^3}$$

or

$$M(r) = M_E\left(\frac{r}{R}\right)^3$$

Substitute in equation (1) and simplify to obtain:

$$F_g = \frac{GM_E\left(\frac{r}{R}\right)^3 m}{r^2} = \frac{GM_E m}{R^2}\frac{r}{R}$$

Apply Newton's law of gravity to yourself at the surface of the earth to obtain:

$$mg = \frac{GM_E m}{R^2}$$

or

$$g = \frac{GM_E}{R^2}$$

where g is the magnitude of free-fall acceleration at the surface of the earth.

Substitute to obtain:

$$F_g = \boxed{\frac{mg}{R}r}$$

i.e., the force of gravity on you is proportional to your distance from the center of the earth.

(*b*) Apply Newton's 2nd law to your body to obtain:

$$F_N - mg\frac{r}{R} = -mr\omega^2$$

Solve for your "effective weight" (i.e., what a spring scale will measure) F_N:

$$F_N = \frac{mg}{R}r - mr\omega^2 = \boxed{\left(\frac{mg}{R} - m\omega^2\right)r}$$

Note that this equation tells us that your effective weight increases linearly with distance from the center of the earth. The second term can be interpreted as a "centrifugal force" pushing out, which increases the farther you get from the center of the earth.

(c) We can decide whether the change in mass with distance from the center of the earth or the rotational effect is more important by examining the ratio of the two terms in the expression for your effective weight:

$$\frac{\frac{mg}{R}r}{mr\omega^2} = \frac{\frac{g}{R}}{\omega^2} = \frac{g}{R\left(\frac{2\pi}{T}\right)^2} = \frac{gT^2}{4\pi^2 R}$$

$$= \frac{\left(9.81\,\text{m/s}^2\right)\left(24\,\text{h} \times \frac{3600\,\text{s}}{\text{h}}\right)^2}{4\pi^2(6370\,\text{km})}$$

$$= 291$$

The change in the mass between you and the center of the earth as you move away from the center is 291 times more important than the rotational effect.

*77 ••• A small diameter hole is drilled into the sphere of Problem 76 toward the center of the sphere to a depth of 2 m below the sphere's surface. A small mass is dropped from the surface into the hole. Determine the speed of the small mass as it strikes the bottom of the hole.

Picture the Problem We can use conservation of energy to relate the work done by the gravitational field to the speed of the small object as it strikes the bottom of the hole. Because we're given the mass of the sphere, we can find C by expressing the mass of the sphere in terms of C. We can then use its definition to find the gravitational field of the sphere inside its surface. The work done by the field equals the negative of the change in the potential energy of the system as the small object falls in the hole.

Use conservation of energy to relate the work done by the gravitational field to the speed of the small object as it strikes the bottom of the hole:

$K_f - K_i + \Delta U = 0$

or, because $K_i = 0$ and $W = -\Delta U$,

$W = \frac{1}{2}mv^2$

where v is the speed with which the object strikes the bottom of the hole and W is the work done by the gravitational field.

Solve for v:

$$v = \sqrt{\frac{2W}{m}} \qquad (1)$$

Express the mass of a differential element of the sphere:

$$dm = \rho \, dV = \rho\left(4\pi r^2 dr\right)$$

Integrate to express the mass of the sphere in terms of C:

$$M = 4\pi C \int_{0}^{5\,\mathrm{m}} r \, dr = \left(50\,\mathrm{m}^2\right)\pi C$$

Solve for and evaluate C:

$$C = \frac{M}{\left(50\,\mathrm{m}^2\right)\pi} = \frac{1011\,\mathrm{kg}}{\left(50\,\mathrm{m}^2\right)\pi}$$

$$= 6.436\,\mathrm{kg/m}^2$$

Use its definition to express the gravitational field of the sphere at a distance from its center less than its radius:

$$g = G\frac{\int_{0}^{r} 4\pi r^2 \rho \, dr}{r^2} = G\frac{\int_{0}^{r} 4\pi r^2 \frac{C}{r} \, dr}{r^2}$$

$$= G\frac{4\pi C \int_{0}^{r} r \, dr}{r^2} = 2\pi GC$$

Express the work done on the small object by the gravitational force acting on it:

$$W = -\int_{5\,\mathrm{m}}^{3\,\mathrm{m}} mg \, dr = \left(2\,\mathrm{m}\right)mg$$

Substitute in equation (1) and simplify to obtain:

$$v = \sqrt{\frac{2\left(2\,\mathrm{m}\right)m\left(2\pi GC\right)}{m}} = \sqrt{\left(8\,\mathrm{m}\right)\pi GC}$$

Substitute numerical values and evaluate v:

$$v = \sqrt{\left(8\,\mathrm{m}\right)\pi\left(6.6726\times10^{-11}\,\mathrm{N\cdot m^2/kg^2}\right)\left(6.436\,\mathrm{kg/m^2}\right)} = \boxed{0.104\,\mathrm{mm/s}}$$

*79 ••• Two identical spherical cavities are made in a lead sphere of radius R. The cavities have a radius $R/2$. They touch the outside surface of the sphere and its center as in Figure 11-26. The mass of the solid uniform lead sphere is M. (a) Find the force of attraction of a point particle of mass m at the position shown in the figure to the cavitated lead sphere. (b) What is the attractive force on the point particle if it is located on the x axis at $x = R$?

Figure 11-26 Problem 79

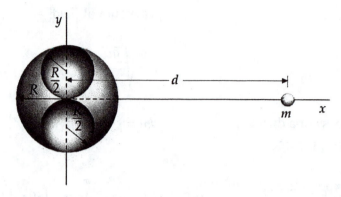

Picture the Problem The force of attraction of the small sphere of mass m to the lead sphere is the sum of the forces due to the solid sphere ($\vec{F}_S$) and the cavities ($\vec{F}_C$) of negative mass.

(a) Express the force of attraction:

$$\vec{F} = \vec{F}_S + \vec{F}_C \qquad (1)$$

Use the law of gravity to express the force due to the solid sphere:

$$\vec{F}_S = -\frac{GMm}{d^2}\hat{i}$$

Express the magnitude of the force acting on the small sphere due to one cavity:

$$F_C = \frac{GM'm}{d^2 + \left(\dfrac{R}{2}\right)^2}$$

where M' is the negative mass of a cavity.

Relate the negative mass of a cavity to the mass of the sphere before hollowing:

$$M' = -\rho V = -\rho\left[\tfrac{4}{3}\pi\left(\frac{R}{2}\right)^3\right]$$

$$= -\tfrac{1}{8}\left(\tfrac{4}{3}\pi\rho R^3\right) = -\tfrac{1}{8}M$$

Letting θ be the angle between the x axis and the line joining the center of the small sphere to the center of either cavity, use the law of gravity to express the force due to the two cavities:

$$\vec{F}_C = 2\frac{GMm}{8\left(d^2 + \dfrac{R^2}{4}\right)}\cos\theta\,\hat{i}$$

because, by symmetry, the y components add to zero.

Express $\cos\theta$:

$$\cos\theta = \frac{d}{\sqrt{d^2 + \dfrac{R^2}{4}}}$$

Substitute to obtain:

$$\vec{F}_C = \frac{GMm}{4\left(d^2 + \dfrac{R^2}{4}\right)}\frac{d}{\sqrt{d^2 + \dfrac{R^2}{4}}}\hat{i}$$

$$= \frac{GMmd}{4\left(d^2 + \dfrac{R^2}{4}\right)^{3/2}}\hat{i}$$

Substitute in equation (1) and simplify:

$$\vec{F} = -\frac{GMm}{d^2}\hat{i} + \frac{GMmd}{4\left(d^2 + \dfrac{R^2}{4}\right)^{3/2}}\hat{i}$$

$$\boxed{= -\frac{GMm}{d^2}\left[1 - \frac{\dfrac{d^3}{4}}{\left\{d^2 + \dfrac{R^2}{4}\right\}^{3/2}}\right]\hat{i}}$$

(b) Evaluate $\vec{F}$ at $d = R$:

$$\vec{F}(R) = -\frac{GMm}{R^2}\left[1 - \frac{\dfrac{R^3}{4}}{\left\{R^2 + \dfrac{R^2}{4}\right\}^{3/2}}\right]\hat{i}$$

$$\boxed{= -0.821\frac{GMm}{R^2}\hat{i}}$$

General Problems

*81 •　　The mean distance of Pluto from the sun is 39.5 AU. Find the period of Pluto.

Picture the Problem We can use Kepler's 3rd law to relate Pluto's period to its mean distance from the sun.

Using Kepler's 3rd law, relate the period of Pluto to its mean distance from the sun:

$$T^2 = Cr^3$$

$$\text{where } C = \frac{4\pi^2}{GM_s} = 2.973 \times 10^{-19} \, s^2/m^3 \, .$$

Solve for T:

$$T = \sqrt{Cr^3}$$

Substitute numerical values and evaluate T:

$$T = \sqrt{\left(2.973 \times 10^{-19} \, s^2/m^3\right)\left(39.5 \, \text{AU} \times \frac{1.50 \times 10^{11} \, m}{\text{AU}}\right)^3}$$

$$= 7.864 \times 10^9 \, s \times \frac{1h}{3600s} \times \frac{1d}{24h} \times \frac{1y}{365.25d}$$

$$= \boxed{249 \, y}$$

*88 ••   In the novel "A Voyage to the Moon" by Jules Verne, astronauts were launched by a giant cannon from the earth to the moon. (*a*) If we estimate the velocity needed to reach the moon as the earth's escape velocity, and the length of the cannon as 900 ft (as stated in the book), what is the probability that the astronauts would have survived "liftoff"? (*b*) In the book, the astronauts in their ship feel the effects of the earth's gravity until they reach the balance point where the moon's gravitational pull on the ship equals that of the earth. At this point, everything flips around and what was the ceiling in the ship becomes the floor. How far away from the center of the earth is this balance point? (*c*) Is this description of what the astronauts experienced reasonable? Is this what actually happened to the Apollo astronauts on their visit to the moon?

Picture the Problem If we assume the astronauts experience a constant acceleration in the barrel of the cannon, we can use a constant-acceleration equation to relate their exit speed (the escape speed from the earth) to the acceleration they would need to undergo in order to reach that speed. We can use conservation of energy to express their escape speed in terms of the mass and radius of the earth and then substitute in the constant-acceleration equation to find their acceleration. To find the balance point between the earth and the moon we can equate the gravitational forces exerted by the earth and the moon at that point.

(*a*) Assuming constant acceleration down the cannon barrel, relate the ship's speed as it exits the barrel to the length of the barrel and the acceleration required to get the ship to escape speed:

$$v_e^2 = 2a\Delta\ell$$

where ℓ is the length of the cannon.

Solve for the acceleration:

$$a = \frac{v_e^2}{2\Delta\ell} \qquad (1)$$

Use conservation of energy to relate the initial energy of astronaut's ship to its energy when it has escaped the earth's gravitational field:

$$\Delta K + \Delta U = 0$$
or
$$K_f - K_i + U_f - U_i = 0$$

When the ship has escaped the earth's gravitational field:

$$K_f = U_f = 0$$
and
$$-K_i - U_i = 0$$
or
$$-\tfrac{1}{2}mv_e^2 - \left(-\frac{GM_E m}{R}\right) = 0$$
where m is the mass of the spaceship.

Solve for v_e^2 to obtain:

$$v_e^2 = \frac{2GM_E}{R}$$

Substitute in equation (1) to obtain:

$$a = \frac{GM_E}{\Delta\ell R}$$

Substitute numerical values and evaluate a:

$$a = \left(6.673 \times 10^{-11}\,\text{N}\cdot\text{m}^2/\text{kg}^2\right)$$
$$\times \frac{\left(5.98 \times 10^{24}\,\text{kg}\right)}{(274\,\text{m})(6370\,\text{km})}$$
$$= 2.29 \times 10^5\,\text{m/s}^2$$
$$\approx 23{,}300g$$

Survival is extremely unlikely!

(b) Let the distance from the center of the earth to the center of the moon be R, and the distance from the center of the spaceship to the earth be x. If M is the mass of the earth and m the mass of the moon, the forces will balance out when:

$$\frac{GM}{x^2} = \frac{Gm}{(R-x)^2}$$
or
$$\frac{x}{\sqrt{M}} = \frac{R-x}{\sqrt{m}}$$
where we've ignored the negative solution, as it doesn't indicate a point between the two bodies.

Solve for x to obtain:

$$x = \frac{R}{1 + \sqrt{\dfrac{m}{M}}}$$

Substitute numerical values and evaluate x:

$$x = \frac{3.84 \times 10^8 \text{ m}}{1 + \sqrt{\dfrac{7.36 \times 10^{22} \text{ kg}}{5.98 \times 10^{24} \text{ kg}}}}$$

$$= \boxed{3.46 \times 10^8 \text{ m}}$$

(c) $\boxed{\text{No it is not. During the entire trip, the astronauts would be in free-fall, and so would not seem to weigh anything.}}$

*91 •• Four identical planets are arranged in a square as shown in Figure 11-27. If the mass of each planet is M and the edge length of the square is a, what must be their speed if they are to orbit their common center under the influence of their mutual attraction?

Figure 11-27 Problem 91

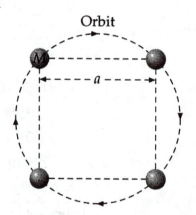

Picture the Problem We can find the orbital speeds of the planets from their distance from the center of mass of the system and the period of their motion. Application of Kepler's 3rd law will allow us to express the period of their motion T in terms of the effective mass of the system ... which we can find from its definition.

Express the orbital speeds of the planets in terms of their period T:

$$v = \frac{2\pi R}{T}$$

where R is the distance to the center of mass of the four-planet system.

Apply Kepler's 3rd law to express the period of the planets:

$$T = \sqrt{\frac{4\pi^2}{GM_{eff}} R^3}$$

where M_{eff} is the effective mass of the four planets.

Substitute to obtain:

$$v = \frac{2\pi R}{\sqrt{\dfrac{4\pi^2}{GM_{\text{eff}}}R^3}} = \sqrt{\frac{GM_{\text{eff}}}{R}}$$

The distance of each planet from the effective mass is:

$$R = \frac{a}{\sqrt{2}}$$

Find M_{eff} from its definition:

$$\frac{1}{M_{\text{eff}}} = \frac{1}{M} + \frac{1}{M} + \frac{1}{M} + \frac{1}{M}$$

and

$$M_{\text{eff}} = \tfrac{1}{4}M$$

Substitute for R and M_{eff} to obtain:

$$\boxed{v = \sqrt{\frac{\sqrt{2}GM}{4a}}}$$

*98 •••   A uniform sphere of mass M is located near a thin, uniform rod of mass m and length L as in Figure 11-29. Find the gravitational force of attraction exerted by the sphere on the rod.

Figure 11-29 Problem 98

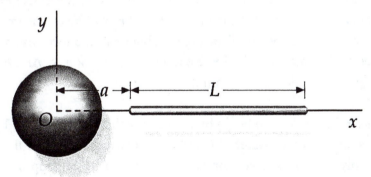

Picture the Problem Choose a mass element dm of the rod of thickness dx at a distance x from the origin. All such elements of the rod experience a gravitational force dF due to presence of the sphere centered at the origin. We can find the total gravitational force of attraction experienced by the rod by integrating dF from $x = a$ to $x = a + L$.

Express the gravitational force dF acting on the element of the

$$dF = \frac{GM\,dm}{x^2}$$

rod of mass *dm*:

Express *dm* in terms of the mass *m* and length *L* of the rod:

$$dm = \frac{m}{L}dx$$

Substitute to obtain:

$$dF = \frac{GMm}{L}\frac{dx}{x^2}$$

Integrate *dF* from $x = a$ to $x = a + L$ to find the total gravitational force acting on the rod:

$$F = \frac{GMm}{L}\int_a^{a+L} x^{-2}dx = -\frac{GMm}{L}\left[\frac{1}{x}\right]_a^{a+L}$$

$$= \boxed{\frac{GMm}{a(a+L)}}$$

*100 •••   Both the sun and the moon exert gravitational forces on the oceans of the earth, causing tides. (*a*) Show that the ratio of the force exerted on a point particle on earth by the sun to that exerted by the moon is $M_S r_m^2/M_m r_S^2$, where M_S and M_m are the masses of the sun and moon and r_S and r_m are the distances from the earth to the sun and to the moon. Evaluate this ratio. (*b*) Even though the sun exerts a much greater force on the oceans than does the moon, the moon has a greater effect on the tides because it is the difference in the force from one side of the earth to the other that is important. Differentiate the expression $F = Gm_1m_2/r^2$ to calculate the change in *F* due to a small change in *r*. Show that $dF/F = (-2\,dr)/r$. (*c*) During one full day, the rotation of the earth can cause the distance from the sun or moon to an ocean to change by, at most, the diameter of the earth. Show that for a small change in distance, the change in the force exerted by the sun is related to the change in the force exerted by the moon by $\Delta F_S/F_m \approx (M_S r_m^3)/(M_m r_S^3)$ and calculate this ratio.

Picture the Problem We can begin by expressing the forces exerted by the sun and the moon on a body of water of mass *m* and taking the ratio of these forces. In (*b*) we'll simply follow the given directions and in (*c*) we can approximate differential quantities with finite quantities to establish the given ratio.

(*a*) Express the force exerted by the sun on a body of water of mass *m*:

$$F_s = \frac{GM_s m}{r_s^2}$$

Express the force exerted by the moon on a body of water of mass *m*:

$$F_m = \frac{GM_m m}{r_m^2}$$

Divide the first of these equations by the second and simplify to obtain:

$$\frac{F_S}{F_m} = \boxed{\frac{M_S r_m^2}{M_m r_S^2}}$$

Substitute numerical values and evaluate this ratio:

$$\frac{F_S}{F_m} = \frac{(1.99\times10^{30}\text{ kg})(3.84\times10^8\text{ m})^2}{(7.36\times10^{22}\text{ kg})(1.50\times10^{11}\text{ m})^2}$$

$$= \boxed{177}$$

(b) Find $\dfrac{dF}{dr}$:

$$\frac{dF}{dr} = -\frac{2Gm_1 m_2}{r^3} = -2\frac{F}{r}$$

Solve for the ratio $\dfrac{dF}{F}$:

$$\frac{dF}{F} = \boxed{-2\frac{dr}{r}}$$

(c) Express the change in force ΔF for a small change in distance Δr:

$$\Delta F = -2\frac{F}{r}\Delta r$$

Express ΔF_S:

$$\Delta F_S = -2\frac{\dfrac{GmM_S}{r_S^2}}{r_S}\Delta r_S$$

$$= -2\frac{GmM_S}{r_S^3}\Delta r_S$$

Express ΔF_m:

$$\Delta F_m = -2\frac{GmM_m}{r_m^3}\Delta r_m$$

Divide the first of these equations by the second and simplify:

$$\frac{\Delta F_S}{\Delta F_m} = \frac{\dfrac{M_S}{r_S^3}\Delta r_S}{\dfrac{M_m}{r_m^3}\Delta r_m} = \frac{M_S r_m^3}{M_m r_S^3}\frac{\Delta r_S}{\Delta r_m}$$

$$= \boxed{\frac{M_S r_m^3}{M_m r_S^3}}$$

because $\dfrac{\Delta r_S}{\Delta r_m} = 1.$

Substitute numerical values and
evaluate this ratio:

$$\frac{\Delta F_S}{\Delta F_m} = \frac{\left(1.99\times10^{30}\text{ kg}\right)\left(3.84\times10^{8}\text{ m}\right)^3}{\left(7.36\times10^{22}\text{ kg}\right)\left(1.50\times10^{11}\text{ m}\right)^3}$$

$$= \boxed{0.454}$$

Chapter 12
Static Equilibrium and Elasticity

Conceptual Problems

*6 • Is it possible to climb a ladder placed against a wall where the ground is frictionless but the wall is not? Explain.

Determine the Concept No. Because the floor can exert no horizontal force, neither can the wall. Consequently, the friction force between the wall and the ladder is zero regardless of the coefficient of friction between the wall and the ladder.

*9 • The horizontal bar in Figure 12-23 will remain horizontal if (a) $L_1 = L_2$ and $R_1 = R_2$, (b) $L_1 = L_2$ and $M_1 = M_2$, (c) $R_1 = R_2$ and $M_1 = M_2$, (d) $L_1M_1 = L_2M_2$, (e) $R_1L_1 = R_2L_2$.

Figure 12-23 Problem 9

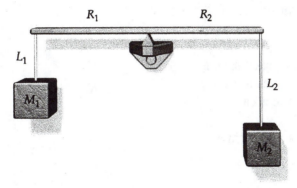

Determine the Concept The condition that the bar is in rotational equilibrium is that the net torque acting on it be zero; i.e., $R_1M_1 = R_2M_2$. This condition is satisfied provided $R_1 = R_2$ and $M_1 = M_2$. $\boxed{(c) \text{ is correct.}}$

*11 •• The great engineering feats of the ancient worlds (Roman arch bridges, the great cathedrals, and the pyramids, to name a few) all have two things in common: they are made of stone, and they are all compressive structures - that is, they are built so that all strains in the structure are compressive rather than tensile in nature. Look up the tensile and compressive strengths of stone and cement to give an explanation why this is true.

Determine the Concept The tensile strengths of stone and concrete are at least an order of magnitude lower than their compressive strengths, so you want to build compressive structures to match their properties.

371

Estimation and Approximation

*13 ••   Consider an atomic model for Young's modulus: assume that we have a large number of atoms arranged in a cubic array separated by distance a.   Imagine that each atom is attached to its six nearest neighbors by little springs with spring constant k. (Atoms are not really attached by springs, but the forces between them act enough like springs to make this a good model.) (*a*) Show that this material, if stretched, will have a Young's modulus $Y = k/a$. (*b*) From Table 12-1, and assuming that $a \approx 1$ nm, estimate a typical value for the "atomic spring constant" k in a metal.

Picture the Problem We can derive this expression by imagining that we pull on an area A of the given material, expressing the force each spring will experience, finding the fractional change in length of the springs, and substituting in the definition of Young's modulus.

(*a*) Express Young's modulus:

$$Y = \frac{F/A}{\Delta L/L} \tag{1}$$

Express the elongation ΔL of each spring:

$$\Delta L = \frac{F_s}{k} \tag{2}$$

Express the force F_s each spring will experience as a result of a force F acting on the area A:

$$F_s = \frac{F}{N}$$

Express the number of springs N in the area A:

$$N = \frac{A}{a^2}$$

Substitute to obtain:

$$F_s = \frac{Fa^2}{A}$$

Substitute in equation (2) to obtain, for the extension of one spring:

$$\Delta L = \frac{Fa^2}{kA}$$

Assuming that the springs extend/compress linearly, express the fractional extension of the springs:

$$\frac{\Delta L_{tot}}{L} = \frac{\Delta L}{a} = \frac{1}{a}\frac{Fa^2}{kA} = \frac{Fa}{kA}$$

Substitute in equation (1) and simplify:

$$Y = \frac{\dfrac{F}{A}}{\dfrac{Fa}{kA}} = \boxed{\frac{k}{a}}$$

(*b*) From our result in part (*a*):

$$k = Ya$$

From Table 12-1: $Y = 200\text{GN/m}^2 = 2 \times 10^{11}\,\text{N/m}^2$

Assuming that $a \sim 1$ nm, $k = (2 \times 10^{11}\,\text{N/m}^2)(10^{-9}\,\text{m}) = \boxed{200\,\text{N/m}}$
evaluate k:

Conditions for Equilibrium

*16 •   Misako wants to measure the strength of her biceps muscle by exerting a
force on a test strap as shown in Figure 12-26. The strap is 28 cm from the pivot
point at the elbow, and her biceps muscle is attached at a point 5 cm from the pivot
point. If the scale reads 18 N when she exerts her maximum force, what force is
exerted by the biceps muscle?

Figure 12-26 Problem 16

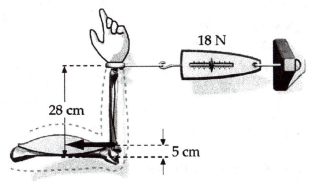

Picture the Problem Let F represent the force exerted by Misako's biceps. To find
F we apply the condition for rotational equilibrium about a pivot chosen at the tip
of her elbow.

Apply $\sum \vec{\tau} = 0$ about the pivot: $(5\,\text{cm})F - (28\,\text{cm})(18\,\text{N}) = 0$

Solve for F: $F = \dfrac{(28\,\text{cm})(18\,\text{N})}{5\,\text{cm}} = \boxed{101\text{N}}$

The Center of Gravity

*19 •   Each of the objects shown in Figure 12-28 is suspended from the ceiling
by a thread attached to the point marked × on the object. Describe the orientation of
each suspended object with a diagram.

Figure 12-28 Problem 19

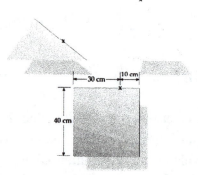

Picture the Problem The figures are shown on the right. The center of mass for each is indicated by a small +. At static equilibrium, the center of gravity is directly below the point of support.

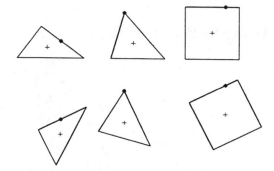

Some Examples of Static Equilibrium

*25 ••   A gravity board for locating the center of gravity of a person consists of a horizontal board supported by a fulcrum at one end and by a scale at the other end. A physics student lies horizontally on the board with the top of his head above the fulcrum point as shown in Figure 12-34. The scale is 2 m from the fulcrum. The student has a mass of 70 kg, and when he is on the gravity board, the scale advances 250 N. Where is the center of gravity of the student?

Figure 12-34 Problem 25

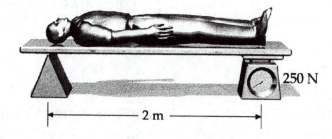

Picture the Problem The diagram shows $\vec{w}$, the weight of the student, $\vec{F}_p$, the force exerted by the board at the pivot, and $\vec{F}_s$, the force exerted by the scale, acting on the student. Because the student is in equilibrium, we can apply the condition for rotational equilibrium to the student to find the location of his center of gravity.

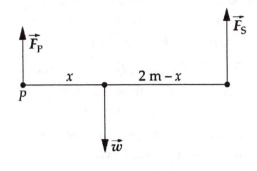

Apply $\sum \vec{\tau} = 0$ at the pivot point P:

$$F_s(2\,\mathrm{m}) - wx = 0$$

Solve for x:

$$x = \frac{(2\,\mathrm{m})F_s}{w}$$

Substitute numerical values and evaluate x:

$$x = \frac{(2\,\mathrm{m})(250\,\mathrm{N})}{(70\,\mathrm{kg})(9.81\,\mathrm{m/s^2})} = \boxed{0.728\,\mathrm{m}}$$

*27 •   A cylinder of weight W is supported by a frictionless trough formed by a plane inclined at 30° to the horizontal on the left and one inclined at 60° on the right as shown in Figure 12-36. Find the force exerted by each plane on the cylinder.

Figure 12-36 Problem 27

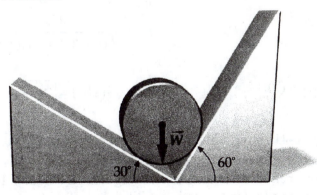

Picture the Problem The planes are frictionless; therefore, the force exerted by each plane must be perpendicular to that plane. Let $\vec{F}_1$ be the force exerted by the 30° plane, and let $\vec{F}_2$ be the force exerted by the 60° plane. Choose a coordinate system in which the positive x direction is to the right and the positive y direction is upward. Because the cylinder is in equilibrium, we can use the conditions for translational equilibrium to find the magnitudes of $\vec{F}_1$ and $\vec{F}_2$.

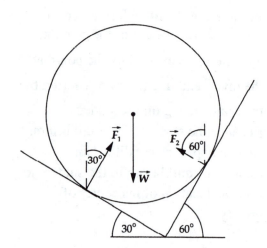

Apply $\sum F_x = 0$ to the cylinder:

$$F_1 \sin 30° - F_2 \sin 60° = 0 \qquad (1)$$

Apply $\sum F_y = 0$ to the cylinder:

$$F_1 \cos 30° + F_2 \cos 60° - W = 0 \qquad (2)$$

Solve equation (1) for F_1:

$$F_1 = \sqrt{3} F_2 \qquad (3)$$

Substitute in equation (2) to obtain:

$$\sqrt{3} F_2 \cos 30° + F_2 \cos 60° - W = 0$$

Solve for F_2:

$$\left(\sqrt{3} \cos 30° + \cos 60°\right) F_2 = W$$

or

$$F_2 = \frac{W}{\sqrt{3} \cos 30° + \cos 60°} = \boxed{\tfrac{1}{2} W}$$

Substitute in equation (3):

$$F_1 = \sqrt{3}\left(\tfrac{1}{2} W\right) = \boxed{\tfrac{\sqrt{3}}{2} W}$$

*37 •• Figure 12-41 shows a hand holding an epee, a weapon used in the sport of fencing. The center of mass of the epee, indicated in Figure 12-41, is 24 cm from the pommel; its total mass is 0.700 kg and its length is 110 cm. (*a*) Apply one of the conditions for static equilibrium to find the (total) force exerted by the hand on epee. (*b*) Apply the other condition for static equilibrium to find the torque exerted by the hand on epee. (*c*) Model the forces exerted by the hand as two oppositely directed forces whose lines of action are separated by the width of the fencer's hand ($\approx$ 10 cm). What are the magnitudes and directions of these two forces?

Figure 12-41 Problem 37

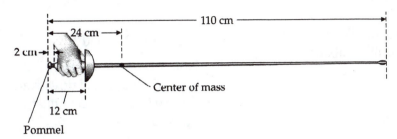

24 cm

110 cm

2 cm

12 cm

Pommel

Center of mass

Picture the Problem The diagram shows the forces F_1 and F_2 that the fencer's hand exerts on the epee. We can use a condition for translational equilibrium to find the upward force the fencer must exert on the epee when it is in equilibrium and the definition of torque to determine the total torque exerted. In part (c) we can use the conditions for translational and rotational equilibrium to obtain two equations in F_1 and F_2 that we can solve simultaneously. In part (d) we can apply Newton's 2nd law in rotational form and the condition for translational equilibrium to obtain two equations in F_1 and F_2 that, again, we can solve simultaneously.

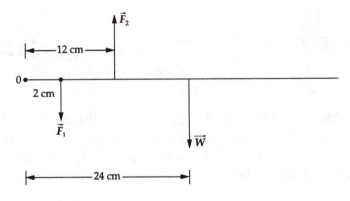

$\vec{F}_2$

12 cm

0

2 cm

$\vec{F}_1$

$\vec{W}$

24 cm

(a) Letting the upward force exerted by the fencer's hand be F, apply $\sum F_y = 0$ to the epee to obtain:

$$F - W = 0$$

Solve for and evaluate F:

$$F = mg = (0.7\,\text{kg})(9.81\,\text{m/s}^2) = \boxed{6.87\,\text{N}}$$

(b) Express the torque due to the weight about the left end of the epee:

$$\tau = \ell w = (0.24\,\text{m})(6.87\,\text{N}) = \boxed{1.65\,\text{N}\cdot\text{m}}$$

(c) Apply $\sum F_y = 0$ to the epee to obtain:

$$-F_1 + F_2 - 6.87\,\text{N} = 0 \qquad\qquad (1)$$

Apply $\sum \tau_0 = 0$ to obtain:

$$-(0.02\,\text{m})F_1 + (0.12\,\text{m})F_2 - 1.65\,\text{N}\cdot\text{m} = 0$$

Solve these equations simultaneously to obtain:

$F_1 = \boxed{8.26\,\text{N}}$ and $F_2 = \boxed{15.1\,\text{N}}$.

Note that the force nearest the butt of the epee is directed downward and the force nearest the hand guard is directed upward.

*41 •• A boat is moored at the end of a dock in a rapidly flowing river by a chain 5 m long, as shown in Figure 12-45. To give the chain some flexibility, a 100-N weight is attached in the center of the chain, to allow for variations in the force pulling the boat away from the dock. (*a*) If the drag force on the boat is 50 N, what is the tension in the chain? (*b*) How far will the chain sag? Ignore the weight of the chain itself. (*c*) How far is the boat from the dock? (*d*) If the maximum tension that the chain can support is 500 N, what is the maximum value of the force that the river can exert on the boat?

Figure 12-45 Problem 41

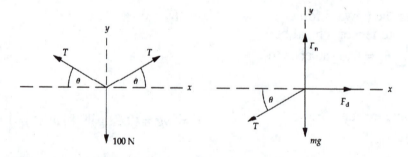

Picture the Problem The free-body diagram shown to the left below is for the weight and the diagram to the right is for the boat. Because both are in equilibrium under the influences of the forces acting on them, we can apply a condition for translational equilibrium to find the tension in the chain.

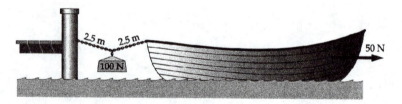

(*a*) Apply $\sum F_x = 0$ to the boat:

$$F_d - T\cos\theta = 0$$

Solve for T:

$$T = \frac{F_d}{\cos\theta}$$

Apply $\sum F_y = 0$ to the weight:

$$2T\sin\theta - 100\,\text{N} = 0 \qquad (1)$$

Substitute for T to obtain:

$$2F_d \tan\theta - 100\,\text{N} = 0$$

Solve for θ:

$$\theta = \tan^{-1}\frac{100\,\text{N}}{2F_d}$$

Substitute for F_d and evaluate θ:

$$\theta = \tan^{-1}\frac{100\,\text{N}}{2(50\,\text{N})} = 45°$$

Solve equation (1) for T:

$$T = \frac{100\,\text{N}}{2\sin\theta}$$

Substitute for θ and evaluate T:

$$T = \frac{100\,\text{N}}{2\sin 45°} = \boxed{70.7\,\text{N}}$$

(b) Use the diagram to the right to relate the sag Δy in the chain to the angle θ the chain makes with the horizontal:

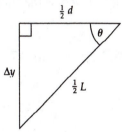

$$\sin\theta = \frac{\Delta y}{\frac{1}{2}L}$$

where L is the length of the chain.

Solve for Δy:

$$\Delta y = \tfrac{1}{2}L\sin\theta$$

Because the horizontal and vertical forces in the chain are equal, $\theta = 45°$ and:

$$\Delta y = \tfrac{1}{2}(5\,\text{m})\sin 45° = \boxed{1.77\,\text{m}}$$

(c) Relate the distance d of the boat from the dock to the angle θ the chain makes with the horizontal:

$$\cos\theta = \frac{\frac{1}{2}d}{\frac{1}{2}L} = \frac{d}{L}$$

Solve for and evaluate d:

$$d = L\cos\theta = (5\,\text{m})\cos 45° = \boxed{3.54\,\text{m}}$$

(d) Relate the resultant tension in the chain to the vertical component of the tension F_v and the maximum drag force exerted on the boat by the water $F_{d,\text{max}}$:

$$F_v^2 + F_{d,\text{max}}^2 = (500\,\text{N})^2$$

Solve for $F_{d,max}$:

$$F_{d,max} = \sqrt{(500\,N)^2 - F_v^2}$$

Because the vertical component of
the tension is 50 N:

$$F_{d,max} = \sqrt{(500\,N)^2 - (50\,N)^2} = \boxed{497\,N}$$

*46 ••   A section of a cathedral wall is shown in Figure 12-48. The arch
attached to the wall exerts a force of 2×10^4 N directed at an angle of 30° below the
horizontal at a point 10 m above the ground and the mass of the wall itself is 30,000
kg. The coefficient of static friction between the wall and the ground is $\mu_k = 0.8$
and the base of the wall is 1.25 m long. (*a*) Calculate the effective normal force, the
frictional force, and the point at which the effective normal force acts on the wall.
(This point is called the *thrust point* by architects and civil engineers.) (*b*) If the
thrust point ever moves outside the base of the wall, the wall will overturn. Apart
from aesthetics, placing a heavy statue on top of the wall has good engineering
practicality: it will move the thrust point toward the center of the wall. Explain
why.

Figure 12-48 Problems 46, 47

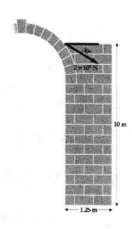

Picture the Problem Choose the coordinate system shown in the diagram and let x be the coordinate of the thrust point. The diagram to the right shows the forces acting on the wall. The normal force must balance out the weight of the wall and the vertical component of the thrust from the arch and the frictional force must balance out the horizontal component of the thrust. We can apply the conditions for translational equilibrium to find f and F_n and the condition for rotational equilibrium to find the distance x from the origin of our coordinate system at which F_n acts.

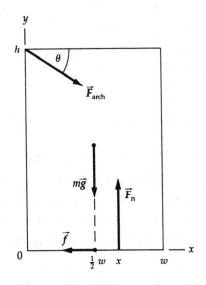

(*a*) Apply the conditions for translational equilibrium to the wall to obtain:

$$\sum F_x = -f + F_{arch}\cos\theta = 0 \qquad (1)$$
and
$$\sum F_y = F_n - mg - F_{arch}\sin\theta = 0 \qquad (2)$$

Solve equation (1) for and evaluate f:

$$f = F_{arch}\cos\theta = (2\times10^4\,\text{N})\cos30°$$
$$= \boxed{17.3\,\text{kN}}$$

Solve equation (2) for F_n:

$$F_n = mg + F_{arch}\sin\theta$$

Substitute numerical values and evaluate F_n:

$$F_n = (3\times10^4\,\text{kg})(9.81\,\text{m/s}^2)$$
$$+ (2\times10^4\,\text{N})\sin30°$$
$$= \boxed{304\,\text{kN}}$$

Apply $\sum \tau_{z\,axis} = 0$ to the to the wall:
Solve for x:

$$xF_n - \tfrac{1}{2}wmg - hF_{arch}\cos\theta = 0$$

$$x = \frac{\tfrac{1}{2}wmg + hF_{arch}\cos\theta}{F_n}$$

Substitute numerical values and evaluate x:

$$x = \frac{\tfrac{1}{2}(1.25\,\text{m})(3\times10^4\,\text{kg})(9.81\,\text{m/s}^2) + (10\,\text{m})(2\times10^4\,\text{N})\cos30°}{304\,\text{kN}} = \boxed{0.570\,\text{m}}$$

(b) If there were no thrust on the side of the wall, the normal force would act through the center of mass, so making the weight larger compared to the thrust must move the point of action of the normal force closer to the center.

Ladder Problems

*48 •• Romeo takes a uniform 10-m ladder and leans it against the smooth (frictionless) wall of the Capulet residence. The ladder's mass is 22.0 kg and the bottom rests on the ground 2.8 m from the wall. When Romeo, whose mass is 70 kg, gets 90 percent of the way to the top, the ladder begins to slip. What is the coefficient of static friction between the ground and the ladder?

Picture the Problem The ladder and the forces acting on it at the critical moment of slipping are shown in the diagram. Use the coordinate system shown. Because the ladder is in equilibrium, we can apply the conditions for translational and rotational equilibrium.

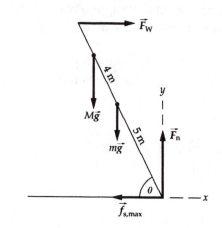

Using its definition, express μ_s:

$$\mu_s = \frac{f_{s,max}}{F_n} \tag{1}$$

Apply $\sum \vec{\tau} = 0$ about the bottom of the ladder:

$$[(9\,\text{m})\cos\theta]Mg + [(5\,\text{m})\cos\theta]mg \\ -[(10\,\text{m})\sin\theta]F_W = 0$$

Solve for F_W:

$$F_W = \frac{(9\,\text{m})M + (5\,\text{m})m}{(10\,\text{m})\sin\theta}g\cos\theta$$

Find the angle θ:

$$\theta = \cos^{-1}\left(\frac{2.8\,\text{m}}{10\,\text{m}}\right) = 73.74°$$

Evaluate F_W:

$$F_W = \frac{(9\,\text{m})(70\,\text{kg}) + (5\,\text{m})(22\,\text{kg})}{(10\,\text{m})\sin 73.74^\circ}$$
$$\times (9.81\,\text{m/s}^2)\cos 73.74^\circ$$
$$= 211.7\,\text{N}$$

Apply $\sum F_x = 0$ to the ladder and solve for $f_{s,max}$:

$F_W - f_{s,max} = 0$
and
$f_{s,max} = F_W = 211.7\,\text{N}$

Apply $\sum F_y = 0$ to the ladder:

$F_n - Mg - mg = 0$

Solve for F_n:

$F_n = (M + m)g$
$= (70\,\text{kg} + 22\,\text{kg})(9.81\,\text{m/s}^2)$
$= 902.5\,\text{N}$

Substitute numerical values in equation (1) and evaluate μ_s:

$$\mu_s = \frac{211.7\,\text{N}}{902.5\,\text{N}} = \boxed{0.235}$$

Stress and Strain

*54 •   A 50-kg ball is suspended from a steel wire of length 5 m and radius 2 mm. By how much does the wire stretch?

Picture the Problem L is the unstretched length of the wire, F is the force acting on it, and A is its cross-sectional area. The stretch in the wire ΔL is related to Young's modulus by $Y = (F/A)/(\Delta L/L)$. We can use Table 12-1 to find the numerical value of Young's modulus for steel.

Find the amount the wire is stretched from Young's modulus:

$$Y = \frac{F/A}{\Delta L/L}$$

Solve for ΔL:

$$\Delta L = \frac{FL}{YA}$$

Substitute for F and A to obtain:

$$\Delta L = \frac{mgL}{Y\pi r^2}$$

Substitute numerical values and evaluate ΔL:

$$\Delta L = \frac{(50\,\text{kg})(9.81\,\text{m/s}^2)(5\,\text{m})}{2\pi \times 10^{11}\,\text{N/m}^2\,(2\times 10^{-3}\,\text{m})^2}$$

$$= \boxed{0.976\,\text{mm}}$$

*57 • As a runner's foot pushes off on the ground, the shearing force acting on an 8-mm-thick sole is shown in Figure 12-50. If the force of 25 N is distributed over an area of 15 cm², find the angle of shear θ, given that the shear modulus of the sole is $1.9 \times 10^5\,\text{N/m}^2$.

Figure 12-50 Problem 57

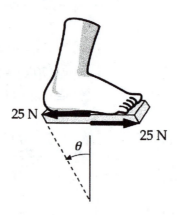

Picture the Problem The shear stress, defined as the ratio of the shearing force to the area over which it is applied, is related to the shear strain through the definition of the shear modulus; $M_s \equiv \dfrac{\text{shear stress}}{\text{shear strain}} = \dfrac{F_s/A}{\tan\theta}$.

Using the definition of shear modulus, relate the angle of shear, θ to the shear force and shear modulus:

$$\tan\theta = \frac{F_s}{M_s A}$$

Solve for θ:

$$\theta = \tan^{-1}\left[\frac{F_s}{M_s A}\right]$$

Substitute numerical values and evaluate θ:

$$\theta = \tan^{-1}\left[\frac{25\,\text{N}}{(1.9\times 10^5\,\text{N/m}^2)(15\times 10^{-4}\,\text{m}^2)}\right]$$

$$= \boxed{5.01^\circ}$$

*61 •• When a rubber strip with a cross section of 3 mm × 1.5 mm is suspended vertically and various masses are attached to it, a student obtains the following data for length versus load:

Load, g	0	100	200	300	400	500
Length, cm	5.0	5.6	6.2	6.9	7.8	10.0

(*a*) Find Young's modulus for the rubber strip for small loads. (*b*) Find the energy stored in the strip when the load is 0.150 kg. (See Problem 59.)

Picture the Problem The table to the right summarizes the ratios $\Delta L/F$ for the student's data. Note that this ratio is constant, to three significant figures, for loads less than or equal to 200 g. We can use this ratio to calculate Young's modulus for the rubber strip.

Load	F	ΔL	$\Delta L/F$
(g)	(N)	(m)	(m/N)
100	0.981	0.006	6.12×10^{-3}
200	1.962	0.012	6.12×10^{-3}
300	2.943	0.019	6.46×10^{-3}
400	3.924	0.028	7.14×10^{-3}
500	4.905	0.05	10.2×10^{-3}

(*a*) Referring to the table, we see that for loads ≤ 200 g:

$$\frac{\Delta L}{F} = 6.12 \times 10^{-3} \text{ m/N}$$

Use the definition of Young's modulus to express Y:

$$Y = \frac{FL}{A\Delta L} = \frac{L}{A\frac{\Delta L}{F}}$$

Substitute numerical values and evaluate Y:

$$Y = \frac{5 \times 10^{-2} \text{ m}}{\left(3 \times 10^{-3} \text{ m}\right)\left(1.5 \times 10^{-3} \text{ m}\right)\left(6.12 \times 10^{-3} \text{ m/N}\right)} = \boxed{1.82 \times 10^{6} \text{ N/m}^2}$$

(*b*) Interpolate to determine the stretch when the load is 150 g, and use the expression from Problem 58, to express the energy stored in the strip:

$$U = \tfrac{1}{2}F\Delta L$$
$$= \tfrac{1}{2}(0.15\,\text{kg})\left(9.81\,\text{m/s}^2\right)\left(9 \times 10^{-3}\,\text{m}\right)$$
$$= \boxed{6.62\,\text{mJ}}$$

*65 •• An elevator cable is to made of a new type of composite developed by Acme Laboratories. In the lab, a sample of the cable that is 2 m long and has a cross-sectional area of 0.2 mm² fails under a load of 1000 N. The cable in the elevator will be 20 m long and have a cross-sectional area of 1.2 mm². It will need to support a load of 20,000 N safely. Will it?

Picture the Problem We can use the definition of stress to calculate the failing stress of the cable and the stress on the elevator cable. Note that the failing stress of the composite cable is the same as the failing stress of the test sample.

Express the stress on the elevator cable:

$$\text{Stress}_{\text{cable}} = \frac{F}{A} = \frac{20\,\text{kN}}{1.2\times10^{-6}\,\text{m}^2}$$
$$= 1.67\times10^{10}\,\text{N/m}^2$$

Express the failing stress of the sample:

$$\text{Stress}_{\text{failing}} = \frac{F}{A} = \frac{1\,\text{kN}}{0.2\times10^{-6}\,\text{m}^2}$$
$$= 0.500\times10^{10}\,\text{N/m}^2$$

Because $\text{Stress}_{\text{failing}} < \text{Stress}_{\text{cable}}$, it will not support the elevator.

*66 •••   When a material is stretched in one direction, if its density remains constant, then (because its total volume remains constant), its length must decrease in one or both of the other directions. Take a rectangular block of length x, width y, and depth z, and pull on it so that it's new length $x'' = x + \Delta x$. If $\Delta x \ll x$ and $\Delta y/y = \Delta z/z$, show that $\Delta y/y = -\frac{1}{2}\Delta x/x$.

Picture the Problem Let the length of the sides of the rectangle be x, y and z. Then the volume of the rectangle will be $V = xyz$ and we can express the new volume V' resulting from the pulling in the x direction and the change in volume ΔV in terms of Δx, Δy, and Δz. Discarding the higher order terms in ΔV and dividing our equation by V and using the given condition that $\Delta y/y = \Delta z/z$ will lead us to the given expression for $\Delta y/y$.

Express the new volume of the rectangular box when its sides change in length by Δx, Δy, and Δz:

$$V' = (x+\Delta x)(y+\Delta y)(z+\Delta z)$$
$$= xyz + \Delta x(yz) + \Delta y(xz) + \Delta z(xy)$$
$$+ \{z\Delta x\Delta y + y\Delta x\Delta z + x\Delta y\Delta z + \Delta x\Delta y\Delta z\}$$

where the terms in brackets are very small (i.e., second order or higher).

Discard the second order and higher terms to obtain:

$$V' = V + \Delta x(yz) + \Delta y(xz) + \Delta z(xy)$$
or
$$\Delta V = V' - V = \Delta x(yz) + \Delta y(xz) + \Delta z(xy)$$

Because $\Delta V = 0$:

$$\Delta x(yz) = -\left[\Delta y(xz) + \Delta z(xy)\right]$$

Divide both sides of this equation by $V = xyz$ to obtain:

$$\frac{\Delta x}{x} = -\left[\frac{\Delta y}{y} + \frac{\Delta z}{z}\right]$$

Because $\Delta y/y = \Delta z/z$, our equation becomes:

$$\frac{\Delta x}{x} = -2\frac{\Delta y}{y} \text{ or } \frac{\Delta y}{y} = \boxed{-\frac{1}{2}\frac{\Delta x}{x}}$$

*68 ••• For most materials listed in Table 12-1, the tensile strength is two to three orders of magnitude lower than Young's modulus. Consequently, most of these materials will break before their strain exceeds 1 percent. Of man-made materials, nylon has about the greatest extensibility - it can take strains of about 0.2 before breaking. But spider silk beats anything man-made. Certain forms of spider silk can take strains on the order of 10 before breaking! (*a*) If such a thread has a circular cross-section of radius r_0 and unstretched length L_0, find its new radius r when stretched to a length $L=10L_0$. (*b*) If the Young's modulus of the spider thread is Y, calculate the tension needed to break the thread in terms of Y and r_0.

Picture the Problem Because the volume of the thread remains constant during the stretching process, we can equate the initial and final volumes to express r_0 in terms of r. We can also use Young's modulus to express the tension needed to break the thread in terms of Y and r_0.

(*a*) Express the conservation of volume during the stretching of the spider's silk:

$$\pi r^2 L = \pi r_0^2 L_0$$

Solve for r:

$$r = r_0\sqrt{\frac{L_0}{L}}$$

Substitute for L to obtain:

$$r = r_0\sqrt{\frac{L_0}{10L_0}} = \boxed{0.316 r_0}$$

(*b*) Express Young's modulus in terms of the breaking tension T:

$$Y = \frac{T/A}{\Delta L/L} = \frac{T/\pi r^2}{\Delta L/L} = \frac{10T/\pi r_0^2}{\Delta L/L}$$

Solve for T to obtain:

$$T = \frac{1}{10}\pi r_0^2 Y \frac{\Delta L}{L}$$

Because $\Delta L/L = 9$:

$$T = \boxed{\frac{9\pi r_0^2 Y}{10}}$$

General Problems

*71 • Figure 12-54 shows a mobile consisting of four weights hanging on three rods of negligible mass. Find the value of each of the unknown weights if the mobile is to balance. *Hint: Find the weight w₁ first.*

Figure 12-54 Problem 71

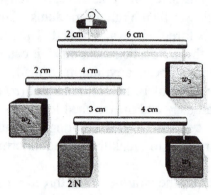

Picture the Problem We can apply the balance condition $\sum \vec{\tau} = 0$ successively, starting with the lowest part of the mobile, to find the value of each of the unknown weights.

Apply $\sum \vec{\tau} = 0$ about an axis through the point of suspension of the lowest part of the mobile:

$$(3\,\text{cm})(2\,\text{N}) - (4\,\text{cm})w_1 = 0$$

Solve for and evaluate w_1:

$$w_1 = \frac{(3\,\text{cm})(2\,\text{N})}{4\,\text{cm}} = \boxed{1.50\,\text{N}}$$

Apply $\sum \vec{\tau} = 0$ about an axis through the point of suspension of the middle part of the mobile:

$$(2\,\text{cm})w_2 - (4\,\text{cm})(2\,\text{N} + 1.5\,\text{N}) = 0$$

Solve for and evaluate w_2:

$$w_2 = \frac{(4\,\text{cm})(2\,\text{N} + 1.5\,\text{N})}{2\,\text{cm}} = \boxed{7.00\,\text{N}}$$

Apply $\sum \vec{\tau} = 0$ about an axis through the point of suspension of the top part of the mobile:

$$(2\,\text{cm})(10.5\,\text{N}) - (6\,\text{cm})w_3 = 0$$

Solve for and evaluate w_3:

$$w_3 = \frac{(2\,\text{cm})(10.5\,\text{N})}{6\,\text{cm}} = \boxed{3.50\,\text{N}}$$

*77 ••   A cube of mass M leans against a frictionless wall making an angle of θ with the floor as shown in Figure 12-58. Find the minimum coefficient of static friction μ_s between the cube and the floor that allows the cube to stay at rest.

Figure 12-58 Problem 77

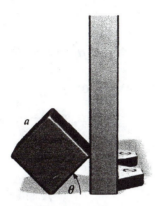

Picture the Problem The figure shows the location of the cube's center of mass and the forces acting on the cube. The opposing couple is formed by the friction force $f_{s,max}$ and the force exerted by the wall. Because the cube is in equilibrium, we can use the condition for translational equilibrium to establish that $f_{s,max} = F_W$ and $F_n = Mg$ and the condition for rotational equilibrium to relate the opposing couples.

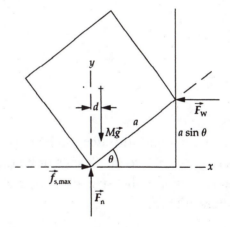

Apply $\sum \vec{F} = 0$ to the cube:

$$\sum F_y = F_n - Mg = 0 \Rightarrow F_n = Mg$$
and
$$\sum F_x = f_s - F_W = 0 \Rightarrow F_W = f_s$$

Noting that $\vec{f}_{s,max}$ and $\vec{F}_W$ form a couple, as do $\vec{F}_n$ and $M\vec{g}$, apply $\sum \vec{\tau} = 0$ about an axis though the center of mass of the cube:

$$f_{s,max}\, a \sin\theta - Mgd = 0$$

Referring to the diagram to the right, note that

$$d = \frac{a}{\sqrt{2}}\sin(45° + \theta):$$

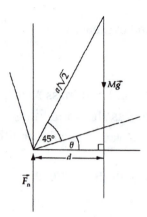

Substitute for d and $f_{s,max}$ to obtain:

$$\mu_s Mga\sin\theta - Mg\frac{a}{\sqrt{2}}\sin(45° + \theta) = 0$$

or

$$\mu_s\sin\theta - \frac{1}{\sqrt{2}}\sin(45° + \theta) = 0$$

Solve for μ_s and simplify to obtain:

$$\mu_s = \frac{1}{\sqrt{2}\sin\theta}\sin(45° + \theta)$$

$$= \frac{1}{\sqrt{2}\sin\theta}(\sin 45°\cos\theta + \cos 45°\sin\theta)$$

$$= \frac{1}{\sqrt{2}\sin\theta}\left(\frac{1}{\sqrt{2}}\cos\theta + \frac{1}{\sqrt{2}}\sin\theta\right)$$

$$= \boxed{\frac{1}{2}(\cot\theta + 1)}$$

*80 •• A uniform cube can be moved along a horizontal plane either by pushing the cube so that it slips or by turning it over ("rolling"). What coefficient of kinetic friction μ_k between the cube and the floor makes both ways equal in terms of the work needed?

Picture the Problem To "roll" the cube one must raise its center of mass from $y = a/2$ to $y = \sqrt{2}a/2$, where a is the cube length. During this process the work done is the change in the gravitational potential energy of the cube. No additional work is done on the cube as it "flops" down. We can also use the definition of work to express the work done in sliding the cube a distance a along a horizontal surface and then

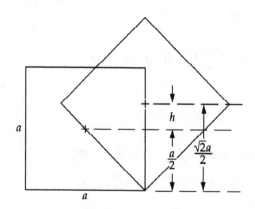

equate the two expressions to determine μ_k.

Express the work done in moving the cube a distance a by raising its center of mass from $y = a/2$ to $y = \sqrt{2}a/2$ and then letting the cube flop down:	$W = mg\left(\dfrac{\sqrt{2}a}{2} - \dfrac{a}{2}\right) = \dfrac{mga}{2}\left(\sqrt{2} - 1\right)$ $= 0.207 mga$
Letting f_k represent the kinetic friction force, express the work done in dragging the cube a distance a along the surface at constant speed:	$W = f_k a = \mu_k mga$
Equate these two expressions to obtain:	$\mu_k = \boxed{0.207}$

*84 ••   A thin rod 60 cm long is balanced 20 cm from one end when an object whose mass is $2m + 2$ g is at the end nearest the pivot and an object of mass m is at the opposite end (Figure 12-63a). Balance is again achieved if the object whose mass is $2m + 2$ g is replaced by the object of mass m and no object is placed at the other end (Figure 12-63b). Determine the mass of the rod.

Figure 12-63 Problem 84

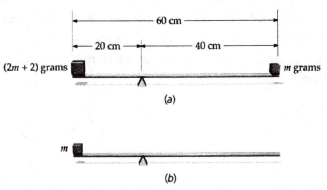

(a)

(b)

Picture the Problem Let the mass of the rod be represented by M. Because the rod is in equilibrium, we can apply the condition for rotational equilibrium to relate the masses of the objects placed on it to its mass.

Apply $\sum \vec{\tau} = 0$ about an axis through the pivot for the initial	$(20\,\text{cm})(2m+2\text{g}) - (40\,\text{cm})m$ $-(10\,\text{cm})M = 0$

condition:

Solve for and evaluate M:

$$M = \frac{(20\,\text{cm})(2m + 2\,\text{g}) - (40\,\text{cm})m}{10\,\text{cm}}$$

$$= \boxed{4.00\,\text{g}}$$

Apply $\sum \vec{\tau} = 0$ about an axis through the pivot for the second condition:

$$(20\,\text{cm})m - (10\,\text{cm})M = 0$$

Solve for and evaluate m:

$$m = \frac{(10\,\text{cm})M}{20\,\text{cm}} = \tfrac{1}{2}M = \boxed{2.00\,\text{g}}$$

*85 ••   If you balance a meter stick across two fingers (one from each hand) and slowly bring your hands together, the fingers will always meet in the middle of the stick, no matter where they are initially placed. (*a*) Explain why this is true. (*b*) As you move your fingers together, first one will move, and then the other; both will not move at the same time. Explain quantitatively why this is true; assume that the coefficient of static friction between the meterstick and your fingers is μ_s and the coefficient of kinetic friction is μ_k.

Picture the Problem Let the distance from the center of the meterstick of either finger be x_1 and x_2 and W the weight of the stick. Because the meterstick is in equilibrium, we can apply the condition for rotational equilibrium to obtain expressions for the forces one's fingers exert on the meterstick as functions of the distances x_1 and x_2 and the weight of the meterstick W. We can then explain the stop-and-start motion of one's fingers as they are brought closer together by considering the magnitudes of these forces in relationship the coefficients of static and kinetic friction.

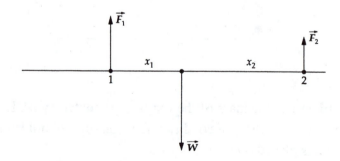

(a)

The stick remains balanced as long as the center of mass is between the two fingers. For a balanced stick the normal force exerted by the finger nearest the center of mass is greater than that exerted by the other finger. Consequently, a larger static-frictional force can be exerted by the finger closer to the center of mass, which means the slipping occurs at the other finger.

(b) Apply $\sum \vec{\tau} = 0$ about an axis through point 1 to obtain:

$$F_2(x_1 + x_2) - Wx_1 = 0$$

Solve for F_2 to obtain:

$$F_2 = W \frac{x_1}{x_1 + x_2}$$

Apply $\sum \vec{\tau} = 0$ about an axis through point 2 to obtain:

$$-F_1(x_1 + x_2) + Wx_2 = 0$$

Solve for F_1 to obtain:

$$F_1 = W \frac{x_2}{x_1 + x_2}$$

The finger farthest from the center of mass will slide inward until the normal force it exerts on the stick is sufficiently large to produce a kinetic-frictional force exceeding the maximum static-frictional force exerted by the other finger. At that point the finger that was not sliding begins to slide, the finger that was sliding stops sliding, and the process is reversed. When one finger is is slipping the other is not.

*89 ••• We have a large number of identical uniform bricks, each of length L. If we stack one on top of another lengthwise (see Figure 12-64), the maximum offset that will allow the top brick to rest on the bottom brick is $L/2$. (a) Show that if we place this two-brick stack on top of a third brick, the maximum offset of the second brick on the third is $L/4$. (b) Show that, in general, if we have a stack of N bricks, the maximum offset of the nth brick (counting down from the top) on the nth brick is $L/2n$, where $n \leq N$. (c) Write a spreadsheet program to calculate total offset (the sum of the individual offsets) for a stack of N bricks, and calculate this for $L = 1$ m and $N = 5$, 10, and 100. (d) Does the sum of the individual offsets approach a finite limit as $N \to \infty$? If so, what is that limit?

Picture the Problem Let the mass of each brick be m and number them as shown in the diagrams for 3 bricks and 4 bricks below. Let ℓ denote the maximum offset of the nth brick. Choose the coordinate system shown and apply the condition for rotational equilibrium about an axis parallel to the z axis and passing through the point P at the supporting edge of the nth brick.

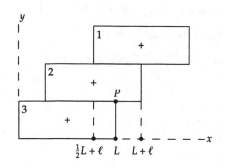

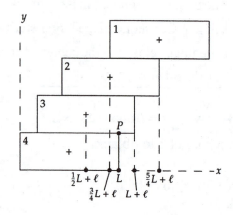

(*a*) Apply $\sum \vec{\tau} = 0$ about an axis through P and parallel to the z axis to bricks 1 and 2 for the 3-brick arrangement shown above on the left:

$$mg\left[L - \left(\tfrac{1}{2}L + \ell\right)\right] - mg\ell = 0$$

Solve for ℓ to obtain:

$$\ell = \boxed{\tfrac{1}{4}L}$$

(*b*) Apply $\sum \vec{\tau} = 0$ about an axis through P and parallel to the z axis to bricks 1 and 2 for the 4-brick arrangement shown above on the right:

$$mg\left[L - \left(\tfrac{1}{2}L + \ell\right)\right] + mg\left[L - \left(\tfrac{3}{4}L + \ell\right)\right] - mg\left(\tfrac{5}{4}L + \ell - L\right) = 0$$

Solve for ℓ to obtain:

$$\ell = \tfrac{1}{6}L$$

Continuing in this manner we obtain, as the successive offsets, the sequence:

$$\boxed{\dfrac{L}{2}, \dfrac{L}{4}, \dfrac{L}{6}, \dfrac{L}{8}, \cdots \dfrac{L}{2n}}$$

where $n = 1, 2, 3, \ldots N$.

(c) Express the offset of the (n +1)st brick in terms of the offset of the nth brick:

$$\ell_{n+1} = \ell_n + \dfrac{L}{2n}$$

A spreadsheet program to calculate the sum of the offsets as a function of n is shown below. The formulas used to calculate the quantities in the columns are as follows:

Cell	Formula/Content	Algebraic Form
B5	B4+1	$n+1$
C5	C4+$B$1/(2*B5)	$\ell_n + \dfrac{L}{2n}$

	A	B	C	D
1	L=	1	m	
2				
3		n	offset	
4		1	0.500	
5		2	0.750	
6		3	0.917	
7		4	1.042	
8		5	1.142	
9		6	1.225	
10		7	1.296	
11		8	1.359	
12		9	1.414	
13		10	1.464	
98		95	2.568	
99		96	2.573	
100		97	2.579	
101		98	2.584	
102		99	2.589	

From the table we see that $\ell_5 = \boxed{1.142\,\text{m}}$, $\ell_{10} = \boxed{1.464\,\text{m}}$, and $\ell_{100} = \boxed{2.594\,\text{m}}$.

(*d*) Increasing N in the spreadsheet solution suggests that the sum of the individual offsets continues to grow as N increases without bound. The series is, in fact, divergent and the stack of bricks has no maximum offset or length.

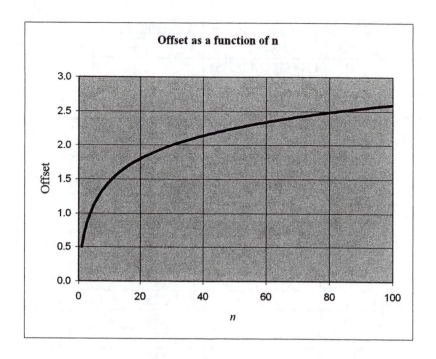

Offset as a function of n

*94 •••   A solid cube of side length a balanced atop a cylinder of diameter d is in unstable equilibrium if $d \ll a$ and is in stable equilibrium if $d \gg a$ (Figure 12-68). Determine the minimum value of d/a for which the cube is in stable equilibrium.

Figure 12-68 Problem 94

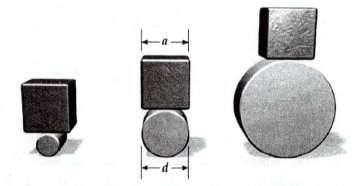

Picture the Problem Consider a small rotational displacement, $\delta\theta$ of the cube from equilibrium. This shifts the point of contact between cube and cylinder by $R\,\delta\theta$, where $R = d/2$. As a result of that motion, the cube itself is rotated through the same angle $\delta\theta$, and so its center is shifted in the same direction by the amount $(a/2)\,\delta\theta$, neglecting higher order terms in $\delta\theta$.

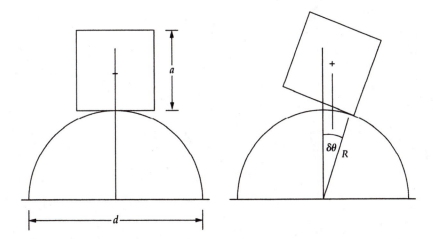

If the displacement of the cube's center of mass is less than that of the point of contact, the torque about the point of contact is a restoring torque, and the cube will return to its equilibrium position. If, on the other hand, $(a/2)\,\delta\theta > (d/2)\,\delta\theta$, then the torque about the point of contact due to mg is in the direction of $\delta\theta$, and will cause the displacement from equilibrium to increase. We see that the minimum value of d/a for stable equilibrium is $d/a = 1$.

Chapter 13
Fluids

Conceptual Problems

*2 •   Does Archimedes' principle hold in a satellite orbiting the earth in a circular orbit? Explain.

Determine the Concept No. In an environment where $g_{eff} = F_g - m\dfrac{v^2}{r} = 0$, there is no buoyant force; there is no "up" or "down."

*6 ••   Two objects are balanced as in Figure 13-24. The objects have identical volumes but different masses. Will the equilibrium be disturbed if the entire system is completely immersed in water? Explain.

Figure 13-24 Problem 6

Determine the Concept Yes. Because the volumes of the two objects are equal, the downward force on each side is reduced by the same amount when they are submerged, not in proportion to their masses. That is, if $m_1L_1 = m_2L_2$ and $L_1 \neq L_2$, then $(m_1 - c)L_1 \neq (m_2 - c)L_2$.

*11 •   True or false: The buoyant force on a submerged object depends on the shape of the object.

Determine the Concept False. The buoyant force on a submerged object depends on the weight of the displaced fluid which, in turn, depends on the volume of the displaced fluid.

*18 ••   You are sitting in a boat floating on a very small pond. You take the anchor out of the boat and drop it into the water. What happens to the water level in the pond?

Determine the Concept The water level in the pond will drop slightly. When the anchor is in the boat, the boat displaces enough water so that the buoyant force on it equals the sum of the weight of the boat, your weight, and the weight of the anchor. When you put the anchor overboard, it will displace its volume and the volume of water displaced by the boat will decrease.

*20 •   Three bottles of different shapes are filled to the same level with water, as shown in Figure 13-27. The area of the bottom of each bottle is the same. The hydrostatic pressure is the same on the bottom of each of the bottles, but the total force must be different, as each bottle has a different amount of liquid in it. Explain this apparent paradox.

Figure 13-27 Problem 20

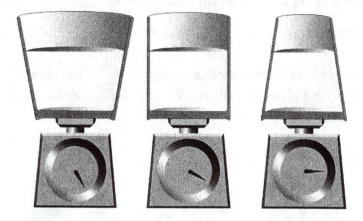

Determine the Concept The diagram that follows shows the forces exerted by the pressure of the liquid on the two cups to the left.

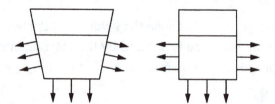

Because the force is normal to the surface of the cup, there is a larger downward component to the net force on the cup on the left.   Similarly, there will be less total force exerted by the fluid in the cup on the far right in the diagram in the problem statement.

Density

*24 •   A 60-mL flask is filled with mercury at 0°C (Figure 13-28). When the temperature rises to 80°C, 1.47 g of mercury spill out of the flask. Assuming that the volume of the flask is constant, find the density of mercury at 80°C if its density at 0°C is 13,645 kg/m^3.

Figure 13-28 Problem 24

Picture the Problem Let ρ_0 represent the density of mercury at 0°C and ρ' its density at 80°C, and let m represent the mass of our sample at 0°C and m' its mass at 80°C. We can use the definition of density to relate its value at the higher temperature to its value at the lower temperature and the amount spilled.

Using its definition, express the density of the mercury at 80°C:	$\rho' = \dfrac{m'}{V}$
Express the mass of the mercury at 80°C in terms of its mass at 0°C and the amount spilled at the higher temperature:	$\rho' = \dfrac{m - \Delta m}{V} = \dfrac{m}{V} - \dfrac{\Delta m}{V}$ $= \rho_0 - \dfrac{\Delta m}{V}$
Substitute numerical values and evaluate ρ':	$\rho' = 1.3645 \times 10^4 \text{ kg/m}^3 - \dfrac{1.47 \times 10^{-3} \text{ kg}}{60 \times 10^{-6} \text{ m}^3}$ $= \boxed{1.3621 \times 10^4 \text{ kg/m}^3}$

Pressure

*27 • A hydraulic lift is used to raise an automobile of mass 1500 kg. The radius of the shaft of the lift is 8 cm and that of the piston is 1 cm. How much force must be applied to the piston to raise the automobile?

Picture the Problem The pressure applied to an enclosed liquid is transmitted undiminished to every point in the fluid and to the walls of the container. Hence we can equate the pressure produced by the force applied to the piston to the pressure due to the weight of the automobile and solve for F.

Express the pressure the weight of the automobile exerts on the shaft of the lift:	$$P_{auto} = \frac{w_{auto}}{A_{shaft}}$$
Express the pressure the force applied to the piston produces:	$$P = \frac{F}{A_{piston}}$$
Because the pressures are the same, we can equate them to obtain:	$$\frac{w_{auto}}{A_{shaft}} = \frac{F}{A_{piston}}$$
Solve for F:	$$F = w_{auto} \frac{A_{piston}}{A_{shaft}} = m_{auto} g \frac{A_{piston}}{A_{shaft}}$$
Substitute numerical values and evaluate F:	$$F = (1500\,kg)(9.81\,m/s^2)\left(\frac{1\,cm}{8\,cm}\right)^2$$ $$= \boxed{230\,N}$$

*29 •   What pressure increase is required to compress the volume of 1 kg of water from 1.00 L to 0.99 L?

Picture the Problem The required pressure ΔP is related to the change in volume ΔV and the initial volume V through the definition of the bulk modulus B;
$$B = -\frac{\Delta P}{\Delta V/V}.$$

Using the definition of the bulk modulus, relate the change in volume to the initial volume and the required pressure:	$$B = -\frac{\Delta P}{\Delta V/V}$$
Solve for ΔP:	$$\Delta P = -B\frac{\Delta V}{V}$$
Substitute numerical values and evaluate ΔP:	$$\Delta P = -2.0\times10^9\,Pa\times\left(\frac{-0.01\,L}{1\,L}\right)$$ $$= 2.00\times10^7\,Pa\times\left(\frac{1\,atm}{101.325\,kPa}\right)$$ $$= \boxed{198\,atm}$$

*36 ••• The volume of a cone of height h and base radius r is $V = \pi r^2 h/3$. A conical vessel of height 25 cm resting on its base of radius 15 cm is filled with water. (*a*) Find the volume and weight of the water in the vessel. (*b*) Find the force exerted by the water on the base of the vessel. Explain how this force can be greater than the weight of the water.

Picture the Problem The weight of the water in the vessel is the product of its mass and the gravitational field. Its mass, in turn, is related to its volume through the definition of density. The force the water exerts on the base of the container can be determined from the product of the pressure it creates and the area of the base.

(*a*) Using the definition of density, relate the weight of the water to the volume it occupies:

$$w = mg = \rho V g$$

Substitute for V to obtain:

$$w = \tfrac{1}{3}\pi\rho r^2 h g$$

Substitute numerical values and evaluate w:

$$w = \tfrac{1}{3}\pi\left(10^3\,\text{kg/m}^3\right)\left(15\times10^{-2}\,\text{m}\right)^2\left(25\times10^{-2}\,\text{m}\right)\left(9.81\,\text{m/s}^2\right)= \boxed{57.8\,\text{N}}$$

(*b*) Using the definition of pressure, relate the force exerted by the water on the base of the vessel to the pressure it exerts and the area of the base:

$$F = PA = \rho g h \pi r^2$$

Substitute numerical values and evaluate F:

$$F = \left(10^3\,\text{kg/m}^3\right)\left(9.81\,\text{m/s}^2\right)\left(25\times10^{-2}\,\text{m}\right)\pi\left(15\times10^{-2}\,\text{m}\right)^2 = \boxed{173\,\text{N}}$$

> This occurs in the same way that the force on Pascal's barrel is much greater than the weight of the water in the tube. The downward force on the base is also the result of the downward component of the force exerted by the slanting walls of the cone on the water.

Buoyancy

*37 •   A 500-g piece of copper (specific gravity 9.0) is suspended from a spring scale and is submerged in water (Figure 13-31). What force does the spring scale read?

Figure 13-31 Problem 37

Picture the Problem The scale's reading will be the difference between the weight of the piece of copper in air and the buoyant force acting on it.

Express the apparent weight w' of the piece of copper:

$$w' = w - B$$

Using the definition of density and Archimedes' principle, substitute for w and B to obtain:

$$w' = \rho_{Cu} V g - \rho_w V g$$
$$= (\rho_{Cu} - \rho_w) V g$$

Express w in terms of ρ_{Cu} and V and solve for Vg:

$$w = \rho_{Cu} V g \Rightarrow V g = \frac{w}{\rho_{Cu}}$$

Substitute to obtain:

$$w' = (\rho_{Cu} - \rho_w) \frac{w}{\rho_{Cu}} = \left(1 - \frac{\rho_w}{\rho_{Cu}}\right) w$$

Substitute numerical values and evaluate w':

$$w' = \left(1 - \frac{1}{9}\right)(0.5\,\text{kg})(9.81\,\text{m/s}^2)$$

$$= \boxed{4.36\,\text{N}}$$

*42 ••   A 5-kg iron block is suspended from a spring scale and is submerged in a fluid of unknown density. The spring scale reads 6.16 N. What is the density of the fluid?

Picture the Problem We can use Archimedes' principle to find the density of the unknown object. The difference between the weight of the block in air and in the fluid is the buoyant force acting on the block.

Apply Archimedes' principle to obtain:

$$B = w_f = m_f g = \rho_f V_f g$$

Solve for ρ_f:

$$\rho_f = \frac{B}{V_f g}$$

Because $V_f = V_{\text{Fe block}}$:

$$\rho_f = \frac{B}{V_{\text{Fe block}} g} = \frac{B}{m_{\text{Fe block}} g} \rho_{\text{Fe}}$$

Substitute numerical values and evaluate ρ_f:

$$\rho_f = \frac{(5\,\text{kg})(9.81\,\text{m/s}^2) - 6.16\,\text{N}}{(5\,\text{kg})(9.81\,\text{m/s}^2)} (7.96 \times 10^3\,\text{kg/m}^3) = \boxed{6.96 \times 10^3\,\text{kg/m}^3}$$

*45 •• An object has neutral buoyancy when its density equals that of the liquid in which it is submerged, which means that it neither floats nor sinks. If the average density of an 85-kg diver is 0.96 kg/L, what mass of lead should be added to give him neutral buoyancy?

Picture the Problem Let V = volume of diver, ρ_D the density of the diver, V_{Pb} the volume of added lead, and m_{Pb} the mass of lead. The diver is in equilibrium under the influence of his weight, the weight of the lead, and the buoyant force of the water.

Apply $\sum F_y = 0$ to the diver:

$$B - w_D - w_{Pb} = 0$$

Substitute to obtain:

$$\rho_w V_{D+Pb} g - \rho_D V_D g - m_{Pb} g = 0$$

or

$$\rho_w V_D + \rho_w V_{Pb} - \rho_D V_D - m_{Pb} = 0$$

Rewrite this expression in terms of masses and densities:

$$\rho_w \frac{m_D}{\rho_D} + \rho_w \frac{m_{Pb}}{\rho_{Pb}} - \rho_D \frac{m_D}{\rho_D} - m_{Pb} = 0$$

Solve for the mass of the lead:

$$m_{Pb} = \frac{\rho_{Pb}(\rho_w - \rho_D) m_D}{\rho_D(\rho_{Pb} - \rho_w)}$$

Substitute numerical values and evaluate m_{Pb}:

$$m_{Pb} = \frac{\left(11.3 \times 10^3 \text{ kg/m}^3\right)\left(10^3 \text{ kg/m}^3 - 0.96 \times 10^3 \text{ kg/m}^3\right)\left(85 \text{ kg}\right)}{\left(0.96 \times 10^3 \text{ kg/m}^3\right)\left(11.3 \times 10^3 \text{ kg/m}^3 - 10^3 \text{ kg/m}^3\right)} = \boxed{3.89 \text{ kg}}$$

*48 ••• The hydrometer shown in Figure 13-34 is a device for measuring the specific gravity of liquids. The bulb contains lead shot, and the specific gravity can be read directly from the liquid level on the stem after the hydrometer has been calibrated. The volume of the bulb is 20 mL, the stem is 15 cm long and has a diameter of 5.00 mm, and the mass of the glass is 6.0 g. (*a*) What mass of lead shot must be added so that the lowest specific gravity of liquid that can be measured is 0.9? (*b*) What is the maximum specific gravity of liquid that can be measured?

Figure 13-34 Problem 48

Picture the Problem For minimum liquid density, the bulb and its stem will be submerged. For maximum liquid density, only the bulb is submerged. In both cases the hydrometer will be in equilibrium under the influence of its weight and the buoyant force exerted by the liquids.

(*a*) Apply $\sum F_y = 0$ to the hydrometer:

$B - w = 0$

Using Archimedes' principle to express B, substitute to obtain:

$\rho_{min} V g - m_{tot} g = 0$

or

$\rho_{min}\left(V_{bulb} + V_{stem}\right) = m_{glass} + m_{Pb}$

Solve for m_{Pb}:

$m_{Pb} = \rho_{min}\left(V_{bulb} + V_{stem}\right) - m_{glass}$

Substitute numerical values and evaluate m_{Pb}:

$$m_{Pb} = (0.9\,kg/L)\left[0.020\,L + \frac{\pi}{4}(0.15\,m)(0.005\,m)^2\left(\frac{1\,L}{10^{-3}\,m^3}\right)\right] - (6\times10^{-3}\,kg)$$

$$= \boxed{14.7\,g}$$

(b) Apply $\sum F_y = 0$ to the hydrometer:

$$\rho_{max}Vg - m_{tot}g = 0$$

or

$$\rho_{max}V_{bulb} = m_{glass} + m_{Pb}$$

Solve for ρ_{max}:

$$\rho_{max} = \frac{m_{glass} + m_{Pb}}{V_{bulb}}$$

Substitute numerical values and evaluate ρ_{max}:

$$\rho_{max} = \frac{6\,g + 14.7\,g}{20\,mL} = \boxed{1.04\,kg/L}$$

Continuity and Bernoulli's Equation

*51 ••   Water exits a circular tap moving straight down with a flow rate of 10.5 cm³/s. (a) If the diameter of the tap is 1.2 cm, what is the speed of the water? (b) As the fluid falls from the tap, the stream of water narrows. Find the new diameter of the stream at a point 7.5 cm below the tap. Assume that the stream still has a circular cross section and neglect any drag forces acting on the water. (c) If turbulent flows are characterized by Reynolds numbers above 2300 or so, how far does the water have to fall before it becomes turbulent? Does this match everyday experience?

Picture the Problem Let J represent the flow rate of the water. Then we can use $J = Av$ to relate the flow rate to the cross-sectional area of the circular tap and the velocity of the water. In (b) we can use the equation of continuity to express the diameter of the stream 7.5 cm below the tap and a constant-acceleration equation to find the velocity of the water at this distance. In (c) we can use a constant-acceleration equation to express the distance-to-turbulence in terms of the velocity of the water at turbulence v_t and the definition of Reynolds number N_R to relate v_t to N_R.

(a) Express the flow rate of the water in terms of the cross-sectional area A of the circular tap and the velocity v of the water:

$$J = Av = \pi r^2 v = \tfrac{1}{4}\pi d^2 v \qquad\qquad (1)$$

Solve for v:

$$v = \frac{J}{\tfrac{1}{4}\pi d^2}$$

Substitute numerical values and evaluate v:

$$v = \frac{10.5\,\text{cm}^3/\text{s}}{\frac{1}{4}\pi(1.2\,\text{cm})^2} = \boxed{9.28\,\text{cm/s}}$$

(b) Apply the equation of continuity to the stream of water:

$$v_f A_f = v_i A_i = vA_i$$

or

$$v_f \frac{\pi}{4}d_f^2 = v\frac{\pi}{4}d_i^2$$

Solve for d_f:

$$d_f = \sqrt{\frac{v}{v_f}}d_i \qquad (2)$$

Use a constant-acceleration equation to relate v_f and v to the distance Δh fallen by the water:

$$v_f^2 = v^2 + 2g\Delta h$$

Solve for v_f to obtain:

$$v_f = \sqrt{v^2 + 2g\Delta h}$$

Substitute numerical values and evaluate v_f:

$$v_f = \sqrt{(9.28\,\text{cm/s})^2 + 2(981\,\text{cm/s}^2)(7.5\,\text{cm})}$$
$$= 122\,\text{cm/s}$$

Substitute in equation (2) and evaluate d_f:

$$d_f = (1.2\,\text{cm})\sqrt{\frac{9.28\,\text{cm/s}}{122\,\text{cm/s}}} = \boxed{0.331\,\text{cm}}$$

(c) Using a constant-acceleration equation, relate the fall-distance-to-turbulence Δd to its initial speed v and its speed v_t when its flow becomes turbulent:

$$v_t^2 = v^2 + 2g\Delta d$$

Solve for Δd to obtain:

$$\Delta d = \frac{v_t^2 - v^2}{2g} \qquad (3)$$

Express Reynolds number N_R for turbulent flow:

$$N_R = \frac{2r\rho v_t}{\eta}$$

From equation (1):

$$r = \sqrt{\frac{J}{\pi v_t}}$$

Substitute to obtain:

$$N_R = \frac{2\rho v_t}{\eta}\sqrt{\frac{J}{\pi v_t}}$$

Solve for v_t:

$$v_t = \frac{\pi N_R^2 \eta^2}{4\rho^2 J}$$

Substitute numerical values (see Figure 13-1 for the density of water and Table 13-1 for the coefficient of viscosity for water) and evaluate v_t:

$$v_t = \frac{\pi (2300)^2 (1.8 \times 10^{-3} \text{ Pa} \cdot \text{s})^2}{4(10^3 \text{ kg/m}^3)^2 (10.5 \text{ cm}^3/\text{s})}$$

$$= 1.28 \text{ m/s}$$

Substitute in equation (3) and evaluate the fall-distance-to turbulence:

$$\Delta d = \frac{(128 \text{ cm/s})^2 - (9.28 \text{ cm/s})^2}{2(981 \text{ cm/s}^2)}$$

$$= \boxed{8.31 \text{ cm}}$$

in reasonable agreement with everyday experience.

*55 •• Blood flows in an aorta of radius 9 mm at 30 cm/s. (*a*) Calculate the volume flow rate in liters per minute. (*b*) Although the cross-sectional area of a capillary is much smaller than that of the aorta, there are many capillaries, so their total cross-sectional area is much larger. If all the blood from the aorta flows into the capillaries and the speed of flow through the capillaries is 1.0 mm/s, calculate the total cross-sectional area of the capillaries.

Picture the Problem We can use the definition of the volume flow rate to find the volume flow rate of blood in an aorta and to find the total cross-sectional area of the capillaries.

(*a*) Use the definition of the volume flow rate to find the volume flow rate through an aorta:

$$I_V = Av$$

Substitute numerical values and evaluate I_V:

$$I_V = \pi (9 \times 10^{-3} \text{ m}^3)(0.3 \text{ m/s})$$

$$= 7.63 \times 10^{-5} \frac{\text{m}^3}{\text{s}} \times \frac{60 \text{ s}}{\text{min}} \times \frac{1 \text{L}}{10^{-3} \text{ m}^3}$$

$$= \boxed{4.58 \text{ L/min}}$$

(*b*) Use the definition of the volume flow rate to express the volume flow rate through the capillaries:

$$I_V = A_{cap} v_{cap}$$

Solve for the total cross-sectional area of the capillaries:

$$A_{cap} = \frac{I_V}{v_{cap}}$$

Substitute numerical values and evaluate A_{cap}:

$$A_{cap} = \frac{7.63 \times 10^{-5} \text{ m}^3/\text{s}}{0.001 \text{ m/s}}$$

$$= 7.63 \times 10^{-2} \text{ m}^2 = \boxed{763 \text{ cm}^2}$$

*58 ••   Water flows through a Venturi meter like that in Example 13-10 with a pipe diameter of 9.5 cm and a constriction diameter of 5.6 cm. The U-tube manometer is partially filled with mercury. Find the volume flow rate of the water if the difference in the mercury level in the U-tube is 2.40 cm.

Picture the Problem We'll use its definition to relate the volume flow rate in the pipe to the velocity of the water and the result of Example 13-9 to find the velocity of the water.

Using its definition, express the volume flow rate:

$$I_V = A_1 v_1 = \pi r^2 v_1$$

Using the result of Example 13-9, find the velocity of the water upstream from the Venturi meter:

$$v_1 = \sqrt{\frac{2\rho_{Hg}gh}{\rho_w\left(\dfrac{R_1^2}{R_2^2} - 1\right)}}$$

Substitute numerical values and evaluate v_1:

$$v_1 = \sqrt{\frac{2\left(13.6 \times 10^3 \text{ kg/m}^3\right)\left(9.81 \text{ m/s}^2\right)\left(0.024 \text{ m}\right)}{\left(10^3 \text{ kg/m}^3\right)\left[\left(\dfrac{0.095 \text{ m}}{0.056 \text{ m}}\right)^2 - 1\right]}} = 1.847 \text{ m/s}$$

Substitute numerical values and evaluate I_V:

$$I_V = \frac{\pi}{4}\left(0.095 \text{ m}\right)^2\left(1.847 \text{ m/s}\right)$$

$$= 1.309 \times 10^{-2} \text{ m}^3/\text{s}$$

$$= \boxed{13.1 \text{ L/s}}$$

*62 ••   Figure 13-37 shows a Pitot-static tube, a device used for measuring the velocity of a gas. The inner pipe faces the incoming fluid, while the ring of holes in the outer tube is parallel to the gas flow. Show that the speed of the gas is given by $v^2 = 2gh\,(\rho - \rho_g)/\rho_g$, where ρ is the density of the liquid used in the manometer and ρ_g is the density of the gas.

Figure 13-37 Problem 62

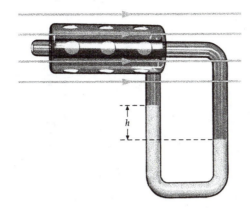

Picture the Problem Let the numeral 1 denote the opening in the end of the inner pipe and the numeral 2 to one of the holes in the outer tube. We can apply Bernoulli's principle at these locations and solve for the pressure difference between them. By equating this pressure difference to the pressure difference due to the height h of the liquid column we can express v as a function of ρ, ρ_g, g, and h.

Apply Bernoulli's principle at locations 1 and 2 to obtain:	$P_1 + \frac{1}{2}\rho_g v_1^2 = P_2 + \frac{1}{2}\rho_g v_2^2$ where we've ignored the difference in elevation between the two openings.
Solve for the pressure difference $\Delta P = P_1 - P_2$:	$\Delta P = P_1 - P_2 = \frac{1}{2}\rho_g v_2^2 - \frac{1}{2}\rho_g v_1^2$
Express the velocity of the gas at 1:	$v_1 = 0$ because the gas is brought to a halt (i.e., is stagnant) at the opening to the inner pipe.
Express the velocity of the gas at 2:	$v_2 = v$ because the gas flows freely past the holes in the outer ring.
Substitute to obtain:	$\Delta P = \frac{1}{2}\rho_g v^2$
Letting A be the cross-sectional area of the tube, express the pressure at a depth h in the column of liquid whose density is ρ_l:	$P_1 = P_2 + \dfrac{w_{\text{displaced liquid}}}{A} - \dfrac{B}{A}$ where $B = \rho_g A h g$ is the buoyant force acting on the column of liquid of height h.

Substitute to obtain:

$$P_1 = P_2 + \frac{\rho\,ghA}{A} - \frac{\rho_g ghA}{A}$$
$$= P_2 + \left(\rho - \rho_g\right)gh$$

or

$$\Delta P = P_1 - P_2 = \left(\rho - \rho_g\right)gh$$

Equate these two expressions for ΔP: $\frac{1}{2}\rho_g v^2 = \left(\rho - \rho_g\right)gh$

Solve for v^2 to obtain:

$$\boxed{v^2 = \frac{2gh\left(\rho - \rho_g\right)}{\rho_g}}$$

Note that the correction for buoyant force due to the displaced gas is very small and that, to a good approximation, $v = \sqrt{\dfrac{2gh\rho}{\rho_g}}$

Remarks: Pitot tubes are used to measure the airspeed of airplanes.

Viscous Flow

*66 •   Blood takes about 1.0 s to pass through a 1-mm-long capillary of the human circulatory system. If the diameter of the capillary is 7 μm and the pressure drop is 2.60 kPa, find the viscosity of blood. Assume laminar flow.

Picture the Problem We can apply Poiseuille's law to relate the pressure drop across the capillary tube to the radius and length of the tube, the rate at which blood is flowing through it, and the viscosity of blood.

Using Poiseuille's law, relate the pressure drop to the length and diameter of the capillary tube, the volume flow rate of the blood, and the viscosity of the blood:

$$\Delta P = \frac{8\eta L}{\pi\,r^4} I_v$$

Solve for the viscosity of the blood:

$$\eta = \frac{\pi\,r^4 \Delta P}{8 L I_v}$$

Using its definition, express the volume flow rate of the blood:

$$I_v = A_{cap} v = \pi\,r^2 v$$

Substitute and simplify:

$$\eta = \frac{r^2 \Delta P}{8Lv}$$

Substitute numerical values to obtain:

$$\eta = \frac{\left(3.5 \times 10^{-6}\,\text{m}\right)^2 \left(2.60\,\text{kPa}\right)}{8\left(10^{-3}\,\text{m}\right)\left(\dfrac{10^{-3}\,\text{m}}{1\text{s}}\right)}$$

$$= \boxed{3.98\,\text{mPa} \cdot \text{s}}$$

*67 •   At Reynolds numbers of about 3×10^5 there is an abrupt transition, where the drag on a sphere abruptly decreases. Estimate the velocity at which this drag crisis occurs for a baseball, and comment on whether or not it should play a role in the physics of the game.

Picture the Problem We can use the definition of Reynolds number to find the velocity of a baseball at which the drag crisis occurs.

Using its definition, relate Reynolds number to the velocity v of the baseball:

$$N_R = \frac{2r\rho v}{\eta}$$

Solve for v:

$$v = \frac{\eta N_R}{2r\rho}$$

Substitute numerical values (see Figure 13-1 for the density of air and Table 13-1 for the coefficient of viscosity for air) and evaluate v:

$$v = \frac{\left(0.018\,\text{mPa} \cdot \text{s}\right)\left(3 \times 10^5\right)}{2\left(0.05\,\text{m}\right)\left(1.293\,\text{kg/m}^3\right)}$$

$$= 41.8\,\text{m/s} \times \frac{1\,\text{mi/h}}{0.447\,\text{m/s}}$$

$$= \boxed{93.4\,\text{mi/h}}$$

> Since most major league pitchers can throw a fastball in the low - to mid - 90s, this drag crisis may very well play a role in the game.

Remarks: This is a topic which has been fiercely debated by people who study the physics of baseball.

General Problems

*69 ••   Very roughly speaking, the mass of a person should increase as the cube of his or her height–that is, $M = C\rho h^3$, where M is the mass, h the height, ρ is body density, and C is a person's "coefficient of roundness." Estimate C for an adult male and female, using "typical" values for height and weight. Assume

$\rho = 1000 \text{ kg/m}^3$.

Picture the Problem We can solve the given equation for the coefficient of roundness C and substitute estimates/assumptions of typical masses and heights for adult males and females.

Express the mass of a person as a function of C, ρ, and h:

$$M = C\rho h^3$$

Solve for C:

$$C = \frac{M}{\rho h^3}$$

Assuming that a "typical" adult male stands 5' 10" (1.78 m) and weighs 170 lbs (77 kg), then:

$$C = \frac{77\,\text{kg}}{\left(10^3\,\text{kg/m}^3\right)\left(1.78\,\text{m}\right)^3} = \boxed{0.0137}$$

Assuming that a "typical" adult female stands 5' 4" (1.63 m) and weighs 110 lbs (50 kg), then:

$$C = \frac{50\,\text{kg}}{\left(10^3\,\text{kg/m}^3\right)\left(1.63\,\text{m}\right)^3} = \boxed{0.0115}$$

*75 •   When submerged in water, a block of copper has an apparent weight of 56 N. What fraction of this copper block will be submerged if it is floated on a pool of mercury?

Picture the Problem When the copper block is floating on a pool of mercury, it is in equilibrium under the influence of its weight and the buoyant force acting on it. We can apply the condition for translational equilibrium to relate these forces. We can find the fraction of the block that is submerged by applying Archimedes' principle and the definition of density to express the forces in terms of the volume of the block and the volume of the displaced mercury. Let V represent the volume of the copper block, V' the volume of the displaced mercury. Then the fraction submerged when the material is floated on water is V'/V. Choose the upward direction to be the positive y direction.

Apply $\sum F_y = 0$ to the block:

$B - w = 0$, where B is the buoyant force and w is the weight of the block.

Apply Archimedes' principle and the definition of density to obtain:

$$\rho_{\text{Hg}}V'g - \rho_{\text{Cu}}Vg = 0$$

Solve for V'/V:

$$\frac{V'}{V} = \frac{\rho_{\text{Cu}}}{\rho_{\text{Hg}}}$$

Substitute numerical values and evaluate V'/V:

$$\frac{V'}{V} = \frac{8.93 \times 10^3 \text{ kg/m}^3}{13.6 \times 10^3 \text{ kg/m}^3} = 0.657 = \boxed{65.7\%}$$

*80 •• A Styrofoam cube, 25 cm on an edge, is placed on one pan of a balance. The balance is in equilibrium when a 20-g mass of brass is placed on the opposite pan of the balance. Find the mass of the Styrofoam cube.

Picture the Problem The true mass of the Styrofoam cube is greater than that indicated by the balance due to the buoyant force acting on it. The balance is in rotational equilibrium under the influence of the buoyant and gravitational forces acting on the Styrofoam cube and the brass masses. Neglect the buoyancy of the brass masses. Let m and V represent the mass and volume of the cube and L the lever arm of the balance.

Apply $\sum \vec{\tau} = 0$ to the balance:

$$(mg - B)L - m_{\text{brass}}gL = 0$$

Use Archimedes' principle to express the buoyant force on the Styrofoam cube as a function of volume and density of the air it displaces:

$$B = \rho_{\text{air}}Vg$$

Substitute and simplify to obtain:

$$m - \rho_{\text{air}}V - m_{\text{brass}} = 0$$

Solve for m:

$$m = \rho_{\text{air}}V + m_{\text{brass}}$$

Substitute numerical values and evaluate m:

$$m = (1.293 \text{ kg/m}^3)(0.25 \text{ m})^3 + 20 \times 10^{-3} \text{ kg}$$
$$= 4.02 \times 10^{-2} \text{ kg} = \boxed{40.2 \text{ g}}$$

*83 •• Crude oil has a viscosity of about 0.8 Pa·s at normal temperature. A 50-km pipeline is to be constructed from an oil field to a tanker terminal. The pipeline is to deliver oil at the terminal at a rate of 500 L/s and the flow through the pipeline is to be laminar to minimize the pressure needed to push the fluid through the pipeline. Assuming that the density of crude oil is 700 kg/m³, estimate the diameter of the pipeline that should be used.

Picture the Problem We can use the definition of Reynolds number and assume a value for N_R of 1000 (well within the laminar flow range) to obtain a trial value for the radius of the pipe. We'll then use Poiseuille's law to determine the pressure difference between the ends of the pipe that would be required to maintain a volume flow rate of 500 L/s.

Use the definition of Reynolds number to relate N_R to the radius of the pipe:	$N_R = \dfrac{2r\rho v}{\eta}$
Use the definition of I_V to relate the volume flow rate of the pipe to its radius:	$I_V = Av = \pi r^2 v \Rightarrow v = \dfrac{I_V}{\pi r^2}$
Substitute to obtain:	$N_R = \dfrac{2\rho I_V}{\eta \pi r}$
Solve for r:	$r = \dfrac{2\rho I_V}{\eta \pi N_R}$
Substitute numerical values and evaluate r:	$r = \dfrac{2(700\,\text{kg/m}^3)(0.500\,\text{m}^3/\text{s})}{\pi(0.8\,\text{Pa}\cdot\text{s})(1000)} = 27.9\,\text{cm}$
Using Poiseuille's law, relate the pressure difference between the ends of the pipe to its radius:	$\Delta P = \dfrac{8\eta L}{\pi r^4} I_V$

Substitute numerical values and evaluate ΔP:

$$\Delta P = \frac{8(0.8\,\text{Pa}\cdot\text{s})(50\,\text{km})}{\pi(0.279\,\text{m})^4}(0.500\,\text{m}^3/\text{s})$$

$$= 8.41\times10^6\,\text{Pa}$$

$$= 8.41\times10^6\,\text{Pa}\times\frac{1\,\text{atm}}{1.01325\times10^5\,\text{Pa}}$$

$$= 83.0\,\text{atm}$$

This pressure is too large to maintain in the pipe.

Evaluate ΔP for a pipe of 50 cm radius:

$$\Delta P = \frac{8(0.8\,\text{Pa}\cdot\text{s})(50\,\text{km})}{\pi(0.50\,\text{m})^4}(0.500\,\text{m}^3/\text{s})$$

$$= 8.15\times10^5\,\text{Pa}$$

$$= 8.15\times10^5\,\text{Pa}\times\frac{1\,\text{atm}}{1.01325\times10^5\,\text{Pa}}$$

$$= 8.04\,\text{atm}$$

1 m is a reasonable diameter for the pipeline.

*86 •• Figure 13-40 is a sketch of an *aspirator*, a simple device that can be used to achieve a partial vacuum in a reservoir connected to the vertical tube at B. An aspirator attached to the end of a garden hose may be used to deliver soap or fertilizer from the reservoir. Suppose that the diameter at A is 2.0 cm and at C, where the water exits to the atmosphere, it is 1.0 cm. If the flow rate is 0.5 L/s and the gauge pressure at A is 0.187 atm, what diameter of the constriction at B will achieve a pressure of 0.1 atm in the container? Assume laminar nonviscous flow.

Figure 13-40 Problem 86

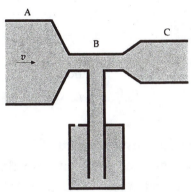

Picture the Problem Because it is not given, we'll neglect the difference in height between the centers of the pipes at A and B. We can use the definition of the volume flow rate to find the speed of the water at A and Bernoulli's equation for constant elevation to find its speed at B. Once we know the speed of the water at B, we can use the equation of continuity to find the diameter of the constriction at B.

Use the definition of the volume flow rate to find v_A:

$$v_A = \frac{I_V}{A_A} = \frac{0.5 \times 10^{-3} \text{ m}^3/\text{s}}{\frac{\pi}{4}(0.02 \text{ m})^2} = 1.59 \text{ m/s}$$

Use Bernoulli's equation for constant elevation to relate the pressures and velocities at A and B:

$$P_B + \tfrac{1}{2}\rho v_B^2 = P_A + \tfrac{1}{2}\rho v_A^2$$

Solve for v_B^2:

$$v_B^2 = \frac{2(P_A - P_B)}{\rho} + v_A^2$$

Substitute numerical values and evaluate v_B^2 :

$$v_B^2 = \frac{2\left[(1.187-0.1)\text{atm}\left(1.01\times10^5\text{ Pa/atm}\right)\right]}{10^3\text{ kg/m}^3} + (1.59\text{ m/s})^2 = 222\text{ m}^2/\text{s}^2$$

Using the continuity equation, relate the volume flow rate to the radius at B:

$$I_V = A_B v_B = \pi r_B^2 v_B$$

Solve for r_B:

$$r_B = \sqrt{\frac{I_V}{\pi v_B}}$$

Substitute numerical values and evaluate r_B and then d_B:

$$r_B = \sqrt{\frac{0.5\times10^{-3}\text{ m}^3/\text{s}}{\pi(14.9\text{ m/s})}} = 3.27\text{ mm}$$

and

$$d_B = 2r_B = \boxed{6.54\text{ mm}}$$

*93 ••   A helium balloon can just lift a load of 750 N. The skin of the balloon has a mass of 1.5 kg. (a) What is the volume of the balloon? (b) If the volume of the balloon were twice that found in Part (a), what would be the initial acceleration of the balloon when it carried a load of 900 N?

Picture the Problem Because the balloon is in equilibrium under the influence of the buoyant force exerted by the air, the weight of its basket and load w, the weight of the skin of the balloon, and the weight of the helium. Choose upward to be the positive y direction and apply the condition for translational equilibrium to relate these forces. Archimedes' principle relates the buoyant force on the balloon to the density of the air it displaces and the volume of the balloon.

(a) Apply $\sum F_y = 0$ to the balloon:

$$B - m_{\text{skin}}g - m_{\text{He}}g - w = 0$$

Letting V represent the volume of the balloon, use Archimedes' principle to express the buoyant force:

$$\rho_{\text{air}}Vg - m_{\text{skin}}g - m_{\text{He}}g - w = 0$$

Substitute for m_{He}:

$$\rho_{\text{air}}Vg - m_{\text{skin}}g - \rho_{\text{He}}Vg - w = 0$$

Solve for V:

$$V = \frac{m_{skin}g + w}{(\rho_{air} - \rho_{He})g}$$

Substitute numerical values and evaluate V:

$$V = \frac{(1.5\,kg)(9.81\,m/s^2) + 750\,N}{(1.293 - 0.1786)(kg/m^3)(9.81\,m/s^2)}$$

$$= \boxed{70.0\,m^3}$$

(b) Apply $\sum F_y = ma$ to the balloon:

$$B - m_{tot}g = m_{tot}a$$

Solve for a:

$$a = \frac{B}{m_{tot}} - g$$

Assuming that the mass of the skin has not changed and letting V' represent the doubled volume of the balloon, express m_{tot}:

$$m_{tot} = m_{load} + m_{He} + m_{skin}$$

$$= \frac{w_{load}}{g} + \rho_{He}V' + m_{skin}$$

Substitute numerical values and evaluate m_{tot}:

$$m_{tot} = \frac{900\,N}{9.81\,m/s^2} + (0.1786\,kg/m^3)(140\,m^3) + 1.5\,kg = 118\,kg$$

Express the buoyant force acting on the balloon:

$$B = w_{displaced\ fluid} = \rho_{air}V'g$$

Substitute numerical values and evaluate B:

$$B = (1.293\,kg/m^3)(140\,m^3)(9.81\,m/s^2)$$

$$= 1.78\,kN$$

Substitute and evaluate a:

$$a = \frac{1.78\,kN}{118\,kg} - 9.81\,m/s^2 = \boxed{5.27\,m/s^2}$$

*95 ••   As mentioned in the discussion of *the law of atmospheres*, the fractional decrease in atmospheric pressure is proportional to the change in altitude. Expressed as a differential equation we have $dP/P = -C\,dh$, where C is a constant. (a) Show that $P(h) = P_0 e^{-Ch}$ is a solution of the differential equation. (b) Show that if $\Delta h \ll h_0$ where $h_0 = 1/C$, then $P(h + \Delta h) \approx P(h)(1 - \Delta h/h_0)$. (c) Given that the pressure at $h = 5.5$ km is half that at sea level, find the constant C.

Picture the Problem We can differentiate the function $P(h)$ to show that it satisfies the differential equation $dP/P = -C \, dh$ and in part (b) we can use the approximation $e^{-x} \approx 1 - x$ and $\Delta h \ll h_0$ to establish the given result.

(a) Differentiate $P(h) = P_0 e^{-Ch}$:

$$\frac{dP}{dh} = -C P_0 e^{-Ch}$$

$$= -CP$$

Separate variables to obtain:

$$\boxed{\frac{dP}{P} = -C dh}$$

(b) Express $P(h + \Delta h)$:

$$P(h + \Delta h) = P_0 e^{-C(h + \Delta h)}$$

$$= P_0 e^{-Ch} e^{-C\Delta h}$$

$$= P(h) e^{-C\Delta h}$$

For $\Delta h \ll h_0$:

$$\frac{\Delta h}{h_0} \ll 1$$

Let $h_0 = 1/C$. Then:

$$C\Delta h \ll 1$$
and
$$e^{-C\Delta h} \approx 1 - C\Delta h = 1 - \frac{\Delta h}{h_0}$$

Substitute to obtain:

$$P(h + \Delta h) = \boxed{P(h)\left(1 - \frac{\Delta h}{h_0}\right)}$$

(c) Take the logarithm of both sides of the function $P(h)$:

$$\ln P = \ln P_0 e^{-Ch} = \ln P_0 + \ln e^{-Ch}$$

$$= \ln P_0 - Ch$$

Solve for C:

$$C = \frac{1}{h} \ln \frac{P_0}{P}$$

Substitute numerical values and evaluate C:

$$C = \frac{1}{5.5 \, \text{km}} \ln \frac{P_0}{\frac{1}{2} P_0} = \frac{1}{5.5 \, \text{km}} \ln 2$$

$$= \boxed{0.126 \, \text{km}^{-1}}$$

Chapter 14
Oscillations

Conceptual Problems

*4 • If the amplitude of a simple harmonic oscillator is tripled, by what factor is the energy changed?

Determine the Concept The energy of a simple harmonic oscillator varies as the square of the amplitude of its motion. Hence, tripling the amplitude increases the energy by a factor of 9.

*9 •• Two identical carts on a frictionless air track are attached by a spring. One is suddenly struck a blow that sends it moving away from the other cart. The motion of the carts is seen to be very jerky-first one cart moves, then stops as the other cart moves and stops in its turn. Explain the motion in a qualitative way.

Determine the Concept Assume that the first cart is given an initial velocity v by the blow. After the initial blow, there are no external forces acting on the carts, so their center of mass moves at a constant velocity $v/2$. The two carts will oscillate about their center of mass in simple harmonic motion where the amplitude of their velocity is $v/2$. Therefore, when one cart has velocity $v/2$ with respect to the center of mass, the other will have velocity $-v/2$. The velocity with respect to the laboratory frame of reference will be $+v$ and 0, respectively. Half a period later, the situation is reversed; one cart will move as the other stops, and vice-versa.

*10 •• The length of the string or wire supporting a pendulum increases slightly when its temperature is raised. How would this affect a clock operated by a simple pendulum?

Determine the Concept The period of a simple pendulum depends on the reciprocal of the length of the pendulum. Increasing the length of the pendulum will decrease its period and the clock would run slow.

*15 • The effect of the mass of a spring on the motion of an object attached to it is usually neglected. Describe qualitatively its effect when it is not neglected.

Determine the Concept We can use the expression for the frequency of a spring-and-mass oscillator to determine the effect of the mass of the spring.

If m represents the mass of the object attached to the spring in a spring-and-mass oscillator, the

$$f = \frac{1}{2\pi}\sqrt{\frac{k}{m}}$$

frequency is given by:

If the mass of the spring is taken into account, the effective mass is greater than the mass of the object alone.	$f' = \dfrac{1}{2\pi}\sqrt{\dfrac{k}{m_{eff}}}$

Divide the second of these equations by the first and simplify to obtain:	$\dfrac{f'}{f} = \dfrac{\dfrac{1}{2\pi}\sqrt{\dfrac{k}{m_{eff}}}}{\dfrac{1}{2\pi}\sqrt{\dfrac{k}{m}}} = \sqrt{\dfrac{m}{m_{eff}}}$

Solve for f':	$f' = f\sqrt{\dfrac{m}{m_{eff}}}$

> Because f' varies inversely with the square root of m, taking into account the effective mass of the spring predicts that the frequency will be reduced.

Estimation and Approximation

*22 •• (a) Estimate the natural period of oscillation for swinging your arms as you walk, with your hands empty. (b) Now estimate it when carrying a heavy briefcase. Look around at other people as they walk by-do these two estimates seem on target?

Picture the Problem Assume that an average length for an arm is about 0.8 m, and that it can be treated as a uniform stick, pivoted at one end. We can use the expression for the period of a physical pendulum to derive an expression for the period of the swinging arm. When carrying a heavy briefcase, the mass is concentrated mostly at the end of the pivot (i.e., in the briefcase), so we can treat the arm-plus-briefcase as a simple pendulum.

(a) Express the period of a uniform rod pivoted at one end:	$T = 2\pi\sqrt{\dfrac{I}{MgD}}$
	where I is the moment of inertia of the stick about an axis through one end, M is the mass of the stick, and $D\,(= L/2)$ is the distance from the end of the stick to its center of mass.

Express the moment of inertia of	$I = \tfrac{1}{3}ML^2$

the stick with respect to an axis through its end:

Substitute the values for I and D to find T:

$$T = 2\pi\sqrt{\frac{\frac{1}{3}ML^2}{Mg(\frac{1}{2}L)}} = 2\pi\sqrt{\frac{2L}{3g}}$$

Substitute numerical values and evaluate T:

$$T = 2\pi\sqrt{\frac{2(0.8\,\text{m})}{3(9.81\,\text{m/s}^2)}} = \boxed{1.47\,\text{s}}$$

(b) Express the period of a simple pendulum:

$$T' = 2\pi\sqrt{\frac{L'}{g}}$$

where L' is slightly longer than the arm length due to the size of the briefcase.

Assuming $L' = 1$ m, evaluate the period of the simple pendulum:

$$T' = 2\pi\sqrt{\frac{1\,\text{m}}{9.81\,\text{m/s}^2}} = \boxed{2.01\,\text{s}}$$

From observation of people as they walk, these estimates seem reasonable.

Simple Harmonic Motion

*25 •   A particle of mass m begins at rest from $x = +25$ cm and oscillates about its equilibrium position at $x = 0$ with a period of 1.5 s. Write equations for (a) the position x as a function of t, (b) the velocity v as a function of t, and (c) the acceleration a as a function of t.

Picture the Problem The position of the particle as a function of time is given by $x = A\cos(\omega t + \delta)$. Its velocity as a function of time is given by $v = -A\omega\sin(\omega t + \delta)$ and its acceleration by $a = -A\omega^2\cos(\omega t + \delta)$. The initial position and velocity give us two equations from which to determine the amplitude A and phase constant δ.

(a) Express the position, velocity, and acceleration of the particle as a function of t:

$$x = A\cos(\omega t + \delta) \qquad (1)$$
$$v = -A\omega\sin(\omega t + \delta) \qquad (2)$$
$$a = -A\omega^2\cos(\omega t + \delta) \qquad (3)$$

Find the angular frequency of the particle's motion:

$$\omega = \frac{2\pi}{T} = \frac{4\pi}{3}\,\text{s}^{-1} = 4.19\,\text{s}^{-1}$$

Relate the initial position and

$$x_0 = A\cos\delta$$

velocity to the amplitude and phase constant:

and

$$v_0 = -\omega A \sin \delta$$

Divide these equations to eliminate A:

$$\frac{v_0}{x_0} = \frac{-\omega A \sin \delta}{A \cos \delta} = -\omega \tan \delta$$

Solve for δ and substitute numerical values to obtain:

$$\delta = \tan^{-1}\left(-\frac{v_0}{x_0 \omega}\right) = \tan^{-1}\left(-\frac{0}{x_0 \omega}\right) = 0$$

Substitute in equation (1) to obtain:

$$x = (25\,\text{cm})\cos\left[\left(\frac{4\pi}{3}\,\text{s}^{-1}\right)t\right]$$

$$= \boxed{(25\,\text{cm})\cos\left[\left(4.19\,\text{s}^{-1}\right)t\right]}$$

(b) Substitute in equation (2) to obtain:

$$v = -(25\,\text{cm})\left(\frac{4\pi}{3}\,\text{s}^{-1}\right)\sin\left[\left(\frac{4\pi}{3}\,\text{s}^{-1}\right)t\right]$$

$$= \boxed{-(105\,\text{cm/s})\sin\left[\left(4.19\,\text{s}^{-1}\right)t\right]}$$

(c) Substitute in equation (3) to obtain:

$$a = -(25\,\text{cm})\left(\frac{4\pi}{3}\,\text{s}^{-1}\right)^2\cos\left[\left(\frac{4\pi}{3}\,\text{s}^{-1}\right)t\right]$$

$$= \boxed{-(439\,\text{cm/s}^2)\cos\left[\left(4.19\,\text{s}^{-1}\right)t\right]}$$

*30 •• Military specifications often call for electronic devices to be able to withstand accelerations of $10g = 98.1\ \text{m/s}^2$. To make sure that their products meet this specification, manufacturers test them using a shaking table that can vibrate a device at various specified frequencies and amplitudes. If a device is given a vibration of amplitude 1.5 cm, what should its frequency be in order to test for compliance with the 10g military specification?

Picture the Problem We can use the expression for the maximum acceleration of an oscillator to relate the 10g military specification to the compliance frequency.

Express the maximum acceleration of an oscillator:

$$a_{max} = A\omega^2$$

Express the relationship between the angular frequency and the frequency of the vibrations:

$$\omega = 2\pi f$$

Substitute to obtain:

$$a_{max} = 4\pi^2 A f^2$$

Solve for f:

$$f = \frac{1}{2\pi}\sqrt{\frac{a_{max}}{A}}$$

Substitute numerical values and evaluate f:

$$f = \frac{1}{2\pi}\sqrt{\frac{98.1\,\text{m/s}^2}{1.5\times10^{-2}\,\text{m}}} = \boxed{12.9\,\text{Hz}}$$

*32 •• (a) Show that $A_0 \cos(\omega t + \delta)$ can be written as $A_s \sin(\omega t) + A_c \cos(\omega t)$, and determine A_s and A_c in terms of A_0 and δ. (b) Relate A_c and A_s to the initial position and velocity of a particle undergoing simple harmonic motion.

Picture the Problem We can use the formula for the cosine of the sum of two angles to write $x = A_0 \cos(\omega t + \delta)$ in the desired form. We can then evaluate x and dx/dt at $t = 0$ to relate A_c and A_s to the initial position and velocity of a particle undergoing simple harmonic motion.

(a) Apply the trigonometric identity $\cos(\omega t + \delta) = \cos\omega t\cos\delta - \sin\omega t\sin\delta$ to obtain:

$$x = A_0\cos(\omega t + \delta) = A_0[\cos\omega t\cos\delta - \sin\omega t\sin\delta]$$
$$= -A_0\sin\delta\sin\omega t + A_0\cos\delta\cos\omega t$$
$$= \boxed{A_s\sin\omega t + A_c\cos\omega t}$$

provided
$$A_s = -A_0\sin\delta \text{ and } A_c = A_0\cos\delta$$

(b) At $t = 0$:

$$x(0) = \boxed{A_0\cos\delta = A_c}$$

Evaluate dx/dt:

$$v = \frac{dx}{dt} = \frac{d}{dt}[A_s\sin\omega t + A_c\cos\omega t]$$
$$= A_s\omega\cos\omega t - A_c\omega\sin\omega t$$

Evaluate $v(0)$ to obtain:

$$v(0) = \omega A_s = \boxed{-\omega A_0\sin\delta}$$

Simple Harmonic Motion and Circular Motion

*34 • A particle moves in a circle of radius 15 cm, making 1 revolution every 3 s. (a) What is the speed of the particle? (b) What is its angular velocity ω? (c) Write an equation for the x component of the position of the particle as a function of time t, assuming that the particle is on the positive x axis at time $t = 0$.

Picture the Problem We can find the period of the motion from the time required for the particle to travel completely around the circle. The angular frequency of the motion is 2π times the reciprocal of its period and the x-component of the particle's position is given by $x = A\cos(\omega t + \delta)$.

(a) Use the definition of speed to express and evaluate the speed of the particle:

$$v = \frac{2\pi r}{T} = \frac{2\pi(15\,\text{cm})}{3\,\text{s}} = \boxed{31.4\,\text{cm/s}}$$

(b) Express the angular velocity of the particle:

$$\omega = \frac{2\pi}{T} = \boxed{\frac{2\pi}{3}\,\text{rad/s}}$$

(c) Express the x component of the position of the particle:

$$x = A\cos(\omega t + \delta)$$

Assuming that the particle is on the positive x axis at time $t = 0$:

$$A = A\cos\delta \Rightarrow \delta = \cos^{-1}1 = 0$$

Substitute to obtain:

$$x = \boxed{(15\,\text{cm})\cos\left(\frac{2\pi}{3}\,\text{s}^{-1}\right)t}$$

Energy in Simple Harmonic Motion

*40 ••   A 3-kg object oscillates on a spring with an amplitude of 8 cm. Its maximum acceleration is 3.50 m/s². Find the total energy.

Picture the Problem The total energy of the object is given, in terms of its maximum kinetic energy, by $E_{tot} = \frac{1}{2}mv_{max}^2$. We can express v_{max} in terms of A and ω and, in turn, express ω in terms of a_{max} to obtain an expression for E_{tot} in terms of a_{max}.

Express the total energy of the object in terms of its maximum kinetic energy:

$$E_{tot} = \frac{1}{2}mv_{max}^2$$

Relate the maximum speed of the object to its angular frequency:

$$v_{max} = A\omega$$

Substitute to obtain:

$$E_{tot} = \frac{1}{2}m(A\omega)^2 = \frac{1}{2}mA^2\omega^2$$

Relate the maximum acceleration

$$a_{max} = A\omega^2$$

of the object to its angular frequency:

or

$$\omega^2 = \frac{a_{max}}{A}$$

Substitute and simplify to obtain:

$$E_{tot} = \tfrac{1}{2}mA^2 \frac{a_{max}}{A} = \tfrac{1}{2}mAa_{max}$$

Substitute numerical values and evaluate E_{tot}:

$$E_{tot} = \tfrac{1}{2}(3\,\mathrm{kg})(0.08\,\mathrm{m})(3.50\,\mathrm{m/s}^2)$$
$$= \boxed{0.420\,\mathrm{J}}$$

Springs

*44 •   An 85-kg person steps into a car of mass 2400 kg, causing it to sink 2.35 cm on its springs. Assuming no damping, with what frequency will the car and passenger vibrate on the springs?

Picture the Problem We can find the frequency of vibration of the car-and-passenger system using $f = \frac{1}{2\pi}\sqrt{\frac{k}{M}}$, where M is the total mass of the system. The spring constant can be determined from the compressing force and the amount of compression.

Express the frequency of the car-and-passenger system:

$$f = \frac{1}{2\pi}\sqrt{\frac{k}{M}}$$

Express the spring constant:

$$k = \frac{F}{\Delta x} = \frac{mg}{\Delta x}$$

where m is the person's mass.

Substitute to obtain:

$$f = \frac{1}{2\pi}\sqrt{\frac{mg}{M\Delta x}}$$

Substitute numerical values and evaluate f:

$$f = \frac{1}{2\pi}\sqrt{\frac{(85\,\mathrm{kg})(9.81\,\mathrm{m/s}^2)}{(2485\,\mathrm{kg})(2.35\times10^{-2}\,\mathrm{m})}}$$
$$= \boxed{0.601\,\mathrm{Hz}}$$

*48 ••   An object of mass m is supported by a vertical spring of force constant 1800 N/m. When pulled down 2.5 cm from equilibrium and released from rest, the object oscillates at 5.5 Hz. (*a*) Find m. (*b*) Find the amount the spring is stretched from its natural length when the object is in equilibrium. (*c*) Write expressions for the displacement *x*, the velocity *v*, and the acceleration *a* as functions of time *t*.

Picture the Problem Choose a coordinate system in which upward is the positive *y* direction. We can find the mass of the object using $m = k/\omega^2$. We can apply a condition for translational equilibrium to the object when it is at its equilibrium position to determine the amount the spring has stretched from its natural length. Finally, we can use the initial conditions to determine *A* and δ and express *x*(*t*) and then differentiate this expression to obtain *v*(*t*) and *a*(*t*).

(*a*) Express the angular frequency of the system in terms of the mass of the object fastened to the vertical spring and solve for the mass of the object:	$\omega^2 = \dfrac{k}{m} \Rightarrow m = \dfrac{k}{\omega^2}$
Express ω^2 in terms of *f*:	$\omega^2 = 4\pi^2 f^2$
Substitute to obtain:	$m = \dfrac{k}{4\pi^2 f^2}$
Substitute numerical values and evaluate *m*:	$m = \dfrac{1800\,\text{N/m}}{4\pi^2 (5.5\,\text{s}^{-1})^2} = \boxed{1.51\,\text{kg}}$
(*b*) Letting Δx represent the amount the spring is stretched from its natural length when the object is in equilibrium, apply $\sum F_y = 0$ to the object when it is in equilibrium:	$k\Delta x - mg = 0$
Solve for Δx:	$\Delta x = \dfrac{mg}{k}$
Substitute numerical values and evaluate Δx:	$\Delta x = \dfrac{(1.51\,\text{kg})(9.81\,\text{m/s}^2)}{1800\,\text{N/m}} = \boxed{8.23\,\text{mm}}$
(*c*) Express the position of the object as a function of time:	$x = A\cos(\omega t + \delta)$

Use the initial conditions
($x_0 = -2.5$ cm and $v_0 = 0$) to find
δ:

$$\delta = \tan^{-1}\left(-\frac{v_0}{\omega x_0}\right) = \tan^{-1} 0 = \pi$$

Evaluate ω:

$$\omega = \sqrt{\frac{k}{m}} = \sqrt{\frac{1800\,\text{N/m}}{1.51\,\text{kg}}} = 34.5\,\text{rad/s}$$

Substitute to obtain:

$$x = (2.5\,\text{cm})\cos\left[(34.5\,\text{rad/s})t + \pi\right]$$
$$= \boxed{-(2.5\,\text{cm})\cos\left[(34.5\,\text{rad/s})t\right]}$$

Differentiate $x(t)$ to obtain v:

$$v = \boxed{(86.4\,\text{cm/s})\sin\left[(34.5\,\text{rad/s})t\right]}$$

Differentiate $v(t)$ to obtain a:

$$a = \boxed{(29.8\,\text{m/s}^2)\cos\left[(34.5\,\text{rad/s})t\right]}$$

*52 •• A suitcase of mass 20 kg is hung from two bungie cords, as shown in Figure 14-26. Each cord is stretched 5 cm when the suitcase is in equilibrium. If the suitcase is pulled down a little and released, what will be its oscillation frequency?

Figure 14-26 Problem 52

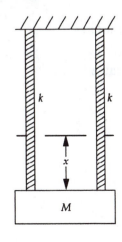

Picture the Problem The diagram shows the stretched bungie cords supporting the suitcase under equilibrium conditions. We can use

$$f = \frac{1}{2\pi}\sqrt{\frac{k_{\text{eff}}}{M}}$$ to express the frequency

of the suitcase in terms of the effective "spring" constant k_{eff} and apply a condition for translational equilibrium to the suitcase to find k_{eff}.

Express the frequency of the suitcase oscillator:	$f = \dfrac{1}{2\pi}\sqrt{\dfrac{k_{\text{eff}}}{M}}$

Apply $\sum F_y = 0$ to the suitcase to obtain:

$kx + kx - Mg = 0$

or

$2kx - Mg = 0$

or

$k_{\text{eff}}x - Mg = 0$

where $k_{\text{eff}} = 2k$

Solve for k_{eff} to obtain:

$$k_{\text{eff}} = \frac{Mg}{x}$$

Substitute to obtain:

$$f = \frac{1}{2\pi}\sqrt{\frac{g}{x}}$$

Substitute numerical values and evaluate f:

$$f = \frac{1}{2\pi}\sqrt{\frac{9.81\,\text{m/s}^2}{0.05\,\text{m}}} = \boxed{2.23\,\text{Hz}}$$

*56 •• A winch cable has a cross-sectional area of $1.5\ \text{cm}^2$ and a length of 2.5 m. Young's modulus for the cable is $150\ \text{GN/m}^2$. A 950-kg engine block is hung from the end of the cable. (*a*) By what length does the cable stretch? (*b*) Treating the cable as a simple spring, what is the oscillation frequency of the engine block at the end of the cable?

Picture the Problem We can relate the elongation of the cable to the load on it using the definition of Young's modulus and use the expression for the frequency of a spring and mass oscillator to find the oscillation frequency of the engine block at the end of the wire.

(*a*) Using the definition of Young's modulus, relate the elongation of the cable to the applied stress:

$$Y = \frac{\text{stress}}{\text{strain}} = \frac{F/A}{\Delta \ell / \ell}$$

Solve for $\Delta \ell$:

$$\Delta \ell = \frac{F\ell}{AY} = \frac{Mg\ell}{AY}$$

Substitute numerical values and evaluate $\Delta\ell$:

$$\Delta\ell = \frac{(950\,\text{kg})(9.81\,\text{m/s}^2)(2.5\,\text{m})}{(1.5\,\text{cm}^2)(150\,\text{GN/m}^2)}$$

$$= \boxed{1.04\,\text{mm}}$$

(b) Express the oscillation frequency of the wire-engine block system:

$$f = \frac{1}{2\pi}\sqrt{\frac{k_{\text{eff}}}{M}}$$

Express the effective "spring" constant of the cable:

$$k_{\text{eff}} = \frac{F}{\Delta\ell} = \frac{Mg}{\Delta\ell}$$

Substitute to obtain:

$$f = \frac{1}{2\pi}\sqrt{\frac{g}{\Delta\ell}}$$

Substitute numerical values and evaluate f:

$$f = \frac{1}{2\pi}\sqrt{\frac{9.81\,\text{m/s}^2}{1.04\,\text{mm}}} = \boxed{15.5\,\text{Hz}}$$

Energy of an Object on a Vertical Spring

*59 ••   A 1.2-kg object hanging from a spring of force constant 300 N/m oscillates with a maximum speed of 30 cm/s. (a) What is its maximum displacement? When the object is at its maximum displacement, find (b) the total energy of the system, (c) the gravitational potential energy, and (d) the potential energy in the spring.

Picture the Problem We can find the amplitude of the motion by relating it to the maximum speed of the object. Let the origin of our coordinate system be at y_0, where y_0 is the equilibrium position of the object and let $U_g = 0$ at this location. Because $F_{\text{net}} = 0$ at equilibrium, the extension of the spring is then $y_0 = mg/k$, and the potential energy stored in the spring is $U_s = \frac{1}{2}ky_0^2$. A further extension of the spring by an amount y increases U_s to
$$\frac{1}{2}k(y + y_0)^2 = \frac{1}{2}ky^2 + kyy_0 + \frac{1}{2}ky_0^2 = \frac{1}{2}ky^2 + mgy + \frac{1}{2}ky_0^2.$$
Consequently, if we set $U = U_g + U_s = 0$, a further extension of the spring by y increases U_s by $\frac{1}{2}ky^2 + mgy$ while decreasing U_g by mgy. Therefore, if $U = 0$ at the equilibrium position, the change in U is given by $\frac{1}{2}k(y')^2$, where $y' = y - y_0$.

(a) Relate the maximum speed of the object to the amplitude of its motion:

$$v_{\text{max}} = A\omega$$

Solve for A:

$$A = \frac{v_{max}}{\omega} = v_{max}\sqrt{\frac{m}{k}}$$

Substitute numerical values and evaluate A:

$$A = (0.3\,\text{m/s})\sqrt{\frac{1.2\,\text{kg}}{300\,\text{N/m}}} = \boxed{1.90\,\text{cm}}$$

(b) Express the energy of the object at maximum displacement:

$$E = \tfrac{1}{2}kA^2$$

Substitute numerical values and evaluate E:

$$E = \tfrac{1}{2}(300\,\text{N/m})(0.019\,\text{m})^2 = \boxed{0.0542\,\text{J}}$$

(c) At maximum displacement from equilibrium:

$$U_g = -mgA$$

Substitute numerical values and evaluate U_g:

$$U_g = -(1.2\,\text{kg})(9.81\,\text{m/s}^2)(0.019\,\text{m})$$
$$= \boxed{-0.224\,\text{J}}$$

(d) Express the potential energy in the spring when the object is at its maximum downward displacement:

$$U_s = \tfrac{1}{2}kA^2 + mgA$$

Substitute numerical values and evaluate U_s:

$$U_s = \tfrac{1}{2}(300\,\text{N/m})(0.019\,\text{m})^2$$
$$+ (1.2\,\text{kg})(9.81\,\text{m/s}^2)(0.019\,\text{m})$$
$$= \boxed{0.278\,\text{J}}$$

Simple Pendulums

*63 • A pendulum set up in the stairwell of a 10-story building consists of a heavy weight suspended on a 34.0-m wire. If $g = 9.81$ m/s², what is the period of oscillation?

Picture the Problem We can use $T = 2\pi\sqrt{L/g}$ to find the period of this pendulum.

Express the period of a simple pendulum:

$$T = 2\pi\sqrt{\frac{L}{g}}$$

Substitute numerical values and evaluate T:

$$T = 2\pi\sqrt{\frac{34\,\text{m}}{9.81\,\text{m/s}^2}} = \boxed{11.7\,\text{s}}$$

Physical Pendulums

*70 •• The pendulum bob of a large town-hall clock has a length of 4 m. (a) What is its period of oscillation? Treat it as a simple pendulum with small amplitude oscillations. (b) To regulate the period of the pendulum there is a tray attached to its shaft, halfway up. The tray holds a stack of coins. To change the period by a little bit, coins are added or removed from the tray. Explain in detail why this works. Will adding coins increase or decrease the period of the pendulum?

Picture the Problem We can use the expression for the period of a simple pendulum to find the period of the clock.

(a) Express the period of a simple pendulum:

$$T = 2\pi\sqrt{\frac{\ell}{g}}$$

Substitute numerical values and evaluate T:

$$T = 2\pi\sqrt{\frac{4\,\text{m}}{9.81\,\text{m/s}^2}} = \boxed{4.01\,\text{s}}$$

(b) | By effectively raising the center of mass of the pendulum, placing coins in the tray shortens the period.

*73 •• You are given a meter stick and asked to drill a narrow hole in it so that, when the stick is pivoted about the hole, the period of the pendulum will be a minimum. Where should you drill the hole?

Picture the Problem Let x be the distance of the pivot from the center of the meter stick, m the mass of the meter stick, and L its length. We'll express the period of the meter stick as a function of the distance x and then differentiate this expression with respect to x to determine where the hole should be drilled to minimize the period.

Express the period of a physical pendulum:

$$T = 2\pi\sqrt{\frac{I}{MgD}} \qquad (1)$$

Express the moment of inertia of the meter stick with respect to its center of mass:

$$I_{cm} = \tfrac{1}{12}mL^2$$

Using the parallel-axis theorem, express the moment of inertia of the meter stick with respect to the pivot point:

$$I = I_{cm} + mx^2$$
$$= \tfrac{1}{12}mL^2 + mx^2$$

Substitute in equation (1) to obtain:

$$T = 2\pi\sqrt{\frac{\tfrac{1}{12}mL^2 + mx^2}{mgx}}$$

$$= \frac{2\pi}{\sqrt{g}}\sqrt{\frac{\tfrac{1}{12}L^2 + x^2}{x}}$$

$$= C\sqrt{\frac{\tfrac{1}{12}L^2 + x^2}{x}}$$

where $C = \dfrac{2\pi}{\sqrt{g}}$

Set $dT/dx = 0$ to find the condition for minimum T:

$$\frac{dT}{dx} = C \times \frac{d}{dx}\sqrt{\frac{\tfrac{1}{12}L^2 + x^2}{x}} = 0 \text{ for extrema}$$

Evaluate the derivative to obtain:

$$\frac{2x^2 - \left(\tfrac{1}{12}L^2 + x^2\right)}{x^2\sqrt{\dfrac{\tfrac{1}{12}L^2 + x^2}{x}}} = 0$$

Because the denominator of this expression cannot be zero, it follows that:

$$2x^2 - \left(\tfrac{1}{12}L^2 + x^2\right) = 0$$

Solve for and evaluate x to obtain:

$$x = \frac{L}{\sqrt{12}} = \frac{100\,\text{cm}}{\sqrt{12}} = 28.9\,\text{cm}$$

The hole should be drilled at a distance:

$$d = 50\,\text{cm} - 28.9\,\text{cm} = \boxed{21.1\,\text{cm}}$$

from the center of the meter stick.

*78 ••   A pendulum clock loses 48 s/d when the amplitude of the pendulum is 8.4°. What should be the amplitude of the pendulum so that the clock keeps perfect time?

Picture the Problem The period of a simple pendulum depends on its amplitude ϕ_0 according to $T = 2\pi \sqrt{\dfrac{L}{g}} \left[1 + \dfrac{1}{2^2} \sin^2 \dfrac{1}{2}\phi_0 + \dfrac{1}{2^2}\left(\dfrac{3}{4}\right)^2 \sin^4 \dfrac{1}{2}\phi_0 + ... \right]$. We can

approximate T to the second-order term and express $\Delta T/T = (T_{\text{slow}} - T_{\text{accurate}})/T$. Equating this expression to $\Delta T/T$ calculated from the fractional daily loss of time will allow us to solve for and evaluate the amplitude of the pendulum that corresponds to keeping perfect time.

Express the fractional daily loss of time:

$$\frac{\Delta T}{T} = \frac{48\,\text{s}}{\text{day}} \times \frac{1\,\text{day}}{24\,\text{h}} \times \frac{1\,\text{h}}{3600\,\text{s}} = \frac{48}{86400}$$

Approximate the period of the clock to the 2nd-order term:

$$T = 2\pi \sqrt{\frac{L}{g}} \left[1 + \frac{1}{2^2} \sin^2 \frac{1}{2}\phi_0 \right]$$

Express the difference in the periods of the slow and accurate clocks:

$$\Delta T = T_{\text{slow}} - T_{\text{accurate}}$$

$$= 2\pi \sqrt{\frac{L}{g}} \left\{ \left[1 + \frac{1}{2^2} \sin^2 \frac{1}{2}(8.4°) \right] \right.$$

$$\left. - \left[1 + \frac{1}{2^2} \sin^2 \frac{1}{2}\phi_0 \right] \right\}$$

$$= 2\pi \sqrt{\frac{L}{g}} \left[\frac{1}{2^2} \sin^2 \frac{1}{2}(8.4°) \right.$$

$$\left. - \frac{1}{2^2} \sin^2 \frac{1}{2}\phi_0 \right]$$

Divide both sides of this equation by T to obtain:

$$\frac{\Delta T}{T} = \frac{1}{4} \sin^2 4.2° - \frac{1}{4} \sin^2 \frac{1}{2}\phi_0$$

Substitute for $\dfrac{\Delta T}{T}$ and simplify to obtain:

$$\frac{1}{4} \sin^2 4.2° - \frac{1}{4} \sin^2 \frac{1}{2}\phi_0 = \frac{48}{86400}$$

and

$$\sin \frac{1}{2}\phi_0 = 0.05605$$

Solve for ϕ_0:

$$\phi_0 = \boxed{6.43°}$$

Damped Oscillations

*87 •• It has been stated that the vibrating earth has a resonance period of 54 min and a Q factor of about 400 and that after a large earthquake, the earth "rings" (continues to vibrate) for about 2 months. (*a*) Find the percentage of the energy of vibration lost to damping forces during each cycle. (*b*) Show that after n periods the energy is $E_n = (0.984)^n E_0$, where E_0 is the original energy. (*c*) If the original energy of vibration of an earthquake is E_0, what is the energy after 2 d?

Picture the Problem We can find the fractional loss of energy per cycle from the physical interpretation of Q for small damping. We will also find a general expression for the earth's vibrational energy as a function of the number of cycles it has completed. We can then solve this equation for the earth's vibrational energy after any number of days.

(*a*) Express the fractional change in energy as a function of Q:

$$\frac{\Delta E}{E} = \frac{2\pi}{Q} = \frac{2\pi}{400} = \boxed{1.57\%}$$

(*b*) Express the energy of the damped oscillator after one cycle:

$$E_1 = E_0\left(1 - \frac{\Delta E}{E}\right)$$

Express the energy after two cycles:

$$E_2 = E_1\left(1 - \frac{\Delta E}{E}\right) = E_0\left(1 - \frac{\Delta E}{E}\right)^2$$

Generalizing to n cycles:

$$E_n = E_0\left(1 - \frac{\Delta E}{E}\right)^n = E_0(1 - 0.0157)^n$$

$$= \boxed{E_0(0.9843)^n}$$

(*c*) Express 2 d in terms of the number of cycles; i.e., the number of vibrations the earth will have experienced:

$$2\,d = 2\,d \times \frac{24\,h}{d} \times \frac{60\,m}{h}$$

$$= 2880\,min \times \frac{1T}{54\,min}$$

$$= 53.3T$$

Evaluate $E(2\,d)$:

$$E(2\,d) = E_0(0.9843)^{53.3} = \boxed{0.430 E_0}$$

General Problems

*99 ••   A small particle of mass m slides without friction in a spherical bowl
of radius r. (*a*) Show that the motion of the particle is the same as if it were
attached to a string of length r. (*b*) Figure 14-36 shows a particle of mass m_1 that
is displaced a small distance s_1 from the bottom of the bowl, where s_1 is much
smaller than r. A second particle of mass m_2 is displaced in the opposite direction
a distance $s_2 = 3s_1$, where s_2 is also much smaller than r. If the particles are
released at the same time, where do they meet? Explain.

Figure 14-36 Problems 99 and 100

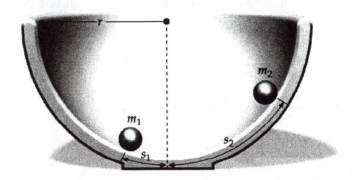

Picture the Problem Compare the
forces acting on the particle to the right
in Figure 14-36 with the forces shown
acting on the bob of the simple
pendulum shown in the free-body
diagram to the right. Because there is
no friction, the only forces acting on
the particle are mg and the normal force
acting radially inward.
In (*b*), we can think of the particles as
the bobs of simple pendulums of equal
length.

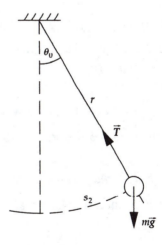

(*a*) The normal force is identical to the tension in a string of length r that keeps
the particle moving in a circular path and a component of mg provides, for small
displacements θ_0 or s_2, the linear restoring force required for oscillatory motion.

(*b*) The particles meet at the bottom. Because s_1 and s_2 are both much smaller
than r, the particles behave like the bobs of simple pendulums of equal length;
therefore, they have the same periods.

*104 •• A wooden cube with edge length a and mass m floats in water with one of its faces parallel to the water surface. The density of the water is ρ. Find the period of oscillation in the vertical direction if the cube is pushed down slightly.

Picture the Problem Choose a coordinate system in which the direction the cube is initially displaced (downward) is the positive y direction. The figure shows the forces acting on the cube when it is in equilibrium floating in the water and when it has been pushed down a small distance y. We can find the period of its oscillatory motion from its angular frequency. By applying Newton's 2nd law to the cube we can obtain its equation of motion; from this equation we can determine the angular frequency of the cube's small-amplitude oscillations.

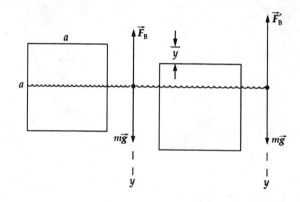

Express the period of oscillation in terms of the angular frequency of the oscillations:

$$T = \frac{2\pi}{\omega} \qquad (1)$$

Apply $\sum F_y = 0$ to the cube when it is floating in the water:

$$mg - F_B = 0$$

Apply $\sum T_y = ma_y$ to the cube when it is pushed down a small distance y:

$$mg - F_B' = ma_y$$

Eliminate mg between these equations to obtain:

$$F_B - F_B' = ma_y$$

or

$$\Delta F_B = F_B - F_B' = ma_y$$

For $y \ll 1$:

$$\Delta F_B \approx dF_B = -\rho V g = -a^2 \rho g y = m \frac{d^2 y}{dt^2}$$

Rewrite the equation of motion as:

$$m\frac{d^2 y}{dt^2} = -a^2 \rho g y$$

or

$$\frac{d^2 y}{dt^2} = -\frac{a^2 \rho g}{m} y$$

where $\dfrac{d^2 y}{dt^2} = -\omega^2 y$

Solve for ω:

$$\omega = a\sqrt{\frac{\rho g}{m}}$$

Substitute in equation (1) to obtain:

$$T = \frac{2\pi}{a\sqrt{\dfrac{\rho g}{m}}} = \boxed{\dfrac{2\pi}{a}\sqrt{\dfrac{m}{\rho g}}}$$

*112 •• A small block of mass m_1 rests on a piston that is vibrating vertically with simple harmonic motion given by $y = A\sin\omega t$. (a) Show that the block will leave the piston if $\omega^2 A > g$. (b) If $\omega^2 A = 3g$ and $A = 15$ cm, at what time will the block leave the piston?

Picture the Problem If the displacement of the block is $y = A\sin\omega t$, its acceleration is $a = -\omega^2 A\sin\omega t$.

(a) At maximum upward extension, the block is momentarily at rest. Its downward acceleration is g. The downward acceleration of the piston is $\omega^2 A$. Therefore, if $\omega^2 A > g$, the block will separate from the piston.

(b) Express the acceleration of the small block:

$$a = -A\omega^2 \sin\omega t$$

For $\omega^2 A = 3g$ and $A = 15$ cm:

$$a = -3g\sin\omega t = -g$$

Solve for t:

$$t = \frac{1}{\omega}\sin^{-1}\left(\frac{1}{3}\right) = \sqrt{\frac{A}{3g}}\sin^{-1}\left(\frac{1}{3}\right)$$

Substitute numerical values and evaluate t:

$$t = \sqrt{\frac{0.15\,\text{m}}{3\left(9.81\,\text{m/s}^2\right)}}\sin^{-1}\frac{1}{3} = \boxed{0.0243\,\text{s}}$$

*117 ••• If we attach two blocks of masses m_1 and m_2, to either end of a spring of spring constant k and set them into oscillation, show that the oscillation frequency $\omega = \left(k/\mu\right)^{1/2}$, where $\mu = m_1 m_2/(m_1 + m_2)$ is the reduced mass of the system.

Picture the Problem The pictorial representation shows the two blocks connected by the spring and displaced from their equilibrium positions. We can apply Newton's 2nd law to each of these coupled oscillators and solve the resulting equations simultaneously to obtain the differential equation of motion of the coupled oscillators. We can then compare this differential equation and its solution to the differential equation of motion of the simple harmonic oscillator and its solution to show that the oscillation frequency is $\omega = \left(k/\mu\right)^{1/2}$ where $\mu = m_1 m_2/(m_1 + m_2)$ is the reduced mass of the system.

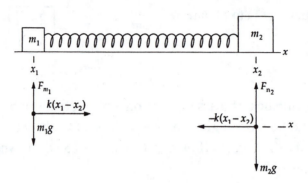

Apply $\sum \vec{F} = m\vec{a}$ to the block whose mass is m_1 and solve for its acceleration:

$$k(x_1 - x_2) = m_1 a_1 = m_1 \frac{d^2 x_1}{dt^2}$$

or

$$a_1 = \frac{d^2 x_1}{dt^2} = \frac{k}{m_1}(x_1 - x_2)$$

Apply $\sum \vec{F} = m\vec{a}$ to the block whose mass is m_2 and solve for its acceleration:

$$-k(x_1 - x_2) = m_2 a_2 = m_1 \frac{d^2 x_2}{dt^2}$$

or

$$a_2 = \frac{d^2 x_2}{dt^2} = \frac{k}{m_2}(x_2 - x_1)$$

Subtract the first equation from the second to obtain:

$$\frac{d^2(x_2 - x_1)}{dt^2} = \frac{d^2 x}{dt^2} = -k\left(\frac{1}{m_1} + \frac{1}{m_2}\right)x$$

where $x = x_2 - x_1$

Define the reduced mass of the system to be:

$$\frac{1}{\mu} = \frac{1}{m_1} + \frac{1}{m_2} \text{ or } \mu = \frac{m_1 m_2}{m_1 + m_2}$$

Substitute to obtain:

$$\frac{d^2x}{dt^2} = -\frac{k}{\mu}x \qquad (1)$$

Compare this differential equation with the differential equation of the simple harmonic oscillator:

$$\frac{d^2x}{dt^2} = -\frac{k}{m}x$$

The solution to this equation is:

$$x = x_0 \cos(\omega t + \delta)$$

$$\text{where } \omega = \sqrt{\frac{k}{m}}$$

Express the solution to equation (1):

$$x = x_0 \cos(\omega t + \delta)$$

$$\text{where } \omega = \boxed{\sqrt{\frac{k}{\mu}}}$$

*122 ••• A straight tunnel is dug through the earth as shown in Figure 14-45. Assume that the walls of the tunnel are frictionless. (a) The gravitational force exerted by the earth on a particle of mass m at a distance r from the center of the earth when $r < R_E$ is $F_r = -\left(GmM_E / R_E^3\right)r$, where M_E is the mass of the earth and R_E is its radius. Show that the net force on a particle of mass m at a distance x from the middle of the tunnel is given by $F_x = -\left(GmM_E / R_E^3\right)x$, and that the motion of the particle is therefore simple harmonic motion. (b) Show that the period of the motion is given by $T = 2\pi\sqrt{R_E / g}$ and find its value in minutes. (This is the same period as that of a satellite orbiting near the surface of the earth and is independent of the length of the tunnel.)

Figure 14-45 Problem 122

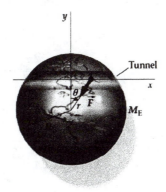

Picture the Problem The net force acting on the particle as it moves in the tunnel is the x-component of the gravitational force acting on it. We can find the period of the particle from the angular frequency of its motion. We can apply Newton's 2nd law to the particle in order to express ω in terms of the radius of the earth and the acceleration due to gravity at the surface of the earth.

(a) From the figure we see that:

$$F_x = F_r \sin\theta = -\frac{GmM_E}{R_E^3} r \frac{x}{r}$$

$$= \boxed{-\frac{GmM_E}{R_E^3} x}$$

Because this force is a linear restoring force, the motion of the particle is simple harmonic motion.

(b) Express the period of the particle as a function of its angular frequency:

$$T = \frac{2\pi}{\omega} \qquad (1)$$

Apply $\sum F_x = ma_x$ to the particle:

$$-\frac{GmM_E}{R_E^3} x = ma$$

Solve for a:

$$a = -\frac{GM_E}{R_E^3} x = -\omega^2 x$$

where

$$\omega = \sqrt{\frac{GM_E}{R_E^2}}$$

Use $GM_E = gR_E^2$ to simplify ω:

$$\omega = \sqrt{\frac{gR_E^2}{R_E^3}} = \sqrt{\frac{g}{R_E}}$$

Substitute in equation (1) to obtain:

$$T = \frac{2\pi}{\sqrt{\dfrac{g}{R_E}}} = \boxed{2\pi\sqrt{\frac{R_E}{g}}}$$

Substitute numerical values and evaluate T:

$$T = 2\pi\sqrt{\frac{6.37\times10^6\,\text{m}}{9.81\,\text{m/s}^2}} = 5.06\times10^3\,\text{s}$$

$$= \boxed{84.4\,\text{min}}$$

*125 ••• In this problem, you will derive the expression for the average power delivered by a driving force to a driven oscillator (Figure 14-24, page 450). (*a*) Show that the instantaneous power input of the driving force is given by $P = Fv = -A\omega F_0 \cos\omega t \sin(\omega t - \delta)$. (*b*) Use the trigonometric identity $\sin(\theta_1 - \theta_2) = \sin\theta_1 \cos\theta_2 - \cos\theta_1 \sin\theta_2$ to show that the equation in (*a*) can be written $P = A\omega F_0 \sin\delta \cos^2\omega t - A\omega F_0 \cos\delta \cos\omega t \sin\omega t$. (*c*) Show that the average value of the second term in your result for (*b*) over one or more periods is zero and that therefore $P_{av} = \frac{1}{2}A\omega F_0 \sin\delta$. (*d*) From Equation 14-54 for $\tan\delta$, construct a right triangle in which the side opposite the angle δ is $b\omega$ and the side adjacent is $m(\omega_0^2 - \omega^2)$, and use this triangle to show that

$$\sin\delta = \frac{b\omega}{\sqrt{m^2(\omega_0^2 - \omega^2)^2 + b^2\omega^2}} = \frac{b\omega A}{F_0}$$

(*e*) Use your result for (*d*) to eliminate ωA from your result for (*c*) so that the average power input can be written

$$P_{av} = \frac{1}{2}\frac{F_0^2}{b}\sin^2\delta = \frac{1}{2}\left[\frac{b\omega^2 F_0^2}{m^2(\omega_0^2 - \omega^2)^2 + b^2\omega^2}\right] \qquad 14\text{-}55$$

Picture the Problem We can follow the step-by-step instructions provided in the problem statement to obtain the desired results.

(*a*) Express the average power delivered by a driving force to a driven oscillator:

$P = \vec{F}\cdot\vec{v} = Fv\cos\theta$
or, because θ is 0°,
$P = Fv$

Express F as a function of time:

$F = F_0 \cos\omega t$

Express the position of the driven oscillator as a function of time:

$x = A\cos(\omega t - \delta)$

Differentiate this expression with respect to time to express the velocity of the oscillator as a function of time:

$v = -A\omega \sin(\omega t - \delta)$

Substitute to express the average power delivered to the driven oscillator:

$P = (F_0 \cos\omega t)[-A\omega \sin(\omega t - \delta)]$
$= \boxed{-A\omega F_0 \cos\omega t \sin(\omega t - \delta)}$

(b) Expand $\sin(\omega t - \delta)$ to obtain:

$$\sin(\omega t - \delta) = \sin \omega t \cos \delta - \cos \omega t \sin \delta$$

Substitute in your result from (a) and simplify to obtain:

$$P = -A\omega F_0 \cos \omega t (\sin \omega t \cos \delta - \cos \omega t \sin \delta)$$

$$= \boxed{\begin{array}{l} A\omega F_0 \sin \delta \cos^2 \omega t \\ - A\omega F_0 \cos \delta \cos \omega t \sin \omega t \end{array}}$$

(c) Integrate $\sin \theta \cos \theta$ over one period to determine $\langle \sin \theta \cos \theta \rangle$:

$$\langle \sin \theta \cos \theta \rangle = \frac{1}{2\pi} \left[\int_0^{2\pi} \sin \theta \cos \theta d\theta \right]$$

$$= \frac{1}{2\pi} \left[\frac{1}{2} \sin^2 \theta \Big|_0^{2\pi} \right]$$

$$= 0$$

Integrate $\cos^2 \theta$ over one period to determine $\langle \cos^2 \theta \rangle$:

$$\langle \cos^2 \theta \rangle = \frac{1}{2\pi} \int_0^{2\pi} \cos^2 \theta d\theta$$

$$= \frac{1}{2\pi} \left[\frac{1}{2} \int_0^{2\pi} (1 + \cos 2\theta) d\theta \right]$$

$$= \frac{1}{2\pi} \left[\frac{1}{2} \int_0^{2\pi} d\theta + \frac{1}{2} \int_0^{2\pi} \cos 2\theta d\theta \right]$$

$$= \frac{1}{2\pi} (\pi + 0) = \frac{1}{2}$$

Substitute and simplify to express P_{av}:

$$P_{av} = A\omega F_0 \sin \delta \langle \cos^2 \omega t \rangle - A\omega F_0 \cos \delta \langle \cos \omega t \sin \omega t \rangle$$

$$= \tfrac{1}{2} A\omega F_0 \sin \delta - A\omega F_0 \cos \delta (0)$$

$$= \boxed{\tfrac{1}{2} A\omega F_0 \sin \delta}$$

(d) Construct a triangle that is consistent with

$$\tan \delta = \frac{b\omega}{m(\omega_0^2 - \omega^2)} :$$

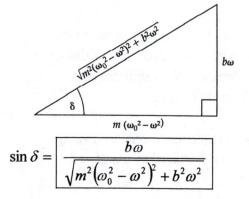

Using the triangle, express $\sin\delta$:

$$\sin \delta = \boxed{\frac{b\omega}{\sqrt{m^2 (\omega_0^2 - \omega^2)^2 + b^2 \omega^2}}}$$

Using equation 14-53, reduce this expression to the simpler form:

$$\sin \delta = \boxed{\dfrac{b\omega A}{F_0}}$$

(e) Solve $\sin \delta = \dfrac{b\omega A}{F_0}$ for ω:

$$\omega = \dfrac{F_0}{bA}\sin \delta$$

Substitute in the expression for P_{av} to eliminate ω:

$$P_{av} = \boxed{\dfrac{F_0^2}{2b}\sin^2 \delta}$$

Substitute for $\sin \delta$ from (d) to obtain Equation 14-55:

$$P_{av} = \dfrac{1}{2}\boxed{\left[\dfrac{b\omega^2 F_0^2}{m^2\left(\omega_0^2 - \omega^2\right)^2 + b^2\omega^2}\right]}$$

Chapter 15
Wave Motion

Conceptual Problems

*1 • A rope hangs vertically from the ceiling. Do waves on the rope move faster, slower, or at the same speed as they move from bottom to top? Explain.

Determine the Concept The speed of a transverse wave on a rope is given by $v = \sqrt{F/\mu}$ where F is the tension in the rope and μ is its linear density. The waves on the rope move faster as they move up because the tension increases due to the weight of the rope below.

*5 • The crack of a bullwhip is caused by the speed of the tip breaking the sound barrier. Explain how the tapered shape of the whip helps the tip move much faster than the hand holding the whip.

Determine the Concept The speed of the wave v on the bullwhip varies with the tension F in the whip and its linear density μ according to $v = \sqrt{F/\mu}$. As the whip tapers, the wave speed in the tapered end increases due to the decrease in the mass density, so the wave travels faster.

*9 •• Stars often occur in pairs revolving around their common center of mass. If one of the stars is a black hole, it is invisible. Explain how the existence of such a black hole might be inferred from the light observed from the other, visible star.

Determine the Concept The light from the companion star will be shifted about its mean frequency periodically due to the relative approach to and recession from the earth of the companion star as it revolves about the black hole.

*13 • While out on patrol, the battleship *Rodger Young* hits a mine and begins to burn, ultimately exploding. Sailor Abel jumps into the water and begins swimming away from the doomed ship, while Sailor Baker gets into a life raft. Comparing their experiences later, Abel tells Baker, "I was swimming underwater, and heard a big explosion from the ship. When I surfaced, I heard a second explosion. What do you think it could be?" Baker says, "I think it was your imagination–I only heard one explosion." Explain why Baker only heard one explosion, while Abel heard two.

Determine the Concept There was only one explosion. Sound travels faster in water than air. Abel heard the sound wave in the water first, then, surfacing, heard the sound wave traveling through the air, which took longer to reach him.

*17 •• The explosion of a depth charge beneath the surface of the water is recorded by a helicopter hovering above its surface, as shown in Figure 15-29. Along which path, A, B, or C, will the sound wave take the least time to reach the helicopter?

Figure 15-29 Problem 17

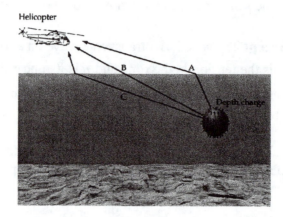

Helicopter

B   A

C   Depth charge

Determine the Concept Path C. Because the wave speed is highest in the water, and more of path C is underwater than A or B, the sound wave will spend the least time on path C.

Estimation and Approximation

*20 •• Estimate the speed of the bullet as it passes through the helium balloon in Figure 15-30 from the angle of its shock cone.

Figure 15-30 Problem 20

Picture the Problem You can use a protractor to measure the angle of the shock cone and then estimate the speed of the bullet using $\sin \theta = v/u$. The speed of sound in helium at room temperature (293 K) is 977 m/s.

Relate the speed of the bullet u to the speed of sound v in helium

$$\sin \theta = \frac{v}{u}$$

and the angle of the shock cone θ:

Solve for u:

$$u = \frac{v}{\sin\theta}$$

Measure θ to obtain:

$$\theta \approx 70°$$

Substitute numerical values and evaluate u:

$$u = \frac{977 \text{ m/s}}{\sin 70°} = \boxed{1.04 \text{ km/s}}$$

Speed of Waves

*23 •   Calculate the speed of sound waves in hydrogen gas at $T = 300$ K. (Take $M = 2$ g/mol and $\gamma = 1.4$.)

Picture the Problem The speed of sound in a gas is given by $v = \sqrt{\gamma RT/M}$ where R is the gas constant, T is the absolute temperature, M is the molecular mass of the gas, and γ is a constant that is characteristic of the particular molecular structure of the gas. Because hydrogen gas is diatomic, $\gamma = 1.4$.

Express the dependence of the speed of sound in hydrogen gas on the absolute temperature:

$$v = \sqrt{\frac{\gamma RT}{M}}$$

Substitute numerical values and evaluate v:

$$v = \sqrt{\frac{1.4(8.314 \text{ J/mol}\cdot\text{K})(300 \text{ K})}{2\times10^{-3} \text{ kg/mol}}}$$

$$= \boxed{1.32 \text{ km/s}}$$

*26 •   A wave pulse propagates along a wire in the positive x direction at 20 m/s. What will be the pulse velocity if we (*a*) double the length of the wire but keep the tension and mass per unit length constant? (*b*) double the tension while holding the length and mass per unit length constant? (*c*) double the mass per unit length while holding the other variables constant?

Picture the Problem The speed of a wave pulse on a wire is given by $v = \sqrt{F/\mu}$ where F is the tension in the wire, m is its mass, L is its length, and μ is its mass per unit length.

(*a*) Doubling the length while

$$v = \boxed{20 \text{ m/s}}$$

keeping the mass per unit length
constant does not change the
linear density:

(b) Because v depends on $\sqrt{F}$,
doubling the tension increases v
by a factor of $\sqrt{2}$:

$$v = \sqrt{2}(20\,\text{m/s}) = \boxed{28.3\,\text{m/s}}$$

(c) Because v depends on $1/\sqrt{\mu}$,
doubling μ reduces v by a factor
of $\sqrt{2}$:

$$v = \frac{20\,\text{m/s}}{\sqrt{2}} = \boxed{14.1\,\text{m/s}}$$

*29 •• (a) Compute the derivative of the speed of a wave on a string with
respect to the tension dv/dF, and show that the differentials dv and dF obey
$dv/v = \frac{1}{2}dF/F$. (b) A wave moves with a speed of 300 m/s on a wire that is under
a tension of 500 N. Using dF to approximate a change in tension, determine how
much the tension must be changed to increase the speed to 312 m/s.

Picture the Problem The speed of a transverse wave on a string is given by
$v = \sqrt{F/\mu}$ where F is the tension in the wire and μ is its linear density. We can
differentiate this expression with respect to F and then separate the variables to
show that the differentials satisfy $dv/v = \frac{1}{2}dF/F$. We'll approximate the
differential quantities to determine by how much the tension must be changed to
increase the speed of the wave to 312 m/s.

(a) Evaluate dv/dF:

$$\frac{dv}{dF} = \frac{d}{dF}\left[\sqrt{\frac{F}{\mu}}\right] = \frac{1}{2}\sqrt{\frac{1}{F\mu}} = \frac{1}{2}\frac{v}{F}$$

Separate the variables to obtain:

$$\boxed{\frac{dv}{v} = \frac{1}{2}\frac{dF}{F}}$$

(b) Solve for dF:

$$dF = 2F\frac{dv}{v}$$

Approximate dF with ΔF and dv
with Δv to obtain:

$$\Delta F = 2F\frac{\Delta v}{v}$$

Substitute numerical values and
evaluate ΔF:

$$\Delta F = 2(500\,\text{N})\frac{12\,\text{m/s}}{300\,\text{m/s}} = \boxed{40.0\,\text{N}}$$

*34 ••• Weather station Beta is located 0.75 mi due east of weather station Alpha. Observers at the two stations see a lightning strike to the north of the stations; observers at station Alpha hear the thunder 3.4 s after seeing the strike, while observers at Beta hear it 2.5 s after seeing the strike. Locate the coordinates of the lightning strike relative to the position of station Alpha.

Picture the Problem Choose a coordinate system in which station Alpha is at the origin and the axes are oriented as shown in the pictorial representation. Because 0.75 mi = 1.21 km, Alpha's coordinates are (0,0), Beta's are (1.21 km,0), and those of the lightning strike are (x,y). We can relate the distances from the stations to the speed of sound in air and the times required to hear the thunder at the two stations.

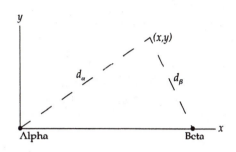

Relate the distance d_α to the position coordinates of Alpha and the lightning strike:

$$x^2 + y^2 = d_\alpha^2 \qquad (1)$$

Relate the distance d_β to the position coordinates of Beta and the lightning strike:

$$(x - 1.21\,\text{km})^2 + y^2 = d_\beta^2 \qquad (2)$$

Relate the distance d_α to the speed of sound in air v and the time that elapses between seeing the lightning at Alpha and hearing the thunder:

$$d_\alpha = v\Delta t_\alpha = (340\,\text{m/s})(3.4\,\text{s}) = 1156\,\text{m}$$

Relate the distance d_β to the speed of sound in air v and the time that elapses between seeing the lightning at Beta and hearing the thunder:

$$d_\beta = v\Delta t_\beta = (340\,\text{m/s})(2.5\,\text{s}) = 850\,\text{m}$$

Substitute in equations (1) and (2) to obtain:

$$x^2 + y^2 = (1156\,\text{m})^2 = 1.336\,\text{km}^2 \qquad (3)$$
and
$$(x - 1.21\,\text{km})^2 + y^2 = (850\,\text{m})^2$$
$$= 0.7225\,\text{km}^2 \qquad (4)$$

Subtract equation (4) from equation (3) to obtain:	$x^2 - (x-1.21\,\text{km})^2 = 1.336\,\text{km}^2$ $-0.7225\,\text{km}^2$

or

$$(2.42\,\text{km})x - (1.21\,\text{km})^2 = 0.6135\,\text{km}^2$$

Solve for x to obtain:	$x = 0.855\,\text{km}$
Substitute in equation (3) to obtain:	$(0.855\,\text{km})^2 + y^2 = 1.336\,\text{km}^2$

Solve for y, keeping the positive root because the lightning strike is to the north of the stations, to obtain:	$y = 0.778\,\text{km}$

The coordinates of the lightning strike are:

$$\boxed{(0.855\,\text{km},\, 0.778\,\text{km})}$$

or

$$\boxed{(0.531\,\text{mi},\, 0.484\,\text{mi})}$$

The Wave Equation

*37 •   Show that the function $y = A\sin kx \cos\omega t$ satisfies the wave equation.

Picture the Problem The general wave equation is $\dfrac{\partial^2 y}{\partial x^2} = \dfrac{1}{v^2}\dfrac{\partial^2 y}{\partial t^2}$. To show that $y = A\sin kx \cos\omega t$ satisfies this equation, we'll need to find the first and second derivatives of y with respect to x and t and then substitute these derivatives in the wave equation.

Find the first two spatial derivatives of $y = A\sin kx \cos\omega t$:

$$\frac{\partial y}{\partial x} = Ak\cos kx \cos\omega t$$

and

$$\frac{\partial^2 y}{\partial x^2} = -Ak^2 \sin kx \cos\omega t \qquad (1)$$

Find the first two temporal derivatives of $y = A\sin kx \cos\omega t$:

$$\frac{\partial y}{\partial t} = -\omega A \sin kx \sin\omega t$$

and

$$\frac{\partial^2 y}{\partial t^2} = -\omega^2 A \sin kx \cos\omega t \qquad (2)$$

Express the ratio of equation (1) to equation (2):

$$\frac{\dfrac{\partial^2 y}{\partial x^2}}{\dfrac{\partial^2 y}{\partial t^2}} = \frac{-Ak^2 \sin kx \cos \omega t}{-A\omega^2 \sin kx \cos \omega t} = \frac{k^2}{\omega^2}$$

$$= \boxed{\frac{1}{v^2}}$$

confirming that $y = A\sin kx \cos \omega t$ satisfies the general wave equation.

Harmonic Waves on a String

*40 • Equation 15-10 applies to all types of periodic waves, including electromagnetic waves such as light waves and microwaves, which travel at 3×10^8 m/s in a vacuum. (*a*) The range of wavelengths of light to which the eye is sensitive is about 4×10^{-7} to 7×10^{-7} m. What are the frequencies that correspond to these wavelengths? (*b*) Find the frequency of a microwave that has a wavelength of 3 cm.

Picture the Problem We can use $f = c/\lambda$ to express the frequency of any periodic wave in terms of its wavelength and velocity.

(*a*) Find the frequency of light of wavelength 4×10^{-7} m:

$$f = \frac{c}{\lambda} = \frac{2.998 \times 10^8 \text{ m/s}}{4 \times 10^{-7} \text{ m}} = 7.50 \times 10^{14} \text{ Hz}$$

Find the frequency of light of wavelength 7×10^{-7} m:

$$f = \frac{c}{\lambda} = \frac{2.998 \times 10^8 \text{ m/s}}{7 \times 10^{-7} \text{ m}} = 4.28 \times 10^{14} \text{ Hz}$$

Therefore the range of frequencies is:

$$\boxed{4.28 \times 10^{14} \text{ Hz} \le f \le 7.50 \times 10^{14} \text{ Hz}}$$

(*b*) Use the same relationship to calculate the frequency of these microwaves:

$$f = \frac{c}{\lambda} = \frac{2.998 \times 10^8 \text{ m/s}}{3 \times 10^{-2} \text{ m}}$$

$$= \boxed{1.00 \times 10^{10} \text{ Hz}}$$

*46 •• In a real string, a wave loses some energy as it travels down the string. Such a situation can be described by a wave function whose amplitude $A(x)$ depends on x: $y = A(x)\sin(kx - \omega t) = \left(A_0 e^{-bx}\right)\sin(kx - \omega t)$ (*a*) What is the power transported by the wave at the origin? (*b*) What is the power transported by the wave at point x, where $x > 0$?

Picture the Problem The power propagated along the rope by a harmonic wave is $P = \frac{1}{2}\mu\omega^2 A^2 v$, where v is the velocity of the wave, and μ, ω, and A are the linear density of the string, the angular frequency of the wave, and the amplitude of the wave, respectively. We can use the wave function $y = \left(A_0 e^{-bx}\right)\sin\left(kx - \omega t\right)$ to determine the amplitude of the wave at $x = 0$ and at point x.

(a) Express the power associated with the wave at the origin:

$$P = \frac{1}{2}\mu\omega^2 A^2 v$$

Evaluate the amplitude at $x = 0$:

$$A(0) = \left(A_0 e^0\right) = A_0$$

Substitute to obtain:

$$P(0) = \boxed{\frac{1}{2}\mu\omega^2 A_0^2 v}$$

(b) Express the amplitude of the wave at x:

$$A(x) = \left(A_0 e^{-bx}\right)$$

Substitute to obtain:

$$P(x) = \frac{1}{2}\mu\omega^2\left(A_0 e^{-bx}\right)^2 v$$
$$= \boxed{\frac{1}{2}\mu\omega^2 A_0^2 v e^{-2bx}}$$

*48 •••   Two very long strings are tied together at the point $x = 0$. In the region $x < 0$, the wave speed is v_1, while in the region $x > 0$, the speed is v_2. A sinusoidal wave is incident from the left ($x < 0$); part of the wave is reflected and part is transmitted. For $x < 0$, the displacement of the wave is describable by $y(x,t) = A \sin(k_1 x - \omega t) + B \sin(k_1 x + \omega t)$, while for $x > 0$, $y(x,t) = C \sin(k_2 x - \omega t)$, where $\omega/k_1 = v_1$ and $\omega/k_2 = v_2$. (a) If we assume that both the wave function y and its first spatial derivative $\partial y/\partial x$ must be continuous at $x = 0$, show that $C/A = 2/(1 + v_1/v_2)$, and that $B/A = -(1 - v_1/v_2)/(1 + v_1/v_2)$. (b) Show that $B^2 + (v_1/v_2)C^2 = A^2$.

Picture the Problem We can use the assumption that both the wave function and its first spatial derivative are continuous at $x = 0$ to establish equations relating A, B, C, k_1, and k_2. Then, we can solve these simultaneous equations to obtain expressions for B and C in terms of A, v_1, and v_2.

(a) Let $y_1(x,t)$ represent the wave function in the region $x < 0$, and $y_2(x,t)$ represent the wave function in the region $x > 0$. Express the continuity of the two wave functions at $x = 0$:

$$y_1(0,t) = y_2(0,t)$$

and

$$A \sin\left[k_1(0) - \omega t\right] + B \sin\left[k_1(0) + \omega t\right]$$
$$= C \sin\left[k_2(0) - \omega t\right]$$

or

$$A \sin\left(-\omega t\right) + B \sin \omega t = C \sin\left(-\omega t\right)$$

Because the sine function is odd;

$$-A \sin \omega t + B \sin \omega t = -C \sin \omega t$$

i.e., $\sin(-\theta) = -\sin\theta$:

and
$$A - B = C \tag{1}$$

Differentiate the wave functions with respect to x to obtain:

$$\frac{\partial y_1}{\partial x} = Ak_1 \cos(k_1 x - \omega t)$$
$$+ Bk_1 \cos(k_1 x + \omega t)$$

and

$$\frac{\partial y_2}{\partial x} = Ck_2 \cos(k_2 x - \omega t)$$

Express the continuity of the slopes of the two wave functions at $x = 0$:

$$\left.\frac{\partial y_1}{\partial x}\right|_{x=0} = \left.\frac{\partial y_2}{\partial x}\right|_{x=0}$$

and

$$Ak_1 \cos[k_1(0) - \omega t] + Bk_1 \cos[k_1(0) + \omega t]$$
$$= Ck_2 \cos[k_2(0) - \omega t]$$

or

$$Ak_1 \cos(-\omega t) + Bk_1 \cos\omega t$$
$$= Ck_2 \cos(-\omega t)$$

Because the cosine function is even; i.e., $\cos(-\theta) = \cos\theta$:

$$Ak_1 \cos\omega t + Bk_1 \cos\omega t = Ck_2 \cos\omega t$$

and

$$k_1 A + k_1 B = k_2 C \tag{2}$$

Multiply equation (1) by k_1 and add it to equation (2) to obtain:

$$2k_1 A = (k_1 + k_2)C$$

Solve for C:

$$C = \frac{2k_1}{k_1 + k_2} A = \frac{2}{1 + k_2/k_1} A$$

Solve for C/A and substitute ω/v_1 for k_1 and ω/v_2 for k_2 to obtain:

$$\frac{C}{A} = \frac{2}{1 + k_2/k_1} = \boxed{\frac{2}{1 + v_1/v_2}}$$

Substitute in equation (1) to obtain:

$$A - B = \left(\frac{2}{1 + v_1/v_2}\right)A$$

Solve for B/A to obtain:

$$\frac{B}{A} = \boxed{-\frac{1 - v_1/v_2}{1 + v_1/v_2}}$$

(b) We wish to show that

$$B^2 + (v_1/v_2)C^2 = A^2$$

Use the results of (a) to obtain the expressions $B = -[(1 - \alpha)/(1 + \alpha)]A$ and $C = 2A/(1 + \alpha)$, where $\alpha = v_1/v_2$.

Substitute these expressions into

$$B^2 + (v_1/v_2)C^2 = A^2$$

and check to see if the resulting equation is an identity:

$$B^2 + \frac{v_1}{v_2}C^2 = A^2$$

$$\left(\frac{1-\alpha}{1+\alpha}\right)^2 A^2 + \alpha\left(\frac{2}{1+\alpha}\right)^2 A^2 = A^2$$

$$\left(\frac{1-\alpha}{1+\alpha}\right)^2 + \alpha\left(\frac{2}{1+\alpha}\right)^2 = 1$$

$$\frac{(1-\alpha)^2 + 4\alpha}{(1+\alpha)^2} = 1$$

$$\frac{1 - 2\alpha + \alpha^2 + 4\alpha}{(1+\alpha)^2} = 1$$

$$\frac{1 + 2\alpha + \alpha^2}{(1+\alpha)^2} = 1$$

$$\frac{(1+\alpha)^2}{(1+\alpha)^2} = 1$$

$$1 = 1$$

The equation is an identity:

Therefore, $\boxed{B^2 + \dfrac{v_1}{v_2}C^2 = A^2}$

Remarks: Our result in (a) can be checked by considering the limit of B/A as $v_2/v_1 \to 0$. This limit gives B/A = +1 which tells us that the transmitted wave has zero amplitude and the incident and reflected waves superpose to gave a standing wave with a node at x = 0.

Harmonic Sound Waves

*49 •   A sound wave in air produces a pressure variation given by

$$p(x,t) = 0.75\cos\frac{\pi}{2}(x - 340t)$$

where p is in pascals, x is in meters, and t is in seconds. Find (a) the pressure amplitude of the sound wave, (b) the wavelength, (c) the frequency, and (d) the speed.

Picture the Problem The pressure variation is of the form $p(x,t) = p_0 \cos k(x - vt)$ where $k = \pi/2$ and $v = 340\,\text{m/s}$. We can find λ from k and f from ω and k.

(a) By inspection of the equation: $p_0 = \boxed{0.750\,\text{Pa}}$

(b) Because $k = \dfrac{2\pi}{\lambda} = \dfrac{\pi}{2}$: $\lambda = \boxed{4.00\,\text{m}}$

(c) Solve $v = \dfrac{\omega}{k} = \dfrac{2\pi f}{k}$ for f to obtain:

$$f = \frac{kv}{2\pi} = \frac{\dfrac{\pi}{2}(340\,\text{m/s})}{2\pi} = \boxed{85.0\,\text{Hz}}$$

(d) By inspection of the equation: $v = \boxed{340\,\text{m/s}}$

*54 •   An octave represents a change in frequency by a factor of two. Over how many octaves can a typical person hear?

Picture the Problem A human can hear sounds between roughly 20 Hz and 20 kHz; a factor of 1000. An octave represents a change in frequency by a factor of 2. We can evaluate $2^N = 1000$ to find the number of octaves heard by a person who can hear this range of frequencies.

Relate the number of octaves to the difference between 20 kHz and 20 Hz:

$2^N = 1000$

Take the logarithm of both sides of the equation to obtain:

$\log 2^N = \log 10^3$

or

$N \log 2 = 3$

Solve for and evaluate N:

$$N = \frac{3}{\log 2} = 9.97 \approx \boxed{10}$$

Waves in Three Dimensions: Intensity

*57 •   A loudspeaker at a rock concert generates 10^{-2} W/m² at 20 m at a frequency of 1 kHz. Assume that the speaker spreads its energy uniformly in three dimensions. (a) What is the total acoustic power output of the speaker? (b) At what distance will the intensity be at the pain threshold of 1 W/m²? (c) What is the intensity at 30 m?

Picture the Problem Because the power radiated by the loudspeaker is the product of the intensity of the sound and the surface area over which it is distributed, we can use this relationship to find the average power, the intensity of the radiation, or the distance to the speaker for a given intensity or average power.

(a) Use $P_{av} = 4\pi r^2 I$ to find the total acoustic power output of the speaker:

$$P_{av} = 4\pi(20\,\text{m})^2(10^{-2}\,\text{W/m}^2)$$
$$= \boxed{50.3\,\text{W}}$$

(b) Relate the intensity of the sound at 20 m to the distance from the speaker:

$$10^{-2}\,\text{W/m}^2 = \frac{P_{av}}{4\pi(20\,\text{m})^2}$$

Relate the threshold-of-pain intensity to the distance from the speaker:

$$1\,\text{W/m}^2 = \frac{P_{av}}{4\pi r^2}$$

Divide the first of these equations by the second; solve for and evaluate r:

$$r = \sqrt{10^{-2}(20\,\text{m})^2} = \boxed{2.00\,\text{m}}$$

(c) Use $I = \dfrac{P_{av}}{4\pi r^2}$ to find the intensity at 30 m:

$$I(30\,\text{m}) = \frac{50.3\,\text{W}}{4\pi(30\,\text{m})^2}$$
$$= \boxed{4.45\times10^{-3}\,\text{W/m}^2}$$

Intensity Level

*61 •   The sound level of a dog's bark is 50 dB. The intensity of a rock concert is 10,000 times that of the dog's bark. What is the sound level of the rock concert?

Picture the Problem The intensity level of a sound wave β, measured in decibels, is given by $\beta = (10\,\text{dB})\log(I/I_0)$, where $I_0 = 10^{-12}\,\text{W/m}^2$ is defined to be the threshold of hearing.

Express the sound level of the rock concert:

$$\beta_{concert} = (10\,\text{dB})\log\left(\frac{I_{concert}}{I_0}\right) \qquad (1)$$

Express the sound level of the dog's bark:

$$50\,\text{dB} = (10\,\text{dB})\log\left(\frac{I_{dog}}{I_0}\right)$$

Solve for the intensity of the dog's bark:

$$I_{dog} = 10^5 I_0 = 10^5(10^{-12}\,\text{W/m}^2)$$
$$= 10^{-7}\,\text{W/m}^2$$

Express the intensity of the rock concert in terms of the intensity of the dog's bark:

$$I_{concert} = 10^4 I_{dog} = 10^4 \left(10^{-7} \text{ W/m}^2\right)$$
$$= 10^{-3} \text{ W/m}^2$$

Substitute in equation (1) and evaluate $\beta_{concert}$:

$$\beta_{concert} = (10\,\text{dB})\log\left(\frac{10^{-3} \text{ W/m}^2}{10^{-12} \text{ W/m}^2}\right)$$
$$= (10\,\text{dB})\log 10^9$$
$$= \boxed{90.0\,\text{dB}}$$

*64 •   What fraction of the acoustic power of a noise would have to be eliminated to lower its sound intensity level from 90 to 70 dB?

Picture the Problem We can express the intensity levels at both 90 dB and 70 dB in terms of the intensities of the sound at those levels. By subtracting the two expressions, we can solve for the ratio of the intensities at the two levels and then find the fractional change in the intensity that corresponds to a decrease in intensity level from 90 dB to 70 dB.

Express the intensity level at 90 dB:

$$90\,\text{dB} = (10\,\text{dB})\log\left(\frac{I_{90}}{I_0}\right)$$

Express the intensity level at 70 dB:

$$70\,\text{dB} = (10\,\text{dB})\log\left(\frac{I_{70}}{I_0}\right)$$

Express $\Delta\beta = \beta_{90} - \beta_{70}$:

$$\Delta\beta = 20\,\text{dB}$$
$$= (10\,\text{dB})\log\left(\frac{I_{90}}{I_0}\right) - (10\,\text{dB})\log\left(\frac{I_{70}}{I_0}\right)$$
$$= (10\,\text{dB})\log\left(\frac{I_{90}}{I_{70}}\right)$$

Solve for I_{90}:

$$I_{90} = 100 I_{70}$$

Express the fractional change in the intensity from 90 dB to 70 dB:

$$\frac{I_{90} - I_{70}}{I_{90}} = \frac{100 I_{70} - I_{70}}{100 I_{70}} = \boxed{99\%}$$

*70 ••   If you double the distance between a source of sound and a receiver, the intensity at the receiver drops by approximately (*a*) 2 dB, (*b*) 3 dB, (*c*) 6 dB, (*d*) Amount cannot be determined from the information given.

Picture the Problem Let P be the power radiated by the source of sound, and r be the initial distance from the source to the receiver. We can use the definition of intensity to find the ratio of the intensities before and after the distance is doubled and then use the definition of the decibel level to find the change in its level.

Relate the change in decibel level to the change in the intensity level:

$$\Delta\beta = 10\log\frac{I}{I'}$$

Using its definition, express the intensity of the sound from the source as a function of P and r:

$$I = \frac{P}{4\pi r^2}$$

Express the intensity when the distance is doubled:

$$I' = \frac{P}{4\pi(2r)^2} = \frac{P}{16\pi r^2}$$

Evaluate the ratio of I to I':

$$\frac{I}{I'} = \frac{\dfrac{P}{4\pi r^2}}{\dfrac{P}{16\pi r^2}} = 4$$

Substitute to obtain:

$$\Delta\beta = 10\log 4 = 6.02\,\text{dB and}$$

$$\boxed{(c) \text{ is correct.}}$$

*72 •••   When a violinist pulls the bow across a string, the force with which the bow is pulled is fairly small, about 0.6 N. Suppose the bow travels across the A string, which vibrates at 440 Hz, at 0.5 m/s. A listener 35 m from the performer hears a sound of 60 dB intensity. With what efficiency is the mechanical energy of bowing converted to sound energy? (Assume that the sound radiates uniformly in all directions.)

Picture the Problem Let η represent the efficiency with which mechanical energy is converted to sound energy. Because we're given information regarding the rate at which mechanical energy is delivered to the string and the rate at which sound energy arrives at the location of the listener, we'll take the efficiency to be the ratio of the sound power delivered to the listener divided by the power delivered to the string. We can calculate the power input directly from the given data. We'll calculate the intensity of the sound at 35 m from its intensity level at that distance and use this result to find the power output.

Express the efficiency of the conversion of mechanical energy to sound energy:	$\eta = \dfrac{P_{out}}{P_{in}}$

Find the power delivered by the bow to the string:	$P_{in} = Fv = (0.6\,N)(0.5\,m/s) = 0.3\,W$

Using $\beta = (10\,dB)\log(I/I_0)$, find the intensity of the sound at 35 m:	$60\,dB = (10\,dB)\log\dfrac{I_{35\,m}}{I_0}$ and $I_{35\,m} = 10^6 I_0 = 10^{-6}\,W/m^2$

Find the power of the sound emitted:	$P_{out} = IA = 4\pi(10^{-6}\,W/m^2)(35\,m)^2$ $= 0.0154\,W$

Substitute numerical values and evaluate η:	$\eta = \dfrac{0.0154\,W}{0.3\,W} = \boxed{5.13\%}$

The Doppler Effect

*83 ••   The Doppler effect is routinely used to measure the speed of winds in storm systems. A weather station uses a Doppler radar system of frequency $f = 625$ MHz to bounce a radar pulse off of the raindrops in a swirling thunderstorm system 50 km away; the reflected radar pulse is found to be up-shifted in frequency by 325 Hz. Assuming the wind is headed directly toward the radar antenna, how fast are the winds in the storm system moving? (The radar system can only measure the radial component of the velocity.)

Picture the Problem Because the radial component u of the velocity of the raindrops is small compared to the speed $v = c$ of the radar pulse, we can approximate the fractional change in the frequency of the reflected radar pulse to find the speed of the winds carrying the raindrops in the storm system.

Express the shift in frequency when the speed of the source (the storm system) u is much smaller than the wave speed $v = c$:	$\dfrac{\Delta f}{f_s} \approx \dfrac{u}{v}$

Solve for u:	$u = v\dfrac{\Delta f}{f_s}$

Substitute numerical values and evaluate u:

$$u = (2.998 \times 10^8 \text{ m/s}) \frac{325 \text{ Hz}}{625 \text{ MHz}}$$

$$= 156 \text{ m/s} \times \frac{0.6215 \text{ mi/h}}{0.2778 \text{ m/s}}$$

$$= \boxed{349 \text{ mi/h}}$$

*92 •• A small speaker radiating sound at 1000 Hz is tied to one end of an 0.8-m-long rod that is free to rotate about its other end. The rod rotates in the horizontal plane at 4.0 rad/s. Derive an expression for the frequency heard by a stationary observer far from the rotating speaker.

Picture the Problem The frequency heard by the stationary observer will vary with time as the speaker rotates on its support arm. We can use a Doppler equation to express the frequency heard by the observer as a function of the velocity of the source and find the velocity of the source from the expression for the tangential velocity of an object moving in a circular path.

Express the frequency f_r heard by a stationary observer:

$$f_r = \frac{1}{1 - u_s/v} f_s = (1 - u_s/v)^{-1} f_s$$

Expand $(1 - u_s/v)^{-1}$ to obtain:

$$(1 - u_s/v)^{-1} \approx 1 + u_s/v$$

because $u_s/v \ll 1$

Substitute in the expression for f_r:

$$f_r = (1 + u_s/v) f_s \qquad (1)$$

Express the speed of the source as a function of time:

$$u_s = r\omega \sin \omega t$$
$$= (0.8 \text{ m})(4 \text{ rad/s}) \sin[(4 \text{ rad/s})t]$$
$$= (3.2 \text{ m/s}) \sin[(4 \text{ rad/s})t]$$

Substitute in equation (1) to obtain:

$$f_r = \left(1 + \frac{3.2 \text{ m/s}}{v} \sin[(4 \text{ rad/s})t]\right) f_s$$

Substitute for v and simplify:

$$f_r = \left(1 + \frac{3.2 \text{ m/s}}{340 \text{ m/s}} \sin[(4 \text{ rad/s})t]\right)(1000 \text{ s}^{-1})$$

$$= \boxed{1000 \text{ Hz} + (9.41 \text{ Hz}) \sin[(4 \text{ rad/s})t]}$$

*97 •• The Hubble space telescope has been used to determine the existence of planets orbiting around distant stars. The planet orbiting the star will cause the star to "wobble" with the same period as the planet's orbit; because of this, light from the star will be Doppler-shifted up and down periodically. Estimate the maximum and minimum wavelengths of light of nominal wavelength 500 nm emitted by the sun that is Doppler-shifted by the motion of the sun due to the planet Jupiter.

Picture the Problem The sun and Jupiter orbit about their effective mass located at their common center of mass. We can apply Newton's 2nd law to the sun to obtain an expression for its orbital speed about the sun-Jupiter center of mass and then use this speed in the Doppler shift equation to estimate the maximum and minimum wavelengths resulting from the Jupiter-induced motion of the sun.

Letting v be the orbital speed of the sun about the center of mass of the sun-Jupiter system, express the Doppler shift of the light due to this motion when the sun is approaching the earth:

$$f' = \frac{c}{\lambda'} = f\sqrt{\frac{1+v/c}{1-v/c}} = \frac{c}{\lambda}\sqrt{\frac{1+v/c}{1-v/c}}$$

Solve for λ':

$$\lambda' = \lambda\sqrt{\frac{1-v/c}{1+v/c}}$$

$$= \lambda\sqrt{(1-v/c)(1+v/c)^{-1}}$$

$$= \lambda(1-v/c)^{1/2}(1+v/c)^{-1/2}$$

Because $v \ll c$, we can expand $(1-v/c)^{1/2}$ and $(1+v/c)^{-1/2}$ binomially to obtain:

$$(1-v/c)^{1/2} \approx 1 - \frac{v}{2c}$$

and

$$(1+v/c)^{-1/2} \approx 1 - \frac{v}{2c}$$

Substitute to obtain:

$$\sqrt{\frac{1-v/c}{1+v/c}} = \left(1-\frac{v}{2c}\right)^2 \approx 1 - \frac{v}{c}$$

When the sun is receding from the earth:

$$\sqrt{\frac{1+v/c}{1-v/c}} = \left(1+\frac{v}{2c}\right)^2 \approx 1 + \frac{v}{c}$$

Hence the motion of the sun will give an observed Doppler shift of:

$$\lambda' \approx \lambda\left(1 \pm \frac{v}{c}\right) \qquad (1)$$

Apply Newton's 2nd law to the sun:

$$\frac{GM_sM_{\text{eff}}}{r_{\text{cm}}^2} = M_s\frac{v^2}{r_{\text{cm}}}$$

Solve for v to obtain:

$$v = \sqrt{\frac{GM_{\text{eff}}}{r_{\text{cm}}}}$$

Measured from the center of the sun, the distance to the center of mass of the sun-Jupiter system is:

$$r_{\text{cm}} = \frac{(0)M_S + r_{\text{s-J}}M_J}{M_s + M_J} = \frac{r_{\text{s-J}}M_J}{M_s + M_J}$$

The effective mass is related to the masses of the sun and Jupiter according to:

$$\frac{1}{M_{\text{eff}}} = \frac{1}{M_s} + \frac{1}{M_J}$$

or

$$M_{\text{eff}} = \frac{M_s M_J}{M_s + M_J}$$

Substitute for M_{eff} and r_{cm} to obtain:

$$v = \sqrt{\frac{G \dfrac{M_s M_J}{M_s + M_J}}{\dfrac{r_{\text{s-J}}M_J}{M_s + M_J}}} = \sqrt{\frac{GM_s}{r_{\text{s-J}}}}$$

Using $r_{\text{s-J}} = 7.78 \times 10^{11}$ m as the mean orbital radius of Jupiter, substitute numerical values and evaluate v:

$$v = \sqrt{\frac{\left(6.673 \times 10^{-11} \, \text{N} \cdot \text{m}^2 / \text{kg}^2\right)\left(1.99 \times 10^{30} \, \text{kg}\right)}{7.78 \times 10^{11} \, \text{m}}} = 1.306 \times 10^4 \, \text{m/s}$$

Substitute in equation (1) to obtain:

$$\lambda' \approx (500 \, \text{nm})\left(1 \pm \frac{1.306 \times 10^4 \, \text{m/s}}{2.998 \times 10^8 \, \text{m/s}}\right)$$

$$= (500 \, \text{nm})\left(1 \pm 4.36 \times 10^{-5}\right)$$

The maximum and minimum wavelengths are:

$$\lambda_{\text{max}} = \boxed{(500 \, \text{nm})\left(1 + 4.36 \times 10^{-5}\right)}$$

and

$$\lambda_{\text{min}} = \boxed{(500 \, \text{nm})\left(1 - 4.36 \times 10^{-5}\right)}$$

General Problems

*108 •• Find the speed of a car the tone of whose horn will drop by 10 percent as it passes you.

Picture the Problem Let the frequency of the car's horn be f_s, the frequency you hear as the car approaches f_r, and the frequency you hear as the car recedes f_r'. We can use $f_r = \dfrac{v \pm u_r}{v \pm u_s} f_s$ to express the frequencies heard as the car approaches and recedes and then use these frequencies to express the fractional change in frequency as the car passes you.

Express the fractional change in frequency as the car passes you:	$\dfrac{\Delta f}{f_r} = 0.1$
Relate the frequency heard as the car approaches to the speed of the car:	$f_r = \dfrac{1}{1 - u_s/v} f_s$
Express the frequency heard as the car recedes in terms of the speed of the car:	$f_r' = \dfrac{1}{1 + u_s/v} f_s$
Divide the second of these frequency equations by the first to obtain:	$\dfrac{f_r'}{f_r} = \dfrac{1 - u_s/v}{1 + u_s/v}$ and $\dfrac{f_r}{f_r} - \dfrac{f_r'}{f_r} = \dfrac{\Delta f}{f_r} = 1 - \dfrac{1 - u_s/v}{1 + u_s/v} = 0.1$
Solve u_s:	$u_s = \dfrac{0.1}{1.9} v$
Substitute numerical values and evaluate u_s:	$u_s = \dfrac{0.1}{1.9}(340\,\text{m/s})$ $= 17.89 \dfrac{\text{m}}{\text{s}} \times \dfrac{1\,\text{km}}{10^3} \times \dfrac{3600\,\text{s}}{\text{h}}$ $= \boxed{64.4\,\text{km/h}}$

*115 •• Laser ranging to the moon is done routinely to accurately determine the earth-moon distance. However, to determine the distance accurately, corrections must be made for the speed of light in the earth's atmosphere, which is 99.997 percent of the speed of light in vacuum. Assuming that the earth's atmosphere is effectively 8 km high, estimate the length of the correction.

Picture the Problem Let d be the distance to the moon, h be the height of earth's atmosphere, and v be the speed of light in earth's atmosphere. We can express d', the distance measured when the earth's atmosphere is ignored, in terms of the time for a pulse of light to make a round-trip from the earth to the moon and solve this equation for the length of correction $d' - d$.

Express the roundtrip time for a pulse of light to reach the moon and return:

$$t = t_{\text{earth's atmosphere}} + t_{\text{out of earth's atmosphere}}$$

$$= 2\frac{h}{v} + 2\frac{d-h}{c}$$

Express the "measured" distance d' when we do not account for the atmosphere:

$$d' = \frac{1}{2}ct = \frac{1}{2}c\left(2\frac{h}{v} + 2\frac{d-h}{c}\right)$$

$$= \frac{c}{v}h + d - h$$

Solve for the length of correction $d' - d$:

$$d' - d = h\left(\frac{c}{v} - 1\right)$$

Substitute numerical values and evaluate $d' - d$:

$$d' - d = (8\,\text{km})\left(\frac{c}{0.99997c} - 1\right)$$

$$= \boxed{24.0\,\text{cm}}$$

Remarks: This is larger than the accuracy of the measurements, which is about 3 to 4 cm.

*119 •••   A heavy rope 3 m long is attached to the ceiling and is allowed to hang freely. (*a*) Show that the speed of transverse waves on the rope is independent of its mass and length but does depend on the distance y from the bottom according to the formula $v = \sqrt{gy}$. (*b*) If the bottom end of the rope is given a sudden sideways displacement, how long does it take the resulting wave pulse to go to the ceiling, reflect, and return to the bottom of the rope?

Picture the Problem We can relate the speed of the pulse to the tension in the rope and its linear density. Because the rope hangs vertically, the tension in it varies linearly with the distance from its bottom. Once we've established the result in part (*a*), we can integrate the resulting velocity equation to find the time for the pulse to travel the length of the rope and then double this time to get the round-trip time.

(*a*) Relate the speed of transverse waves to tension and linear density:

$$v = \sqrt{\frac{F}{\mu}}$$

Express the force acting on a segment of the rope of length y:

$$F = mg = \mu y g$$

Substitute to obtain:

$$v = \sqrt{\frac{\mu y g}{\mu}} = \boxed{\sqrt{gy}}$$

(b) Because the speed of the pulse varies with the distance from the bottom of the rope, express v as dy/dt and solve for dt:

$$\frac{dy}{dt} = \sqrt{gy} \quad \text{and} \quad dt = \frac{1}{\sqrt{g}} \frac{dy}{\sqrt{y}}$$

Integrate the left side of the equation from 0 to t and the right side from 0 to 3 m:

$$\int_0^t dt' = \frac{1}{\sqrt{g}} \int_0^{3\,\text{m}} \frac{dy}{\sqrt{y}}$$

and

$$t = \frac{1}{\sqrt{g}} \left(2\sqrt{y}\right)_0^{3\,\text{m}} = \frac{2\sqrt{3\,\text{m}}}{\sqrt{9.81\,\text{m/s}^2}}$$

$$= 1.106\,\text{s}$$

The time for the pulse to make the round trip is:

$$t_{\text{round trip}} = 2t = 2(1.106\,\text{s}) = \boxed{2.21\,\text{s}}$$

Chapter 16
Superposition and Standing Waves

Conceptual Problems

*1 •• Two rectangular wave pulses are traveling in opposite directions along a string. At $t = 0$, the two pulses are as shown in Figure 16-27. Sketch the wave functions for $t = 1, 2,$ and 3 s.

Figure 16-27 Problem 1, 2

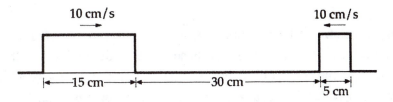

Picture the Problem We can use the speeds of the pulses to determine their positions at the given times.

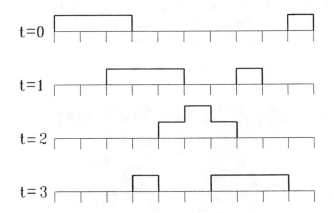

*6 • The resonant frequencies of a violin string are all integer multiples of the fundamental frequency, while the resonant frequencies of a circular drumhead are irregularly spaced. Given this information, explain the difference in the sounds of a violin and a drum.

Determine the Concept Our ears and brain find frequencies which are small-integer multiples of one another pleasing when played in combination. In particular, the ear hears frequencies related by a factor of 2 (one octave) as identical. Thus, a violin sounds much more "musical" than the sound of a drum.

*10 •   When two waves moving in opposite directions superimpose as in Figure 16-1, does either impede the progress of the other?

Determine the Concept Because the two waves move independently, neither impedes the progress of the other.

*15 ••   When the tension on a piano wire is increased, which of the following occurs? (*a*) Its wavelength decreases. (*b*) Its wavelength remains the same while its frequency increases. (*c*) Its wavelength and frequency increase. (*d*) None of the above occur.

Determine the Concept Increasing the tension on a piano wire increases the speed of the waves. The wavelength of these waves is determined by the length of the wire. Because the speed of the waves is the product of their wavelength and frequency, the wavelength remains the same and the frequency increases.

$\boxed{(b) \text{ is correct.}}$

*18 ••   Figure 16-28 is a photograph of two pieces of very finely woven silk placed one on top of the other. Where the pieces overlap, a series of light and dark lines are seen. This moiré pattern can also be seen when a scanner is used to copy photos from a book or newspaper. What causes the moiré pattern, and how is it similar to the phenomenon of interference?

Figure 16-28 Problem 18

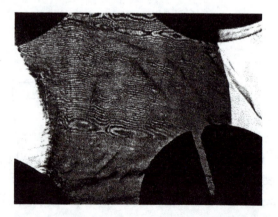

Determine the Concept The light is being projected up from underneath the silk, so you will see light where there is a gap and darkness where two threads overlap. Because the two weaves have almost the same spatial period but not exactly identical (because the two are stretched unequally), there will be places where, for large sections of the cloth, the two weaves overlap in phase, leading to brightness, and large sections where the two overlap 90° out of phase (i.e., thread on gap and vice versa), leading to darkness. This is exactly the same idea as in the interference of two waves.

Estimation and Approximation

*20 •   The shortest pipes used in organs are about 7.5 cm long. (*a*) What is the fundamental frequency of a pipe this long that is open at both ends? (*b*) For such a pipe, what is the highest harmonic that is within the audible range? (The normal range of hearing is about 20 to 20,000 Hz.)

Picture the Problem We can use $v = f_1 \lambda_1$ to express the resonance frequencies in the organ pipes in terms of their wavelengths and $L = n\frac{\lambda_n}{2}, n = 1, 2, 3, \ldots$ to relate the length of the pipes to the resonance wavelengths.

(*a*) Relate the fundamental frequency of the pipe to its wavelength and the speed of sound:	$$f_1 = \frac{v}{\lambda_1}$$
Express the condition for constructive interference in a pipe that is open at both ends:	$$L = n\frac{\lambda_n}{2}, n = 1, 2, 3, \ldots \qquad (1)$$
Solve for λ_1:	$$\lambda_1 = 2L$$
Substitute and evaluate f_1:	$$f_1 = \frac{v}{2L} = \frac{340\,\text{m/s}}{2(7.5 \times 10^{-2}\,\text{m})} = \boxed{2.27\,\text{kHz}}$$
(*b*) Relate the resonance frequencies of the pipe to their wavelengths and the speed of sound:	$$f_n = \frac{v}{\lambda_n}$$
Solve equation (2) for λ_n:	$$\lambda_n = \frac{2L}{n}$$
Substitute to obtain:	$$f_n = n\frac{v}{2L} = n\frac{340\,\text{m/s}}{2(7.5 \times 10^{-2}\,\text{m})}$$ $$= n(2.27\,\text{kHz})$$
Set $f_n = 20$ kHz and evaluate n:	$$n = \frac{20\,\text{kHz}}{2.27\,\text{kHz}} = 8.81$$

> The eighth harmonic is within the range defined as audible. The ninth harmonic might be heard by a person with very good hearing.

Superposition and Interference

*24 •   Two sound sources oscillate in phase with the same amplitude A. They are separated in space by $\lambda/3$. What is the amplitude of the resultant wave formed from the two sources at a point that is on the line that passes through the sources but is not between the sources?

Picture the Problem The phase shift in the waves generated by these two sources is due to their separation of $\lambda/3$. We can find the phase difference due to the path difference from $\delta = 2\pi \dfrac{\Delta x}{\lambda}$ and then the amplitude of the resultant wave from $A = 2y_0 \cos\frac{1}{2}\delta$.

Evaluate the phase difference δ:

$$\delta = 2\pi \frac{\Delta x}{\lambda} = 2\pi \frac{\lambda/3}{\lambda} = \frac{2}{3}\pi$$

Find the amplitude of the resultant wave:

$$A_{\text{res}} = 2y_0 \cos\tfrac{1}{2}\delta = 2A\cos\frac{1}{2}\left(\frac{2}{3}\pi\right)$$

$$= 2A\cos\frac{\pi}{3} = \boxed{A}$$

*26 •   With a compass, draw circular arcs representing wave crests originating from each of two point sources a distance $d = 6$ cm apart for $\lambda = 1$ cm. Connect the intersections corresponding to points of constant path difference and label the path difference for each line. (See Figure 16-8.)

Picture the Problem The diagram is shown below. Lines of constructive interference are shown for path differences of 0, λ, 2λ, and 3λ.

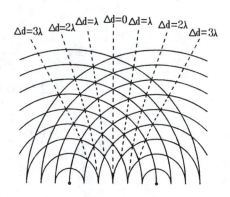

*30 •• Show that, if the separation between two sound sources radiating coherently in phase is less than half a wavelength, complete destructive interference will not be observed in any direction.

Picture the Problem The drawing shows a generic point P located a distance r_1 from source S_1 and a distance r_2 from source S_2. The sources are separated by a distance d and we're given that $d < \lambda/2$. Because the condition for destructive interference is that $\delta = n\pi$ where $n = 1, 2, 3,...$, we'll show that, with $d < \lambda/2$, this condition cannot be satisfied.

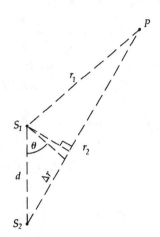

Relate the phase shift to the path difference and the wavelength of the sound:

$$\delta = 2\pi \frac{\Delta r}{\lambda}$$

Relate Δr to d and θ:

$$\Delta r < d \sin\theta \le d$$

Substitute to obtain:

$$\delta < 2\pi \frac{d \sin\theta}{\lambda} \le 2\pi \frac{d}{\lambda}$$

Because $d < \lambda/2$:

$$\delta < 2\pi \frac{\lambda/2}{\lambda} = \pi$$

Express the condition for destructive interference:

$$\delta = n\pi$$
$$\text{where } n = 1, 2, 3,...$$

Because $\delta < \pi$, there is no complete destructive interference in any direction.

*35 •• Two harmonic water waves of equal amplitudes but different frequencies, wave vectors, *and velocities* are superposed on each other. The total displacement of the wave can be written as $y(x,t) = A[\cos(k_1 x - \omega_1 t) + \cos(k_2 x - \omega_2 t)]$, where $\omega_1/k_1 = v_1$ (the speed of the first wave) and $\omega_2/k_2 = v_2$ (the speed of the second wave). (*a*) Show that $y(x,t)$ can be written in the form $y(x,t) = 2A\cos[(\Delta k/2)x - (\Delta\omega/2)t)]$ where $\omega_{av} = (\omega_1 + \omega_2)/2$. $k_{av} = (k_1 + k_2)/2$, $\Delta\omega = \omega_1 - \omega_2$, and $\Delta k = k_1 - k_2$. The factor $2A\cos[(\Delta k/2)x - (\Delta\omega/2)t]$ is referred to as the *envelope* of the wave. (*b*) Using a spreadsheet program or graphing calculator, make a graph of $y(x,t)$ for $A = 1$,

$\omega_1 = 1$ rad/s, $k_1 = 1$ m^{-1}, $\omega_2 = 0.9$ rad/s, and $k_2 = 0.8$ m^{-1} at $t = 0$ s, 0.5 s, and 1 s for x between 0 m and 50 m. (*c*) What is the speed at which the envelope moves?

Picture the Problem We can use the trigonometric identity

$\cos A + \cos B = 2\cos\left(\dfrac{A+B}{2}\right)\cos\left(\dfrac{A-B}{2}\right)$ to derive the expression given in (*a*) and

the speed of the envelope can be found from the second factor in this expression; i.e., from $\cos\left((\Delta k/2)x - (\Delta\omega/2)t\right)$.

(*a*) Express the amplitude of the resultant wave function $y(x,t)$:

$$y(x,t) = A\left(\cos(k_1 x - \omega_1 t) + \cos(k_2 x - \omega_2 t)\right)$$

Use the trigonometric identity $\cos A + \cos B = 2\cos\left(\dfrac{A+B}{2}\right)\cos\left(\dfrac{A-B}{2}\right)$ to obtain:

$$y(x,t) = 2A\left[\cos\frac{k_1 x - \omega_1 t + k_2 x - \omega_2 t}{2}\cos\frac{k_1 x - \omega_1 t - k_2 x + \omega_2 t}{2}\right]$$

$$= 2A\left[\cos\left(\frac{k_1 + k_2}{2}x - \frac{\omega_1 + \omega_2}{2}t\right)\cos\left(\frac{k_1 - k_2}{2}x + \frac{\omega_2 - \omega_1}{2}t\right)\right]$$

Substitute $\omega_{ave} = (\omega_1 + \omega_2)/2$, $k_{ave} = (k_1 + k_2)/2$, $\Delta\omega = \omega_1 - \omega_2$ and $\Delta k = k_1 - k_2$ to obtain:

$$\boxed{y(x,t) = 2A\left[\cos(k_{ave}x - \omega_{ave}t)\cos\left(\frac{\Delta k}{2}x - \frac{\Delta\omega}{2}t\right)\right]}$$

(*b*) A spreadsheet program to calculate $y(x,t)$ between 0 m and 50 m at $t = 0$, 0.5 s, and 1 s follows. The constants and cell formulas used are shown in the table.

Cell	Content/Formula	Algebraic Form
B11	B10+0.25	$x + \Delta x$
C10	COS($B$3*B10−$B$5*$C$9) + COS($B$4*B10−$B$6*$C$9)	$y(x,0)$
D10	COS($B$3*B10−$B$5*$D$9) + COS($B$4*B10−$B$6*$D$9)	$y(x,0.5\,\text{s})$
E10	COS($B$3*B10−$B$5*$E$9)	$y(x,1\,\text{s})$

	A	B	C	D	E
1					
2					
3	k1=	1	m^{-1}		

4	k2=	0.8	m^{-1}		
5	w1=	1	rad/s		
6	w2=	0.9	rad/s		
7		x	y(x,0)	y(x,0.5 s)	y(x,1 s)
8		(m)			
9			0.000	2.000	4.000
10		0.00	2.000	−0.643	−1.550
11		0.25	1.949	−0.207	−1.787
12		0.50	1.799	0.241	−1.935
13		0.75	1.557	0.678	−1.984
14		1.00	1.237	1.081	−1.932
206		49.00	0.370	−0.037	0.021
207		49.25	0.397	0.003	−0.024
208		49.50	0.397	0.065	−0.075
209		49.75	0.364	0.145	−0.124
210		50.00	0.298	0.237	−0.164

The solid line is the graph of $y(x,0)$, the dashed line that of $y(x,0.5$ s$)$, and the dotted line is the graph of $y(x,1$ s$)$.

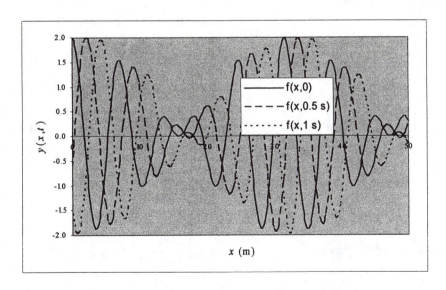

(c) Express the speed of the envelope:

$$v_{envelope} = \frac{\Delta\omega}{\Delta k} = \frac{\omega_1 - \omega_2}{k_1 - k_2}$$

Substitute numerical values and evaluate $v_{envelope}$:

$$v_{envelope} = \frac{1\,\text{rad/s} - 0.9\,\text{rad/s}}{1\,\text{m}^{-1} - 0.8\,\text{m}^{-1}} = \boxed{0.500\,\text{m/s}}$$

*38 ••• Two loudspeakers are driven in phase by an audio amplifier at a frequency of 600 Hz. The speakers are on the y axis, one at $y = +1.00$ m and the other at $y = -1.00$ m. A listener begins at $y = 0$ a very large distance D away and walks along a line parallel to the y axis. (See Problem 36.) (*a*) At what angle θ will she first hear a minimum in the sound intensity? (*b*) At what angle will she first hear a maximum (after $\theta = 0$)? (*c*) How many maxima can she possibly hear if she keeps walking in the same direction?

Picture the Problem Because the speakers are driven in phase and the path difference is 0 at her initial position, the listener will hear a maximum at $(D, 0)$. As she walks along a line parallel to the y axis she will hear a minimum wherever it is true that the path difference is an odd multiple of a half wavelength. She will hear an intensity maximum wherever the path difference is an integral multiple of a wavelength. We'll apply the condition for destructive interference in part (*a*) to determine the angular location of the first minimum and, in part (*b*), the condition for constructive interference find the angle at which she'll hear the first maximum after the one at 0°. In part (*c*), we can apply the condition for constructive interference to determine the number of maxima she can hear as keeps walking parallel to the y axis.

(*a*) Express the condition for destructive interference:

$$d \sin \theta_m = m \frac{\lambda}{2}$$

where $m = 1, 3, 5, \ldots$, or

$$\theta_m = \sin^{-1} \left(\frac{m\lambda}{2d} \right)$$

Evaluate this expression for $m = 1$:

$$\theta_1 = \sin^{-1} \left(\frac{v}{2fd} \right) = \sin^{-1} \left[\frac{340 \text{ m/s}}{2(600 \text{ s}^{-1})(2 \text{ m})} \right]$$

$$= \boxed{8.14°}$$

(*b*) Express the condition for additional intensity maxima:

$$d \sin \theta_m = m\lambda$$

where $m = 0, 1, 2, 3, \ldots$, or

$$\theta_m = \sin^{-1} \left(\frac{m\lambda}{d} \right)$$

Evaluate this expression for $m = 1$:

$$\theta_1 = \sin^{-1} \left(\frac{v}{fd} \right) = \sin^{-1} \left[\frac{340 \text{ m/s}}{(600 \text{ s}^{-1})(2 \text{ m})} \right]$$

$$= \boxed{16.5°}$$

(c) Express the limiting condition on $\sin\theta$:

$$\sin\theta_m = m\frac{\lambda}{d} \le 1$$

Solve for m to obtain:

$$m \le \frac{d}{\lambda} = \frac{fd}{v} = \frac{\left(600\,s^{-1}\right)\left(2\,m\right)}{340\,m/s} = 3.53$$

Because m must be an integer:

$$m = \boxed{3}$$

Standing Waves

*44 •   A string fixed at both ends is 3 m long. It resonates in its second harmonic at a frequency of 60 Hz. What is the speed of transverse waves on the string?

Picture the Problem We can use $v = f\lambda$ to relate the second-harmonic frequency to the wavelength of the standing wave for the second harmonic.

Relate the speed of transverse waves on the string to their frequency and wavelength:

$$v = f_2\lambda_2$$

Express λ_2 in terms of the length L of the string:

$$\lambda_2 = L$$

Substitute for λ_2 and evaluate v:

$$v = f_2 L = \left(60\,s^{-1}\right)\left(3\,m\right) = \boxed{180\,m/s}$$

*50 •   What is the greatest length that an organ pipe can have in order to have its fundamental note in the audible range (20 to 20,000 Hz) if (a) the pipe is closed at one end and (b) it is open at both ends?

Picture the Problem Because the frequency and wavelength of sounds waves are inversely proportional, the greatest length of the organ pipe corresponds to the lowest frequency in the normal hearing range. We can relate wavelengths to the length of the pipes using the expressions for the resonance frequencies for pipes that are open at both ends and open at one end.

Find the wavelength of a 20-Hz note:

$$\lambda_{max} = \frac{v}{f_{lowest}} = \frac{340\,m/s}{20\,s^{-1}} = 17\,m$$

(a) Relate the length L of a closed-at-one-end organ pipe to the wavelengths of its standing waves:

$$L = n\frac{\lambda_n}{4}, n = 1, 3, 5, \ldots$$

Solve for and evaluate λ_1:

$$L = \frac{\lambda_{max}}{4} = \frac{17\,m}{4} = \boxed{4.25\,m}$$

(b) Relate the length L of an open organ pipe to the wavelengths of its standing waves:

$$L = n\frac{\lambda_n}{2}, n = 1, 2, 3, \ldots$$

Solve for and evaluate λ_1:

$$L = \frac{\lambda_{max}}{2} = \frac{17\,m}{2} = \boxed{8.50\,m}$$

*54 ••   Three successive resonance frequencies for a certain string are 75, 125, and 175 Hz. (a) Find the ratios of each pair of successive resonance frequencies. (b) How can you tell that these frequencies are for a string fixed at one end only rather than for a string fixed at both ends? (c) What is the fundamental frequency? (d) Which harmonics are these resonance frequencies? (e) If the speed of transverse waves on this string is 400 m/s, find the length of the string.

Picture the Problem Whether these frequencies are for a string fixed at one end only rather than for a string fixed at both ends can be decided by determining whether they are integral multiples or odd-integral multiples of a fundamental frequency. The length of the string can be found from the wave speed and the wavelength of the fundamental frequency using the standing-wave condition for a string with one end free.

(a) Letting the three frequencies be represented by f', f'', and f''', find the ratio of the first two frequencies:

$$\frac{f'}{f''} = \frac{75\,Hz}{125\,Hz} = \boxed{\frac{3}{5}}$$

Find the ratio of the second and third frequencies:

$$\frac{f''}{f'''} = \frac{125\,Hz}{175\,Hz} = \boxed{\frac{5}{7}}$$

(b) $\boxed{\text{There are no even harmonics, so the string must be fixed at one end only.}}$

(c) Express the resonance

$$f_n = nf_1, n = 1, 3, 5, \ldots$$

frequencies in terms of the fundamental frequency:
Noting that the frequencies are multiples of 25 Hz, we can conclude that:

$$f_1 = \frac{f_3}{3} = \frac{75\,\text{Hz}}{3} = \boxed{25\,\text{Hz}}$$

(d) | Because the frequencies are 3, 5, and 7 times the fundamental frequency, they are the third, fifth, and seventh harmonics.

(e) Express the length of the string in terms of the standing-wave condition for a string fixed at one end:

$$L = n\frac{\lambda_n}{4}, n = 1, 3, 5, \ldots$$

Using $v = f_1\lambda_1$, find λ_1:

$$\lambda_1 = \frac{v}{f_1} = \frac{400\,\text{m/s}}{25\,\text{s}^{-1}} = 16\,\text{m}$$

Evaluate L for $\lambda_1 = 16$ m and $n = 1$:

$$L = \frac{\lambda_1}{4} = \frac{16\,\text{m}}{4} = \boxed{4.00\,\text{m}}$$

*57 ••   At 16°C, the fundamental frequency of an organ pipe is 440.0 Hz. What will be the fundamental frequency of the pipe if the temperature increases to 32°C? Would it be better to construct the pipe with a material that expands substantially as the temperature increases or should the pipe be made of material that maintains the same length at all normal temperatures?

Picture the Problem We can use $v = f\lambda$ to express the fundamental frequency of the organ pipe in terms of the speed of sound and $v = \sqrt{\dfrac{\gamma RT}{M}}$ to relate the speed of sound and the fundamental frequency to the absolute temperature.

Express the fundamental frequency of the organ pipe in terms of the speed of sound:

$$f = \frac{v}{\lambda}$$

Relate the speed of sound to the temperature:

$$v = \sqrt{\frac{\gamma RT}{M}}$$

where γ and R are constants, M is the molar mass, and T is the absolute temperature.

Substitute to obtain:

$$f = \frac{1}{\lambda}\sqrt{\frac{\gamma RT}{M}}$$

Using primed quantities to represent the higher temperature, express the new frequency as a function of T:

$$f' = \frac{1}{\lambda'}\sqrt{\frac{\gamma RT'}{M}}$$

As we have seen, λ is proportional to the length of the pipe. For the first question, we assume the length of the pipe does not change, so $\lambda = \lambda'$. Then the ratio of f' to f is:

$$\frac{f'}{f} = \sqrt{\frac{T'}{T}}$$

Solve for and evaluate f' with $T' = 305$ K and $T = 289$ K:

$$f' = f_{305\,K} = f_{289\,K}\sqrt{\frac{305\,K}{289\,K}}$$

$$= (440.0\,Hz)\sqrt{\frac{305\,K}{289\,K}}$$

$$= \boxed{452\,Hz}$$

It would be better to have the pipe expand so that v/L, where L is the length of the pipe, is independent of temperature.

*67 ••   Show that the standing wave function $A'\sin kx\cos(\omega t + \delta)$ can be written as the sum of two harmonic wave functions—one for a wave of amplitude A traveling in the positive x direction and the other for a wave of the same amplitude A traveling in the negative x direction. The two traveling waves each have the same wave number and angular frequency as the standing wave.

Picture the Problem Let the wave function for the wave traveling to the right be $y_R(x,t) = A\sin(kx - \omega t - \delta)$ and the wave function for the wave traveling to the lef be $y_L(x,t) = A\sin(kx + \omega t + \delta)$ and use the identity $\sin\alpha + \sin\beta = 2\sin\left(\frac{\alpha + \beta}{2}\right)\cos\left(\frac{\alpha - \beta}{2}\right)$ to show that the sum of the wave functions can be written in the form $y(x,t) = A'\sin kx\cos(\omega t + \delta)$.

Express the sum of the traveling waves of equal amplitude moving in opposite directions:

$$y(x,t) = y_R(x,t) + y_L(x,t) = A\sin(kx - \omega t - \delta) + A\sin(kx + \omega t + \delta)$$

Use the trigonometric identity to obtain:

$$y(x,t) = 2A\sin\left(\frac{kx - \omega t - \delta + kx + \omega t + \delta}{2}\right)\cos\left(\frac{kx - \omega t - \delta - kx - \omega t - \delta}{2}\right)$$

$$= 2A\sin kx\cos(-\omega t - \delta)$$

Because the cosine function is even; i.e., $\cos(-\theta) = \cos\theta$:

$$y(x,t) = 2A\sin kx\cos(\omega t + \delta)$$
$$= A'\sin kx\cos(\omega t + \delta)$$
where $A' = 2A$

Thus, we have:

$$y(x,t) = \boxed{A'\sin kx\cos(\omega t + \delta)}$$
provided $A' = 2A$.

*69 ••   A commonly used physics experiment that examines resonances of transverse waves on a wire is shown in Figure 16-30. A weight is attached to the end of a string draped over a pulley; the other end of the string is attached to a mechanical oscillator that moves the string up and down at a set frequency f. The length L between the oscillator and the pulley is fixed. For certain values of the weight the string resonates. If $L = 1$ m, $f = 80$ Hz, and the mass density of the string is $\mu = 0.415$ g/m, what weights are needed for each of the first two modes (standing waves) of the string?

Figure 16-30 Problem 69

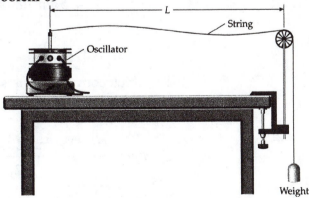

Picture the Problem We can equate the expression for the velocity of a wave on a string and the expression for the velocity of a wave in terms of its frequency and wavelength to obtain an expression for the weight that must be suspended from the end of the string in order to produce a given standing wave pattern. By using the condition on the wavelength that must be satisfied at resonance, we can express the weight on the end of the string in terms of μ, f, L, and an integer n and then evaluate this expression for $n = 1, 2,$ and 3 for the first three standing wave patterns.

Express the velocity of a wave on the string in terms of the tension T in the string and its linear density μ:

$$v = \sqrt{\frac{T}{\mu}} = \sqrt{\frac{mg}{\mu}}$$

where mg is the weight of the object suspended from the end of the string.

Express the wave speed in terms of its wavelength λ and frequency f:

$$v = f\lambda$$

Eliminate v to obtain:

$$f\lambda = \sqrt{\frac{mg}{\mu}}$$

Solve for mg:

$$mg = \mu f^2 \lambda^2$$

Express the condition on λ that corresponds to resonance:

$$\lambda = \frac{2L}{n}, n = 1, 2, 3, \ldots$$

Substitute to obtain:

$$mg = \mu f^2 \left(\frac{2L}{n}\right)^2, n = 1, 2, 3, \ldots$$

or

$$mg = \frac{4\mu f^2 L^2}{n^2}, n = 1, 2, 3, \ldots$$

Evaluate mg for $n = 1$:

$$mg = \frac{4(0.415\,\text{g/m})(80\,\text{s}^{-1})^2(0.2\,\text{m})^2}{(1)^2}$$

$$= \boxed{0.425\,\text{N}}$$

which corresponds, at sea level, to a mass of 43.3 g

Evaluate mg for $n = 2$:

$$mg = \frac{4(0.415\,\text{g/m})(80\,\text{s}^{-1})^2(0.2\,\text{m})^2}{(2)^2}$$

$$= \boxed{0.106\,\text{N}}$$

which corresponds, at sea level, to a mass of 10.8 g

*71 •   A tuning fork of frequency f_0 begins vibrating at time $t = 0$ and is stopped after a time interval Δt. The waveform of the sound at some later time is shown as a function of x. Let N be the (approximate) number of cycles in this waveform. (a) How are N, f_0, and Δt related? (b) If Δx is the length in space of this wave packet, what is the wavelength in terms of Δx and N? (c) What is the wave number k in terms of N and Δx? (d) The number N is uncertain by about ±1 cycle. Use Figure 16-31 to explain why. (e) Show that the uncertainty in the wave number due to the uncertainty in N is $2\pi/\Delta x$.

Figure 16-31 Problem 71

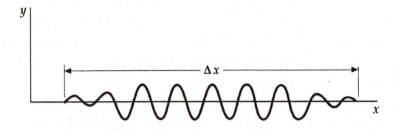

Picture the Problem We can approximate the duration of the pulse from the product of the number of cycles in the interval and the period of each cycle and the wavelength from the number of complete wavelengths in Δx. We can use its definition to find the wave number k from the wavelength λ.

(a) Relate the duration of the pulse to the number of cycles in the interval and the period of each cycle:

$$\Delta t \approx NT = \boxed{\frac{N}{f_0}}$$

(b) There are about N complete wavelengths in Δx; hence:

$$\lambda \approx \boxed{\frac{\Delta x}{N}}$$

(c) Use its definition to express the wave number k:

$$k = \frac{2\pi}{\lambda} = \boxed{\frac{2\pi N}{\Delta x}}$$

(d) | N is uncertain because the waveform dies out gradually rather than stopping abruptly at some time; hence, where the pulse starts and stops is not well defined.

(e) Using our result in part (c), express the uncertainty in k:

$$\Delta k = \frac{2\pi \Delta N}{\Delta x} = \boxed{\frac{2\pi}{\Delta x}}$$

because $\Delta N = \pm 1$

General Problems

*81 ••   In an early method used to determine the speed of sound in gases, powder was spread along the bottom of a horizontal, cylindrical glass tube. One end of the tube was closed by a piston that oscillated at a known frequency f. The other end was closed by a movable piston whose position was adjusted until resonance occurred. At resonance, the powder collected in equally spaced piles along the bottom of the tube. (*a*) Explain why the powder collects in this way. (*b*) Derive a formula that gives the speed of sound in the gas in terms of f and the distance between the piles of powder. (*c*) Give suitable values for the frequency f and the distance between the piles of powder. (*d*) Give suitable values for the frequency f and the length L of the tube for which the speed of sound could be measured in either air or helium.

Picture the Problem We can use $v = f\lambda$ to relate the speed of sound in the gas to the distance between the piles of powder in the glass tube.

(*a*)
> At resonance, standing waves are set up in the tube. At a displacement antinode, the powder is moved about; at a node, the powder is stationary, and so it collects at the nodes.

(*b*) Relate the speed of sound to its frequency and wavelength:	$v = f\lambda$
Letting D = distance between nodes, relate the distance between the nodes to the wavelength of the sound:	$\lambda = 2D$
Substitute to obtain:	$v = \boxed{2fD}$
(*c*) If we let the length L of the tube be 1.2 m and assume that $v_{air} = 344$ m/s (the speed of sound in air at 20°C), then the tenth harmonic corresponds to $D = 25.3$ cm and a driving frequency of:	$f_{air} = \dfrac{v_{air}}{2D} = \dfrac{344\,\text{m/s}}{2(0.253\,\text{m})} = \boxed{680\,\text{Hz}}$

> If $f = 2$ kHz and $v_{He} = 1008$ m/s (the speed of sound in helium at 20°C),
> then D for the 10th harmonic in helium would 25.3 cm and D for the tenth
> harmonic in air would be 8.60 cm. Hence, neglecting end effects at the
> (d) driven end, a tube whose length is the least common multiple of 8.60 cm
> and 25.3 cm (218 cm) would work well for the measurement of the
> speed of sound in either air or helium.

*89 ••   The speed of sound is proportional to the square root of the absolute
temperature T (Equation 15-5). (a) Show that if the temperature changes by a
small amount ΔT, the fundamental frequency of an organ pipe changes by
approximately Δf, where $\Delta f / f = \frac{1}{2} \Delta T / T$. (b) Suppose that an organ pipe that is
closed at one end has a fundamental frequency of 200 Hz when the temperature is
20°C. What will be its fundamental frequency when the temperature is 30°C?
(Ignore any change in the length of the pipe due to thermal expansion.)

Picture the Problem We can express the fundamental frequency of the organ
pipe as a function of the air temperature and differentiate this expression with
respect to the temperature to express the rate at which the frequency changes with
respect to temperature. For changes in temperature that are small compared to the
temperature, we can approximate the differential changes in frequency and
temperature with finite changes to complete the derivation of $\Delta f/f = \frac{1}{2}\Delta T/T$. In
part (b), we'll use this relationship and the data for the frequency at 20°C to find
the frequency of the fundamental at 30°C.

(a) Express the fundamental frequency of an organ pipe in terms of its wavelength and the speed of sound:	$f = \dfrac{v}{\lambda}$
Relate the speed of sound in air to the absolute temperature:	$v = \sqrt{\dfrac{\gamma R T}{M}} = C\sqrt{T}$ where $C = \sqrt{\dfrac{\gamma R}{M}} = \text{constant}$
Defining a new constant C', substitute to obtain:	$f = \dfrac{C}{\lambda}\sqrt{T} = C'\sqrt{T}$ because λ is constant for the fundamental frequency we ignore any change in the length of the pipe.

Differentiate this expression with respect to T:

$$\frac{df}{dT} = \frac{1}{2}C'T^{-1/2} = \frac{f}{2T}$$

Separate the variables to obtain:

$$\frac{df}{f} = \frac{1}{2}\frac{dT}{T}$$

For $\Delta T \ll T$, we can approximate df by Δf and dT by ΔT to obtain:

$$\boxed{\frac{\Delta f}{f} = \frac{1}{2}\frac{\Delta T}{T}}$$

(b) Express the fundamental frequency at 30°C in terms of its frequency at 20°C:

$$f_{30} = f_{20} + \Delta f$$

Solve our result in (a) for Δf:

$$\Delta f = \tfrac{1}{2}f\frac{\Delta T}{T}$$

Substitute numerical values and evaluate f_{30}:

$$f_{30} = 200\,\text{Hz} + \tfrac{1}{2}(200\,\text{Hz})\frac{10\,\text{K}}{293\,\text{K}}$$

$$= \boxed{203\,\text{Hz}}$$

*94 •• Two sources of harmonic waves have a phase difference that is proportional to time: $\delta_s = Ct$, where C is a constant. The amplitude of the wave from each source at some point P is A_0. (a) Write the wave functions for each of the two waves at point P, assuming this point to be a distance x_1 from one source and $x_1 + \Delta x$ from the other. (b) Find the resultant wave function and show that its amplitude is $2A_0\cos(\delta + \delta_s)$, where δ is the phase difference at P due to the path difference. (c) Using a spreadsheet program or graphing calculator, graph the intensity at point P versus time for a zero path difference. (Let I_0 be the intensity due to each wave separately.) What is the time average of the intensity? (d) Make the same graph for the intensity at a point for which the path difference is $\lambda/2$.

Picture the Problem Let the sources be denoted by the numerals 1 and 2. The phase difference between the two waves at point P is the sum of the phase difference due to the sources δ_0 and the phase difference due to the path difference δ.

(a) Write the wave function due to source 1:

$$f_1(x,t) = \boxed{A_0\cos(kx_1 - \omega t)}$$

Write the wave function due to

$$f_2(x,t) = \boxed{A_0\cos(k(x_1 + \Delta x) - \omega t + \delta_s)}$$

source 2:

(b) Express the sum of the two wave functions:

$$f(x,t) = f_1(x,t) + f_2(x,t) = A_0 \cos(kx_1 - \omega t) + A_0 \cos(k(x_1 + \Delta x) - \omega t + \delta_s)$$
$$= A_0 [\cos(kx_1 - \omega t)\cos(k(x_1 + \Delta x) - \omega t + \delta_s)]$$

Use $\cos\alpha + \cos\beta = 2\cos\left(\dfrac{\alpha+\beta}{2}\right)\cos\left(\dfrac{\alpha-\beta}{2}\right)$ to obtain:

$$f(x,t) = 2A_0\left[\cos\left(\frac{k\Delta x}{2} + \frac{\delta_s}{2}\right)\cos\left(k\left(x + \frac{\Delta x}{2}\right) - \omega t + \frac{\delta_s}{2}\right)\right]$$

Express the phase difference δ in terms of the path difference Δx and the wave number k:

$$\frac{\delta}{\Delta x} = \frac{2\pi}{\lambda} = k \text{ or } k\Delta x = \delta$$

Substitute to obtain:

$$\boxed{f(x,t) = 2A_0\left[\cos\left(\frac{\delta+\delta_s}{2}\right)\cos\left(k\left(x + \frac{\Delta x}{2}\right) - \omega t + \frac{\delta_s}{2}\right)\right]}$$

The amplitude of the resultant wave function is the coefficient of the time-dependent factor:

$$A = \boxed{2A_0 \cos\tfrac{1}{2}(\delta + \delta_s)}$$

(c) Express the intensity at an arbitrary point P:

$$I_P = C'A^2$$
$$= C'[2A_0 \cos\tfrac{1}{2}(\delta + \delta_s)]^2$$
$$= C'[4A_0^2 \cos^2 \tfrac{1}{2}(\delta + \delta_s)]$$

Evaluate I for $\delta = 0$ and $\delta_s = Ct$:

$$I = C'[4A_0^2 \cos^2 \tfrac{1}{2}(Ct)]$$

Because the average value of $\cos^2\theta$ over a complete period is ½:

$$I_{ave} \propto 2A_0^2 = 2I_0$$
and
$$I \propto \boxed{4I_0 \cos^2 \tfrac{1}{2}(Ct)}$$

(d) Evaluate I for $\Delta x = \tfrac{1}{2}\lambda$ and $\delta_s = Ct$:

$$\Delta x = \tfrac{1}{2}\lambda \Rightarrow \delta = \pi$$
$$\therefore I = C'[4A_0^2 \cos^2 \tfrac{1}{2}(\pi + Ct)]$$

and at $t = 0$, $I = 0$. i.e., the waves

interfere destructively.

A spreadsheet program to calculate the intensity at point P as a function of time for a zero path difference and a path difference of λ follows. The constants and cell formulas used are shown in the table.

Cell	Content/Formula	Algebraic Form
B1	1	C
B7	B6+0.1	$t + \Delta t$
C6	COS($B$6*B6/2)^2	$\cos^2 \frac{1}{2}(Ct)$
D6	COS($B$6*B6/2–PI()/2)^2	$\cos^2 \frac{1}{2}(\pi + Ct)$

	A	B	C	D
1		C= 1	s^{-1}	
2				
3				
4		t	I	I
5		(s)	(W/m^2)	(W/m^2)
6		0.00	1.000	0.000
7		0.10	0.998	0.002
8		0.20	0.990	0.010
9		0.30	0.978	0.022
103		9.70	0.019	0.981
104		9.80	0.035	0.965
105		9.90	0.055	0.945

The solid curve is the graph of $\cos^2 \frac{1}{2}(Ct)$ and the dashed curve is the graph of $\cos^2 \frac{1}{2}(\pi + Ct)$.

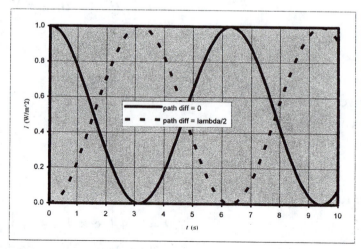

Chapter 17
Temperature and the Kinetic Theory of Gases

Conceptual Problems

*1 •   True or false:

(*a*) Two objects in thermal equilibrium with each other must be in thermal equilibrium with a third object.
(*b*) The Fahrenheit and Celsius temperature scales differ only in the choice of the zero temperature.
(*c*) The kelvin is the same size as the Celsius degree.
(*d*) All thermometers give the same result when measuring the temperature of a particular system.

(*a*) False. If two objects are in thermal equilibrium with a third, then they are in thermal equilibrium with each other.

(*b*) False. The Fahrenheit and Celsius temperature scales differ in the number of intervals between the ice-point temperature and the steam-point temperature.

(*c*) True.

(*d*) False. The result one obtains for the temperature of a given system is thermometer-dependent.

*6 ••   Figure 17-18 shows a plot of pressure versus temperature for a process that takes an ideal gas from point A to point B. What happens to the volume of the gas?

Figure 17-18 Problem 6

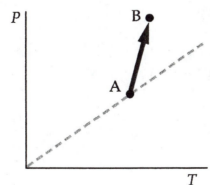

Determine the Concept From the ideal-gas law, we have $V = nRT/P$. In the process depicted, both the temperature and the pressure increase, but the pressure increases faster than does the temperature. Hence, the volume decreases.

*13 •   The temperature of the interior of the sun is said to be about 10^7 degrees. Do you think that this is degrees Celsius or kelvins, or does it matter?

Determine the Concept Because $10^7 \gg 273$, it does not matter.

*21 ••   Explain in terms of molecular motion why the pressure on the walls of a container increases when the volume of a gas is reduced at constant temperature.

Determine the Concept Because the temperature remains constant, the average speed of the molecules remains constant. When the volume decreases, the molecules travel less distance between collisions, so the pressure increases because the frequency of collisions increases.

Estimation and Approximation

*24 ••   A stoppered test tube that has a volume of 10 mL has 1 mL of water at its bottom and is at a temperature of 100°C and initially at a pressure of 1 atm $(1.01 \times 10^5 \text{ N/m}^2)$. Estimate the pressure inside the test tube when the water is completely boiled away.

Picture the Problem Assuming the steam to be an ideal gas at a temperature of 373 K, we can use the ideal-gas law to estimate the pressure inside the test tube when the water is completely boiled away.

Using the ideal-gas law, relate the pressure inside the test tube to its volume and the temperature:

$$P = \frac{NkT}{V}$$

Relate the number of particles N to the mass of water, its molar mass M, and Avogadro's number N_A:

$$\frac{m}{N} = \frac{M}{N_A}$$

Solve for N:

$$N = m\frac{N_A}{M}$$

Relate the mass of 1 mL of water to its density:

$$m = \rho V = \left(10^3 \text{ kg/m}^3\right)\left(10^{-6} \text{ m}^3\right) = 1\text{g}$$

Substitute for m, N_A, and M (18 g/mol) and evaluate N:

$$N = (1\text{g})\frac{6.022 \times 10^{23} \text{ particles/mol}}{18\text{g/mol}}$$

$$= 3.35 \times 10^{22} \text{ particles}$$

Substitute numerical values and evaluate P:

$$P = \frac{(3.35 \times 10^{22} \text{ particles})(1.381 \times 10^{-23} \text{ J/K})(373\text{K})}{10 \times 10^{-6} \text{ m}^3}$$

$$= 172 \times 10^5 \text{ N/m}^2 \times \frac{1\text{atm}}{1.01 \times 10^5 \text{ N/m}^2}$$

$$= \boxed{171\text{atm}}$$

*27 •• Repeat Problem 26 for Jupiter, whose escape velocity is 60 km/s and whose temperature is typically −150°C.

Picture the Problem We can use $v_{rms} = \sqrt{3RT/M}$ to calculate the rms speeds of H_2, O_2, and CO_2 at 123 K and then compare these speeds to 20% of the escape velocity on Jupiter to decide the likelihood of finding these gases in the atmosphere of Jupiter.

Express the rms speed of an atom as a function of the temperature:

$$v_{rms} = \sqrt{\frac{3RT}{M}}$$

(*a*) Substitute numerical values and evaluate v_{rms} for H_2:

$$v_{rms,H_2} = \sqrt{\frac{3(8.314 \text{ J/mol} \cdot \text{K})(123\text{K})}{2 \times 10^{-3} \text{ kg/mol}}}$$

$$= \boxed{1.24\text{km/s}}$$

(*b*) Evaluate v_{rms} for O_2:

$$v_{rms,O_2} = \sqrt{\frac{3(8.314 \text{ J/mol} \cdot \text{K})(123\text{K})}{32 \times 10^{-3} \text{ kg/mol}}}$$

$$= \boxed{310\text{m/s}}$$

(*c*) Evaluate v_{rms} for CO_2:

$$v_{rms,CO_2} = \sqrt{\frac{3(8.314 \text{ J/mol} \cdot \text{K})(123\text{K})}{44 \times 10^{-3} \text{ kg/mol}}}$$

$$= \boxed{264\text{m/s}}$$

(*d*) Calculate 20% of v_{esc} for

$$v = \tfrac{1}{5}v_{esc} = \tfrac{1}{5}(60\text{km/s}) = 12\text{km/s}$$

Jupiter:

> Because v is greater than v_{rms} for O_2, CO_2, and H_2, O_2, CO_2, and H_2 should be found on Jupiter.

Temperature Scales

*30 •   What is the Celsius temperature corresponding to the normal temperature of the human body, 98.6°F?

Picture the Problem We can use the Fahrenheit-Celsius conversion equation to express the temperature of the human body on the Celsius scale.

Convert 98.6°F to the equivalent Celsius temperature:	$t_C = \frac{5}{9}(t_F - 32°) = \frac{5}{9}(98.6° - 32°)$ $= \boxed{37.0°C}$

*35 •   A constant-volume gas thermometer reads 50 torr at the triple point of water. (*a*) What will be the pressure when the thermometer measures a temperature of 300 K? (*b*) What ideal-gas temperature corresponds to a pressure of 678 torr?

Picture the Problem We can use the information that the thermometer reads 50 torr at the triple point of water to calibrate it. We can then use the direct proportionality between the absolute temperature and the pressure to either the pressure at a given temperature or the temperature for a given pressure.

Using the ideal-gas temperature scale, relate the temperature to the pressure:	$T = \dfrac{273.16\,\text{K}}{P_3}P = \dfrac{273.16\,\text{K}}{50\,\text{torr}}P$ $= (5.463\,\text{K/torr})P$
(*a*) Solve for and evaluate P when $T = 300$ K:	$P = (0.1830\,\text{torr/K})T$ $= (0.1830\,\text{torr/K})(300\,\text{K})$ $= \boxed{54.9\,\text{torr}}$
(*b*) Find T when the pressure is 678 torr:	$T = (5.463\,\text{K/torr})(678\,\text{torr})$ $= \boxed{3704\,\text{K}}$

*41 •••   A thermistor is a solid-state device whose resistance varies greatly with temperature. Its temperature dependence is given approximately by $R = R_0 e^{B/T}$, where R is in ohms (Ω), T is in kelvins, and R_0 and B are constants that can be determined by measuring R at calibration points such as the ice point and the steam point. (*a*) If $R = 7360 \, \Omega$ at the ice point and $153 \, \Omega$ at the steam point, find R_0 and B. (*b*) What is the resistance of the thermistor at $t = 98.6°F$? (*c*) What is the rate of change of the resistance with temperature (dR/dT) at the ice point and the steam point? (*d*) At which temperature is the thermistor most sensitive?

Picture the Problem We can use the temperature dependence of the resistance of the thermistor and the given data to determine R_0 and B. Once we know these quantities, we can use the temperature-dependence equation to find the resistance at any temperature in the calibration range. Differentiation of R with respect to T will allow us to express the rate of change of resistance with temperature at both the ice point and the steam point temperatures.

(*a*) Express the resistance at the ice point as a function of temperature of the ice point:	$7360 \, \Omega = R_0 e^{B/273 \, \mathrm{K}}$	(1)
Express the resistance at the steam point as a function of temperature of the steam point:	$153 \, \Omega = R_0 e^{B/373 \, \mathrm{K}}$	(2)
Divide equation (1) by equation (2) to obtain:	$\dfrac{7360 \, \Omega}{153 \, \Omega} = 48.10 = e^{B/273 \, \mathrm{K} - B/373 \, \mathrm{K}}$	

Solve for B by taking the logarithm of both sides of the equation:

$$\ln 48.1 = B \left(\frac{1}{273} - \frac{1}{373} \right) \mathrm{K}^{-1}$$

and

$$B = \frac{\ln 48.1}{\left(\dfrac{1}{273} - \dfrac{1}{373} \right) \mathrm{K}^{-1}} = \boxed{3.94 \times 10^3 \, \mathrm{K}}$$

Solve equation (1) for R_0 and substitute for B:

$$R_0 = \frac{7360 \, \Omega}{e^{B/273 \, \mathrm{K}}} = (7360 \, \Omega) e^{-B/273 \, \mathrm{K}}$$

$$= (7360 \, \Omega) e^{-3.94 \times 10^3 \, \mathrm{K}/273 \, \mathrm{K}}$$

$$= \boxed{3.97 \times 10^{-3} \, \Omega}$$

(*b*) From (*a*) we have:

$$R = (3.97 \times 10^{-3} \, \Omega) e^{3.94 \times 10^3 \, \mathrm{K}/T}$$

Convert 98.6°F to kelvins to

$$T = 310 \mathrm{K}$$

obtain:

Substitute to obtain:

$$R = (3.97 \times 10^{-3} \, \Omega) e^{3.94 \times 10^3 \, \mathrm{K}/310 \, \mathrm{K}}$$

$$= \boxed{1.31 \, \mathrm{k}\Omega}$$

(c) Differentiate R with respect to T to obtain:

$$\frac{dR}{dT} = \frac{d}{dT}\left(R_0 e^{B/T}\right) = R_0 e^{B/T} \frac{d}{dT}\left(\frac{B}{T}\right)$$

$$= \frac{-B}{T^2} R_0 e^{B/T} = -\frac{RB}{T^2}$$

Evaluate dR/dT at the ice point:

$$\left(\frac{dR}{dT}\right)_{\text{ice point}} = -\frac{(7360 \, \Omega)(3.94 \times 10^3 \, \mathrm{K})}{(273.16 \, \mathrm{K})^2}$$

$$= \boxed{-389 \, \Omega/\mathrm{K}}$$

Evaluate dR/dT at the steam point:

$$\left(\frac{dR}{dT}\right)_{\text{steam point}} = -\frac{(153 \, \Omega)(3.94 \times 10^3 \, \mathrm{K})}{(373.16 \, \mathrm{K})^2}$$

$$= \boxed{-4.33 \, \Omega/\mathrm{K}}$$

(d) | The thermistor is more sensitive; i.e., it has greater sensitivity at lower temperatures.

The Ideal-Gas Law

*48 ••   The boiling point of helium at one atmosphere is 4.2 K. What is the volume occupied by helium gas due to evaporation of 10 g of liquid helium at 1 atm pressure and a temperature of (a) 4.2 K, and (b) 293 K?

Picture the Problem Let the subscript 1 refer to helium gas at 4.2 K and the subscript 2 to the gas at 293 K. We can apply the ideal-gas law to find the volume of the gas at 4.2 K and a fixed amount of gas to find its volume at 293 K.

(a) Apply the ideal-gas law to the helium gas and solve for its volume:

$$V_1 = \frac{nRT_1}{P_1}$$

Substitute numerical values to obtain:

$$V_1 = n\frac{(0.08206 \, \mathrm{L \cdot atm/mol \cdot K})(4.2 \, \mathrm{K})}{1 \, \mathrm{atm}}$$

$$= (0.3447 \, \mathrm{L/mol})n$$

Find the number of moles in 10 g of helium:	$n = \dfrac{10\,\text{g}}{4\,\text{g/mol}} = 2.5\,\text{mol}$
Substitute for n to obtain:	$V_1 = (0.3447\,\text{L/mol})(2.5\,\text{mol})$
	$= \boxed{0.862\,\text{L}}$

(b) Apply the ideal-gas law for a fixed amount of gas and solve for the volume of the helium gas at 293 K:	$\dfrac{P_2 V_2}{T_2} = \dfrac{P_1 V_1}{T_1}$ and, because $P_1 = P_2$, $V_2 = \dfrac{T_2}{T_1} V_1$
Substitute numerical values and evaluate V_2:	$V_2 = \dfrac{293\,\text{K}}{4.2\,\text{K}}(0.862\,\text{L}) = \boxed{60.1\,\text{L}}$

*50 •• An automobile tire is filled to a gauge pressure of 200 kPa when its temperature is 20°C. (Gauge pressure is the difference between the actual pressure and atmospheric pressure.) After the car has been driven at high speeds, the tire temperature increases to 50°C. (*a*) Assuming that the volume of the tire does not change and that air behaves as an ideal gas, find the gauge pressure of the air in the tire. (*b*) Calculate the gauge pressure if the volume of the tire expands by 10 percent.

Picture the Problem Let the subscript 1 refer to the tire when its temperature is 20°C and the subscript 2 to conditions when its temperature is 50°C. We can apply the ideal-gas law for a fixed amount of gas to relate the temperatures to the pressures of the air in the tire.

(*a*) Apply the ideal-gas law for a fixed amount of gas and solve for pressure at the higher temperature:	$\dfrac{P_2 V_2}{T_2} = \dfrac{P_1 V_1}{T_1}$ (1) and $P_2 = \dfrac{T_2}{T_1} P_1$ because $V_1 = V_2$.
Substitute numerical values to obtain:	$P_2 = \dfrac{323\,\text{K}}{293\,\text{K}}(200\,\text{kPa} + 101\,\text{kPa})$ $= 332\,\text{kPa}$ and

$$P_{2,\text{gauge}} = 332\,\text{kPa} - 101\,\text{kPa}$$

$$= \boxed{231\,\text{kPa}}$$

(b) Solve equation (1) for P_2 with $V_2 = 1.1\,V_1$ and evaluate P_2:

$$P_2 = \frac{V_1 T_2}{V_2 T_1} P_1$$

$$= \frac{323\,\text{K}}{1.1(293\,\text{K})}(200\,\text{kPa} + 101\,\text{kPa})$$

$$= 302\,\text{kPa}$$

and

$$P_{2,\text{gauge}} = 302\,\text{kPa} - 101\,\text{kPa} = \boxed{201\,\text{kPa}}$$

Kinetic Theory of Gases

*55 •   (a) Find v_{rms} for an argon atom if 1 mol of the gas is confined to a 1-L container at a pressure of 10 atm. (For argon, $M = 40 \times 10^{-3}$ kg/mol.) (b) Compare this with v_{rms} for a helium atom under the same conditions. (For helium, $M = 4 \times 10^{-3}$ kg/mol.)

Picture the Problem We can express the rms speeds of argon and helium atoms by combining $PV = nRT$ and $v_{\text{rms}} = \sqrt{3RT/M}$ to obtain an expression for v_{rms} in terms of P, V, and M.

Express the rms speed of an atom as a function of the temperature:

$$v_{\text{rms}} = \sqrt{\frac{3RT}{M}}$$

From the ideal-gas law we have:

$$RT = \frac{PV}{n}$$

Substitute to obtain:

$$v_{\text{rms}} = \sqrt{\frac{3PV}{nM}}$$

(a) Substitute numerical values and evaluate v_{rms} for an argon atom:

$$v_{\text{rms}}(\text{Ar}) = \sqrt{\frac{3(10\,\text{atm})(101.3\,\text{kPa/atm})(10^{-3}\,\text{m}^3)}{(1\,\text{mol})(40 \times 10^{-3}\,\text{kg/mol})}} = \boxed{276\,\text{m/s}}$$

(b) Substitute numerical values and evaluate v_{rms} for a helium atom:

$$v_{rms}(He) = \sqrt{\frac{3(10\,atm)(101.3\,kPa/atm)(10^{-3}\,m^3)}{(1\,mol)(4 \times 10^{-3}\,kg/mol)}} = \boxed{872\,m/s}$$

*58 •   In one model of a solid, the material is assumed to consist of a regular array of atoms in which each atom has a fixed equilibrium position and is connected by springs to its neighbors. Each atom can vibrate in the x, y, and z directions. The total energy of an atom in this model is

$$E = \tfrac{1}{2}mv_x^2 + \tfrac{1}{2}mv_y^2 + \tfrac{1}{2}mv_z^2 + \tfrac{1}{2}kx^2 + \tfrac{1}{2}ky^2 + kz^2$$

What is the average energy of an atom in the solid when the temperature is T? What is the total energy of one mole of such a solid?

Picture the Problem Because there are 6 squared terms in the expression for the total energy of an atom in this model, we can conclude that there are 6 degrees of freedom. Because the system is in equilibrium, we can conclude that there is energy of $\tfrac{1}{2}kT$ per molecule or $\tfrac{1}{2}RT$ per mole associated with each degree of freedom.

Express the average energy per atom in the solid in terms of its temperature and the number of degrees of freedom:

$$\frac{E_{av}}{atom} = N(\tfrac{1}{2}kT) = 6(\tfrac{1}{2}kT) = \boxed{3kT}$$

Relate the total energy of one mole to its temperature and the number of degrees of freedom:

$$\frac{E_{tot}}{mole} = N(\tfrac{1}{2}RT) = 6(\tfrac{1}{2}RT) = \boxed{3RT}$$

*61 ••   Oxygen (O_2) is confined to a cubic container 15 cm on a side at a temperature of 300 K. Compare the average kinetic energy of a molecule of the gas to the change in its gravitational potential energy if it falls from the top of the container to the bottom.

Picture the Problem We can use $K = \tfrac{3}{2}kT$ and $\Delta U = mgh = Mgh/N_A$ to express the ratio of the average kinetic energy of a molecule of the gas to the change in its gravitational potential energy if it falls from the top of the container to the bottom.

Express the average kinetic energy of a molecule of the gas as a function of its temperature:

$$K = \tfrac{3}{2}kT$$

Letting h represent the height of the container, express the change in the potential energy of a molecule as it falls from the top of the container to the bottom:

$$\Delta U = mgh = \frac{Mgh}{N_A}$$

Express the ratio of K to ΔU and simplify to obtain:

$$\frac{K}{\Delta U} = \frac{\frac{3}{2}kT}{\dfrac{Mgh}{N_A}} = \frac{3N_A kT}{2Mgh}$$

Substitute numerical values and evaluate $K/\Delta U$:

$$\frac{K}{\Delta U} = \frac{3(6.022\times10^{23})(1.381\times10^{-23}\ \text{J/K})(300\,\text{K})}{2(32\times10^{-3}\,\text{kg})(9.81\,\text{m/s}^2)(0.15\,\text{m})} = \boxed{7.95\times10^4}$$

The Distribution of Molecular Speeds

*63 •• $f(v)$ is defined in Equation 17.37. Because $f(v)dv$ gives the fraction of molecules that have speeds in the range dv, the integral of $f(v)dv$ over all the possible ranges of speeds must equal 1. Given the integral

$$\int_0^\infty v^2 e^{-av^2}\,dv = \frac{\sqrt{\pi}}{4}a^{-3/2}$$

show that $\int_0^\infty f(v)dv = 1$, where $f(v)$ is given by Equation 17-37.

Picture the Problem We can show that $f(v)$ is normalized by using the given integral to integrate it over all possible speeds

Express the integral of Equation 17-37:

$$\int_0^\infty f(v)dv = \frac{4}{\sqrt{\pi}}\left(\frac{m}{2kT}\right)^{3/2}\int_0^\infty v^2 e^{-mv^2/2kT}\,dv$$

Let $a = m/2kT$ to obtain:

$$\int_0^\infty f(v)dv = \frac{4}{\sqrt{\pi}}a^{3/2}\int_0^\infty v^2 e^{-av^2}\,dv$$

Use the given integral to obtain:

$$\int_0^\infty f(v)dv = \frac{4}{\sqrt{\pi}}a^{3/2}\left(\frac{\sqrt{\pi}}{4}a^{-3/2}\right) = \boxed{1}$$

i.e., $f(v)$ is normalized.

*65 ••   Current experiments in atomic trapping and cooling can create low-density gases of Rubidium and other atoms with temperatures in the nanokelvin (10^{-9} K) range. These atoms are trapped and cooled using magnetic fields and lasers in ultrahigh vacuum chambers. One method that is used to measure the temperature of a trapped gas is to turn the trap off and measure the time it takes for molecules of the gas to fall a given distance! Consider a gas of Rubidium atoms at a temperature of 120 nK. Calculate how long it would take an atom traveling at the rms speed of the gas to fall a distance of 10 cm if (*a*) it were initially moving directly downward and (*b*) if it were initially moving directly upward. Assume that the atom doesn't collide with any others along its trajectory.

Picture the Problem Choose a coordinate system in which downward is the positive direction. We can use a constant-acceleration equation to relate the fall distance to the initial velocity of the molecule, the acceleration due to gravity, the fall time, and $v_{rms} = \sqrt{3kT/m}$ to find the initial velocity of the molecules.

(*a*) Using a constant-acceleration equation, relate the fall distance to the initial velocity of a molecule, the acceleration due to gravity, and the fall time:

$$y = v_0 t + \tfrac{1}{2} g t^2 \qquad (1)$$

Express the rms speed of the atom to its temperature and mass:

$$v_{rms} = \sqrt{\frac{3kT}{m}}$$

Substitute numerical values and evaluate v_{rms}:

$$v_{rms} = \sqrt{\frac{3\left(1.381\times10^{-23}\ \text{J/K}\right)\left(120\,\text{nK}\right)}{\left(85.47\,\text{u}\right)\left(1.660\times10^{-27}\ \text{kg/u}\right)}}$$
$$= 5.92\times10^{-3}\ \text{m/s}$$

Letting $v_{rms} = v_0$, substitute in equation (1) to obtain:

$$0.1\,\text{m} = \left(5.92\times10^{-3}\ \text{m/s}\right) t + \tfrac{1}{2}\left(9.81\,\text{m/s}^2\right) t$$

Solve this equation to obtain:

$$t = \boxed{0.142\,\text{s}}$$

(*b*) If the atom is initially moving upward:

$$v_{rms} = v_0 = -5.92\times10^{-3}\ \text{m/s}$$

Substitute in equation (1) to obtain:

$$0.1\,\text{m} = \left(-5.92\times10^{-3}\ \text{m/s}\right) t + \tfrac{1}{2}\left(9.81\,\text{m/s}^2\right) t^2$$

Solve for t to obtain: $t = \boxed{0.143\,\text{s}}$

General Problems

*69 ••   Water, H_2O, can be converted into H_2 and O_2 gases by electrolysis. How many moles of these gases result from the electrolysis of 2 L of water?

Picture the Problem We can use the molar mass of water to find the number of moles in 2 L of water. Because there are two hydrogen atoms in each molecule of water, there must be as many hydrogen molecules in the gas formed by electrolysis as there were molecules of water and, because there is one oxygen atom in each molecule of water, there must be half as many oxygen molecules in the gas formed by electrolysis as there were molecules of water.

Express the electrolysis of water into H_2 and O_2:

$$n(H_2O) \rightarrow n(H_2) + \tfrac{1}{2}n(O_2)$$

Express the number of moles in 2 L of water:

$$n(H_2O) = \frac{2000\,\text{g}}{18\,\text{g/mol}} = 111\,\text{mol}$$

Because there is one hydrogen atom for each water molecule:

$$n(H_2) = \boxed{111\,\text{mol}}$$

Because there are two oxygen atoms for each water molecule:

$$n(O_2) = \tfrac{1}{2}n(H_2O) = \tfrac{1}{2}(111\,\text{mol})$$
$$= \boxed{55.5\,\text{mol}}$$

*72 ••   Three insulated vessels of equal volumes V are connected by thin tubes that can transfer gas but do not transfer heat. Initially all vessels are filled with the same type of gas at a temperature T_0 and pressure P_0. Then the temperature in the first vessel is doubled and the temperature in the second vessel is tripled. The temperature in the third vessel remains unchanged. Find the final pressure P' in the system in terms of the initial pressure P_0.

Picture the Problem Initially, we have $3P_0V = n_0RT_0$. Later, the pressures in the three vessels, each of volume V, are still equal, but the number of moles is not. The total number of moles, however, is constant and equal to the number of moles in the three vessels initially. Applying the ideal-gas law to each of the vessels will allow us to relate the number of moles in each to the final pressure and temperature. Equating this sum n_0 will leave us with an equation in P' and P_0 that we can solve for P'.

Relate the number of moles of gas in the system in the three vessels initially to the number in each vessel when the pressure is P':

$$n_0 = n_1 + n_2 + n_3$$

Relate the final pressure in the first vessel to its temperature and solve for n_1:

$$P' = \frac{n_1 R(2T_0)}{V} \Rightarrow n_1 = \frac{P'V}{2RT_0}$$

Relate the final pressure in the second vessel to its temperature and solve for n_2:

$$P' = \frac{n_2 R(3T_0)}{V} \Rightarrow n_2 = \frac{P'V}{3RT_0}$$

Relate the final pressure in the third vessel to its temperature and solve for n_3:

$$P' = \frac{n_3 RT_0}{V} \Rightarrow n_3 = \frac{P'V}{RT_0}$$

Substitute to obtain:

$$n_0 = \frac{P'V}{2RT_0} + \frac{P'V}{3RT_0} + \frac{P'V}{RT_0}$$

$$= \left(\frac{1}{2} + \frac{1}{3} + 1\right)\frac{P'V}{RT_0} = \frac{11}{6}\left(\frac{P'V}{RT_0}\right)$$

Express the number of moles in the three vessels initially in terms of the initial pressure and total volume:

$$n_0 = \frac{P_0(3V)}{RT_0}$$

Equate the two expressions for n_0 and solve for P' to obtain:

$$\boxed{P' = \frac{18}{11}P_0}$$

*74 •• The mean free path for O_2 molecules at a temperature of 300 K at 1 atmosphere of pressure ($P = 1.01 \times 10^5$ Pa) is $\lambda = 7.1 \times 10^{-8}$ m. Use this data to estimate the size of an O_2 molecule.

Picture the Problem Because the O_2 molecule resembles 2 spheres stuck together, which in cross-section look something like two circles, we can estimate the radius of the molecule from the formula for the area of a circle. We can express the area, and hence the radius, of the circle in terms of the mean free path and the number density of the molecules and use the ideal-gas law to express the number density.

Express the area of two circles of
diameter d that touch each other:

$$A = 2\left(\frac{\pi d^2}{4}\right) = \frac{\pi d^2}{2}$$

Solve for d to obtain:

$$d = \sqrt{\frac{2A}{\pi}} \qquad (1)$$

Relate the mean free path of the
molecules to their number density
and cross-sectional area:

$$\lambda = \frac{1}{n_v A}$$

Solve for A to obtain:

$$A = \frac{1}{n_v \lambda}$$

Substitute in equation (1) to obtain:

$$d = \sqrt{\frac{2}{\pi n_v \lambda}}$$

Use the ideal-gas law to relate
the number density of the O_2
molecules to their temperature
and pressure:

$$PV = NkT \text{ or } n_v = \frac{N}{V} = \frac{P}{kT}$$

Substitute to obtain:

$$d = \sqrt{\frac{2kT}{\pi P \lambda}}$$

Substitute numerical values and
evaluate d:

$$d = \sqrt{\frac{2(1.381 \times 10^{-23}\, \text{J/K})(300\,\text{K})}{\pi(1.01 \times 10^5\, \text{Pa})(7.1 \times 10^{-8}\, \text{m})}}$$

$$= 6.06 \times 10^{-10}\, \text{m} = \boxed{0.606\,\text{nm}}$$

*77 •••   The table below gives values of

$$\left(4/\sqrt{\pi}\right)\int_0^x z^2 e^{-z^2}\, dz$$

for different values of x. Use the table to answer the following questions: (a) For
O_2 gas at 273 K, what fraction of molecules have speeds less than 400 m/s? (b)
For the same gas, what percentage of molecules have speeds between 190 m/s and
565 m/s?

x	$\left(4/\sqrt{\pi}\right)\int_0^x z^2 e^{-z^2}\, dz$	x	$\left(4/\sqrt{\pi}\right)\int_0^x z^2 e^{-z^2}\, dz$
0.1	7.48×10^{-4}	0.7	0.194
0.2	5.88×10^{-3}	0.8	0.266
0.3	0.019	0.9	0.345

0.4	0.044	1	0.438
0.5	0.081	1.5	0.788
0.6	0.132	2	0.954

Picture the Problem We can show that $\int_0^V f(v)dv = I(x)$, where $f(v)$ is the

Maxwell-Boltzmann distribution function, $x = mV^2/2kT$, and $I(x)$ is the integral whose values are tabulated in the problem statement. Then, we can use this table to find the value of x corresponding to the fraction of the gas molecules with speeds less than v by evaluating $I(x)$.

(*a*) The Maxwell-Boltzmann speed distribution $f(x)$ is given by:

$$f(v) = \frac{4}{\sqrt{\pi}}\left(\frac{m}{2kT}\right)^{3/2} v^2 e^{-mv^2/2kT}$$

which means that the fraction of particles with speeds between v and $v + dv$ is $f(v)dv$.

Express the fraction $F(V)$ of particles with speeds less than $V = 400$ m/s:

$$F(V) = \int_0^V f(v)dv$$

$$= \frac{4}{\sqrt{\pi}}\left(\frac{m}{2kT}\right)^{3/2} \int_0^V v^2 e^{-mv^2/2kT} dv$$

Change integration variables by letting $z = v\sqrt{m/2kT}$ so we can use the table of values to evaluate the integral. Then:

$$v = \sqrt{\frac{2kT}{m}} z \Rightarrow dv = \sqrt{\frac{2kT}{m}} dz$$

Substitute in the integrand of $F(V)$ to obtain:

$$v^2 e^{-mv^2/2kT} dv = z^2 \frac{2kT}{m} e^{-z^2} \left(\frac{2kT}{m}\right)^{1/2} dz$$

$$= \left(\frac{2kT}{m}\right)^{3/2} z^2 e^{-z^2} dz$$

Transform the integration limits to correspond to the new integration variable $z = v\sqrt{m/2kT}$:

When $v = 0$, $z = 0$, and when $v = V$, $z = V\sqrt{m/2kT}$

The new lower integration limit is 0. Evaluate $z = V\sqrt{m/2kT}$ to find the upper limit:

$$z = (400\,\text{m/s})\sqrt{\frac{(32\,\text{u})(1.661\times10^{-27}\,\text{kg})}{2(1.381\times10^{-23}\,\text{J/K})(273\,\text{K})}} = 1.06$$

Evaluate $F(400\text{ m/s})$ to obtain:

$$F(400\,\text{m/s}) = \int_0^{400\,\text{m/s}} f(v)\,dv = \frac{4}{\sqrt{\pi}}\left(\frac{m}{2kT}\right)^{3/2}\int_0^{400\,\text{m/s}} v^2 e^{-mv^2/2kT}\,dv = \frac{4}{\sqrt{\pi}}\int_0^{1.06} z^2 e^{-z^2}\,dz$$

$$= I(1.06)$$

where $I(x) = \dfrac{4}{\sqrt{\pi}}\displaystyle\int_0^x z^2 e^{-z^2}\,dz$

Letting r represent the fraction of the molecules with speeds less than 400 m/s, interpolate from the table to obtain:

$$\frac{r-0.438}{1.06-1} = \frac{0.788-0.438}{1.5-1}$$

and

$$r = \boxed{48.0\%}$$

(b) Express the fraction r of the molecules with speeds between $V_1 = 190$ m/s and $V_2 = 565$ m/s:

$$r = F(V_2)-F(V_1) = I(x_2)-I(x_1)$$

where

$$x_1 = V_1\sqrt{m/2kT} \text{ and } x_2 = V_2\sqrt{m/2kT}$$

Evaluate x_1 and x_2 to obtain:

$$x_1 = (190\,\text{m/s})\sqrt{\frac{(32\,\text{u})(1.661\times10^{-27}\,\text{kg})}{2(1.381\times10^{-23}\,\text{J/K})(273\,\text{K})}} = 0.504$$

and

$$x_2 = (565\,\text{m/s})\sqrt{\frac{(32\,\text{u})(1.661\times10^{-27}\,\text{kg})}{2(1.381\times10^{-23}\,\text{J/K})(273\,\text{K})}} = 1.50$$

Substitute to obtain:

$$r = I(1.50)-I(0.504) \tag{1}$$

Using the table, evaluate $I(1.50)$:

$$I(1.50) = 0.788$$

Letting r represent the fraction of the molecules with speeds less than 190 m/s, interpolate from the table to obtain:

$$\frac{r-0.081}{0.504-0.5} = \frac{0.132-0.081}{0.6-0.5}$$

and

$$r = 0.083$$

Substitute in equation (1) to obtain:

$$r = 0.788-0.083 = \boxed{70.5\%}$$

Chapter 18
Heat and the First Law of Thermodynamics

Conceptual Problems

*2 • The temperature change of two blocks of masses M_A and M_B is the same when they absorb equal amounts of heat. It follows that the specific heats are related by (a) $c_A = (M_A/M_B)c_B$, (b) $c_A = (M_B/M_A)c_B$, (c) $c_A = c_B$, (d) none of the above.

Picture the Problem We can use the relationship $Q = mc\Delta T$ to relate the temperature changes of bodies A and B to their masses, specific heats, and the amount of heat supplied to each.

Relate the temperature change of block A to its specific heat and mass:

$$\Delta T_A = \frac{Q}{M_A c_A}$$

Relate the temperature change of block B to its specific heat and mass:

$$\Delta T_B = \frac{Q}{M_B c_B}$$

Equate the temperature changes to obtain:

$$\frac{1}{M_B c_B} = \frac{1}{M_A c_A}$$

Solve for c_A:

$$c_A = \frac{M_B}{M_A} c_B \text{ and } \boxed{(b) \text{ is correct}}$$

*5 • Can a system absorb heat with no change in its internal energy?

Determine the Concept Yes, if the heat absorbed by the system is equal to the work done by the system.

*10 •• Two gas-filled rubber balloons of (initially) equal volume are at the bottom of a dark, cold lake. The top of the lake is warmer than the bottom. One balloon one rises rapidly and expands adiabatically as it rises. The other balloon rises more slowly and expands isothermally. Which balloon is larger when it reaches the surface of the lake?

Determine the Concept The balloon that expands isothermally is larger when it reaches the surface. The balloon that expands adiabatically will be at a lower temperature than the one that expands isothermally. Because each balloon has the same number of gas molecules and are at the same pressure, the one with the higher temperature will be bigger. An analytical argument that leads to the same conclusion is shown below.

Letting the subscript "a" denote the adiabatic process and the subscript "i" denote the isothermal process, express the equation of state for the adiabatic balloon:

$$P_0 V_0^\gamma = P_f V_{f,a}^\gamma \Rightarrow V_{f,a} = V_0 \left(\frac{P_0}{P_f}\right)^{1/\gamma}$$

For the isothermal balloon:

$$P_0 V_0 = P_f V_{f,i} \Rightarrow V_{f,i} = V_0 \left(\frac{P_0}{P_f}\right)$$

Divide the second of these equations by the first and simplify to obtain:

$$\frac{V_{f,i}}{V_{f,a}} = \frac{V_0 \left(\dfrac{P_0}{P_f}\right)}{V_0 \left(\dfrac{P_0}{P_f}\right)^{1/\gamma}} = \left(\frac{P_0}{P_f}\right)^{1-1/\lambda}$$

Because $P_0/P_f > 1$ and $\gamma > 1$:

$$\boxed{V_{f,i} > V_{f,a}}$$

*14 • If a system's volume remains constant while undergoing changes in temperature and pressure, then (*a*) the internal energy of the system is unchanged, (*b*) the system does no work, (*c*) the system absorbs no heat, (*d*) the change in internal energy equals the heat absorbed by the system.

Determine the Concept For a constant-volume process, no work is done on or by the gas. Applying the first law of thermodynamics we obtain $Q_{in} = \Delta E_{int}$. Because the temperature must change during such a process, we can conclude that $\Delta E_{int} \neq 0$ and hence $Q_{in} \neq 0$. $\boxed{(b) \text{ and } (d) \text{ are correct.}}$

*18 •• Which would you expect to have a higher heat capacity per *unit mass*, lead or copper? Why? (Don't look up the heat capacities to answer this question.)

Determine the Concept At room temperature, most solids have a roughly constant heat capacity per *mole* of 6 cal/mol-K (Dulong-Petit law). Because 1 mole of lead is more massive than 1 mole of copper, the heat capacity of lead should be lower than the heat capacity of copper. This is, in fact, the case.

Estimation and Approximation

*20 ••   A simple demonstration to show that heat is a form of energy is to take a bag of lead shot and drop it repeatedly onto a very rigid surface from a small height. The bag's temperature will increase, allowing an estimate of the heat capacity of lead. (*a*) Estimate the temperature increase of a bag filled with 1 kg of lead shot dropped 50 times from a height of 1 m. (*b*) In principle, the change in temperature is independent of the mass of the shot in the bag; in practice, it's better to use a larger mass than a smaller one. Why might this be true?

Picture the Problem The heat capacity of lead is c = 128 J/kg·K. We'll assume that all of the work done in lifting the bag through a vertical distance of 1 m goes into raising the temperature of the lead shot and use conservation of energy to relate the number of drops of the bag and the distance through which it is dropped to the heat capacity and change in temperature of the lead shot.

(*a*) Use conservation of energy to relate the change in the potential energy of the lead shot to the change in its temperature:

$$Nmgh = mc\Delta T$$

where N is the number of times the bag of shot is dropped.

Solve for ΔT to obtain:

$$\Delta T = \frac{Nmgh}{mc} = \frac{Ngh}{c}$$

Substitute numerical values and evaluate ΔT:

$$\Delta T = \frac{50(9.81\,\text{`m/s}^2)(1\,\text{m})}{128\,\text{J/kg}\cdot\text{K}} = \boxed{3.83\,\text{K}}$$

(*b*)   It is better to use a larger mass because the rate at which heat is lost by the lead shot is proportional to its surface area while the rate at which it gains heat is proportional to its mass. The amount of heat lost varies as the surface area of the shot divided by its mass ($L^2/L^3 = L^{-1}$); which decreases as the mass increases.

*24 ••   A certain molecule has vibrational energy levels that are equally spaced by 0.15 eV. Find the critical temperature T_c so for $T \gg T_c$ you would expect the equipartition theorem to hold and for $T \ll T_c$ you would expect the equipartition theorem to fail.

Picture the Problem We can apply the condition for the validity of the equipartition theorem, i.e., that the spacing of the energy levels be large compared to kT, to find the critical temperature T_c:

Express the failure condition for

$$kT_c \approx 0.15\,\text{eV}$$

the equipartition theorem:

Solve for T_c:

$$T_c = \frac{0.15\,\text{eV}}{k}$$

Substitute numerical values and evaluate T_c:

$$T_c = \frac{0.15\,\text{eV} \times \dfrac{1.602 \times 10^{-19}\,\text{J}}{\text{eV}}}{1.381 \times 10^{-23}\,\text{J/K}} = \boxed{1740\,\text{K}}$$

Heat Capacity; Specific Heat; Latent Heat

*25 •   A "typical" adult male consumes about 2500 kcal of food in a day. (a) How many joules is this? (b) If this consumed energy is dissipated over the course of 24 hours, what is his average output power in watts?

Picture the Problem We can use the conversion factor 1 cal = 4.184 J to convert 2500 kcal into joules and the definition of power to find the average output if the consumed energy is dissipated over 24 h.

(a) Convert 2500 kcal to joules:

$$2500\,\text{kcal} = 2500\,\text{kcal} \times \frac{4.184\,\text{J}}{\text{cal}}$$

$$= \boxed{10.5\,\text{MJ}}$$

(b) Use the definition of power to obtain:

$$P_{av} = \frac{\Delta E}{\Delta t} = \frac{1.05 \times 10^7\,\text{J}}{24\,\text{h} \times \dfrac{3600\,\text{s}}{\text{h}}} = \boxed{121\,\text{W}}$$

Remarks: Note that this average power output is essentially that of a widely used light bulb.

*31 ••   A 30-g lead bullet initially at 20°C comes to rest in the block of a ballistic pendulum. Assume that half the initial kinetic energy of the bullet is converted into thermal energy within the bullet. If the speed of the bullet was 420 m/s, what is the temperature of the bullet immediately after coming to rest in the block?

Picture the Problem The temperature of the bullet immediately after coming to rest in the block is the sum of its pre-collision temperature and the change in its temperature as a result of being brought to a stop in the block. We can equate the heat gained by the bullet and half its pre-collision kinetic energy to find the change in its temperature.

Express the temperature of the bullet immediately after coming to rest in terms of its initial temperature and the change in its temperature as a result of being stopped in the block:	$T = T_i + \Delta T$ $= 293\,K + \Delta T$
Relate the heat absorbed by the bullet as it comes to rest to its kinetic energy before the collision:	$Q = \tfrac{1}{2}K$
Substitute for Q and K to obtain:	$m_{Pb}c_{Pb}\Delta T = \tfrac{1}{2}\left(\tfrac{1}{2}m_{Pb}v^2\right)$
Solve for ΔT:	$\Delta T = \dfrac{v^2}{4c_{Pb}}$
Substitute to obtain:	$T = 293\,K + \dfrac{v^2}{4c_{Pb}}$
Substitute numerical values and evaluate T:	$T = 293\,K + \dfrac{(420\,m/s)^2}{4(0.128\,kJ/kg\cdot K)}$ $= 638\,K = \boxed{365°C}$

Calorimetry

*34 •   The specific heat of a certain metal can be determined by measuring the temperature change that occurs when a piece of the metal is heated and then placed in an insulated container made of the same material and containing water. Suppose a piece of metal has a mass of 100 g and is initially at 100°C. The container has a mass of 200 g and contains 500 g of water at an initial temperature of 20.0°C. The final temperature is 21.4°C. What is the specific heat of the metal?

Picture the Problem During this process the water and the container will gain energy at the expense of the piece of metal. We can set the heat out of the metal equal to the heat into the water and the container and solve for the specific heat of the metal.

Apply conservation of energy to the system to obtain:	$\Delta Q = 0$ or $Q_{gained} = Q_{lost}$

Express the heat lost by the metal in terms of its specific heat and temperature change:	$Q_{lost} = m_{metal}c_{metal}\Delta T_{metal}$

Express the heat gained by the water and the container in terms of their specific heats and temperature change:	$Q_{gained} = m_w c_w \Delta T_w + m_{container}c_{metal}\Delta T_w$

Substitute to obtain:

$$m_w c_w \Delta T_w + m_{container}c_{metal}\Delta T_w = m_{metal}c_{metal}\Delta T_{metal}$$

Substitute numerical values:

$$(0.5\,kg)(4.18\,kJ/kg \cdot K)(294.4\,K - 293\,K) + (0.2\,kg)(294.4\,K - 293\,K)c_{metal}$$
$$= (0.1kg)(373\,K - 294.4\,K)c_{metal}$$

Solve for c_{metal}:

$$c_{metal} = \boxed{0.386\,kJ/kg \cdot K}$$

*41 •• A 200-g aluminum calorimeter contains 500 g of water at 20°C. A 100-g piece of ice cooled to –20°C is placed in the calorimeter. (*a*) Find the final temperature of the system, assuming no heat loss. (Assume that the specific heat of ice is 2.0 kJ/kg·K.) (*b*) A second 200-g piece of ice at –20°C is added. How much ice remains in the system after it reaches equilibrium? (*c*) Would you give a different answer for (*b*) if both pieces of ice were added at the same time?

Picture the Problem Assume that the calorimeter is in thermal equilibrium with the water it contains. During this process the ice will gain heat in warming to 0°C and melting, as will the water formed from the melted ice. The water in the calorimeter and the calorimeter will lose heat. We can do a preliminary calculation to determine whether there is enough heat available to melt all of the ice and, if there is, equate the heat the heat lost by the water to the heat gained by the ice and resulting ice water as the system achieves thermal equilibrium.

Find the heat available to melt the ice:

$$Q_{avail} = m_{water}c_{water}\Delta T_{water} + m_{cal}c_{cal}\Delta T_{water}$$
$$= [(0.5\,kg)(4.18\,kJ/kg \cdot K) + (0.2\,kg)(0.9\,kJ/kg \cdot K)](293\,K - 273\,K)$$
$$= 45.40\,kJ$$

Find the heat required to melt all of the ice:

$$Q_{\text{melt ice}} = m_{\text{ice}} c_{\text{ice}} \Delta T_{\text{ice}} + m_{\text{ice}} L_{\text{f}}$$
$$= (0.1\,\text{kg})(2\,\text{kJ/kg} \cdot \text{K})(273\,\text{K} - 253\,\text{K}) + (0.1\,\text{kg})(333.5\,\text{kJ/kg})$$
$$= 37.35\,\text{kJ}$$

(*a*) Because $Q_{\text{avail}} > Q_{\text{melt ice}}$, we know that the final temperature will be greater than 0°C. Apply the conservation of energy to the system to obtain:

$$\Delta Q = 0 \text{ or } Q_{\text{gained}} = Q_{\text{lost}}$$

Express Q_{lost} in terms of the final temperature of the system:

$$Q_{\text{lost}} = (m_{\text{water}} c_{\text{water}} + m_{\text{cal}} c_{\text{cal}}) \Delta T_{\text{water+calorimeter}}$$

Express Q_{gained} in terms of the final temperature of the system:

$$Q_{\text{gained}} = m_{\text{ice}} c_{\text{ice}} \Delta T_{\text{ice}} + m_{\text{ice}} L_{\text{f}}$$

Substitute to obtain:

$$m_{\text{ice}} c_{\text{ice}} \Delta T_{\text{ice}} + m_{\text{ice}} L_{\text{f}} + m_{\text{ice water}} c_{\text{water}} \Delta T_{\text{ice water}} = (m_{\text{water}} c_{\text{water}} + m_{\text{cal}} c_{\text{cal}}) \Delta T_{\text{water+calorimeter}}$$

Substitute numerical values:

$$37.35\,\text{kJ} + (0.1\,\text{kg})(4.18\,\text{kJ/kg} \cdot \text{K})(t_{\text{f}} - 273\,\text{K})$$
$$= [(0.5\,\text{kg})(4.18\,\text{kJ/kg} \cdot \text{K}) + (0.2\,\text{kg})(0.9\,\text{kJ/kg} \cdot \text{K})](293\,\text{K} - t_{\text{f}})$$

Solving for t_f yields:

$$t_{\text{f}} = 276\,\text{K} = \boxed{2.99°\text{C}}$$

(*b*) Find the heat required to raise 200 g of ice to 0°C:

$$Q_{\text{warm ice}} = m_{\text{ice}} c_{\text{ice}} \Delta T_{\text{ice}} = (0.2\,\text{kg})(2\,\text{kJ/kg} \cdot \text{K})(273\,\text{K} - 253\,\text{K}) = 8.00\,\text{kJ}$$

Noting that there are now 600 g of water in the calorimeter, find the heat available from cooling the calorimeter and water from 3°C to 0°C:

$$Q_{\text{avail}} = m_{\text{water}} c_{\text{water}} \Delta T_{\text{water}} + m_{\text{cal}} c_{\text{cal}} \Delta T_{\text{water}}$$
$$= [(0.6\,\text{kg})(4.18\,\text{kJ/kg} \cdot \text{K}) + (0.2\,\text{kg})(0.9\,\text{kJ/kg} \cdot \text{K})](293\,\text{K} - 273\,\text{K})$$
$$= 8.064\,\text{kJ}$$

Express the amount of ice that will melt in terms of the difference between the heat available and the heat required to warm the ice:

$$m_{\text{melted ice}} = \frac{Q_{\text{avail}} - Q_{\text{warm ice}}}{L_f}$$

Substitute numerical values and evaluate $m_{\text{melted ice}}$:

$$m_{\text{melted ice}} = \frac{8.064\,\text{kJ} - 8\,\text{kJ}}{333.5\,\text{kJ/kg}}$$
$$= 0.1919\,\text{g}$$

Find the ice remaining in the system:

$$m_{\text{remaining ice}} = 200\,\text{g} - 0.1919\,\text{g}$$
$$= \boxed{199.8\,\text{g}}$$

(c) | Because the initial and final conditions are the same, the answer would be the same.

First Law of Thermodynamics

*46 • If 400 kcal is added to a gas that expands and does 800 kJ of work, what is the change in the internal energy of the gas?

Picture the Problem We can apply the first law of thermodynamics to find the change in internal energy of the gas during this process.

Apply the first law of thermodynamics to express the change in internal energy of the gas in terms of the heat added to the system and the work done on the gas:

$$\Delta E_{\text{int}} = Q_{\text{in}} + W_{\text{on}}$$

The work done by the gas is the negative of the work done on the gas. Substitute numerical values and evaluate ΔE_{int}:

$$\Delta E_{\text{int}} = 400\,\text{kcal} \times \frac{4.184\,\text{J}}{\text{cal}} - 800\,\text{kJ}$$
$$= \boxed{874\,\text{kJ}}$$

*51 •• On a cold day you can warm your hands by rubbing them together. (*a*) Assume that the coefficient of friction between your hands is 0.5, that the normal force between your hands is 35 N, and that you rub them together at an average speed of 35 cm/s. What is the rate at which heat is generated? (*b*) Assume further that the mass of each of your hands is approximately 350 g, that the specific heat of your hands is about 4 kJ/kg·K, and that all the heat generated goes into raising the temperature of your hands. How long must you rub your hands together to produce a 5°C increase in their temperature?

Picture the Problem We can find the rate at which heat is generated when you rub your hands together using the definition of power and the rubbing time to produce a 5°C increase in temperature from $\Delta Q = (dQ/dt)\Delta t$ and $Q = mc\Delta T$.

(*a*) Express the rate at which heat is generated as a function of the friction force and the average speed of your hands:	$$\frac{dQ}{dt} = P = f_k v = \mu F_n v$$
Substitute numerical values and evaluate dQ/dt:	$$\frac{dQ}{dt} = 0.5(35\,\text{N})(0.35\,\text{m/s}) = \boxed{6.13\,\text{W}}$$
(*b*) Relate the heat required to raise the temperature of your hands 5 K to the rate at which it is being generated:	$$\Delta Q = \frac{dQ}{dt}\Delta t = mc\Delta T$$
Solve for Δt:	$$\Delta t = \frac{mc\Delta T}{dQ/dt}$$
Substitute numerical values and evaluate Δt:	$$\Delta t = \frac{(0.35\,\text{kg})(4\,\text{kJ/kg}\cdot\text{K})(5\,\text{K})}{6.13\,\text{W}}$$ $$= 1143\,\text{s} \times \frac{1\,\text{min}}{60\,\text{s}} = \boxed{19.0\,\text{min}}$$

Work and the *PV* Diagram for a Gas

*54 •• The gas is allowed to expand isothermally until its volume is 3 L and its pressure is 1 atm. It is then heated at constant volume until its pressure is 2 atm. (*a*) Show this process on a *PV* diagram, and calculate the work done by the gas. (*b*) Find the heat added during this process.

Picture the Problem We can find the work done by the gas during this process from the area under the curve. Because no work is done along the constant volume (vertical) part of the path, the work done by the gas is done during its isothermal expansion. We can then use the first law of thermodynamics to find the heat added to the system during this process.

(*a*) The path from the initial state (1) to the final state (2) is shown on the *PV* diagram.

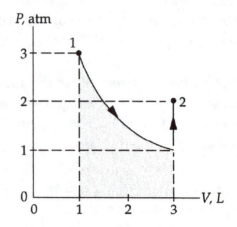

The work done by the gas equals the area under the curve:

$$W_{\text{by gas}} = \int_{V_1}^{V_2} P\,dV = nRT_1 \int_{1L}^{3L} \frac{dV}{V}$$

$$= P_1V_1 \int_{1L}^{3L} \frac{dV}{V} = P_1V_1 \left[\ln V\right]_{1L}^{3L}$$

$$= P_1V_1 \ln 3$$

Substitute numerical values and evaluate $W_{\text{by gas}}$:

$$W_{\text{by gas}} = \left(3\,\text{atm} \times \frac{101.3\,\text{kPa}}{\text{atm}}\right)\left(1\,\text{L} \times \frac{10^{-3}\,\text{m}^3}{\text{L}}\right) \ln 3 = \boxed{334\,\text{J}}$$

(*b*) The work done by the gas is the negative of the work done on the gas. Apply the first law of thermodynamics to the system to obtain:

$$Q_{\text{in}} = \Delta E_{\text{int}} - W_{\text{on}}$$
$$= \left(E_{\text{int},2} - E_{\text{int},1}\right) - \left(-W_{\text{by gas}}\right)$$
$$= \left(E_{\text{int},2} - E_{\text{int},1}\right) + W_{\text{by gas}}$$

Substitute numerical values and evaluate Q_{in}:

$$Q_{\text{in}} = \left(912\,\text{J} - 456\,\text{J}\right) + 334\,\text{J} = \boxed{790\,\text{J}}$$

*58 • An *isobaric* expansion is one carried out at constant pressure. Draw several isobars for an ideal gas on a diagram showing volume as a function of temperature.

Picture the Problem From the ideal gas law, $PV = NkT$, or $V = NkT/P$. Hence, on a VT diagram, isobars will be straight lines with slope $1/P$.

A spreadsheet program was used to plot the following graph. The graph was plotted for 1 mol of gas.

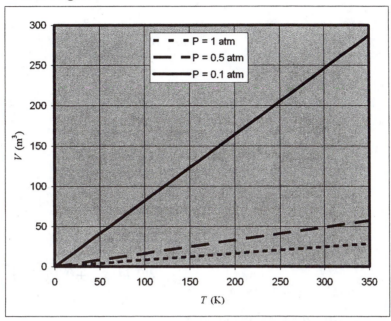

Heat Capacities of Gases and the Equipartition Theorem

*62 •• (a) Calculate the specific heats per unit mass of air at constant volume and constant pressure. Assume a temperature of 300 K and a pressure of 10^5 N/m^2. Assume that air is composed of 74% N_2 (molecular weight 28 g/mole) molecules and 26% O_2 molecules (molecular weight 32 g/mole) and that both components are ideal gases. (b) Compare your answer to the value listed in the *Handbook of Chemistry and Physics* for the heat capacity at constant pressure of 1.032 J/g·K.

Picture the Problem The specific heats of air at constant volume and constant pressure are given by $c_V = C_V/m$ and $c_P = C_P/m$ and the heat capacities at constant volume and constant pressure are given by $C_V = \frac{5}{2}nR$ and $C_P = \frac{7}{2}nR$, respectively.

(a) Express the specific heats per unit mass of air at constant volume and constant pressure:

$$c_V = \frac{C_V}{m} \qquad (1)$$

and

$$c_P = \frac{C_P}{m} \qquad (2)$$

Express the heat capacities of a diatomic gas in terms of the gas constant R, the number of moles n, and the number of degrees of freedom:

$$C_V = \tfrac{5}{2}nR$$
and
$$C_P = \tfrac{7}{2}nR$$

Express the mass of 1 mol of air:

$$m = 0.74M_{N_2} + 0.26M_{O_2}$$

Substitute in equation (1) to obtain:

$$c_V = \frac{5nR}{2\left(0.74M_{N_2} + 0.26M_{O_2}\right)}$$

Substitute numerical values and evaluate c_V:

$$c_V = \frac{5(1\,\text{mol})(8.314\,\text{J/mol}\cdot\text{K})}{2\left[0.74\left(28\times10^{-3}\,\text{kg}\right)+0.26\left(32\times10^{-3}\,\text{kg}\right)\right]} = \boxed{716\,\text{J/kg}\cdot\text{K}}$$

Substitute in equation (2) to obtain:

$$c_P = \frac{7nR}{2\left(0.74M_{N_2} + 0.26M_{O_2}\right)}$$

Substitute numerical values and evaluate c_P:

$$c_P = \frac{7(1\,\text{mol})(8.314\,\text{J/mol}\cdot\text{K})}{2\left[0.74\left(28\times10^{-3}\,\text{kg}\right)+0.26\left(32\times10^{-3}\,\text{kg}\right)\right]} = \boxed{1002\,\text{J/kg}\cdot\text{K}}$$

(b) Express the percent difference between the value from the *Handbook of Chemistry and Physics* and the calculated value:

$$\%\,\text{difference} = \frac{1.032\,\text{J/g}\cdot\text{K} - 1.002\text{J/g}\cdot\text{K}}{1.032\,\text{J/g}\cdot\text{K}} = \boxed{2.91\%}$$

*67 •• Carbon dioxide (CO_2) at 1 atm of pressure and a temperature of $-78.5°C$ sublimates directly from a solid to a gaseous state, without going through a liquid phase. What is the change in the heat capacity (at constant pressure) per mole of CO_2 when it undergoes sublimation? Assume that the gas molecules can rotate but do not vibrate. Is the change in the heat capacity positive or negative on sublimation? The CO_2 molecule is pictured in Figure 18-21.

Figure 18-21 Problem 67

O   C   O

Picture the Problem We can find the change in the heat capacity at constant pressure as CO_2 undergoes sublimation from the energy per molecule of CO_2 in the solid and gaseous states.

Express the change in the heat capacity (at constant pressure) per mole as the CO_2 undergoes sublimation:

$$\Delta C_P = C_{P,gas} - C_{P,solid}$$

Express $C_{p,gas}$ in terms of the number of degrees of freedom per molecule:

$$C_{P,gas} = f\left(\tfrac{1}{2} Nk\right) = \tfrac{5}{2} Nk$$

because each molecule has three translational and two rotational degrees of freedom in the gaseous state.

We know, from the Dulong-Petit Law, that the molar specific heat of most solids is $3R = 3Nk$. This result is essentially a per-atom result as it was obtained for a monatomic solid with six degrees of freedom. Use this result and the fact CO_2 is triatomic to express $C_{p,solid}$:

$$C_{P,solid} = \frac{3Nk}{\text{atom}} \times 3\,\text{atoms} = 9Nk$$

Substitute to obtain:

$$\Delta C_P = \tfrac{5}{2} Nk - \tfrac{18}{2} Nk = \boxed{-\tfrac{13}{2} Nk}$$

Quasi-Static Adiabatic Expansion of a Gas

*70 ••   One mole of an ideal gas ($\gamma = \tfrac{5}{3}$) is expanded adiabatically and quasi-statically from a pressure of 10 atm and a temperature of 0°C to a pressure of 2 atm. Find (*a*) the initial and final volumes, (*b*) the final temperature, and (*c*) the work done by the gas.

Picture the Problem The adiabatic expansion is shown in the *PV* diagram. We can use the ideal-gas law to find the initial volume of the gas and the equation for a quasi-static adiabatic process to find the final volume of the gas. A second application of the ideal-gas law, this time at the final state, will yield the final temperature of the gas. In (*c*) we can use the first law of thermodynamics to find the work done by the gas during this process.

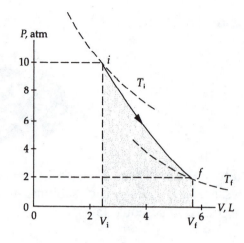

(*a*) Apply the ideal-gas law to express the initial volume of the gas:

$$V_i = \frac{nRT_i}{P_i}$$

Substitute numerical values and evaluate V_i:

$$V_i = \frac{(1\,\text{mol})(8.314\,\text{J/mol}\cdot\text{K})(273\,\text{K})}{10\,\text{atm}\times\dfrac{101.3\,\text{kPa}}{\text{atm}}}$$

$$= 2.24\times10^{-3}\,\text{m}^3 = \boxed{2.24\,\text{L}}$$

Use the relationship between the pressures and volumes for a quasi-static adiabatic process to express V_f:

$$P_i V_i^\gamma = P_f V_f^\gamma \;\Rightarrow\; V_f = V_i\left(\frac{P_i}{P_f}\right)^{1/\gamma}$$

Substitute numerical values and evaluate V_f:

$$V_f = V_i\left(\frac{P_i}{P_f}\right)^{1/\gamma} = (2.24\,\text{L})\left(\frac{10\,\text{atm}}{2\,\text{atm}}\right)^{3/5}$$

$$= \boxed{5.88\,\text{L}}$$

(*b*) Apply the ideal-gas law to express the final temperature of the gas:

$$T_f = \frac{P_f V_f}{nR}$$

Substitute numerical values and evaluate T_f:

$$T_f = \frac{(2\,\text{atm})(5.88\,\text{L})}{8.206\times10^{-2}\,\text{L}\cdot\text{atm/mol}\cdot\text{K}}$$

$$= \boxed{143\,\text{K}}$$

(*c*) Apply the first law of

$$W_{\text{on}} = \Delta E_{\text{int}} - Q_{\text{in}}$$

thermodynamics to express the work done on the gas:

or, because the process is adiabatic,

$$W_{on} = \Delta E_{int} = C_V \Delta T = \tfrac{3}{2} nR\Delta T$$

Substitute numerical values and evaluate W_{on}:

$$W_{on} = \tfrac{3}{2}(1\,\mathrm{mol})(8.314\,\mathrm{J/mol \cdot K})(-130\,\mathrm{K})$$
$$= -1.62\,\mathrm{kJ}$$

Because $W_{by\ the\ gas} = -W_{on}$:

$$W_{by\ gas} = \boxed{1.62\,\mathrm{kJ}}$$

*73 •• Half a mole of an ideal monatomic gas at a pressure of 400 kPa and a temperature of 300 K expands until the pressure has diminished to 160 kPa. Find the final temperature and volume, the work done, and the heat absorbed by the gas if the expansion is (*a*) isothermal and (*b*) adiabatic.

Picture the Problem We can use the ideal-gas law to find the initial volume of the gas. In (*a*) we can apply the ideal-gas law for a fixed amount of gas to find the final volume and the expression (Equation 19-16) for the work done in an isothermal process. Application of the first law of thermodynamics will allow us to find the heat absorbed by the gas during this process. In (*b*) we can use the relationship between the pressures and volumes for a quasi-static adiabatic process to find the final volume of the gas. We can apply the ideal-gas law to find the final temperature and, as in (*a*), apply the first law of thermodynamics, this time to find the work done by the gas.

Use the ideal-gas law to express the initial volume of the gas:

$$V_i = \frac{nRT_i}{P_i}$$

Substitute numerical values and evaluate V_i:

$$V_i = \frac{(0.5\,\mathrm{mol})(8.314\,\mathrm{J/mol \cdot K})(300\,\mathrm{K})}{400\,\mathrm{kPa}}$$
$$= 3.12 \times 10^{-3}\,\mathrm{m}^3 = 3.12\,\mathrm{L}$$

(*a*) Because the process is isothermal:

$$T_f = T_i = \boxed{300\,\mathrm{K}}$$

Use the ideal-gas law for a fixed amount of gas to express V_f:

$$\frac{P_i V_i}{T_i} = \frac{P_f V_f}{T_f}$$

or, because T = constant,

$$V_f = V_i \frac{P_i}{P_f}$$

Substitute numerical values and evaluate V_f:	$$V_f = (3.12\,L)\frac{400\,kPa}{160\,kPa} = \boxed{7.80\,L}$$
Express the work done by the gas during the isothermal expansion:	$$W_{by\ gas} = nRT\ln\frac{V_f}{V_i}$$
Substitute numerical values and evaluate $W_{by\ gas}$:	$$W_{by\ gas} = (0.5\,mol)(8.314\,J/mol\cdot K)$$ $$\times (300\,K)\ln\left(\frac{7.80\,L}{3.12\,L}\right)$$ $$= \boxed{1.14\,kJ}$$
Noting that the work done by the gas during the process equals the negative of the work done on the gas, apply the first law of thermodynamics to find the heat absorbed by the gas:	$$Q_{in} = \Delta E_{int} - W_{on} = 0 - (-1.14\,kJ)$$ $$= \boxed{1.14\,kJ}$$
(b) Using $\gamma = 5/3$ and the relationship between the pressures and volumes for a quasi-static adiabatic process, express V_f:	$$P_iV_i^\gamma = P_fV_f^\gamma \Rightarrow V_f = V_i\left(\frac{P_i}{P_f}\right)^{1/\gamma}$$
Substitute numerical values and evaluate V_f:	$$V_f = (3.12\,L)\left(\frac{400\,kPa}{160\,kPa}\right)^{3/5} = \boxed{5.41\,L}$$
Apply the ideal-gas law to find the final temperature of the gas:	$$T_f = \frac{P_fV_f}{nR}$$
Substitute numerical values and evaluate T_f:	$$T_f = \frac{(160\,kPa)(5.41\times10^{-3}\,m^3)}{(0.5\,mol)(8.314\,J/mol\cdot K)}$$ $$= \boxed{208\,K}$$
For an adiabatic process:	$$Q_{in} = \boxed{0}$$
Apply the first law of thermodynamics to express the	$$W_{on} = \Delta E_{int} - Q_{in} = C_V\Delta T - 0 = \tfrac{3}{2}nR\Delta T$$

work done on the gas during the
adiabatic process:

Substitute numerical values and
evaluate W_{on}:

$$W_{on} = \tfrac{3}{2}(0.5\,\text{mol})(8.314\,\text{J/mol}\cdot\text{K})$$
$$\times(208\,\text{K} - 300\,\text{K})$$
$$= -574\,\text{J}$$

Because the work done by the gas
equals the negative of the work done
on the gas:

$$W_{\text{by gas}} = -(-574\,\text{J}) = \boxed{574\,\text{J}}$$

*76 ••• A hand pump is used to inflate a bicycle tire to a gauge pressure of 482
kPa (about 70 lb/in^2). How much work must be done if each stroke of the pump is
an adiabatic process? Atmospheric pressure is 1 atm, the air temperature is
initially 20°C, and the volume of the air in the tire remains constant at 1 L.

Picture the Problem Consider the process to be accomplished in a single
compression. The initial pressure is 1 atm = 101 kPa. The final pressure is
(101 + 482) kPa = 583 kPa, and the final volume is 1 L. Because air is a mixture
of diatomic gases, $\gamma_{\text{air}} = 1.4$. We can find the initial volume of the air using
$P_iV_i^\gamma = P_fV_f^\gamma$ and use Equation 19-39 to find the work done by the air.

Express the work done in an
adiabatic process:

$$W = \frac{P_iV_i - P_fV_f}{\gamma - 1} \qquad (1)$$

Use the relationship between
pressure and volume for a quasi-
static adiabatic process to express
the initial volume of the air:

$$P_iV_i^\gamma = P_fV_f^\gamma \Rightarrow V_i = V_f\left(\frac{P_f}{P_i}\right)^{\frac{1}{\gamma}}$$

Substitute numerical values and
evaluate V_i:

$$V_i = (1\,\text{L})\left(\frac{583\,\text{kPa}}{101\,\text{kPa}}\right)^{\frac{1}{1.4}} = 3.50\,\text{L}$$

Substitute numerical values in equation (1) and evaluate W:

$$W = \frac{(101\,\text{kPa})(3.5\times10^{-3}\,\text{m}^3) - (583\,\text{kPa})(10^{-3}\,\text{m}^3)}{1.4 - 1} = \boxed{-574\,\text{J}}$$

where the minus sign tells us that work is done on the gas.

Cyclic Processes

*79 •• One mole of an ideal diatomic gas is allowed to expand along the straight line from 1 to 2 in the *PV* diagram (Figure 18-22). It is then compressed back isothermally from 2 to 1. Calculate the total work done on the gas during this cycle.

Figure 18-22 Problem 79

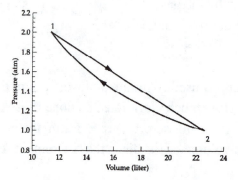

Picture the Problem The total work done as the gas is taken through this cycle is the area bounded by the two processes. Because the process from 1→2 is linear, we can use the formula for the area of a trapezoid to find the work done during this expansion. We can use $W_{\text{isothermal process}} = nRT\ln(V_f/V_i)$ to find the work done on the gas during the process 2→1. The total work is then the sum of these two terms.

Express the net work done per cycle:

$$W_{\text{total}} = W_{1\to2} + W_{2\to1} \qquad (1)$$

Work is done by the gas during its expansion from 1 to 2 and hence is equal to the negative of the area of the trapezoid defined by this path and the vertical lines at $V_1 = 11.5$ L and $V_2 = 23$ L. Use the formula for the area of a trapezoid to express $W_{1\to2}$:

$$W_{1\to2} = -A_{\text{trap}}$$
$$= -\tfrac{1}{2}(23\,\text{L} - 11.5\,\text{L})(2\,\text{atm} + 1\,\text{atm})$$
$$= -17.3\,\text{L}\cdot\text{atm}$$

Work is done on the gas during the isothermal compression from V_2 to V_1 and hence is equal to the area under the curve representing this process. Use the expression

$$W_{2\to1} = nRT\ln\frac{V_f}{V_i}$$

for the work done during an
isothermal process to express
$W_{2\rightarrow1}$:

Apply the ideal-gas law at point 1 to find the temperature along the isotherm
2→1:

$$T = \frac{PV}{nR} = \frac{(2\,\text{atm})(11.5\,\text{L})}{(1\,\text{mol})(8.206\times10^{-2}\,\text{L}\cdot\text{atm/mol}\cdot\text{K})} = 280\,\text{K}$$

Substitute numerical values and evaluate $W_{2\rightarrow1}$:

$$W_{2\rightarrow1} = \left|(1\,\text{mol})(8.206\times10^{-2}\,\text{L}\cdot\text{atm/mol}\cdot\text{K})(280\,\text{K})\ln\left(\frac{11.5\,\text{L}}{23\,\text{L}}\right)\right| = 15.9\,\text{L}\cdot\text{atm}$$

Substitute in equation (1) and evaluate
W_{net}:

$$W_{\text{net}} = -17.3\,\text{L}\cdot\text{atm} + 15.9\,\text{L}\cdot\text{atm}$$

$$= -1.40\,\text{L}\cdot\text{atm} \times \frac{101.325\,\text{J}}{\text{L}\cdot\text{atm}}$$

$$= \boxed{-142\,\text{J}}$$

Remarks: The work done by the gas during each cycle is 142 J.

Figure 18-23 Problems 81 and 82

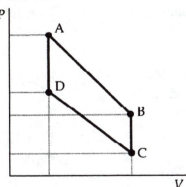

*82 ••• Repeat Problem 81 with a diatomic gas.

Picture the Problem We can find the temperatures, pressures, and volumes at all
points for this ideal diatomic gas (5 degrees of freedom) using the ideal-gas law
and the work for each process by finding the areas under each curve. We can find
the heat exchanged for each process from the heat capacities and the initial and
final temperatures for each process.

Express the total work done by the gas per cycle:	$W_{by\,gas,\,tot} = W_{D \to A} + W_{A \to B} + W_{B \to C} + W_{C \to D}$

1. Use the ideal-gas law to find the volume of the gas at point D:

$$V_D = \frac{nRT_D}{P_D}$$
$$= \frac{(2\,mol)(8.314\,J/mol \cdot K)(360\,K)}{(2\,atm)(101.3\,kPa/atm)}$$
$$= 29.5\,L$$

2. We're given that the volume of the gas at point B is three times that at point D:

$$V_B = V_C = 3V_D$$
$$= 88.6\,L$$

Use the ideal-gas law to find the pressure of the gas at point C:

$$P_C = \frac{nRT_C}{V_C}$$
$$= \frac{(2\,mol)(8.206 \times 10^{-2}\,L \cdot atm/mol \cdot K)}{88.6\,L}$$
$$\times (360\,K)$$
$$= 0.667\,atm$$

We're given that the pressure at point B is twice that at point C:

$$P_B = 2P_C = 2(0.667\,atm) = 1.33\,atm$$

3. Because path DC represents an isothermal process:

$$T_D = T_C = 360\,K$$

Use the ideal-gas law to find the temperatures at points B and A:

$$T_A = T_B = \frac{P_B V_B}{nR}$$
$$= \frac{(1.333\,atm)(88.6\,L)}{(2\,mol)(8.206 \times 10^{-2}\,L \cdot atm/mol \cdot K)}$$
$$= 720\,K$$

Because the temperature at point A is twice that at D and the volumes are the same, we can conclude that:

$$P_A = 2P_D = 4\,atm$$

The pressure, volume, and temperature at points A, B, C, and D are summarized in the table to

Point	P	V	T
	(atm)	(L)	(K)
A	4	29.5	720

the right.

B	1.33	88.6	720
C	0.667	88.6	360
D	2	29.5	360

4. For the path D→A:

$$W_{D \to A} = 0$$

and

$$\begin{aligned}Q_{D \to A} = \Delta U_{D \to A} &= \tfrac{5}{2} nR\Delta T_{D \to A} \\ &= \tfrac{5}{2} nR(T_A - T_D) \\ &= \tfrac{5}{2}(2\,\text{mol})(8.314\,\text{J/mol} \cdot \text{K}) \\ &\quad \times (720\,\text{K} - 360\,\text{K}) \\ &= 15.0\,\text{kJ}\end{aligned}$$

For the path A→B:

$$\begin{aligned}W_{A \to B} = Q_{A \to B} &= nRT_{A,B} \ln\frac{V_B}{V_A} \\ &= (2\,\text{mol})(8.314\,\text{J/mol} \cdot \text{K})(720\,\text{K}) \\ &\quad \times \ln\left(\frac{88.6\,\text{L}}{29.5\,\text{L}}\right) \\ &= 13.2\,\text{kJ}\end{aligned}$$

and, because process A→B is isothermal, $\Delta E_{\text{int}, A \to B} = 0$

For the path B→C:

$$W_{B \to C} = 0$$

and

$$\begin{aligned}Q_{B \to C} = \Delta U_{B \to C} &= C_V \Delta T \\ &= \tfrac{5}{2} nR(T_C - T_B) \\ &= \tfrac{5}{2}(2\,\text{mol})(8.314\,\text{J/mol} \cdot \text{K}) \\ &\quad \times (360\,\text{K} - 720\,\text{K}) \\ &= -15.0\,\text{kJ}\end{aligned}$$

For the path C→D:

$$\begin{aligned}W_{C \to D} &= nRT_{C,D} \ln\frac{V_D}{V_C} \\ &= (2\,\text{mol})(8.314\,\text{J/mol} \cdot \text{K})(360\,\text{K}) \\ &\quad \times \ln\left(\frac{29.5\,\text{L}}{88.6\,\text{L}}\right) \\ &= -6.58\,\text{kJ}\end{aligned}$$

Also, because process A→B is

isothermal, $\Delta E_{int, A \to B} = 0$ and

$$Q_{C \to D} = W_{C \to D} = -6.58 \, \text{kJ}$$

Q_{in}, W_{on}, and ΔE_{int} are summarized for each of the processes in the table to the right.

Process	Q_{in} (kJ)	W_{on} (kJ)	ΔE_{int} (kJ)
D→A	15.0	0	15.0
A→B	13.2	−13.2	0
B→C	−15.0	0	−15.0
C→D	−6.58	6.58	0

Referring to the table and noting that the work done by the gas equals the negative of the work done on the gas, find the total work done by the gas per cycle:

$$W_{by \, gas, \, tot} = W_{D \to A} + W_{A \to B} + W_{B \to C} + W_{C \to D}$$
$$= 0 + 13.2 \, \text{kJ} + 0 - 6.58 \, \text{kJ}$$
$$= \boxed{6.62 \, \text{kJ}}$$

Remarks: Note that ΔE_{int} for the complete cycle is zero and that the total work done is the same for the diatomic gas of this problem and the monatomic gas of problem 81.

General Problems

*86 • What is the number of moles n of the gas in Problem 85?

Picture the Problem We can find the number of moles of the gas from the expression for the work done on or by a gas during an isothermal process.

Express the work done on the gas during the isothermal process:

$$W = nRT \ln \frac{V_f}{V_i}$$

Solve for n:

$$n = \frac{W}{RT \ln \dfrac{V_f}{V_i}}$$

Substitute numerical values and evaluate n:

$$n = \frac{-180\,\text{kJ}}{(8.314\,\text{J/mol}\cdot\text{K})(293\,\text{K})\ln\left(\frac{1}{5}\right)}$$

$$= \boxed{45.9\,\text{mol}}$$

*89 •• Repeat Problem 87 with the gas following path ADC.

Picture the Problem We can use the ideal-gas law to find the temperatures T_A and T_C. Because the process DC is isobaric, we can find the area under this line geometrically. We can use the expression for the work done during an isothermal expansion to find the work done between A and D and the first law of thermodynamics to find Q_{ADC}.

(a) Using the ideal-gas law, find the temperature at point A:

$$T_A = \frac{P_A V_A}{nR}$$

$$= \frac{(4\,\text{atm})(4.01\,\text{L})}{(3\,\text{mol})(8.206\times10^{-2}\,\text{L}\cdot\text{atm/mol}\cdot\text{K})}$$

$$= \boxed{65.2\,\text{K}}$$

Using the ideal-gas law, find the temperature at point C:

$$T_C = \frac{P_C V_C}{nR}$$

$$= \frac{(1\,\text{atm})(20\,\text{L})}{(3\,\text{mol})(8.206\times10^{-2}\,\text{L}\cdot\text{atm/mol}\cdot\text{K})}$$

$$= \boxed{81.2\,\text{K}}$$

(b) Express the work done by the gas along the path ADC:

$$W_{ADC} = W_{AD} + W_{DC}$$

$$= nRT_A \ln\frac{V_D}{V_A} + P_{DC}\Delta V_{DC}$$

Use the ideal-gas law to find the volume of the gas at point D:

$$V_D = \frac{nRT_D}{P_D} = \frac{(3\,\text{mol})(8.206\times10^{-2}\,\text{L}\cdot\text{atm/mol}\cdot\text{K})(65.2\,\text{K})}{1\,\text{atm}} = 16.1\,\text{L}$$

Substitute numerical values and evaluate W_{ADC}:

$$W_{ADC} = (3\,mol)(8.206 \times 10^{-2}\,L \cdot atm/mol \cdot K)(65.2\,K) \ln\left(\frac{16.1\,L}{4.01\,L}\right)$$

$$+ (1\,atm)(20\,L - 16.1\,L)$$

$$= 26.2\,L \cdot atm \times \frac{101.325\,J}{L \cdot atm} = \boxed{2.65\,kJ}$$

(c) Apply the first law of thermodynamics to obtain:

$$Q_{ADC} = W_{ADC} + \Delta E_{int} = W_{ADC} + C_V \Delta T$$
$$= W_{ADC} + \tfrac{3}{2} nR\Delta T$$
$$= W_{ADC} + \tfrac{3}{2} nR(T_C - T_A)$$

Substitute numerical values and evaluate Q_{ADC}:

$$Q_{ADC} = 2.65\,kJ + \tfrac{3}{2}(3\,mol)(8.314\,J/mol \cdot K)(81.2\,K - 65.2\,K) = \boxed{3.25\,kJ}$$

*98 ••   When an ideal gas undergoes a temperature change at constant volume, its energy changes by $\Delta E_{int} = C_V \Delta T$. (a) Explain why this result holds for an ideal gas for any temperature change independent of the process. (b) Show explicitly that this result holds for the expansion of an ideal gas at constant pressure by first calculating the work done and showing that it can be written as $W = nR\Delta T$, and then by using $\Delta E_{int} = Q - W$, where $Q = C_P \Delta T$.

Picture the Problem We can express the work done during an isobaric process as the product of the temperature and the change in volume and relate Q to ΔT through the definition of C_P. Finally, we can use the first law of thermodynamics to show that $\Delta E_{int} = C_V \Delta T$.

(a)

> For an ideal gas, the internal energy is the sum of the kinetic energies of the gas molecules, which is proportional to kT. Consequently, U is a function of T only and $\Delta E_{int} = C_V \Delta T$.

(b) Use the first law of thermodynamics to relate the work done on the gas, the heat entering the gas, and the change in the internal energy of the gas:

$$\Delta E_{int} = Q_{in} + W_{on}$$

At constant pressure:

$$W_{by\,gas} = P(V_f - V_i) = nR(T_f - T_i) = nR\Delta T$$
and
$$W_{on} = -W_{by\,gas} = -nR\Delta T$$

Relate Q_{in} to C_P and ΔT:

$$Q_{in} = C_P \Delta T$$

Substitute to obtain:

$$\Delta E_{int} = C_P \Delta T - nR\Delta T$$
$$= (C_P - nR)\Delta T = \boxed{C_V \Delta T}$$

*100 •• Heat in the amount of 500 J is supplied to 2 mol of an ideal diatomic gas. (*a*) Find the change in temperature if the pressure is kept constant. (*b*) Find the work done by the gas. (*c*) Find the ratio of the final volume of the gas to the initial volume if the initial temperature is 20°C.

Picture the Problem We can use $Q_{in} = C_P \Delta T$ to find the change in temperature during this isobaric process and the first law of thermodynamics to relate W, Q, and ΔE_{int}. We can use $\Delta E_{int} = \frac{5}{2} nR\Delta T$ to find the change in the internal energy of the gas during the isobaric process and the ideal-gas law for a fixed amount of gas to express the ratio of the final and initial volumes.

(*a*) Relate the change in temperature to Q_{in} and C_P and evaluate ΔT:

$$\Delta T = \frac{Q_{in}}{C_P} = \frac{Q_{in}}{\frac{7}{2} nR}$$

$$= \frac{500\,J}{\frac{7}{2}(2\,mol)(8.314\,J/mol \cdot K)}$$

$$= \boxed{8.59\,K}$$

(*b*) Apply the first law of thermodynamics to relate the work done on the gas to the heat supplied and the change in its internal energy:

$$W_{on} = \Delta E_{int} - Q_{in} = C_V \Delta T - Q_{in}$$
$$= \frac{5}{2} nR\Delta T - Q_{in}$$

Substitute numerical values and evaluate W_{on}:

$$W_{on} = \frac{5}{2}(2\,mol)(8.314\,J/mol \cdot K)(8.59\,K)$$
$$- 500\,J$$
$$= -143\,J$$

Because $W_{by\ gas} = -W_{on}$:

$$W_{by\ gas} = \boxed{143\,J}$$

(*c*) Using the ideal-gas law for a fixed amount of gas, relate the initial and final pressures, volumes and temperatures:

$$\frac{P_i V_i}{T_i} = \frac{P_f V_f}{T_f}$$

or, because the process is isobaric,

$$\frac{V_i}{T_i} = \frac{V_f}{T_f}$$

Solve for and evaluate V_f/V_i:

$$\frac{V_f}{V_i} = \frac{T_f}{T_i} = \frac{T_i + \Delta T}{T_i}$$

$$= \frac{293.15\,\text{K} + 8.59\,\text{K}}{293.15\,\text{K}} = \boxed{1.03}$$

*102 •• Two moles of a diatomic ideal gas expand adiabatically. The initial temperature of the gas is 300 K. The work done by the gas during the expansion is 3.5 kJ. What is the final temperature of the gas?

Picture the Problem We know that, for an adiabatic process, $Q_{in} = 0$. Hence, the work done by the expanding gas equals the change in its internal energy. Because we're given the work done by the gas during the expansion, we can express the change in the temperature of the gas in terms of this work and C_V.

Express the final temperature of the gas as a result of its expansion:

$$T_f = T_i + \Delta T$$

Apply the equation for adiabatic work and solve for ΔT:

$$W_{adiabatic} = -C_V \Delta T$$
and
$$\Delta T = -\frac{W_{adiabatic}}{C_V} = -\frac{W_{adiabatic}}{\frac{5}{2}nR}$$

Substitute and evaluate T_f:

$$T_f = T_i - \frac{W_{adiabatic}}{\frac{5}{2}nR}$$

$$= 300\,\text{K} - \frac{3.5\,\text{kJ}}{\frac{5}{2}(2\,\text{mol})(8.314\,\text{J/mol}\cdot\text{K})}$$

$$= \boxed{216\,\text{K}}$$

*107 ••• In an isothermal expansion, an ideal gas at an initial pressure P_0 expands until its volume is twice its initial volume. (a) Find its pressure after the expansion. (b) The gas is then compressed adiabatically and quasi-statically back to its original volume, at which point its pressure is $1.32P_0$. Is the gas monatomic, diatomic, or polyatomic? (c) How does the translational kinetic energy of the gas change in these processes?

Picture the Problem The isothermal expansion followed by an adiabatic compression is shown on the PV diagram. The path $1\rightarrow2$ is isothermal and the path $2\rightarrow3$ is adiabatic. We can apply the ideal-gas law for a fixed amount of gas and an isothermal process to find the pressure at point 2 and the pressure-volume relationship for a quasi-static adiabatic process to determine γ.

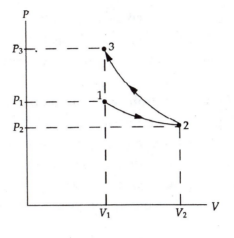

(*a*) Relate the initial and final pressures and volumes for the isothermal expansion and solve for and evaluate the final pressure:

$$P_1V_1 = P_2V_2$$
and
$$P_2 = P_1\frac{V_1}{V_2} = P_0\frac{V_1}{2V_1} = \boxed{\tfrac{1}{2}P_0}$$

(*b*) Relate the initial and final pressures and volumes for the adiabatic compression:

$$P_2V_2^\gamma = P_3V_3^\gamma$$
or
$$\tfrac{1}{2}P_0(2V_0)^\gamma = 1.32P_0V_0^\gamma$$
which simplifies to
$$2^\gamma = 2.64$$

Take the natural logarithm of both sides of this equation and solve for and evaluate γ:

$$\gamma\ln 2 = \ln 2.64$$
and
$$\gamma = \frac{\ln 2.64}{\ln 2} = 1.40$$

$\boxed{\therefore \text{the gas is diatomic.}}$

(*c*)
> In the isothermal process, T is constant, and the translational kinetic energy is unchanged.

> In the adiabatic process, $T_3 = 1.32T_0$, and the translational kinetic energy increases by a factor of 1.32.

*109 ••• Carbon monoxide and oxygen combine to form carbon dioxide with an energy release of 280 kJ/mol of CO according to the reaction $2(CO) + O_2 \rightarrow 2(CO_2)$. Two mol of CO and one mol of O_2 at 300 K are confined in an 80-L container; the combustion reaction is initiated with a spark. (*a*) What is the pressure in the container prior to the reaction? (*b*) If the reaction proceeds adiabatically, what are the final temperature and pressure? (*c*) If the resulting CO_2 gas is cooled to 0°C, what is the pressure in the container?

Picture the Problem In this problem the specific heat of the combustion products depends on the temperature. Although C_P increases gradually from $(9/2)R$ per mol to $(15/2)R$ per mol at high temperatures, we'll assume that $C_P = 4.5R$ below $T = 2000$ K and $C_P = 7.5R$ above $T = 2000$ K. We can find the final temperature following combustion from the heat made available during the combustion and the final pressure by applying the ideal-gas law to the initial and final states of the gases.

(*a*) Apply the ideal-gas law to find the pressure due to 3 mol at 300 K in the container prior to the reaction:

$$P_i = \frac{nRT_i}{V_i}$$

$$= \frac{(3\,\text{mol})(8.314\,\text{J/mol}\cdot\text{K})(300\,\text{K})}{80\,\text{L}}$$

$$= \boxed{93.5\,\text{kPa}}$$

(*b*) Relate the heat available in this adiabatic process to C_V and the change in temperature of the gases:

$$\Delta E_{int} = Q_{available}$$
$$= C_V(T_f - T_i)$$

Because $T > 2000$ K:

$$C_V = C_P - nR = n(7.5R) - nR = 6.5nR$$

Substitute to obtain:

$$Q_{available} = 6.5nR(T_f - T_i)$$

Solve for T_f to obtain:

$$T_f = \frac{Q_{available}}{6.5nR} + T_i \qquad (1)$$

Find Q required to raise 2 mol of CO_2 to 2000 K:

$$Q_{heat\,CO_2} = C_V\Delta T$$

For $T < 2000$ K:

$$C_V = C_P - nR = n(4.5R) - nR = 3.5nR$$

Substitute for C_V and find the heat required to warm to CO_2 to 2000 K:

$$Q_{heat\,CO_2} = 3.5nR\Delta T$$
$$= 3.5(2\,\text{mol})(8.314\,\text{J/mol}\cdot\text{K})$$
$$\times(2000\,\text{K} - 300\,\text{K})$$
$$= 98.94\,\text{kJ}$$

Find Q available to heat 2 mol of CO_2 above 2000 K:

$$Q_{available} = 560\,\text{kJ} - 98.94\,\text{kJ}$$
$$= 461.1\,\text{kJ}$$

Substitute in equation (1) and evaluate T_f:

$$T_f = \frac{461.1\,\text{kJ}}{6.5(2\,\text{mol})(8.314\,\text{J/mol}\cdot\text{K})} + 2000\,\text{K}$$
$$= \boxed{6266\,\text{K}}$$

Apply the ideal-gas law to relate the final temperature, pressure, and volume to the number of moles in the final state:

$$P_f V_f = n_f R T_f$$

Apply the ideal-gas law to relate the initial temperature, pressure, and volume to the number of moles in the initial state:

$$P_i V_i = n_i R T_i$$

Divide the first of these equations by the second and solve for P_f:

$$\frac{P_f V_f}{P_i V_i} = \frac{n_f R T_f}{n_i R T_i}$$

or, because $V_f = V_i$,

$$P_f = P_i\left(\frac{n_f}{n_i}\right)\left(\frac{T_f}{T_i}\right) \qquad (2)$$

Substitute numerical values in equation (2) and evaluate P_f:

$$P_f = (93.53\,\text{kPa})\left(\frac{2\,\text{mol}}{3\,\text{mol}}\right)\left(\frac{6266\,\text{K}}{300\,\text{K}}\right)$$
$$= \boxed{1.30\,\text{MPa}}$$

(c) Substitute numerical values in equation (2) and evaluate P_f for $T_f = 273$ K:

$$P_f = (93.53\,\text{kPa})\left(\frac{2\,\text{mol}}{3\,\text{mol}}\right)\left(\frac{273\,\text{K}}{300\,\text{K}}\right)$$
$$= \boxed{56.7\,\text{kPa}}$$

Chapter 19
The Second Law of Thermodynamics

Conceptual Problems

*2 • Explain why you can't just open your refrigerator to cool your kitchen on a hot day. Why is it that turning on a room air conditioner will cool down a room but opening a refrigerator door will not?

Determine the Concept As described by the second law of thermodynamics, more heat must be transmitted to the outside world than is removed by a refrigerator or air conditioner. The heating coils on a refrigerator are inside the room–the refrigerator actually heats the room it is in. The heating coils on an air conditioner are outside, so the waste heat is vented to the outside.

*5 • In a reversible adiabatic process, (*a*) the internal energy of the system remains constant, (*b*) no work is done by the system, (*c*) the entropy of the system remains constant, (*d*) the temperature of the system remains constant.

Determine the Concept

(*a*) Because the temperature changes during an adiabatic process, the internal energy of the system changes continuously during the process.

(*b*) Both the pressure and volume change during an adiabatic process and hence work is done by the system.

(*c*) $\Delta Q = 0$ during an adiabatic process. Therefore $\Delta S = 0$. $\boxed{(c) \text{ is correct.}}$

(*d*) Because the pressure and volume change during an adiabatic process, so does the temperature.

*8 •• Figure 19-12 shows a thermodynamic cycle on an *ST* diagram. Identify this cycle and sketch it on a *PV* diagram.

Figure 19-12 Problems 8 and 68

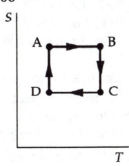

535

Determine the Concept The processes A→B and C→D are adiabatic; the processes B→C and D→A are isothermal. The cycle is therefore the Carnot cycle shown in the adjacent PV diagram.

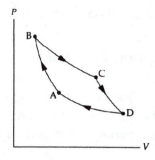

*14 • Which has a greater effect on increasing the efficiency of a Carnot engine, a 5-K increase in the temperature of the hot reservoir or a 5-K decrease in the temperature of the cold reservoir?

Picture the Problem Let ΔT be the change in temperature and $\varepsilon = (T_h - T_c)/T_h$ be the initial efficiency. We can express the efficiencies of the Carnot engine resulting from the given changes in temperature and examine their ratio to decide which has the greater effect on increasing the efficiency.

If T_h is increased by ΔT, ε', the new efficiency is:

$$\varepsilon' = \frac{T_h + \Delta T - T_c}{T_h + \Delta T}$$

If T_c is reduced by ΔT, the efficiency is:

$$\varepsilon'' = \frac{T_h - T_c + \Delta T}{T_h}$$

Divide the second of these equations by the first to obtain:

$$\frac{\varepsilon''}{\varepsilon'} = \frac{\dfrac{T_h - T_c + \Delta T}{T_h}}{\dfrac{T_h + \Delta T - T_c}{T_h + \Delta T}} = \frac{T_h + \Delta T}{T_h} > 1$$

> Therefore, a reduction in the temperature of the cold reservoir by ΔT increases the efficiency more than an equal increase in the temperature of the hot reservoir.

Estimation and Approximation

*16 •• (a) Estimate the highest COP possible for a "typical" household refrigerator. (b) If the refrigerator draws an 600 W of electrical power, estimate the rate at which heat is being drawn from the refrigerator compartment.

Picture the Problem If we assume that the temperature on the inside of the refrigerator is 0°C (273 K) and the room temperature to be about 30°C (303 K), then the refrigerator must be able to maintain a temperature difference of about 30 K. We can use the definition of the COP of a refrigerator and the relationship between the temperatures of the hot and cold reservoir and $|Q_h|$ and Q_c to find an upper limit on the COP of a household refrigerator. In (*b*) we can solve the definition of COP for Q_c and differentiate the resulting equation with respect to time to estimate the rate at which heat is being drawn from the refrigerator compartment.

(*a*) Using its definition, express the COP of a household refrigerator:

$$COP = \frac{Q_c}{W} \qquad (1)$$

Apply the 1st law of thermodynamics to the refrigerator to obtain:

$$W + Q_c = |Q_h|$$

Substitute for W and simplify to obtain:

$$COP = \frac{Q_c}{|Q_h| - Q_c} = \frac{1}{\dfrac{|Q_h|}{Q_c} - 1}$$

Assume, for the sake of finding the upper limit on the COP, that the refrigerator is a Carnot refrigerator and relate the temperatures of the hot and cold reservoirs to $|Q_h|$ and Q_c:

$$\frac{|Q_h|}{Q_c} = \frac{T_h}{T_c}$$

Substitute to obtain:

$$COP_{max} = \frac{1}{\dfrac{T_h}{T_c} - 1}$$

Substitute numerical values and evaluate COP_{max}:

$$COP_{max} = \frac{1}{\dfrac{303\,K}{273\,K} - 1} = \boxed{9.10}$$

(b) Solve equation (1) for Q_c:

$$Q_c = W(COP) \qquad (2)$$

Differentiate equation (2) with respect to time to obtain:

$$\frac{dQ_c}{dt} = (COP)\frac{dW}{dt}$$

Substitute numerical values and evaluate dQ_c/dt:

$$\frac{dQ_c}{dt} = (9.10)(600\,J/s) = \boxed{5.46\,kW}$$

*20 •••   How long, on average, should we have to wait until all of the air molecules in a room rush to one half of the room? (As a friend of mine put it, "Don't hold your breath ...") Assume that the air molecules are contained in a 1m × 1 m × 1m box and that they reshuffle their positions 100 times per second.

Calculate the average time it should take for all the molecules to occupy only one half of the box if there are (*a*) 10 molecules, (*b*) 100 molecules, (*c*) 1000 molecules, and (*d*) 1 mole of molecules in the box. (*e*) The highest vacuums that have been created to date have pressures of about 10^{-12} torr. If a typical vacuum chamber has a capacity of about 1 liter, how long will a physicist have to wait before all of the gas molecules in the vacuum chamber occupy only one half of it? Compare that to the expected lifetime of the universe, which is about 10^{10} years.

Picture the Problem If you had one molecule in a box, it would have a 50% chance of being on one side or the other. We don't care which side the molecules are on as long as they all are on one side, so with one molecule you have a 100% chance of it being on one side or the other. With two molecules, there are four possible combinations (both on one side, both on the other, one on one side and one on the other, and the reverse), so there is a 25% (1 in 4) chance of them both being on a particular side, or a 50% chance of them both being on either side. Extending this logic, the probability of N molecules all being on one side of the box is $P = 2/2^N$, which means that, if the molecules shuffle 100 times a second, the time it would take them to cover all the combinations and all get on one side or the other is $t = \dfrac{2^N}{2(100)}$. In (*e*) we can apply the ideal gas law to find the number of molecules in 1 L of air at a pressure of 10^{-12} torr and an assumed temperature of 300 K.

(*a*) Evaluate t for $N = 10$ molecules:

$$t = \frac{2^{10}}{2(100)} = \boxed{5.12\text{s}}$$

(*b*) Evaluate t for $N = 100$ molecules:

$$t = \frac{2^{100}}{2(100)} = 6.34 \times 10^{27} \text{ s}$$

$$= \boxed{2.01 \times 10^{20} \text{ y}}$$

(*c*) Evaluate t for $N = 1000$ molecules:

$$t = \frac{2^{1000}}{2(100)}$$

To evaluate 2^{1000} let $10^x = 2^{1000}$ and take the logarithm of both sides of the equation to obtain:

$$(1000)\ln 2 = x \ln 10$$

Solve for x to obtain:

$$x = 301$$

Substitute to obtain:

$$t = \frac{10^{301}}{2(100)} = 0.5 \times 10^{299} \text{ s}$$

$$= \boxed{1.58 \times 10^{290} \text{ y}}$$

(d) Evaluate t for
$N = 6.02 \times 10^{23}$ molecules:

$$t = \frac{2^{6.02 \times 10^{23}}}{2(100)}$$

To evaluate $2^{6.02 \times 10^{23}}$ let
$10^x = 2^{6.02 \times 10^{23}}$ and take the
logarithm of both sides of the
equation to obtain:

$$\left(6.02 \times 10^{23}\right) \ln 2 = x \ln 10$$

Solve for x to obtain:

$$x \approx 10^{23}$$

Substitute to obtain:

$$t \approx \frac{10^{10^{23}}}{2(100)} \approx \boxed{10^{10^{23}} \text{ y}}$$

(e) Solve the ideal gas law for the
number of molecules N in the gas:

$$N = \frac{PV}{kT}$$

Assuming the gas to be at room
temperature (300 K), substitute
numerical values and evaluate N:

$$N = \frac{\left(10^{-12} \text{ torr}\right)\left(133.32 \text{ Pa/torr}\right)\left(10^{-3} \text{ m}^3\right)}{\left(1.381 \times 10^{-23} \text{ J/K}\right)\left(300 \text{ K}\right)}$$

$$= 3.22 \times 10^7 \text{ molecules}$$

Evaluate T for $N = 3.22 \times 10^7$
molecules:

$$t = \frac{2^{3.22 \times 10^7}}{2(100)}$$

To evaluate $2^{3.22 \times 10^7}$ let
$10^x = 2^{3.22 \times 10^7}$ and take the
logarithm of both sides of the
equation to obtain:

$$\left(3.22 \times 10^7\right) \ln 2 = x \ln 10$$

Solve for x to obtain:

$$x \approx 10^7$$

Substitute to obtain:

$$T = \frac{10^{10^7}}{2(100)} \approx \boxed{10^{10^7} \text{ y}}$$

Express the ratio of this waiting
time to the lifetime of the
universe $T_{universe}$:

$$\frac{T}{T_{universe}} = \frac{10^{10^7} \text{ y}}{10^{10} \text{ y}} \approx 10^{10^7}$$

or

$$T \approx \boxed{10^{10^7} T_{\text{universe}}}$$

Heat Engines and Refrigerators

*24 •   A refrigerator absorbs 5 kJ of energy from a cold reservoir and rejects 8 kJ to a hot reservoir. (*a*) Find the coefficient of performance of the refrigerator. (*b*) The refrigerator is reversible and is run backward as a heat engine between the same two reservoirs. What is its efficiency?

Picture the Problem We can apply their definitions to find the COP of the refrigerator and the efficiency of the heat engine.

(*a*) Using the definition of the COP, relate the heat absorbed from the cold reservoir to the work done each cycle:	$COP = \dfrac{Q_c}{W}$
Relate the work done per cycle to Q_h and Q_c:	$W = \lvert Q_h \rvert - Q_c$
Substitute to obtain:	$COP = \dfrac{Q_c}{\lvert Q_h \rvert - Q_c}$
Substitute numerical values and evaluate COP:	$COP = \dfrac{5\,\text{kJ}}{\lvert 8\,\text{kJ} \rvert - 5\,\text{kJ}} = \boxed{1.67}$
(*b*) Use the definition of efficiency to relate the work done per cycle to the heat absorbed from the high-temperature reservoir:	$\varepsilon = \dfrac{W}{Q_h}$
Substitute numerical values and evaluate ε:	$\varepsilon = \dfrac{3\,\text{kJ}}{8\,\text{kJ}} = \boxed{37.5\%}$

*28 ••   One mole of an ideal monatomic gas at an initial volume $V_1 = 25$ L follows the cycle shown in Figure 19-15. All the processes are quasi-static. Find (*a*) the temperature of each state of the cycle, (*b*) the heat flow for each part of the cycle, and (*c*) the efficiency of the cycle.

Figure 19-15 Problem 28

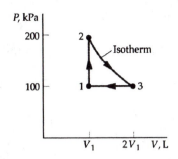

Picture the Problem We can use the ideal-gas law to find the temperatures of each state of the gas and the heat capacities at constant volume and constant pressure to find the heat flow for the constant-volume and isobaric processes. Because the change in internal energy is zero for the isothermal process, we can use the expression for the work done on or by a gas during an isothermal process to find the heat flow during such a process. Finally, we can find the efficiency of the cycle from its definition.

(*a*) Use the ideal-gas law to find the temperature at point 1:

$$T_1 = \frac{P_1 V_1}{nR} = \frac{(100\,\text{kPa})(25\,\text{L})}{(1\,\text{mol})(8.314\,\text{J/mol}\cdot\text{K})}$$

$$= \boxed{301\,\text{K}}$$

Use the ideal-gas law to find the temperatures at points 2 and 3:

$$T_2 = T_3 = \frac{P_2 V_2}{nR} = \frac{(200\,\text{kPa})(25\,\text{L})}{(1\,\text{mol})(8.314\,\text{J/mol}\cdot\text{K})}$$

$$= \boxed{601\,\text{K}}$$

(*b*) Find the heat entering or leaving the system for the constant-volume process from 1 to 2:

$$Q_{1\to2} = C_V \Delta T_{1\to2} = \tfrac{3}{2} R \Delta T_{1\to2}$$
$$= \tfrac{3}{2}(8.314\,\text{J/mol}\cdot\text{K})$$
$$\quad \times (601\,\text{K} - 301\,\text{K})$$
$$= \boxed{3.74\,\text{kJ}}$$

Find the heat entering or leaving the system for the isothermal process from 2 to 3:

$$Q_{2\to3} = nRT_2 \ln\frac{V_3}{V_2}$$
$$= (1\,\text{mol})(8.314\,\text{J/mol}\cdot\text{K})$$
$$\quad \times (601\,\text{K})\ln\left(\frac{50\,\text{L}}{25\,\text{L}}\right)$$
$$= \boxed{3.46\,\text{kJ}}$$

Find the heat entering or leaving the system during the isobaric compression from 3 to 1:

$$Q_{3\to1} = C_p \Delta T_{3\to1} = \tfrac{5}{2} R \Delta T_{3\to1}$$
$$= \tfrac{5}{2}(8.314 \, \text{J/mol} \cdot \text{K})$$
$$\times (301 \, \text{K} - 601 \, \text{K})$$
$$= \boxed{-6.24 \, \text{kJ}}$$

(c) Express the efficiency of the cycle:

$$\varepsilon = \frac{W}{Q_{\text{in}}} = \frac{W}{Q_{1\to2} + Q_{2\to3}}$$

Apply the 1st law of thermodynamics to the cycle:

$$W = \sum Q = Q_{1\to2} + Q_{2\to3} + Q_{3\to1}$$
$$= 3.74 \, \text{kJ} + 3.46 \, \text{kJ} - 6.24 \, \text{kJ}$$
$$= 0.960 \, \text{kJ}$$

because, for the cycle, $\Delta U = 0$.

Substitute and evaluate ε:

$$\varepsilon = \frac{0.960 \, \text{kJ}}{3.74 \, \text{kJ} + 3.46 \, \text{kJ}} = \boxed{13.3\%}$$

*31 •• "As far as we know, Nature has never evolved a heat engine"–Steven Vogel, *Life's Devices*. Princeton University Press (1988). (a) Calculate the efficiency of a heat engine operating between body temperature (98.6°F) and a typical outdoor temperature (70°F), and compare this to the human body's efficiency for converting chemical energy into work (approximately 20%). Does this contradict the Second Law of Thermodynamics? (b) From the result of Part (a), and a general knowledge of the conditions under which most warm-blooded animal life exists, explain why no warm-blooded animals have evolved heat engines to supply their internal energy.

Picture the Problem We can use the efficiency of a Carnot engine operating between reservoirs at body temperature and typical outdoor temperatures to find an upper limit on the efficiency of an engine operating between these temperatures.

(a) Express the maximum efficiency of an engine operating between body temperature and 70°F:

$$\varepsilon_C = 1 - \frac{T_c}{T_h}$$

Use $T = \tfrac{5}{9}(t_F - 32) + 273$ to obtain:

$T_{\text{body}} = 310 \, \text{K}$ and $T_{\text{room}} = 294 \, \text{K}$

Substitute numerical values and evaluate ε_C:

$$\varepsilon_C = 1 - \frac{294 \, \text{K}}{310 \, \text{K}} = \boxed{5.16\%}$$

The fact that this efficiency is considerably less than the actual efficiency of a human body does not contradict the second law of thermodynamics. The application of the second law to chemical reactions such as the ones that supply the body with energy have not been discussed in the text but we can note that don't get our energy from heat swapping between our body and the environment. Rather, we eat food to get the energy that we need.

(b)

Most warm - blooded animals survive under roughly the same conditions as humans. To make a heat engine work with appreciable efficiency, internal body temperatures would have to be maintained at an unreasonably high level.

Second Law of Thermodynamics

*34 •• If two adiabatic curves intersect on a PV diagram, a cycle could be completed by an isothermal path between the two adiabatic curves shown in Figure 19-18. Show that such a cycle could violate the second law of thermodynamics.

Figure 19-18 Problem 34

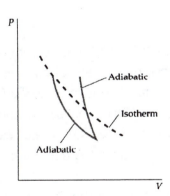

Determine the Concept The work done by the system is the area enclosed by the cycle, where we assume that we start with the isothermal expansion. It is only in this expansion that heat is extracted from a reservoir. There is no heat transfer in the adiabatic expansion or compression. Thus, we would completely convert heat to mechanical energy, without exhausting any heat to a cold reservoir, in violation of the second law.

Carnot Engines

*39 ••   A Carnot engine works between two heat reservoirs as a refrigerator. It does 50 J of work to remove 100 J from the cold reservoir and gives off 150 J to the hot reservoir during each cycle. Its coefficient of performance COP = Q_c /W = (100 J)/(50 J) = 2. (*a*) What is the efficiency of the Carnot engine when it works as a heat engine between the same two reservoirs? (*b*) Show that no other engine working as a refrigerator between the same two reservoirs can have a COP greater than 2.

Picture the Problem We can use the definition of efficiency to find the efficiency of the Carnot engine operating between the two reservoirs.

(*a*) Use its definition to find the efficiency of the Carnot engine:

$$\varepsilon_C = \frac{W}{Q_h} = \frac{50\,J}{150\,J} = \boxed{33.3\%}$$

(*b*) If COP > 2, then 50 J of work will remove more than 100 J of heat from the cold reservoir and put more than 150 J of heat into the hot reservoir. So running engine (*a*) to operate the refrigerator with a COP > 2 will result in the transfer of heat from the cold to the hot reservoir without doing any net mechanical work in violation of the second law.

Heat Pumps

*43 •   A heat pump delivers 20 kW to heat a house. The outside temperature is –10°C and the inside temperature of the hot-air supply for the heating fan is 40°C. (*a*) What is the coefficient of performance of a Carnot heat pump operating between these temperatures? (*b*) What must be the minimum power of the engine needed to run the heat pump? (*c*) If the COP of the heat pump is 60% of the efficiency of an ideal pump, what must the minimum power of the engine be?

Picture the Problem We can use the definition of the COP_{HP} and the Carnot efficiency of an engine to express the maximum efficiency of the refrigerator in terms of the reservoir temperatures. We can apply equation 19-10 and the definition of power to find the minimum power needed to run the heat pump.

(a) Express the COP$_{HP}$ in terms of T_h and T_c:

$$COP_{HP} = \frac{|Q_h|}{W} = \frac{|Q_h|}{|Q_h| - Q_c}$$

$$= \frac{1}{1 - \frac{Q_c}{|Q_h|}} = \frac{1}{1 - \frac{T_c}{T_h}}$$

$$= \frac{T_h}{T_h - T_c}$$

Substitute numerical values and evaluate the COP$_{HP}$:

$$COP_{HP} = \frac{313\,K}{313\,K - 263\,K} = \boxed{6.26}$$

(b) Using its definition, express the power output of the engine:

$$P = \frac{W}{\Delta t}$$

Use equation 19-10 to express the work done by the heat pump:

$$W = \frac{|Q_h|}{1 + COP_{HP}}$$

Substitute and evaluate P:

$$P = \frac{|Q_h|/\Delta t}{1 + COP_{HP}} = \frac{20\,kW}{1 + 6.26} = \boxed{2.75\,kW}$$

(c) Find the minimum power if the COP is 60% of the efficiency of an ideal pump:

$$P_{min} = \frac{|Q_c|/\Delta t}{1 + 0.6(COP_{HP.\,max})} = \frac{20\,kW}{1 + 0.6(6.26)}$$

$$= \boxed{4.21\,kW}$$

Entropy Changes

*47 •• Consider the freezing of 50 g of water by placing it in the freezer compartment of a refrigerator. Assume the walls of the freezer are maintained at −10°C. The water, initially liquid at 0°C, is frozen into ice and cooled to −10°C. Show that even though the entropy of the ice decreases, the net entropy of the universe increases.

Picture the Problem The change in the entropy of the world resulting from the freezing of this water and the cooling of the ice formed is the sum of the entropy changes of the water-ice and the freezer. Note that, while the entropy of the water decreases, the entropy of the freezer increases.

Express the change in entropy of the universe resulting from this freezing and cooling process:

$$\Delta S_u = \Delta S_{water} + \Delta S_{freezer} \qquad (1)$$

Express ΔS_{water}:

$$\Delta S_{water} = \Delta S_{freezing} + \Delta S_{cooling} \qquad (2)$$

Express $\Delta S_{freezing}$:

$$\Delta S_{freezing} = \frac{-Q_{freezing}}{T_{freezing}} \qquad (3)$$

where the minus sign is a consequence of the fact that heat is leaving the water as it freezes.

Relate $Q_{freezing}$ to the latent heat of fusion and the mass of the water:

$$Q_{freezing} = mL_f$$

Substitute in equation (3) to obtain:

$$\Delta S_{freezing} = \frac{-mL_f}{T_{freezing}}$$

Express $\Delta S_{cooling}$:

$$\Delta S_{cooling} = mC_p \ln \frac{T_f}{T_i}$$

Substitute in equation (2) to obtain:

$$\Delta S_{water} = \frac{-mL_f}{T_{freezing}} + mC_p \ln \frac{T_f}{T_i}$$

Noting that the freezer gains heat (at 263 K) from the freezing water and cooling ice, express $\Delta S_{freezer}$:

$$\Delta S_{freezer} = \frac{\Delta Q_{ice}}{T_{freezer}} + \frac{\Delta Q_{cooling\ ice}}{T_{freezer}}$$

$$= \frac{mL_f}{T_{freezer}} + \frac{mC_p \Delta T}{T_{freezer}}$$

Substitute for ΔS_{water} and $\Delta S_{freezer}$ in equation (1):

$$\Delta S_u = \frac{-mL_f}{T_{freezing}} + mC_p \ln \frac{T_f}{T_i} + \frac{mL_f}{T_{freezer}} + \frac{mC_p \Delta T}{T_{freezer}}$$

$$= m \left[\frac{-L_f}{T_{freezing}} + C_p \ln \frac{T_f}{T_i} + \frac{L_f}{T_{freezer}} + \frac{C_p \Delta T}{T_{freezer}} \right]$$

Substitute numerical values and evaluate ΔS_u:

$$\Delta S_u = (0.05\,\text{kg})\left[-\frac{333.5\times10^3\,\text{J/kg}}{273\,\text{K}} + (2100\,\text{J/kg}\cdot\text{K})\ln\left(\frac{263\,\text{K}}{273\,\text{K}}\right) + \frac{333.5\times10^3\,\text{J/kg}}{263\,\text{K}}\right.$$

$$\left. + \frac{(2100\,\text{J/kg}\cdot\text{K})(273\,\text{K} - 263\,\text{K})}{263\,\text{K}}\right]$$

$$= \boxed{2.40\,\text{J/K}}$$

and, because $\Delta S_u > 0$, the entropy of the universe increases.

*53 •• A system absorbs 300 J from a reservoir at 300 K and 200 J from a reservoir at 400 K. It then returns to its original state, doing 100 J of work and rejecting 400 J of heat to a reservoir at a temperature T. (*a*) What is the entropy change of the system for the complete cycle? (*b*) If the cycle is reversible, what is the temperature T?

Picture the Problem We can use the fact that the system returns to its original state to find the entropy change for the complete cycle. Because the entropy change for the complete cycle is the sum of the entropy changes for each process, we can find the temperature T from the entropy changes during the 1st two processes and the heat rejected during the third.

(*a*) Because S is a state function of the system:

$$\Delta S_{\text{complete cycle}} = \boxed{0}$$

(*b*) Relate the entropy change for the complete cycle to the entropy change for each process:

$$\frac{Q_1}{T_1} + \frac{Q_2}{T_2} + \frac{Q_3}{T} = 0$$

Substitute numerical values to obtain:

$$\frac{300\,\text{J}}{300\,\text{K}} + \frac{200\,\text{J}}{400\,\text{K}} + \frac{-400\,\text{J}}{T} = 0$$

Solve for T:

$$T = \boxed{267\,\text{K}}$$

*57 •• A 1-kg block of copper at 100°C is placed in a calorimeter of negligible heat capacity containing 4 L of water at 0°C. Find the entropy change of (*a*) the copper block, (*b*) the water, and (*c*) the universe.

Picture the Problem We can use conservation of energy to find the equilibrium temperature of the water and apply the equations for the entropy change during a constant pressure process to find the entropy changes of the copper block, the water, and the universe.

(a) Using the equation for the entropy change during a constant-pressure process, express the entropy change of the copper block:

$$\Delta S_{\mathrm{Cu}} = m_{\mathrm{Cu}} c_{\mathrm{Cu}} \ln \frac{T_{\mathrm{f}}}{T_{\mathrm{i}}}$$

Apply conservation of energy to obtain:

$$Q_{\mathrm{lost}} = Q_{\mathrm{gained}}$$

or

$$Q_{\mathrm{copper\ block}} = Q_{\mathrm{warming\ water}}$$

Substitute to relate the masses of the block and water to their temperatures, specific heats, and the final temperature t of the water:

$$(1\,\mathrm{kg})(0.386\,\mathrm{kJ/kg \cdot C°})(100°C - t) = (4\,\mathrm{L})(1\,\mathrm{kg/L})(4.184\,\mathrm{kJ/kg \cdot C°})(t)$$

Solve for t and T_{f}:

$t = 2.26°C$ and $T_{\mathrm{f}} = 275.4\,\mathrm{K}$

Substitute numerical values and evaluate ΔS_{Cu}:

$$\Delta S_{\mathrm{Cu}} = (1\,\mathrm{kg})(0.386\,\mathrm{kJ/kg \cdot K})$$
$$\times \ln\left(\frac{275.4\,\mathrm{K}}{373.15\,\mathrm{K}}\right)$$
$$= \boxed{-117\,\mathrm{J/K}}$$

(b) Express the entropy change of the water:

$$\Delta S_{\mathrm{water}} = m_{\mathrm{water}} c_{\mathrm{water}} \ln \frac{T_{\mathrm{f}}}{T_{\mathrm{i}}}$$

Substitute numerical values and evaluate $\Delta S_{\mathrm{water}}$:

$$\Delta S_{\mathrm{water}} = (4\,\mathrm{kg})(4.184\,\mathrm{kJ/kg \cdot K})$$
$$\times \ln\left(\frac{275.4\,\mathrm{K}}{273\,\mathrm{K}}\right)$$
$$= \boxed{146\,\mathrm{J/K}}$$

(c) Substitute for ΔS_{Cu} and $\Delta S_{\mathrm{water}}$ and evaluate the entropy change of the universe:

$$\Delta S_{\mathrm{u}} = \Delta S_{\mathrm{Cu}} + \Delta S_{\mathrm{water}}$$
$$= -117\,\mathrm{J/K} + 146\,\mathrm{J/K}$$
$$= \boxed{29.0\,\mathrm{J/K}}$$

Remarks: The result that $\Delta S_{\mathrm{u}} > 0$ tells us that this process is irreversible.

*60 ••   A box is divided into two identical halves by an impermeable partition through its middle. On one side is 1 mole of ideal gas A; on the other, 1 mole of ideal gas B (which is different from A). (*a*) Calculate the change in entropy when the partition is lifted, and the two gases mix together. (*b*) If we repeat the process with the same type of gas in each side, should the entropy change when the partition is lifted? Explain. (Think carefully about this question.)

Picture the Problem The total change in entropy resulting from the mixing of these gases is the sum of the changes in their entropies.

(*a*) Express the total change in entropy resulting from the mixing of the gases:

$$\Delta S = \Delta S_A + \Delta S_B$$

Express the change in entropy of each of the gases:

$$\Delta S_A = nR \ln \frac{V_{fA}}{V_{iA}}$$

and

$$\Delta S_B = nR \ln \frac{V_{fB}}{V_{iB}}$$

Because the initial and final volumes of the gases are the same and both volumes double:

$$\Delta S = 2nR \ln \frac{V_f}{V_i} = 2nR \ln 2$$

Substitute numerical values and evaluate ΔS:

$$\Delta S = 2(1\,\text{mol})(8.314\,\text{J/mol}\cdot\text{K})\ln 2$$
$$= \boxed{11.5\,\text{J/K}}$$

(*b*) Because the gas molecules are indistinguishable, the entropy doesn't change. A complete description of this phonomenon has been derived using quantum mechanics.

Entropy and Work Lost

*61 •• If 500 J of heat is conducted from a reservoir at 400 K to one at 300 K, (*a*) what is the change in entropy of the universe, and (*b*) how much of the 500 J of heat conducted could have been converted into work using a cold reservoir at 300 K?

Picture the Problem We can find the entropy change of the universe from the entropy changes of the high- and low-temperature reservoirs. The maximum amount of the 500 J of heat that could be converted into work can be found from the maximum efficiency of an engine operating between the two reservoirs.

(*a*) Express the entropy change of the universe:

$$\Delta S_u = \Delta S_h + \Delta S_c = -\frac{Q}{T_h} + \frac{Q}{T_c}$$

$$= -Q\left(\frac{1}{T_h} - \frac{1}{T_c}\right)$$

Substitute numerical values and evaluate ΔS_u:

$$\Delta S_u = (-500\,\text{J})\left(\frac{1}{400\,\text{K}} - \frac{1}{300\,\text{K}}\right)$$

$$= \boxed{0.417\,\text{J/K}}$$

(*b*) Express the heat that could have been converted into work in terms of the maximum efficiency of an engine operating between the two reservoirs:

$$W = \varepsilon_{max} Q_h$$

Express the maximum efficiency of an engine operating between the two reservoir temperatures:

$$\varepsilon_{max} = \varepsilon_C = 1 - \frac{T_c}{T_h}$$

Substitute and evaluate W:

$$W = \left(1 - \frac{T_c}{T_h}\right)Q_h = \left(1 - \frac{300\,\text{K}}{400\,\text{K}}\right)(500\,\text{J})$$

$$= \boxed{125\,\text{J}}$$

General Problems

*66 ••   (*a*) Calvin Cliffs Nuclear Power Plant, located on the Hobbes River, generates 1 GW of power. In this plant liquid sodium circulates between the reactor core and a heat exchanger located in the superheated steam that drives the turbine. Heat is transferred into the liquid sodium in the core, and out of the liquid sodium (and into the superheated steam) in the heat exchanger. The temperature of the superheated steam is 500 K. Waste heat is dumped into the river, which flows by at a temperature of 25°C. (*a*) What is the highest efficiency that this plant can have? (*b*) How much waste heat is dumped into the river every second? (*c*) How much heat must be generated to supply 1 GW of power? (*d*) Assume that new, tough environmental laws have been passed (to preserve the unique wildlife of the river). Because of this, the plant is not allowed to heat the river by more than 0.5°C. What is the minimum flow rate that the Hobbes River must have (in L/s)?

Picture the Problem We can use the expression for the Carnot efficiency of the plant to find the highest efficiency this plant can have. We can then use this efficiency to find the power that must be supplied to the plant to generate 1 GW of power and, from this value, the power that is wasted. The rate at which heat is being delivered to the river is related to the requisite flow rate of the river by $dQ/dt = c\Delta T\rho\, dV/dt$.

(a) Express the Carnot efficiency of a plant operating between temperatures T_c and T_h:

$$\varepsilon_{max} = \varepsilon_C = 1 - \frac{T_c}{T_h}$$

Substitute numerical values and evaluate ε_C:

$$\varepsilon_{max} = 1 - \frac{298\,K}{500\,K} = \boxed{0.404}$$

(c) Find the power that must be supplied, at 40.4% efficiency, to produce an output of 1 GW:

$$P_{supplied} = \frac{P_{output}}{\varepsilon_{max}} = \frac{1\,GW}{0.404} = \boxed{2.48\,GW}$$

(b) Relate the wasted power to the power generated and the power supplied:

$$P_{wasted} = P_{supplied} - P_{generated}$$

Substitute numerical values and evaluate P_{wasted}:

$$P_{wasted} = 2.48\,GW - 1\,GW = \boxed{1.48\,GW}$$

(d) Express the rate at which heat is being dumped into the river:

$$\frac{dQ}{dt} = c\Delta T\frac{dm}{dt} = c\Delta T\frac{d}{dt}(\rho V)$$

$$= c\Delta T\rho\frac{dV}{dt}$$

Solve for the flow rate dV/dt of the river:

$$\frac{dV}{dt} = \frac{dQ/dt}{c\Delta T\rho}$$

Substitute numerical values (see Table 19-1 for the specific heat of water) and evaluate dV/dt:

$$\frac{dV}{dt} = \frac{1.48\times10^9\,J/s}{(4180\,J/kg)(0.5\,K)(10^3\,kg/m^3)}$$

$$= 708\,m^3/s = \boxed{7.08\times10^5\,L/s}$$

*71 •• A heat engine that does the work of blowing up a balloon at a pressure of 1 atm extracts 4 kJ from a hot reservoir at 120°C. The volume of the balloon increases by 4 L, and heat is exhausted to a cold reservoir at a temperature T_c. If the efficiency of the heat engine is 50 percent of the efficiency of a Carnot engine working between the same reservoirs, find the temperature T_c.

Picture the Problem We can express the temperature of the cold reservoir as a function of the Carnot efficiency of an ideal engine and, given that the efficiency of the heat engine is half that of a Carnot engine, relate T_c to the work done by and the heat input to the real heat engine.

Using its definition, relate the efficiency of a Carnot engine working between the same reservoirs to the temperature of the cold reservoir:	$\varepsilon_C = 1 - \dfrac{T_c}{T_h}$

Solve for T_c:	$T_c = T_h(1 - \varepsilon_C)$

Relate the efficiency of the heat engine to that of a Carnot engine working between the same temperatures:	$\varepsilon = \dfrac{W}{Q_{in}} = \frac{1}{2}\varepsilon_C$ or $\varepsilon_C = \dfrac{2W}{Q_{in}}$

Substitute to obtain:	$T_c = T_h\left(1 - \dfrac{2W}{Q_{in}}\right)$

The work done by the gas in expanding the balloon is:	$W = P\Delta V = (1\,\text{atm})(4\,\text{L}) = 4\,\text{atm}\cdot\text{L}$

Substitute numerical values and evaluate T_c:

$$T_c = (393.15\,\text{K})\left(1 - \frac{2\left(4\,\text{atm}\cdot\text{L}\times\dfrac{101.325\,\text{J}}{\text{atm}\cdot\text{L}}\right)}{4\,\text{kJ}}\right) = \boxed{313\,\text{K}}$$

*79 ••• Using the equation for the entropy change of an ideal gas when the volume and temperature change and $TV^{\gamma-1}$ is a constant, show explicitly that the entropy change is zero for a quasi-static adiabatic expansion from state (V_1, T_1) to state (V_2, T_2).

Picture the Problem We can use $nR = C_P - C_V$, $\gamma = C_P/C_V$, and $TV^{\gamma-1} = a$ constant to show that the entropy change for a quasi-static adiabatic expansion that proceeds from state (V_1, T_1) to state (V_2, T_2) is zero.

Express the entropy change for a general process that proceeds from state 1 to state 2:

$$\Delta S = C_V \ln\frac{T_2}{T_1} + nR\ln\frac{V_2}{V_1}$$

For an adiabatic process:

$$\frac{T_2}{T_1} = \left(\frac{V_1}{V_2}\right)^{\gamma-1}$$

Substitute and simplify to obtain:

$$\Delta S = C_V \ln\left(\frac{V_1}{V_2}\right)^{\gamma-1} + nR\ln\left(\frac{V_2}{V_1}\right) = \ln\left(\frac{V_2}{V_1}\right)\left[nR + \frac{C_V \ln\left(\frac{V_1}{V_2}\right)^{\gamma-1}}{\ln\frac{V_2}{V_1}}\right]$$

$$= \ln\left(\frac{V_2}{V_1}\right)\left[nR + \frac{(\gamma-1)C_V \ln\left(\frac{V_1}{V_2}\right)}{-\ln\frac{V_1}{V_2}}\right] = \ln\left(\frac{V_2}{V_1}\right)\left[nR - (\gamma-1)C_V\right]$$

Use the relationship between C_P and C_V to obtain:

$$nR = C_P - C_V$$

Substitute for nR and γ and simplify:

$$\Delta S = \ln\left(\frac{V_2}{V_1}\right)\left[C_P - C_V - \left(\frac{C_P}{C_V}-1\right)C_V\right]$$

$$= \boxed{0}$$

*82 ••• Suppose that each engine in Figure 19-22 is an ideal reversible heat engine. Engine 1 operates between temperatures T_h and T_m and engine 2 operates between T_m and T_c, where $T_h > T_m > T_c$. Show that

$$\varepsilon_{net} = 1 - \frac{T_C}{T_H}$$

This means that two reversible heat engines in series are equivalent to one reversible heat engine operating between the hottest and coldest reservoirs.

Picture the Problem We can express the net efficiency of the two engines in terms of W_1, W_2, and Q_h and then use $\varepsilon_1 = W_1/Q_h$ and $\varepsilon_2 = W_2/Q_m$ to eliminate W_1, W_2, Q_h, and Q_m. Finally, we can substitute the expressions for the efficiencies of the ideal reversible engines to obtain $\varepsilon_{net} = 1 - T_c/T_h$.

Express the efficiencies of ideal reversible engines 1 and 2:	$\varepsilon_1 = 1 - \dfrac{T_m}{T_h}$ (1)
	and
	$\varepsilon_2 = 1 - \dfrac{T_c}{T_m}$ (2)

Express the net efficiency of the two engines connected in series:

$$\varepsilon_{net} = \frac{W_1 + W_2}{Q_h} \qquad (3)$$

Express the efficiencies of engines 1 and 2:

$$\varepsilon_1 = \frac{W_1}{Q_h}$$

and

$$\varepsilon_2 = \frac{W_2}{Q_m}$$

Solve for W_1 and W_2 and substitute in equation (3) to obtain:

$$\varepsilon_{net} = \frac{\varepsilon_1 Q_h + \varepsilon_2 Q_m}{Q_h} = \varepsilon_1 + \frac{Q_m}{Q_h}\varepsilon_2$$

Express the efficiency of engine 1 in terms of Q_m and Q_h:

$$\varepsilon_1 = 1 - \frac{Q_m}{Q_h}$$

Solve for Q_m/Q_h:

$$\frac{Q_m}{Q_h} = 1 - \varepsilon_1$$

Substitute to obtain:

$$\varepsilon_{net} = \varepsilon_1 + (1 - \varepsilon_1)\varepsilon_2$$

Substitute for ε_1 and ε_2 and simplify to obtain:

$$\varepsilon_{net} = 1 - \frac{T_m}{T_h} + \left(\frac{T_m}{T_h}\right)\left(1 - \frac{T_c}{T_m}\right)$$

$$= 1 - \frac{T_m}{T_h} + \frac{T_m}{T_h} - \frac{T_c}{T_h} = \boxed{1 - \frac{T_c}{T_h}}$$

Chapter 20
Thermal Properties and Processes

Conceptual Problems

*1 • Why does the mercury level first decrease slightly when a thermometer is placed in warm water?

Determine the Concept The glass bulb warms and expands first, before the mercury warms and expands.

*5 •• The phase diagram in Figure 20-12 can be interpreted to yield information on how the boiling and melting points of water change with altitude. (*a*) Explain how this information can be obtained. (*b*) How might this information affect cooking procedures in the mountains?

Figure 20-12 Problem 5

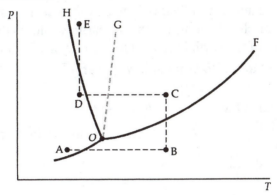

(*a*) With increasing altitude, *P* decreases; from curve OF, *T* of the liquid-gas interface diminishes, so the boiling temperature decreases. Likewise, from curve OH, the melting temperature increases with increasing altitude.

(*b*) Boiling at a lower temperature means that the cooking time will have to be increased.

*7 • In a cool room, a metal or marble table top feels much colder to the touch than does a wood surface does even though they are at the same temperature. Why?

Determine the Concept The thermal conductivity of metal and marble is much greater than that of wood; consequently, heat transfer from the hand is more rapid.

*15 •   In artistic nomenclature, blue is often referred to as a "cool" color, while red is referred to as a "warm" color. In physics, however, red is considered a "cooler" color than blue. Explain why.

Determine the Concept The temperature of an object is inversely proportional to the maximum wavelength at which the object radiates (Wein's displacement law). Because blue light has a shorter wavelength than red light, an object for which the wavelength of the peak of thermal emission is blue is hotter than one that is red.

Estimation and Approximation

*18 ••   Estimate the effective emissivity of the earth, given the following information: the solar constant (the intensity of light incident on the earth from the sun) is 1370 W/m², 70 percent of this light is absorbed by the earth, and the earth's average temperature is 288 K. (Assume that the effective area that is absorbing the light is πR^2, where R is the earth's radius, while the blackbody-emission area is $4\pi R^2$.)

Picture the Problem The amount of heat radiated by the earth must equal the solar flux from the sun, or else the temperature on earth would continually increase. The emissivity of the earth is related to the rate at which it radiates energy into space by the Stefan-Boltzmann law $P_r = e\sigma A T^4$.

Using the Stefan-Boltzmann law, express the rate at which the earth radiates energy as a function of its emissivity e and temperature T:	$P_r = e\sigma A'T^4$ where A' is the surface area of the earth.
Solve for the emissivity of the earth:	$e = \dfrac{P_r}{\sigma A'T^4}$
Use its definition to express the intensity of the radiation received by the earth:	$I = \dfrac{P_{absorbed}}{A}$ where A is the cross-sectional area of the earth.
For 70% absorption of the sun's radiation incident on the earth:	$I = \dfrac{0.7P_r}{A}$
Substitute for P_r and A and simplify to obtain:	$e = \dfrac{0.7AI}{\sigma AT^4} = \dfrac{0.7\pi R^2 I}{4\pi R^2 \sigma T^4} = \dfrac{0.7I}{4\sigma T^4}$

Substitute numerical values and evaluate e:

$$e = \frac{0.7(1370 \, \text{W/m}^2)}{4(5.670 \times 10^{-8} \, \text{W/m}^2 \cdot \text{K}^4)(288 \, \text{K})^4}$$

$$= \boxed{0.615}$$

Thermal Expansion

*24 •• Repeat Problem 23 when the temperature of both the steel shaft and copper collar are raised simultaneously.

Picture the Problem Because the temperatures of both the steel shaft and the copper collar change together, we can find the temperature change required for the collar to fit the shaft by equating their diameters for a temperature increase ΔT. These diameters are related to their diameters at 20°C and the increase in temperature through the definition of the coefficient of linear expansion.

Express the temperature to which the collar and the shaft must be raised in terms of their initial temperature and the increase in their temperature:

$$T = T_i + \Delta T \qquad (1)$$

Express the diameter of the steel shaft when its temperature has been increased by ΔT:

$$D_{\text{steel}} = D_{\text{steel,20°C}}(1 + \alpha_{\text{steel}}\Delta T)$$

Express the diameter of the copper collar when its temperature has been increased by ΔT:

$$D_{\text{Cu}} = D_{\text{Cu,20°C}}(1 + \alpha_{\text{Cu}}\Delta T)$$

If the collar is to fit over the shaft when the temperature of both has been increased by ΔT:

$$D_{\text{Cu,20°C}}(1 + \alpha_{\text{Cu}}\Delta T)$$
$$= D_{\text{steel,20°C}}(1 + \alpha_{\text{steel}}\Delta T)$$

Solve for ΔT to obtain:

$$\Delta T = \frac{D_{\text{steel,20°C}} - D_{\text{Cu,20°C}}}{D_{\text{Cu,20°C}}\alpha_{\text{Cu}} - D_{\text{steel,20°C}}\alpha_{\text{steel}}}$$

Substitute in equation (1) to obtain:

$$T = T_i + \frac{D_{\text{steel,20°C}} - D_{\text{Cu,20°C}}}{D_{\text{Cu,20°C}}\alpha_{\text{Cu}} - D_{\text{steel,20°C}}\alpha_{\text{steel}}}$$

Substitute numerical values and evaluate T:

$$T = 293\,\mathrm{K} + \frac{6.0000\,\mathrm{cm} - 5.9800\,\mathrm{cm}}{(5.98\,\mathrm{cm})(17 \times 10^{-6}\,/\mathrm{K}) - (6.00\,\mathrm{cm})(11 \times 10^{-6}\,/\mathrm{K})} = 854\,\mathrm{K} = \boxed{581°\mathrm{C}}$$

*27 ••   A rookie crew was left to put in the final 1 km of rail for a stretch of railroad track. When they finished, the temperature was 20°C, and they headed to town for some refreshments. After an hour or two, one of the old-timers noticed that the temperature had gone up to 25°C, so he said, "I hope you left some gaps to allow for expansion." From the look on their faces, he knew that they had not, and they all rushed back to the work site. The rail had buckled into an isosceles triangle. How high was the buckle?

Picture the Problem Let L be the length of the rail at 20°C and L' its length at 25°C. The diagram shows these distances and the height h of the buckle. We can use Pythagorean theorem to relate the height of the buckle to the distances L and L' and the definition of the coefficient of linear expansion to relate L and L'.

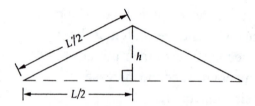

Apply the Pythagorean theorem to obtain:

$$h = \sqrt{\left(\frac{L'}{2}\right)^2 - \left(\frac{L}{2}\right)^2} = \tfrac{1}{2}\sqrt{L'^2 - L^2}$$

Use the definition of the coefficient of linear expansion to relate L and L':

$$L'^2 = L^2(1 + \alpha_{\mathrm{steel}}\Delta T)^2$$
or, because $(\alpha_{\mathrm{steel}}\Delta T)^2 \ll 2\alpha_{\mathrm{steel}}\Delta T$,
$$L'^2 \approx L^2(1 + 2\alpha_{\mathrm{steel}}\Delta T)$$

Substitute to obtain:

$$h = \frac{1}{2}\sqrt{L^2(1 + 2\alpha_{\mathrm{steel}}\Delta T) - L^2}$$
$$= \frac{L}{2}\sqrt{2\alpha_{\mathrm{steel}}\Delta T}$$

Substitute numerical values and evaluate h:

$$h = \frac{1000\,\mathrm{m}}{2}\sqrt{2(11 \times 10^{-6}\,\mathrm{K}^{-1})(5\,\mathrm{K})}$$
$$= \boxed{5.24\,\mathrm{m}}$$

*33 ••• What is the tensile stress in the copper collar of Problem 23 when its temperature returns to 20°C?

Picture the Problem We can use the definition of Young's modulus to express the tensile stress in the copper in terms of the strain it undergoes as its temperature returns to 20°C. We can show that $\Delta L/L$ for the circumference of the collar is the same as $\Delta d/d$ for its diameter.

Using Young's modulus, relate the stress in the collar to its strain:

$$\text{Stress} = Y \times \text{Strain} = Y \frac{\Delta L}{L_{20°C}}$$

where $L_{20°C}$ is the circumference of the collar at 20°C.

Express the circumference of the collar at the temperature at which it fits over the shaft:

$$L_T = \pi d_T$$

Express the circumference of the collar at 20°C:

$$L_{20°C} = \pi d_{20°C}$$

Substitute to obtain:

$$\text{Stress} = Y \frac{\pi d_T - \pi d_{20°C}}{\pi d_{20°C}}$$

$$= Y \frac{d_T - d_{20°C}}{d_{20°C}}$$

Substitute numerical values and evaluate the stress:

$$\text{Stress} = \left(11 \times 10^{-10}\ \text{N/m}^2\right)\left(\frac{0.02\ \text{cm}}{5.98\ \text{cm}}\right)$$

$$= \boxed{3.68 \times 10^{-12}\ \text{N/m}^2}$$

The van der Waals Equation, Liquid-Vapor Isotherms, and Phase Diagrams

*36 •• The van der Waals constants for helium are $a = 0.03412\ \text{L}^2 \cdot \text{atm}/\text{mol}^2$ and $b = 0.0237\ \text{L/mol}$. Use these data to find the volume in cubic centimeters occupied by one helium atom and to estimate the radius of the atom.

Picture the Problem Assume that a helium atom is spherical. Then we can find its radius from $V = \frac{4}{3}\pi r^3$ and its volume from the van der Waals equation.

Express the radius of a spherical atom in terms of its volume:

$$r = \sqrt[3]{\frac{3V}{4\pi}}$$

In the van der Waals equation, b is the volume of 1 mol of molecules. For He, 1 molecule = 1 atom. Use Avogadro's number to express b in cm^3/atom:

$$b = \frac{(0.0237\,L/mol)(10^3\,cm^3/L)}{6.022 \times 10^{23}\,atoms/mol}$$

$$= 3.94 \times 10^{-23}\,cm^3/atom$$

Substitute numerical values and evaluate r:

$$r = \sqrt[3]{\frac{3(3.94 \times 10^{-23}\,cm^3)}{4\pi}}$$

$$= 2.11 \times 10^{-8}\,cm = \boxed{0.211\,nm}$$

Heat Conduction

*44 ••   For a boiler at a power station, heat must be transferred to boiling water at the rate of 3 GW. The boiling water passes through copper pipes having a wall thickness of 4.0 mm and a surface area of 0.12 m^2 per meter length of pipe. Find the total length of pipe (actually there are many pipes in parallel) that must pass through the furnace if the steam temperature is 225°C and the external temperature of the pipes is 600°C.

Picture the Problem We can use the expression for the thermal current to express the thickness of the walls in terms of the thermal conductivity of copper, the area of the walls, and the temperature difference between the inner and outer surfaces. Letting $\Delta A/\Delta x'$ represent the area per unit length of the pipe and L its length, we can eliminate the surface area and solve for and evaluate L.

Write the expression for the thermal current:

$$I = kA\frac{\Delta T}{\Delta x}$$

Solve for A:

$$A = \frac{I\Delta x}{k\Delta T}$$

Express the total surface area of the pipe:

$$A = \frac{\Delta A}{\Delta x'}L$$

Substitute for A and solve for L to obtain:

$$L = \frac{\dfrac{I\Delta x}{k\Delta T}}{\Delta A/\Delta x'}$$

Substitute numerical values and evaluate L:

$$L = \frac{\left[\dfrac{(3\,\text{GW})(4\times10^{-3}\,\text{m})}{(401\,\text{W/m}\cdot\text{K})(873\,\text{K}-498\,\text{K})}\right]}{0.12\,\text{m}}$$

$$= \boxed{665\,\text{m}}$$

General Problems

*51 • A steel tape is placed around the earth at the equator when the temperature is 0°C. What will the clearance between the tape and the ground (assumed to be uniform) be if the temperature of the tape rises to 30°C? Neglect the expansion of the earth.

Picture the Problem The distance by which the tape clears the ground equals the change in the radius of the circle formed by the tape placed around the earth at the equator.

Express the change in the radius of the circle defined by the steel tape:

$\Delta R = R\alpha\Delta T$
where R is the radius of the earth, α is the coefficient of linear expansion of steel, and ΔT is the increase in temperature.

Substitute numerical values and evaluate ΔR.

$\Delta R = (6.37\times10^{6}\,\text{m})(11\times10^{-6}\,\text{K}^{-1})(30\,\text{K})$

$= 2.10\times10^{3}\,\text{m}$

$= \boxed{2.10\,\text{km}}$

*61 •• A hot-water tank of cylindrical shape has an inside diameter of 0.55 m and inside height of 1.2 m. The tank is enclosed with a 5-cm-thick insulating layer of glass wool whose thermal conductivity is 0.035 W/m·K. The metallic interior and exterior walls of the container have thermal conductivities that are much greater than that of the glass wool. How much power must be supplied to this tank in order to maintain the water temperature at 75°C when the external temperature is 1°C?

Picture the Problem We'll do this problem twice. First, we'll approximate the answer by disregarding the fact that the surrounding insulation is cylindrical. In the second solution, we'll obtain the exact answer by taking into account the cylindrical insulation surrounding the side of the tank. In both cases, the power required to maintain the temperature of the water in the tank is equal to the rate at which thermal energy is conducted through the insulation.

1st solution:

Using the thermal current equation, relate the rate at which energy is transmitted through the insulation to the temperature gradient, thermal conductivity of the insulation, and the area of the insulation/tank:

$$I = kA\frac{\Delta T}{\Delta x}$$

Letting d represent the inside diameter of the tank and L its inside height, express and evaluate its surface area:

$$
\begin{aligned}
A &= A_{side} + A_{bases} \\
&= \pi\, dL + 2\left(\frac{\pi d^2}{4}\right) \\
&= \pi\left(dL + \tfrac{1}{2}d^2\right) \\
&= \pi\left[(0.55\,\text{m})(1.2\,\text{m}) + \tfrac{1}{2}(0.55\,\text{m})^2\right] \\
&= 2.55\,\text{m}^2
\end{aligned}
$$

Substitute numerical values and evaluate I:

$$I = (0.035\,\text{W/m}\cdot\text{K})(2.55\,\text{m}^2)\left(\frac{74\,\text{K}}{0.05\,\text{m}}\right)$$

$$= \boxed{132\,\text{W}}$$

2nd solution:

Express the total heat loss as the sum of the losses through the top and bottom and the side of the hot-water tank:

$$I = I_{top\,and\,bottom} + I_{side}$$

Express I through the top and bottom surfaces:

$$
\begin{aligned}
I_{top\,and\,bottom} &= 2\left(kA\frac{\Delta T}{\Delta x}\right) \\
&= \tfrac{1}{2}\pi\, d^2 k \frac{\Delta T}{\Delta x}
\end{aligned}
$$

Substitute numerical values and evaluate $I_{top\,and\,bottom}$:

$$
\begin{aligned}
I_{top\,and\,bottom} &= \tfrac{1}{2}\pi(0.55\,\text{m})^2 \\
&\quad \times \frac{(0.035\,\text{W/m}\cdot\text{K})(74\,\text{K})}{0.05\,\text{m}} \\
&= 24.6\,\text{W}
\end{aligned}
$$

Letting r represent the inside radius of the tank, express the heat current through the cylindrical side:	$I_{side} = -kA\dfrac{dT}{dr} = -2\pi kLr\dfrac{dT}{dr}$ where the minus sign is a consequence of the heat current being opposite the temperature gradient.

Separate the variables:

$$dT = -\frac{I_{side}}{2\pi kL}\frac{dr}{r}$$

Integrate from $r = r_1$ to $r = r_2$ and $T = T_1$ to $T = T_2$:

$$\int_{T_1}^{T_2} dT = -\frac{I_{side}}{2\pi kL}\int_{r_1}^{r_2}\frac{dr}{r}$$

and

$$T_2 - T_1 = -\frac{I_{side}}{2\pi kL}\ln r\Big]_{r_1}^{r_2}$$

$$= -\frac{I_{side}}{2\pi kL}\ln\frac{r_2}{r_1} = \frac{I_{side}}{2\pi kL}\ln\frac{r_1}{r_2}$$

Solve for I_{side} to obtain:

$$I_{side} = \frac{2\pi kL}{\ln\dfrac{r_1}{r_2}}(T_2 - T_1)$$

Substitute numerical values and evaluate I_{side}:

$$I_{side} = \frac{2\pi(0.035\,\text{W/m}\cdot\text{K})(1.2\,\text{m})}{\ln\left(\dfrac{0.325\,\text{m}}{0.275\,\text{m}}\right)}(74\,\text{K})$$

$$= 117\,\text{W}$$

Substitute for I_{side} and evaluate I:

$$I = 24.6\,\text{W} + 117\,\text{W} = \boxed{142\,\text{W}}$$

---

*66 ••• A 200-g copper container holding 0.7 L of water is thermally isolated from its surroundings—except for a 10-cm-long copper rod of cross-sectional area 1.5 cm^2 connecting it to a second copper container filled with an ice and water mixture so its temperature remains at 0°C. The initial temperature of the first container is $T_0 = 60$°C. (Assume the heat capacity of the rod to be negligible.) (a) Show that the temperature T of the first container changes over time t according to

$$T = T_0 e^{-t/RC}$$

where T is in degrees Celsius, R is the thermal resistance of the rod, and C is the total heat capacity of the container plus the water. (Neglect the heat capacity of the rod.) (b) Evaluate R, C, and the "time constant" RC. (c) Show that the total amount of heat Q conducted in time t is

$$Q = CT_0\left(1 - e^{-t/RC}\right)$$

(d) Using a spreadsheet program, graph both $T(t)$ and $Q(t)$; from the graph, find the time it takes for the temperature of the first container to be reduced to 30°C.

Picture the Problem We can use the thermal current equation and the definition of heat capacity to obtain the differential equation describing the rate at which the temperature of the water in the 200-g container is changing. Integrating this equation will yield $T = T_0 e^{-t/RC}$. Substituting for dT/dt in $dQ/dt = -CdT/dt$ and integrating will lead to $Q = CT_0\left(1 - e^{-t/RC}\right)$.

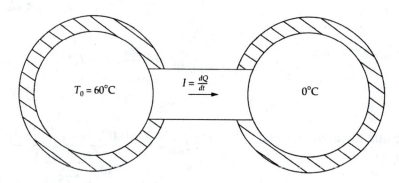

| (a) Use the thermal current equation to express the rate at which heat is conducted from the water at 60°C by the rod: | $I = \dfrac{\Delta T}{R} = \dfrac{T}{R}$ |

because the temperature of the second container is maintained at 0°C.

| Using the definition of heat capacity, relate the thermal current to the rate at which the temperature of the water initially at 60°C is changing: | $I = \dfrac{dQ}{dt} = -C\dfrac{dT}{dt}$ (1) |

| Equate these two expressions to obtain: | $C\dfrac{dT}{dt} = -\dfrac{1}{R}T$, the differential equation |

describing the rate at which the temperature of the water in the 200-g container is changing.

Separate variables to obtain:	$$\frac{dT}{T} = -\frac{1}{RC}dt$$
Integrate dT from T_0 to T and dt from 0 to t:	$$\int_{T_0}^{T}\frac{dT'}{T'} = -\frac{1}{RC}\int_0^t dt'$$ or $$\ln\frac{T}{T_0} = -\frac{1}{RC}t$$
Transform from logarithmic to exponential form and solve for T to obtain:	$$\boxed{T = T_0 e^{-t/RC}} \qquad (2)$$
(b) Use its definition to express the thermal resistance R:	$$R = \frac{\Delta x}{kA}$$
Substitute numerical values (see Table 20-8 for the thermal conductivity of copper) and evaluate R:	$$R = \frac{0.1\,\mathrm{m}}{(401\,\mathrm{W/m\cdot K})(1.5\times10^{-4}\,\mathrm{m^2})}$$ $$= \boxed{1.66\,\mathrm{K/W}}$$
Use its definition to express the heat capacity of the water and the copper container:	$$C = m_c c_c + m_w c_w = m_c c_c + \rho_w V_w c_w$$
Substitute numerical values (see Table 18-1 for the specific heats of water and copper) and evaluate C:	$$C = (0.2\,\mathrm{kg})(386\,\mathrm{kJ/kg\cdot K})$$ $$\quad + (10^3\,\mathrm{kg/m^3})(0.7\,\mathrm{L})(4.18\,\mathrm{kJ/kg\cdot K})$$ $$= \boxed{3.00\,\mathrm{kJ/K}}$$
Evaluate the product of R and C to find the "time constant" τ:	$$\tau = RC = (1.66\,\mathrm{K/W})(3.00\,\mathrm{kJ/K})$$ $$= 4985\,\mathrm{s} = \boxed{1.38\,\mathrm{h}}$$
(c) Solve equation (1) for dQ to obtain:	$$dQ = -C\left(\frac{dT}{dt}\right)dt = -C\,dT$$
Integrate dQ' from $Q = 0$ to Q and dT from T_0 to T:	$$\int_0^Q dQ' = -\int_{T_0}^T C\,dT$$ or $$Q = C(T_0 - T(t))$$
Substitute (equation (2) for $T(t)$ to obtain:	$$Q = C(T_0 - T_0 e^{-t/RC}) = \boxed{CT_0(1 - e^{-t/RC})}$$

A spreadsheet program to evaluate Q as a function of t is shown below. The formulas used to calculate the quantities in the columns are as follows:

Cell	Formula/Content	Algebraic Form
D1	1.35	τ
D2	60	T_0
D3	3000	C
A6	0	t
A7	A6+0.1	$t + \Delta t$
B6	$B$2*EXP(−A6/$B$1)	$T_0 e^{-t/RC}$
C7	$B$3*$B$2*(1−EXP(−A6/$B$1))	$CT_0\left(1-e^{-t/RC}\right)$

	A	B	C	D
1	tau=	1.35	h	
2	T0=	60	deg-C	
3	C=	3000	J/K	
4				
5	t (hr)	T	Q	Q/1000
6	0.0	60.00	0.00E+00	0
7	0.1	55.72	1.29E+04	13
8	0.2	51.74	2.48E+04	24
13	0.7	35.72	7.28E+04	71
14	0.8	33.17	8.05E+04	79
15	0.9	30.81	8.76E+04	86
16	1.0	28.61	9.42E+04	92
33	2.7	8.12	1.56E+05	152
34	2.8	7.54	1.57E+05	154

From the table we can see that the temperature of the container drops to 30°C in a little more than $\boxed{0.9\,\text{h}.}$ If we wanted to know this time to the nearest hundredth of an hour, we could change the step size in the spreadsheet program to 0.01 h. A graph of T as a function of t is shown in the following graph.

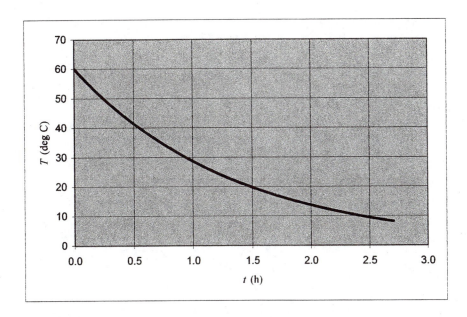

A graph of Q as a function of t is shown below.

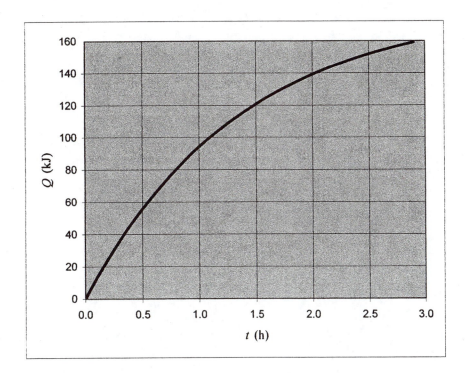